Sturm-Liouville Operators and Applications

Revised Edition

Sturm-Liouville Operators and Applications

Revised Edition

Vladimir A. Marchenko

AMS CHELSEA PUBLISHING
American Mathematical Society • Providence, Rhode Island

2000 *Mathematics Subject Classification.* Primary 34A55, 34B24, 35Q51, 47E05, 47J35.

For additional information and updates on this book, visit
www.ams.org/bookpages/chel-373

Library of Congress Cataloging-in-Publication Data

Marchenko, V. A. (Vladimir Aleksandrovich), 1922–
Sturm-Liouville operators and applications / Vladimir A. Marchenko. — Rev. ed.
p. cm.
Rev. ed. of: Sturm-Liouville operators and applications. 1986.
Includes bibliographical references.
ISBN 978-0-8218-5316-0 (alk. paper)
1. Spectral theory (Mathematics) 2. Transformations (Mathematics) 3. Operator theory. I. Title.

QA320.M286 2011
515′.7222—dc22

2010051019

♾ The paper used in this book is acid-free and falls within the guidelines
established to ensure permanence and durability.
Visit the AMS home page at `http://www.ams.org/`

10 9 8 7 6 5 4 3 2 1 16 15 14 13 12 11

CONTENTS

PREFACE TO THE REVISED EDITION vii

PREFACE ix

Chapter 1 THE STURM-LIOUVILLE EQUATION AND TRANSFORMATION OPERATORS 1
1. Riemann's Formula 1
2. Transformation Operators 7
3. The Sturm-Liouville Boundary Value Problem on a Bounded Interval 26
4. Asymptotic Formulas for Solutions of the Sturm-Liouville Equation 50
5. Asymptotic Formulas for Eigenvalues and Trace Formulas 67

Chapter 2 THE STURM-LIOUVILLE BOUNDARY VALUE PROBLEM ON THE HALF LINE 101
1. Some Information on Distributions 101
2. Distribution-Valued Spectral Functions 117
3. The Inverse Problem 134
4. The Asymptotic Formula for the Spectral Functions of Symmetric Boundary Value Problems and the Equiconvergence Theorem 153

Chapter 3 THE BOUNDARY VALUE PROBLEM OF SCATTERING THEORY 173
1. Auxiliary Propositions 173
2. The Parseval Equality and the Fundamental Equation 200
3. The Inverse Problem of Quantum Scattering Theory 216
4. Inverse Sturm-Liouville Problems on a Bounded Interval 240
5. The Inverse Problem of Scattering Theory on the Full Line 284

Chapter 4 NONLINEAR EQUATIONS 307
1. Transformation Operators of a Special Form 307
2. Rapidly Decreasing Solutions of the Korteweg-de Vries Equation 322
3. Periodic Solutions of the Korteweg-de Vries Equation 332
4. Explicit Formulas for Periodic Solutions of the Korteweg-de Vries Equation 356

Chapter 5 STABILITY OF INVERSE PROBLEMS 363
1. Problem Formulation and Derivation of Main Formulas 363
2. Stability of the Inverse Scattering Problem 370

3. Error Estimate for the Reconstruction of a Boundary Value Problem from its Spectral Function Given on the Set $(-\infty, N^2)$ Only 380

References 389

Preface to the revised edition

In the first edition of this book the main attention was focused on the methods of solving the inverse problem of spectral analysis and on the conditions (necessary and sufficient) which the spectral data must satisfy in order to make it possible to reconstruct the potential of the corresponding Sturm-Liouville operator. These conditions imply that the spectral data (e.g. spectral function or scattering data) must be known for all values of spectral parameter which belong to the spectrum of the operator.

But from the physical meaning of the inverse problem it is obvious that the values of spectral data on the whole spectrum are impossible to obtain from any observations. For example, in the inverse problem of quantum scattering theory the energy of the particles acts as the spectral parameter, and in order to find the values of scattering data on the whole spectrum one has to conduct an experiment with the particles of infinitely large energy. But for big enough values of energy the scattering process is not any more described by Schrödinger equation with potential $q(x)$. Therefore, even allowing, ideally, the possibility to experiment with particles of arbitrarily large energies, we would obtain, starting from a certain energy, data relevant to process, which has certainly nothing to do with the equation that we want to reconstruct. Hence, a principal question is as follows: what information about the potential $q(x)$ can be obtained, if the spectral function or scattering data are known (generally speaking, approximately) only on a finite interval of values of the spectral parameter?

The new Chapter 5, devoted to solving this problem, was added to this edition. The convenient formulae are obtained, which allow to estimate the precision with which the eigenfunctions and potentials of Schrödinger operator can be restored when the scattering data or spectral function are known only on a finite interval of values of spectral parameter.

V. Marchenko

PREFACE

The development of many important directions of mathematics and physics owes a major debt to the concepts and methods which evolved during the investigation of such simple objects as the Sturm-Liouville equation $y'' + q(x)y = zy$ and the allied Sturm-Liouville operator $L = - d^2/dx^2 + q(x)$ (lately L and $q(x)$ are often termed the one-dimensional Schrödinger operator and the potential). These provided a constant source of new ideas and problems in the spectral theory of operators and kindred areas of analysis. This sourse goes back to the first studies of D. Bernoulli and L. Euler on the solution of the equation describing the vibrations of a string, and still remains productive after more than two hundred years. This is confirmed by the recent discovery, made by C. Gardner, J. Green, M. Kruskal, and R. Miura [6], of an unexpected connection between the spectral theory of Sturm-Liouville operators and certain nonlinear partial differential evolution equations.

The methods used (and often invented) during the study of the Sturm-Liouville equation have been constantly enriched. In the 40's a new investigation tool joined the arsenal - that of transformation operators. The latter first appeared in the theory of generalized translation operators of J. Delsarte and B. M. Levitan (see [16]). Transformation operators for arbitrary Sturm-Liouville equations were constructed by A. Ya. Povzner [24], who used them to derive the eigenfunction expansion for a Sturm-Liouville equation with a decreasing potential (it seem that his work is the first in which transformation operators were used in spectral theory). V. A. Marchenko enlisted transformation operators to investigate both inverse problems of spectral analysis [17] and the asymptotic behavior of the spectral function of singular Sturm-Liouville operators [18].

The role of transformation operators in spectral theory became even more important following several discoveries. Specifically, I. M. Gelfand and

B. M. Levitan [8] found that these operators can be used to provide a complete solution to the problem of recovering a Sturm-Liouville equation from its spectral function; B. M. Levitan [15] proved the equiconvergence theorem in its general form; B. Ya. Levin [14] introduced a new type of transformation operators which preserve the asymptotics of the solutions at infinity; and V. A. Marchenko [19] used them to solve the inverse scattering problem.

The main goal of this monograph is to show what can be achieved with the aid of transformation operators in spectral theory, as well as in its recently revealed untraditional applications. We made such an attempt in our book [20], which was published in 1972 and was based on lectures delivered at Khar'kov University. In the years that followed, transformation operators have been applied to an increasing number of problems, and we felt that a more complete discussion of the results in this area was needed. In the present book, aside from traditional topics that are treated roughly in the same way as in our previous monograph, we include new applications of transformation operators and problems connected with the use of spectral theory in the study of nonlinear equations.

In the first chapter transformation operators are used to investigate the boundary value problem generated on a finite interval by the Sturm-Liouville operator with arbitrary nondegenerate boundary conditions. One proves the completeness of the system of eigenfunctions and generalized eigenfunctions. Moreover, one obtains asymptotic formulas for $\lambda \to \infty$ for the solutions of the Sturm-Liouville equation, and then use them to derive asymptotic formulas for the eigenvalues of the boundary value problems under consideration. All these formulas have been known for a long time. However, it turns out that with the aid of transformation operators one can express the principal parts of their remainders explicitly in terms of the Fourier coefficients of the potential $q(x)$. For example, one can establish the exact relationship between the smoothness of a periodic potential and the rate of decay of the lengths of the lacunae in the spectrum of the corresponding Hill operator. The concluding part of the chapter is devoted to the derivation of the Gelfand-Levitan trace formulas [9], which are becoming more and more important.

In the second chapter we discuss the singular boundary value problems generated on the half line $0 \leqslant x < \infty$ by the Sturm-Liouville operator having an arbitrary complex-valued potential $q(x)$ and subject to the boundary conditions $y'(0) - hy(0) = 0$. The notion of a generalized (distribution) spectral function is introduced, and its existence is established for this class of boundary value problems. The Riesz theorem on the form of linear positive functionals shows that the distribution spectral functions are measures whenever $q(x)$ and h are real. In this case the formulas for the expansion in eigenfunctions and the Parseval equality, generated by a distribution spectral function, lead to classical results of H. Weyl. In Section 3 we derive the Gelfand-Levitan integral equation [8]. This enables us to recover the operator from its distribution spectral function. We also find conditions necessary and sufficient for a distribution to be the spectral function of a Sturm-Liouville operator. In the last section the asymptotic formula of Marchenko [18] for the spectral functions of symmetric boundary value problems is obtained (in the sharpened form due to B. M. Levitan [15]) and Levitan's equiconvergence theorem [15] is proved.

The third chapter is devoted to inverse problems in scattering theory and the inverse problem for the Hill equation. The Levin transformation operators [14] are introduced and then used to study the properties of the solutions to a Sturm-Liouville equation whose potential satisfies the constraint $\int_0^\infty x|q(x)|dx < \infty$. Next, we derive Marchenko's integral equation [19], which enables us to recover the potential from the scattering data, and we establish the characteristic properties of these data. In addition, we discuss the results of V. A. Marchenko and I. V. Ostrovskii [22]: we find conditions necessary and sufficient for a given sequence of intervals to equal the set of stability zones of a Hill equation, and show that the set of potentials having a finite number of such zones (known as "finite-zone" potentials) is dense. We also prove a theorem of Gasymov and Levitan [7] which includes a complete solution of the inverse problem in G. Borg's formulation [2]. The last section of Chapter 3 is devoted to the inverse problem of scattering theory for the Sturm-Liouville operator on the full real line. There we prove Faddeev's theorem [5] which gives the characteristic properties of scattering data.

In the last chapter we show how spectral theory can be used to integrate certain nonlinear partial differential equations - a fact which was discovered by C. Gardner, J. Green, M. Kruskal, and R. Miura [6]. Following the publication of Lax's work [12] and of the paper [27] by V. E. Zakharov and A. B. Shabat, based on the ideas of [12], it became clear that this new integration method can be applied to a large number of nonlinear equations which occur in mathematical physics (see the survey paper [3]). Here we discuss in detail only the Korteweg-de Vries equation $\dot{v} = 6vv' - v'''$, and use the exercises to guide the reader's way towards possible generalizations. In Section 1 we give a general presentation of the new integration method which differs somewhat from Lax's scheme, and which permits us to include auxiliary linear operators which depend arbitrarily upon the spectral parameter z. Next, we solve the Cauchy problem for the Korteweg-de Vries equation in the class of rapidly decreasing potentials using the method developed in [6]. The periodic problem for this equation was attacked first in 1974 using different methods, in studies by S. P. Novikov [23], P. Lax [13], and V. A. Marchenko [21]. In Section 3 we discuss the method invented in [21], while Problems 2 and 3 give the proofs of two theorems of Novikov [23], and thereby exhibit a connection between these methods. For a detailed discussion of the results obtained by following the ideas of [23] we refer the interested reader to the survey paper [3].

In 1961 N. I. Akhiezer [1] discovered a relation between the inverse problems for certain Sturm-Liouville operators having a finite number of lacunae in the spectrum and the Jacobi inversion problem for Abelian integrals. Developing Akhiezer's ideas, A. R. Its and V. B. Matveev [10] found an explicit formula for the finite-zone potentials in terms of Riemann's Θ-function. Combining this formula with results of B. A. Dubrovin and S. P. Novikov [4] one finds a simple expression for the finite-zone periodic and almost-periodic solutions of the Korteweg-de Vries equation. We derive this formula in the last section of Chapter 4.

The exercises in the monograph are presented with enough hints so that one can recover the full proofs. This should enable the reader to see possible refinements and generalizations of the material treated in the main text. In particular, the problems include results of M. Crum, M. G. Krein, and V. F. Korop (on degenerate transformation operators and equations with singularities), of M. G. Gasymov, B. M. Levitan, and I. S. Sargsyan (on Dirac systems of equations), of F. S. Rofe-Beketov (on operator Sturm-Liouville equations), and of V. S. Buslaev, M. I. Lomonosov, and L. D. Faddeev (on a continual analog of the trace formula).

Finally, we wish to emphasize that the author did not intend to exhaust all the aspects and methods of spectral theory, and this is why many of its facets are not discussed here. In particular, we do not touch upon deficiency indices, or the character of the spectrum, or the theory of extensions of operators. Nor have we include fundamental results of H. Weyl, E. Titchmarsh, M. V. Keldysh, M. G. Krein, and M. A. Naimark, the majority of which are treated in well-known monographs on the spectral theory of operators. We also omit an analysis of the stability of the inverse problem of spectral theory. This topic is dealt with in detail in the monograph [20].

V. A. Marchenko

CHAPTER 1

THE STURM-LIOUVILLE EQUATION AND TRANSFORMATION OPERATORS

1. RIEMANN'S FORMULA

Let $u(x,y)$ $(-\infty < x < \infty, 0 \leq y \leq \infty)$ be a twice continuously differentiable solution of the Cauchy problem

$$u_{xx}(x,y) - q_1(x)u(x,y) = u_{yy}(x,y) - q_2(y)u(x,y) \tag{1.1.1}$$

$$u(x,0) = \varphi(x) \quad , \quad u_y(x,0) = \psi(x) \ . \tag{1.1.1'}$$

The value of the function $u(x,y)$ at the point (x_0,y_0) may be thought of as the value of a linear functional $T_{x_0}^{y_0}$ on the vector $(\varphi(x),\psi(x))$:

$$u(x_0,y_0) = T_{x_0}^{y_0}[\varphi(x),\psi(x)] \quad . \tag{1.1.2}$$

An expression for this functional was first found by B. Riemann using the following arguments: let $R(x,y\,;x_0,y_0)$ denote the twice continuously differentiable solution of the equation

$$R_{xx} - q_1(x)R = R_{yy} - q_2(y)R \tag{1.1.3}$$

in the domain shown in Figure 1 which takes the value 1 on the characteristics $x - x_0 = (y - y_0)$ of this equation. Multiplying equations (1.1.1) and (1.1.3) by R and u, respectively, and substracting the second from the first we obtain the identity

$$u_{xx}R - uR_{xx} = u_{yy}R - uR_{yy}$$

in D or, equivalently,

$$\frac{\partial}{\partial x}(u_xR - uR_x) - \frac{\partial}{\partial y}(u_yR - uR_y) = 0 .$$

Therefore

$$\iint_D [\frac{\partial}{\partial x}(u_xR - uR_x) - \frac{\partial}{\partial y}(u_yR - uR_y)]dxdy = 0 ,$$

whence, by Green's formula,

$$\int_\Gamma (u_xR - uR_x)dy + (u_yR - uR_y)dx = 0 , \tag{1.1.4}$$

where Γ denotes the oriented boundary of D, consisting of the three anti-clockwise oriented segments I, II, and III (see Figure 1); thus,

$$\int_\Gamma = \int_I + \int_{II} + \int_{III} . \tag{1.1.5}$$

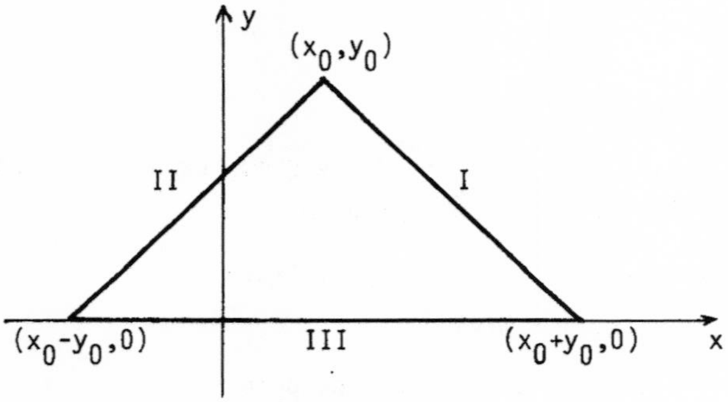

Fig. 1

On segment I, $y = y_0 - (x - x_0)$, $dy = -dx$, and hence

$$\int_I = \int_{x_0+y_0}^{x_0} \{-u_xR + uR_x + u_yR - uR_y\}dx =$$

$$= - \int_{x_0+y_0}^{x_0} \{(u_x - u_y)R + (R_x - R_y)u - 2u(R_x - R_y)\}dx , \tag{1.1.6}$$

where we must take $y_0 - (x - x_0)$ for y in the integrand. Since, for any differentiable function F,

$$\frac{d}{dx} F(x,y_0 - (x - x_0)) = (F_x - F_y)\big|_{y=y_0 - (x - x_0)} ,$$

we have that

$$(u_x - u_y)R + (R_x - R_y)u\big|_{y=y_0 - (x - x_0)} = \frac{d}{dx}(uR\big|_{y=y_0 - (x - x_0)}) ,$$

and

$$(R_x - R_y)\Big|_{y=y_0-(x-x_0)} = \frac{d}{dx} R(x,y_0 - (x-x_0)\,;\,x_0,y_0) = 0$$

because R is identically equal to 1 on the line $y = y_0 - (x - x_0)$. Combining these equalities with formula (1.1.6), we obtain

$$\int_{I} = -\int_{x_0+y_0}^{x_0} \frac{d}{dx}\{uR\big|_{y=y_0-(x-x_0)}\}dx =$$

$$= -\int_{x_0+y_0}^{x_0} \frac{d}{dx} u(x,y_0 - (x-x_0))dx = -u(x,y_0 - (x-x_0))\Big|_{x_0+y_0}^{x_0} =$$

$$= -u(x_0,y_0) + u(x_0 + y_0,0) = -u(x_0,y_0) + \varphi(x_0 + y_0)\ .$$

A similar computation yields the integral on segment II:

$$\int_{II} = \varphi(x_0 - y_0) - u(x_0,y_0)\ .$$

Finally, on segment III, $dy = 0$ and $u(x,y) = \varphi(x), u_y(x,y) = \psi(x)$, and hence

$$\int_{III} = \int_{x_0-y_0}^{x_0+y_0} \{\psi(x)R(x,0\,;\,x_0,y_0) - \varphi(x)R_y(x,0\,;\,x_0,y_0)\}dx\ .$$

Substituting the expressions found above into (1.1.5), we find that

$$0 = -2u(x_0,y_0) + \varphi(x_0+y_0) + \psi(x_0-y_0) +$$
$$+ \int_{x_0-y_0}^{x_0+y_0} \{\psi(x)R(x,0\,;\,x_0,y_0) - \varphi(x)R_y(x,0\,;\,x_0,y_0)\}dx$$

or, equivalently,

$$u(x_0,y_0) = \frac{\varphi(x_0+y_0) + \psi(x_0-y_0)}{2} +$$
$$+ \frac{1}{2}\int_{x_0-y_0}^{x_0+y_0} \{\psi(x)R(x,0\,;\,x_0,y_0) - \varphi(x)R_y(x,0\,;\,x_0,y_0)\}dx\ . \qquad (1.1.7)$$

The function $R(x,y\,;\,x_0,y_0)$ is called the Riemann function, and the expression (1.1.7) for the functional $T_{x_0}^{y_0}$ is known as the Riemann formula. For a rigorous formula we still need to show that the Riemann function with the above properties exists. To this end, we write equation (1.1.3) in the new variables $\xi = x + y$, $\eta = x - y$ $(\xi_0 = x_0 + y_0, \eta_0 = x_0 - y_0)$, setting

$$r(\xi,\eta\,;\xi_0,\eta_0) = R(\frac{\xi+\eta}{2}, \frac{\xi-\eta}{2}; x_0, y_0) .$$

This yields the following equation for the function $r(\xi,\eta) = r(\xi,\eta\,;\xi_0,\eta_0)$ in the domain $D' = \{(\xi,\eta) \mid \eta_0 \leq \eta \leq \xi \leq \xi_0\}$:

$$4r_{\xi\eta} - \{q_1(\frac{\xi+\eta}{2}) - q_2(\frac{\xi-\eta}{2})\}r = 0 \tag{1.1.8}$$

together with the following conditions on the characteristics:

$$r(\xi_0,\eta) = r(\xi,\eta_0) = 1 . \tag{1.1.8'}$$

It is readily verified that if $r(\xi,\eta)$ is a twice continuously differentiable solution of problem (1.1.8)-(1.1.8') (known as the Goursat problem), then the function $r(x+y\,;x-y)$ enjoys all the properties of the Riemann function. Therefore it suffices to prove that problem (1.1.8),(1.1.8') admits a twice continuously differentiable solution in the domain D' . We shall assume that functions $q_1(x)$ and $q_2(y)$ are continuous, so that the function

$$s(\xi,\eta) = \frac{1}{4}\{q_1(\frac{\xi+\eta}{2}) - q_2(\frac{\xi-\eta}{2})\} \tag{1.1.9}$$

is continuous in D' . Problem (1.1.8), (1.1.8') is obviously equivalent to the integral equation

$$r(\xi,\eta) = 1 - \int_{\xi}^{\xi_0} d\alpha \int_{\eta_0}^{\eta} s(\alpha,\beta)r(\alpha,\beta)d\beta . \tag{1.1.10}$$

This equation has a unique continuous solution which can be obtained by the method of successive approximations. In fact, set

$$r_0(\xi,\eta) = 1 , \quad r_n(\xi,\eta) = -\int_{\xi}^{\xi_0} d\alpha \int_{\eta_0}^{\eta} s(\alpha,\beta)r_{n-1}(\alpha,\beta)d\beta$$

and denote $\sup_{D'}|s(\alpha,\beta)| = M$. Then,

$$|r_0(\xi,\eta)| \leq 1 ,$$

$$|r_1(\xi,\eta)| \leq M(\xi_0-\xi)(\eta-\eta_0) ,$$

$$|r_2(\xi,\eta)| \leq \int_{\xi}^{\xi_0} d\alpha \int_{\eta_0}^{\eta} |s(\alpha,\beta)||r_1(\alpha,\beta)|d\beta \leq \frac{M^2(\xi-\xi_0)^2(\eta-\eta_0)^2}{(2!)^2} ,$$

uniformly in D' and, by induction,

$$|r_n(\xi,\eta)| \leq \int_{\xi}^{\xi_0} d\alpha \int_{\eta_0}^{\eta} |s(\alpha,\beta)||r_{n-1}(\alpha,\beta)|d\beta \leq \frac{M^n(\xi-\xi_0)^n(\eta-\eta_0)^n}{(n!)^2} .$$

These estimates show that the series $r(\xi,\eta) = \sum_{n=0}^{\infty} r_n(\xi,\eta)$, of continuous functions r_n , converges absolutely and uniformly in D' and that its sum satisfies equation (1.1.10).

In view of the continuity of the function $s(\alpha,\beta)r(\alpha,\beta)$, it follows from equation (1.1.10) that $r(\xi,\eta)$ is twice continuously differentiable in D' , and that

$$\left.\begin{aligned} r_\xi(\xi,\eta) &= \int_{\eta_0}^{\eta} s(\xi,\beta)r(\xi,\beta)d\beta , \\ r_\eta(\xi,\eta) &= -\int_{\xi}^{\xi_0} s(\alpha,\eta)r(\alpha,\eta)d\alpha . \end{aligned}\right\} \qquad (1.1.11)$$

Therefore, if $q_1(x)$ and $q_2(y)$ (and hence $s(\alpha,\beta)$) are continuously differentiable, then $r(\xi,\eta)$ is twice continuously differentiable. This completes the proof of the existence of the Riemann function in the case where $q_1(x)$ and $q_2(y)$ are continuously differentiable; also,

$$R(x,y\,;x_0,y_0) = r(x+y\,,x-y) , \qquad (1.1.12)$$

where $r(\xi,\eta)$ is the solution of the integral equation (1.1.10).

In the case where the functions $q_1(x)$ and $q_2(y)$ are merely continuous, we use the solution of the integral equation (1.1.10) to define Riemann's function by formula (1.1.12). By the preceding discussion, this yields a continuously differentiable function whose derivatives are calculated by means of formula (1.1.11). Let $q_1^{(n)}(x)$ and $q_2^{(n)}(y)$ be sequences of continuously differentiable functions which converge uniformly to $q_1(x)$ and $q_2(y)$ (in the domain $-\infty < x < \infty$, $0 \leq y < \infty$) as $n \to \infty$. Then, it follows from equation (1.1.10) that the sequence of corresponding Riemann functions $R^{(n)}(x,y\,;x_0,y_0)$ converges uniformly in the domain D to the Riemann function $R(x,y\,;x_0,y_0)$, which was constructed for $q_1(x)$ and $q_2(y)$. Furthermore, by formulas (1.1.11), the sequences of first order partial derivatives $R_x^{(n)}(x,y\,;x_0,y_0)$ and $R_y^{(n)}(x,y\,;x_0,y_0)$ converge uniformly in D to $R_x(x,y\,;x_0,y_0)$ and $R_y(x,y\,;x_0,y_0)$, respectively.

The functions $R^{(n)}(x,y\,;x_0,y_0)$ are twice continuously differentiable and satisfy the equation

$$R_{xx}^{(n)} - q_1^{(n)}(x)R^{(n)} = R_{yy}^{(n)} - q_2^{(n)}R^{(n)} . \tag{1.1.13}$$

Arguing for (1.1.1) and (1.1.13) in exactly the same way we did for equations (1.1.1) and (1.1.3), we obtain the formula

$$u(x_0,y_0) = \frac{\varphi(x_0+y_0) + \varphi(x_0-y_0)}{2} +$$

$$+ \frac{1}{2}\int_{x_0-y_0}^{x_0+y_0} \{\psi(x)R^{(n)}(x,0\,;x_0,y_0) - \varphi(x)R_y^{(n)}(x,0\,;x_0,y_0)\}dx +$$

$$+ \frac{1}{2}\iint_D \{q_1^{(n)}(x) - q_1(x) + q_2^{(n)}(y) - q_2(y)\}R^{(n)}(x,y\,;x_0,y_0)dxdy$$

in place of (1.1.7). The previous arguments show that in these formulas we can let $n\to\infty$ under the integral signs. This proves Riemann's formula (1.1.7) for the case of continuous functions $q_1(x)$ and $q_2(y)$.

We can thus formulate the following result.

THEOREM 1.1.1. *Suppose that the functions* $q_1(x)$ *and* $q_2(y)$ *are continuous. Then every twice continuously differentiable solution of the Cauchy problem* (1.1.1)-(1.1.1') *admits the representation* (1.1.7), *in which* $R(x,y\,;x_0,y_0)$ *is the continuously differentiable function constructed from the solution of the integral equation* (1.1.10) *by formula* (1.1.12).

□

Remark 1. For this theorem to hold true it suffices that the functions $q_1(x)$, $q_2(x)$, and $u(x,y)$ satisfy the conditions listed in its formulation in the domain D , because in deriving formula (1.1.7) we never left D .

Remark 2. In the proof of the theorem it was assumed that $y>0$. It is readily seen, however, that this is not essential. The theorem is valid for all values of y provided that the functions $q_1(x)$, $q_2(x)$, and $u(x,y)$ satisfy the required conditions in the domain D bounded by the lines $x-x_0 = \pm(y-y_0)$ and the abcissa axis $y=0$.

In the next section we shall find the particular case of formula (1.1.7), where $\psi(x) = \varphi'(x)$, useful. Since the Riemann function is continuously differentiable, the term in formula (1.1.7) containing $\psi(x) = \varphi'(x)$ can be integrated by parts once. This leads to the following representation for the twice continuously differentiable solution $u(x,y)$ of the Cauchy problem (1.1.1), (1.1.1') with $\psi(x) = \varphi'(x)$:

$$-\frac{1}{2}\int_{x_0-y_0}^{x_0+y_0} \{R_x(x,0;x_0,y_0) + R_y(x,0;x_0,y_0)\}\varphi(x)dx \quad . \tag{1.1.14}$$

PROBLEMS

1. Show that equation (1.1.10) admits a differentiable solution even when the functions $q_1(x)$ and $q_2(x)$ are merely locally summable.

2. Generalize Theorem 1.1.1 to the case of locally summable functions $q_1(x)$ and $q_2(x)$.

2. TRANSFORMATION OPERATORS

Consider the Sturm-Liouville differential equation

$$y'' - q(x)y + \lambda^2 y = 0 \tag{1.2.1}$$

on the interval $(-a,a)$ (with $a \le \infty$), where $q(x)$ is a complex-valued function which is continuous on this interval and λ is a complex parameter. Henceforth $q(x)$ will be referred to as the potential of equation (1.2.1) or of the allied Sturm-Liouville operator $L = -(d^2/dx^2) + q(x)$. Let $e_0(\lambda,x)$ denote the solution of equation (1.2.1) with initial data

$$e_0(\lambda,0) = 1 \ , \quad e_0'(\lambda,0) = i\lambda \tag{1.2.2}$$

(here the index "0" indicates that the initial data are given at the point 0, while the letter e indicates that these data are the same as for the function $e^{i\lambda x}$, which coincides with $e_0(\lambda,x)$ when $q(x) \equiv 0$). The function $u(x,y) = e^{i\lambda x}e_0(\lambda,y)$ is obviously twice continuously differentiable for $-\infty < x < \infty$, $-a < y < a$, and solves the Cauchy problem

$$\left.\begin{aligned} &u_{xx} = u_{yy} - q(y)u \ , \\ &u(x,0) = e^{i\lambda x} \ , \quad u_y(x,0) = i\lambda e^{i\lambda x} = (e^{i\lambda x})' \ . \end{aligned}\right\} \tag{1.2.3}$$

Hence, formula (1.1.14) applies and yields

$$e^{i\lambda x_0} e_0(\lambda, y_0) = e^{i\lambda(x_0+y_0)} -$$

$$- \frac{1}{2} \int_{x_0-y_0}^{x_0+y_0} \{R_x(x,0\,;x_0,y_0) + R_y(x,0\,;x_0,y_0)\} e^{i\lambda x} dx .$$

Letting $x_0 = 0$ here and denoting

$$K(y_0,x) = -\frac{1}{2} \{R_x(x,0\,;0,y_0) + R_y(x,0\,;0,y_0)\} , \tag{1.2.4}$$

we get

$$e_0(\lambda,y_0) = e^{i\lambda y_0} + \int_{-y_0}^{y_0} K(y_0,x) e^{i\lambda x} dx .$$

We have thus proved the following theorem.

THEOREM 1.2.1. *The solution* $e_0(\lambda,x)$ *of equation* (1.2.1) *with initial data* (1.2.2) *admits the representation*

$$e_0(\lambda,x) = e^{i\lambda x} + \int_{-x}^{x} K(x,t) e^{i\lambda t} dt , \tag{1.2.5}$$

where $K(x,t)$ *is a continuous function which is expressed in terms of the Riemann function of equation* (1.2.3) *by formula* (1.2.4).

□

The integral operator $\mathbb{I} + \mathbb{K}$ defined by the rule

$$(\mathbb{I} + \mathbb{K})f = f(x) + \int_{-x}^{x} K(x,t) f(t) dt$$

is called the transformation operator preserving the initial data at the point $x = 0$. It takes the function $e^{i\lambda x}$ (i.e., the solution of the simplest equation of the form (1.2.1) with initial data (1.2.2)) into the solution of equation (1.2.1) with the same initial data. Actually, since $e^{i\lambda x}$ and $e^{-i\lambda x}$ constitute a fundamental system of solutions for the equation $y'' + \lambda^2 y = 0$, the operator $\mathbb{I} + \mathbb{K}$ transforms any solution of this equation into a solution of equation (1.2.1) with the same initial data at $x = 0$. Therefore, the solution $\omega(\lambda,x\,;h)$ of Equation (1.2.1) with initial data

$$\omega(\lambda,0\,;h) = 1 \quad , \quad \omega'(\lambda,0\,;h) = h \tag{1.2.6}$$

can be expressed as

$$\omega(\lambda,x\,;h) = \cos\lambda x + h\,\frac{\sin\lambda x}{\lambda} + \int_{-x}^{x} K(x,t)\{\cos\lambda t + h\,\frac{\sin\lambda t}{\lambda}\}\,dt =$$
$$= \cos\lambda x + \int_0^x \{h + K(x,t) + K(x,-t)\}\cos\lambda t\,dt +$$
$$+ h\int_0^x [\{K(x,t) - K(x,-t)\}\int_0^t \cos\lambda\xi\,d\xi]\,dt =$$
$$= \cos\lambda x + \int_0^x K(x,t\,;h)\cos\lambda t\,dt\ ,$$

where

$$K(x,t\,;h) = h + K(x,t) + K(x,-t) + h\int_t^x \{K(x,\xi) - K(x,-\xi)\}\,d\xi\ . \tag{1.2.7}$$

Similarly, the solution $\omega(\lambda,x\,;\infty)$ of equation (1.2.1) with initial data

$$\omega(\lambda,0\,;\infty) = 0\quad ,\quad \omega'(\lambda,0\,;\infty) = 1 \tag{1.2.8}$$

admits the representation

$$\omega(\lambda,x\,;\infty) = \frac{\sin\lambda x}{\lambda} + \int_0^x K(x,t\,;\infty)\,\frac{\sin\lambda t}{\lambda}\,dt\ ,$$

where

$$K(x,t\,;\infty) = K(x,t) - K(x,-t)\ . \tag{1.2.9}$$

Thus, Theorem 1.2.1 has the following corollary.

COROLLARY. *The solutions* $\omega(\lambda,x\,;h)$ and $\omega(\lambda,x\,;\infty)$ *of equation* (1.2.1) *with initial data* (1.2.6) *and* (1.2.8) *can be expressed as*

$$\omega(\lambda,x\,;h) = \cos\lambda x + \int_0^x K(x,t\,;h)\cos\lambda t\,dt\ , \tag{1.2.10}$$

$$\omega(\lambda,x\,;\infty) = \frac{\sin\lambda x}{\lambda} + \int_0^x K(x,t\,;\infty)\,\frac{\sin\lambda t}{\lambda}\,dt\ , \tag{1.2.11}$$

respectively, where the continuous functions $K(x,t\,;h)$ *and* $K(x,t\,;\infty)$ *are given in terms of the kernel* $K(x,t)$ *of the operator* (1.2.5) *by formulas* (1.2.7) *and* (1.2.9), *respectively.*

□

The operators $\mathbb{I} + \mathbb{K}$, $\mathbb{I} + \mathbb{K}_h$, and $\mathbb{I} + \mathbb{K}_\infty$ defined by the right-hand sides of the equalities (1.2.5), (1.2.10), and (1.2.11) will be referred to as the transformation operators attached to the point 0. It is clear that instead of 0 we could take any point a , and accordingly replace formulas

(1.2.5), (1.2.10), and (1.2.11) by

$$e_a(\lambda,x) = e^{i\lambda(x-a)} + \int_{-x+2a}^{x} K_a(x,t)e^{i\lambda(t-a)}dt\ , \tag{1.2.5'}$$

$$\omega_a(\lambda,x\,;h) = \cos\lambda(x-a) + \int_a^x K_a(x,t\,;h)\cos\lambda(t-a)dt\ , \tag{1.2.10'}$$

and

$$\omega_a(\lambda,x\,;\infty) = \frac{\sin\lambda(x-a)}{\lambda} + \int_a^x K_a(x,t\,;\infty)\frac{\sin\lambda(t-a)}{\lambda}dt\ , \tag{1.2.11'}$$

respectively, in which $e_a(\lambda,x)$, $\omega_a(\lambda,x\,;h)$, and $\omega(\lambda,x\,;\infty)$ designate the solutions of equation (1.2.1) with the following initial data at the point $x = a : e_a(\lambda,a) = 1$, $e'(\lambda,a) = i\lambda$; $\omega_a(\lambda,a\,;h) = 1$, $\omega_a'(\lambda,a\,;h) = h$; and $\omega_a(\lambda,a\,;\infty) = 0$, $\omega_a'(\lambda,a\,;\infty) = 1$.

Naturally, the operators defined by the right-hand sides of the equalities (1.2.5'), (1.2.10') and (1.2.11') will be referred to as the transformation operators attached to the point a . Since $\mathbb{K}$, $\mathbb{K}_h$ and $\mathbb{K}_\infty$ are Volterra integral operators, the operators $\mathbb{I}+\mathbb{K}$, $\mathbb{I}+\mathbb{K}_h$, and $\mathbb{I}+\mathbb{K}_\infty$ have inverses of the same form, which we denote by $\mathbb{I}+\mathbb{L}$, $\mathbb{I}+\mathbb{L}_h$, and $\mathbb{I}+\mathbb{L}_\infty$, respectively. Thus, we can write along with formulas (1.2.5), (1.2.10), and (1.2.11) the following equalities:

$$e^{i\lambda x} = e_0(\lambda,x) + \int_{-x}^{x} L(x,t)e_0(\lambda,t)dt\ , \tag{1.2.5''}$$

$$\cos\lambda x = \omega(\lambda,x\,;h) + \int_0^x L(x,t\,;h)\omega(\lambda,t\,;h)dt\ , \tag{1.2.10''}$$

and

$$\frac{\sin\lambda x}{\lambda} = \omega(\lambda,x\,;\infty) + \int_0^x L(x,t\,;\infty)\omega(\lambda,t\,;\infty)dt\ . \tag{1.2.11''}$$

The kernels $L(x,t)$, $L(x,t\,;h)$, and $L(x,t\,;\infty)$ are continuous solutions of the corresponding Volterra integral equations.

The transformation operators $\mathbb{I}+K$, $\mathbb{I}+K_h$, and $\mathbb{I}+K_\infty$ and their inverses $\mathbb{I}+\mathbb{L}$, $\mathbb{I}+\mathbb{L}_h$, and $\mathbb{I}+\mathbb{L}_\infty$ play a very important role in the spectral theory of Sturm-Liouville equations. The very fact that they exist suffices to solve many basic problems. However, in some situations it is desirable to have more detailed information about the properties of these operators. For example, estimates for the kernels or for the derivatives of the kernels of these operators are of interest. To provide tools for extracting such information, we next derive some convenient integral equations for the kernels of transformation operators. We remark that since the results obtained above will not be used in what follows, we shall obtain en route a new proof of Theorem 1.2.1, i.e., of the existence of transformation operators.

Let us rewrite equation (1.2.1) in the form

$$y'' + \lambda^2 y = q(x)y$$

and, regarding the right-hand side as known, seek the solution $e_0(\lambda,x)$ of this equation by the method of variation of constants. This yields the equation

$$e_0(\lambda,x) = e^{i\lambda x} + \int_0^x \frac{\sin \lambda(x-t)}{\lambda} q(t)e_0(\lambda,t)dt , \tag{1.2.12}$$

which is equivalent to the problem (1.2.1), (1.2.2). This is an integral equation for the function $e_0(\lambda,x)$ (which is known as the Sturm-Liouville integral equation). We seek its solution in the form (1.2.5). In order for a function of the latter form to satisfy equation (1.2.12), it is necessary that the equality

$$\begin{aligned}\int_{-x}^{x} K(x,t)e^{i\lambda t}dt &= \int_0^x \frac{\sin \lambda(x-t)}{\lambda} q(t)e^{i\lambda t}dt + \\ &+ \int_0^x \frac{\sin \lambda(x-t)}{\lambda} q(t) \int_{-t}^{t} K(t,\xi)e^{i\lambda\xi}d\xi dt\end{aligned} \tag{1.2.13}$$

hold. Conversely, if $K(x,t)$ satisfies this equality, then the function $e_0(\lambda,x)$ satisfies equation (1.2.12), i.e., it solves equation (1.2.1) with initial data (1.2.2).

Next, we express the right-hand side of equality (1.2.13) as a Fourier transform. Since

$$\frac{\sin \lambda(x-1)}{\lambda} e^{i\lambda\xi} = \frac{1}{2} \int_{\xi-(x-t)}^{\xi+(x-t)} e^{i\lambda u}du , \tag{1.2.14}$$

it follows that

$$\int_0^x \frac{\sin \lambda(x-t)}{\lambda} q(t)e^{i\lambda t}dt = \frac{1}{2}\int_0^x q(t) \left\{ \int_{2t-x}^{x} e^{i\lambda u}du \right\} dt =$$

$$= \frac{1}{2} \int_{-x}^{x} e^{i\lambda u}\left\{ \int_0^{\frac{x+u}{2}} q(t)dt \right\} du .$$

Changing the variables of integration we get

$$\int_0^x \frac{\sin\lambda(x-t)}{\lambda} q(t)e^{i\lambda t}dt = \int_{-x}^{x} \frac{1}{2}\left\{ \int_0^{\frac{x+t}{2}} q(\xi)d\xi \right\} e^{i\lambda t}dt . \tag{1.2.15}$$

Using formula (1.2.14) once more, we obtain the equality

$$\int_0^x \frac{\sin\lambda(x-t)}{\lambda} q(t) \int_{-t}^{t} K(t,\xi)e^{i\lambda\xi}d\xi dt =$$

$$= \frac{1}{2}\int_0^x q(t)\left\{ \int_{-t}^{t} K(t,\xi) \int_{\xi-(x-t)}^{\xi+(x-t)} e^{i\lambda u}dud\xi \right\} dt .$$

Next, upon extending the function $K(t,\xi)$ by zero for $|\xi| > |t|$, we can write

$$\int_{-t}^{t} K(t,\xi) \int_{\xi-(x-t)}^{\xi+(x-t)} e^{i\lambda u}dud\xi = \int_{-\infty}^{\infty} K(t,\xi) \int_{\xi-(x-t)}^{\xi+(x-t)} e^{i\lambda u}dud\xi =$$

$$= \int_{-\infty}^{\infty} e^{i\lambda u} \int_{u-(x-t)}^{u+(x-t)} K(t,\xi)d\xi du = \int_{-x}^{x} e^{i\lambda u} \int_{u-(x-t)}^{u+(x-t)} K(t,\xi)d\xi du ,$$

for any choice of $t \in (-x,x)$. Consequently,

$$\int_t^x q(t) \left\{ \int_{-t}^{t} K(t,\xi) \int_{\xi-(x-t)}^{\xi+(x-t)} e^{i\lambda u}dud\xi \right\} dt =$$

$$= \int_{-x}^{x} e^{i\lambda u}\left\{ \int_0^x q(t) \int_{u-(x-t)}^{u+(x-t)} K(t,\xi)d\xi dt \right\} du ,$$

whence, upon relabelling the variables of integration,

$$\int_0^x \frac{\sin\lambda(x-t)}{\lambda} q(t) \int_{-t}^{t} K(t,\xi)e^{i\lambda\xi}d\xi dt =$$

$$= \int_{-x}^{x} e^{i\lambda t} \frac{1}{2}\left\{ \int_0^x q(u) \int_{t-(x-u)}^{t+(x-u)} K(u,\xi)d\xi du \right\} dt . \tag{1.2.16}$$

Formulas (1.2.15) and (1.2.16) show that equality (1.2.13) is equivalent to

$$\int_{-x}^{x} K(x,t)e^{i\lambda t}dt =$$

$$= \frac{1}{2}\int_{-x}^{x}\left\{ \int_0^{\frac{x+t}{2}} q(u)du + \int_0^x q(u) \int_{t-(x-u)}^{t+(x-u)} K(u,\xi)d\xi du \right\} e^{i\lambda t}dt .$$

Therefore, if the function $K(x,t)$ vanishes for $|t| > |x|$ and satisfies the equation

$$K(x,t) = \frac{1}{2} \int_0^{\frac{x+t}{2}} q(u)du + \frac{1}{2} \int_0^x q(u)du \int_{t-(x-u)}^{t+(x-u)} K(u,\xi)d\xi du \ , \qquad (1.2.17)$$

then the function $e(\lambda,x)$ constructed by means of formula (1.2.5) are solutions of equation (1.2.12) for all values of λ, and conversely.

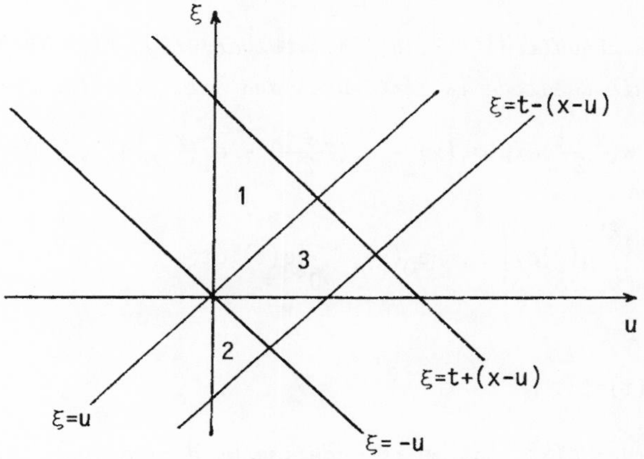

Fig. 2

The domain of integration of the double integral on the right-hand side of formula (1.2.17) is shown in Figure 2. It consists of three regions: 1, 2, and 3. In regions 1 and 2, $|\xi| > |u|$, and hence $K(u,\xi) \equiv 0$ there. Thus, in equation (1.2.17) the integral is actually taken over the rectangle 3. Upon performing the change of variables $u + \xi = 2\alpha$ and $u - \xi = 2\beta$ in this integral, we get the equation

$$K(x,t) = \frac{1}{2} \int_0^{\frac{x+t}{2}} q(y)dy + \int_0^{\frac{x+t}{2}} d\alpha \int_0^{\frac{x-t}{2}} q(\alpha + \beta)K(\alpha+\beta,\alpha - \beta)d\beta \ , \qquad (1.2.18)$$

in which it is already taken into account that $K(x,t) \equiv 0$ for $|t| > |x|$.

Therefore, if the solutions $e_0(\lambda,x)$ of equation (1.2.1) with initial data (1.2.2) admit the representation (1.2.5) for all values of λ , then the kernel $K(x,t)$ must satisfy equation (1.2.18). Conversely, if $K(x,t)$ satisfies (1.2.18), then the right-hand side of formula (1.2.15) is a solution of equation (1.2.1) with initial data (1.2.2) for every value of λ .

Next we set

$$\left.\begin{aligned} &H(\alpha,\beta) = K(\alpha+\beta\,,\,\alpha-\beta)\ , \\ &x+t = 2u \quad , \quad x-t = 2v\ . \end{aligned}\right\} \tag{1.2.19}$$

Then equation (1.2.18) takes the form

$$H(u,v) = \frac{1}{2}\int_0^u q(y)dy + \int_0^u d\alpha \int_0^v q(\alpha+\beta)H(\alpha,\beta)d\beta\ . \tag{1.2.18'}$$

THEOREM 1.2.2. *Equation* (1.2.18) *or, equivalently,* (1.2.18'), *has a unique solution. This solution is continuous and satisfies the inequality*

$$|K(t,x)| \leq \frac{1}{2}\, w\left(\frac{x+t}{2}\right)\exp\left\{\sigma_1(x) - \sigma_1\left(\frac{x+t}{2}\right) - \sigma_1\left(\frac{x-t}{2}\right)\right\}\ , \tag{1.2.20}$$

in which

$$w(u) = \max_{0\leq\xi\leq u}\left|\int_0^\xi q(y)dy\right| \quad , \quad \sigma_0(x) = \int_0^x |q(t)|dt\ ,$$

and (1.2.21)

$$\sigma_1(x) = \int_0^x \sigma_0(t)dt\ .$$

If the function $q(x)$ *has* $n\geq 0$ *continuous derivatives, then the kernel* $K(x,t)$ *has continuous derivatives of order* $n+1$ *with respect to both variables.*

PROOF. We solve equation (1.2.18') by the method of successive approximations, setting

$$H_0(u,v) = \frac{1}{2}\int_0^u q(y)dy\ ,$$

and

$$H_n(u,v) = \int_0^u d\alpha \int_0^v q(\alpha+\beta)H_{n-1}(\alpha,\beta)d\beta\ .$$

If the series $\sum_{n=0}^{\infty} H_n(u,v)$ converges uniformly on some square $0\leq u\ ,\ v\leq a$, then its sum is obviously a continuous solution of equation (1.2.18') in this square. To verify the uniform convergence of the indicated series, we show that

$$|H_n(u,v)| \leq \frac{1}{2}\, w(u)\, \frac{\{\sigma_1(u+v) - \sigma_1(u) - \sigma_1(v)\}^n}{n!}\ , \tag{1.2.22}$$

It follows from the definition of the function $w(x)$ that

$$|H_0(u,v)| = \frac{1}{2}\left|\int_0^u q(y)dy\right| \leqslant \frac{1}{2}w(u) ,$$

so that estimate (1.2.22) holds for $n = 0$. Assuming that it holds for $n - 1$, we conclude that it holds for n too, because

$$|H_n(u,v)| \leqslant$$

$$\leqslant \frac{1}{2}\left|\int_0^u d\alpha \int_0^v |q(\alpha+\beta)|w(\alpha) \frac{\{\sigma_1(\alpha+\beta) - \sigma_1(\alpha) - \sigma_1(\beta)\}^{n-1}}{(n-1)!} d\beta\right| \leqslant$$

$$\leqslant \frac{1}{2}w(u) \int_0^u \left\{\frac{\{\sigma_1(v+\alpha) - \sigma_1(\alpha) - \sigma_1(v)\}^{n-1}}{(n-1)!} \int_0^v |q(\alpha+\beta)|d\beta\right\} d\alpha \leqslant$$

$$\leqslant \frac{1}{2}w(u) \int_0^u \frac{\{\sigma_1(v+\alpha) - \sigma_1(\alpha) - \sigma_1(v)\}^{n-1}}{(n-1)!} \{\sigma_0(v+\alpha) - \sigma_0(\alpha)\}d\alpha =$$

$$= \frac{1}{2}w(u) \frac{\{\sigma_1(v+u) - \sigma_1(u) - \sigma_1(v)\}^n}{n!}$$

(here we used the fact that $w(u)$ and $\sigma_1(u+v) - \sigma_1(u) - \sigma_1(v)$ are nondecreasing functions of u and v).

From the validity of the estimates (1.2.22) it follows that equation (1.2.18') has a continuous solution $H(u,v) = \sum_{n=0}^{\infty} H_n(u,v)$ which satisfies the bound

$$|H(u,v)| \leqslant \sum_{n=0}^{\infty} |H_n(u,v)| \leqslant$$

$$\leqslant \frac{1}{2}w(u)\exp\{\sigma_1(u+v) - \sigma_1(u) - \sigma_1(v)\} . \tag{1.2.23}$$

Thus, the existence of a continuous solution to equation (1.2.18'), and hence to (1.2.18), is established. Moreover, by (1.2.23) the function $K(x,t)$ satisfies inequality (1.2.20). The uniqueness of the solution is proved in the usual manner.

□

Theorem 1.2.2 provides a new proof of the existence of transformation operators as well as an estimate for their kernels. Let us pause to examine equation (1.2.18') in more detail. Since $q(y)$ and $H(\alpha,\beta)$ are continuous, the function $H(u,v)$ is continuously differentiable, and

$$\left.\begin{aligned}\frac{\partial H(u,v)}{\partial u} &= \frac{1}{2}q(u) + \int_0^v q(u+\beta)H(u,\beta)d\beta \ , \\ \frac{\partial H(u,v)}{\partial v} &= \int_0^u q(\alpha+v)H(\alpha,v)d\alpha \ .\end{aligned}\right\} \qquad (1.2.24)$$

From these equalities it follows that $H(u,v)$ has a continuous mixed derivative, and

$$\frac{\partial^2 H(u,v)}{\partial u \partial v} = q(u+v)H(u,v) \ . \qquad (1.2.25)$$

Moreover, equation (1.2.18') implies readily that

$$H(u,0) = \frac{1}{2}\int_0^u q(y)dy \quad \text{and} \quad H(0,v) = 0 \ . \qquad (1.2.26)$$

Therefore, in order for $K(x,t)$ to be the kernel of the transformation operator (1.2.5), it is necessary and sufficient that $H(u,v) = K(u+v,u-v)$ be a solution of the Goursat problem (1.2.25),(1.2.26).

Next, it follows from formulas (1.2.24) that the function $H(u,v)$, and hence also $K(x,t)$, has continuous derivatives of order $n+1$ if the function $q(x)$ has n continuous derivatives. Hence, if $q(x)$ is continuously differentiable, then in equation (1.2.25) we can shift back to the variables $x = u + v$ and $t = u - v$. This transforms the problem (1.2.25), (1.2.26) into

$$\frac{\partial^2 K(x,t)}{\partial x^2} - \frac{\partial^2 K(x,t)}{\partial t^2} = q(x)K(x,t) \ , \qquad (1.2.27)$$

$$K(x,x) = \frac{1}{2}\int_0^x q(\xi)d\xi \ , \quad K(x,-x) = 0 \ . \qquad (1.2.28)$$

(Formulas (1.2.28) are of course valid for arbitrary continuous functions $q(x)$, but equation (1.2.27) can be written only for differentiable functions $q(x)$ if one wishes to avoid using generalized derivatives.)

The kernels of the other transformation operators also satisfy equations analogous to (1.2.27), and these can in fact be derived more directly. We illustrate this by deriving the equation for the kernel $L(x,t) = L(x,t;h)$ of the operator which transforms the solution

$\omega(\lambda,x) = \omega(\lambda,x;h)$ into $\cos \lambda x$. To this end we set

$$z(x) = \omega(\lambda,x) + \int_0^x L(x,t)\omega(\lambda,t)dt \tag{1.2.29}$$

and determine the conditions that the twice continuously differentiable function $L(x,t)$ must obey in order that $z(x) \equiv \cos\ x$. Differentiating both sides of equality (1.2.29) twice with respect to x , we get

$$z''(x) = \omega''(\lambda,x) + [L(x,x)\omega(\lambda,x)]' + L_x(x,t)\omega(\lambda,t)\Big|_{t=x} +$$

$$+ \int_0^x L_{xx}(x,t)\omega(\lambda,t)dt ,$$

whence

$$z''(x) + \lambda^2 z(x) = \omega''(\lambda,x) + \lambda^2\omega(\lambda,x) + [L(x,x)\omega(\lambda,x)]' +$$

$$+ L_x(x,t)\omega(\lambda,t)\Big|_{t=x} + \int_0^x L_{xx}(x,t)\omega(\lambda,t)dt +$$

$$+ \lambda^2 \int_0^x L(x,t)\omega(\lambda,t)dt .$$

Now, upon using the equation $y'' - q(x)y + \lambda^2 y = 0$ satisfied by the function $\omega(\lambda,x)$ and integrating by parts twice, we find that

$$\lambda^2 \int_0^x L(x,t)\omega(\lambda,t)dt = - \int_0^x L(x,t)\{\omega''(\lambda,t) - q(t)\omega(\lambda,t)\}dt =$$

$$= -L(x,x)\omega'(\lambda,x) + L(x,0)\omega'(\lambda,0) + L_t(x,t)\omega(\lambda,t)\Big|_{t=x} -$$

$$- L_t(x,t)\omega(\lambda,t)\Big|_{t=0} - \int_0^x \{L_{tt}(x,t) - q(t)L(x,t)\}\omega(\lambda,t)dt .$$

Consequently,

$$z''(x) + \lambda^2 z(x) = q(x)\omega(\lambda,x) + L'(x,x)\omega(\lambda,x) + L(x,x)\omega'(\lambda,x) +$$

$$+ L_x(x,t)\omega(\lambda,t)\Big|_{t=x} - L(x,x)\omega'(\lambda,x) + L(x,0)h +$$

$$+ L_t(x,t)\omega(\lambda,t)\Big|_{t=x} - L_t(x,0) +$$

$$+ \int_0^x \{L_{xx}(x,t) - L_{tt}(x,t) + q(t)L(x,t)\}\omega(\lambda,t)dt ,$$

which shows that in order for the function $z(x)$ to be a solution of the equation $z'' + \lambda^2 z = 0$ with initial data $z(0) = 1$, $z'(0) = 0$ (i.e., that $z(x) \equiv \cos \lambda x$), it suffices that the kernel $L(x,t)$ satisfy the equation

$$L_{xx}(x,t) = L_{tt}(x,t) - q(t)L(x,t) \tag{1.2.30}$$

and the conditions

$$L(x,x) = -h - \frac{1}{2}\int_0^x q(t)dt \quad , \quad L(x,0)h - L_t(x,0) = 0 \ . \tag{1.2.31}$$

We claim that if the function $q(x)$ is differentiable, then equation (1.2.30) has a solution satisfying conditions (1.2.31).

In fact, passing to the variables $u = x + t$ and $v = x - t$, we obtain the equation

$$4\frac{\partial^2 A(u,v)}{\partial u \partial u} = -q(\frac{u+v}{2})A(u,v) \tag{1.2.32}$$

for the function

$$A(u,v) = L(\frac{u+v}{2}, \frac{u-v}{2}) \ ,$$

and the conditions

$$\left.\begin{aligned} &A(u,0) = -h - \frac{1}{2}\int_0^{\frac{u}{2}} q(t)dt \ , \\ &\{A(u,v)h + A_v(u,v) - A_u(u,v)\}\big|_{u=0} = 0 \ . \end{aligned}\right\} \tag{1.2.33}$$

By successive integration we further obtain

$$A_u(u,v) = -\frac{1}{4}\int_0^v q(\frac{u-\xi}{2})A(u,\xi)d\xi - \frac{1}{4}q(\frac{u}{2}) \ ,$$

$$A(u,v) = -\frac{1}{4}\int_v^u q(\frac{\xi}{2})d\xi - \frac{1}{4}\int_v^u d\eta \int_0^v q(\frac{\eta-\xi}{2})A(\eta,\xi)d\xi + A(v,v) \ ,$$

and

$$2A_u(u,v)\big|_{u=v} = \{A(u,v)h + A_v(u,v) + A_u(u,v)\}\big|_{u=v} =$$

$$= -\frac{1}{2}q(\tfrac{v}{2}) - \frac{1}{2}\int_0^v q(\tfrac{v-\xi}{2})A(v,\xi)d\xi \ .$$

The last equality implies that

$$\{e^{hv}A(v,v)\}' = -\frac{1}{2}e^{hv}\left\{q(\tfrac{v}{2}) + \int_0^v q(\tfrac{v-\xi}{2})A(v,\xi)d\xi\right\}$$

whence

$$A(v,v) = -he^{-hv} - \frac{1}{2}e^{-hv}\int_0^v e^{h\eta}\left\{q(\tfrac{\eta}{2}) + \int_2^\eta q(\tfrac{\eta-\xi}{2})A(\eta,\xi)d\xi\right\}d\eta \ .$$

Therefore, the function $A(u,v)$ must satisfy the integral equation

$$A(u,v) = -he^{-hv} - \frac{1}{4}\int_v^u q(\tfrac{\xi}{2})d\xi - \frac{e^{-hv}}{2}\int_0^v e^{h\eta}q(\tfrac{\eta}{2})d\eta -$$

$$- \frac{e^{-hv}}{2}\int_0^v e^{h\eta}\int_0^\eta q(\tfrac{\eta-\xi}{2})A(\eta,\xi)d\xi d\eta - \frac{1}{4}\int_v^u d\eta\int_0^v q(\tfrac{\eta-\xi}{2})A(\eta,\xi)d\xi \ . \qquad (1.2.34)$$

Conversely, if $A(u,v)$ is a solution of this equation and $q(x)$ is differentiable, then it is a straightforward job to check that the function $L(x,t)$ satisfies equation (1.2.30) and conditions (1.2.31), and hence is the kernel of the sought-for transformation operator.

The existence of a solution to the integral equation (1.2.34) is established by the method of successive approximations, exactly as in Theorem 1.2.2. This automatically yields estimates for the solution and its first derivatives. In the case $\operatorname{Re} h < 0$ it is technically simpler to pass to the new unknown function $B(u,v) = e^{hv}A(u,v)$ in order to avoid unnecessarily crude estimates.

We give the final results for the kernel $L(x,t;h)$:

$L(0,0;h) = -h$,

$$|L(x,t;h)| \le \{|h| + \sigma_0(\tfrac{x+t}{2})\}\, e^{2\sigma_1(\frac{x+t}{2})}\chi(\tfrac{x+t}{2};h) \quad ,$$

and

$$|L_x(x,0;h)| \le \frac{1}{2}|q(\tfrac{x}{2})| + \{|h| + \sigma_0(\tfrac{x}{2})\}\, e^{2\sigma_1(\frac{x}{2})}\chi(\tfrac{x}{2};h) \ ,$$

where

$$\kappa(z;h) = \begin{cases} |e^{-hz}|, & \text{for } \operatorname{Re} h < 0 , \\ 1 , & \text{for } \operatorname{Re} h \geq 0 . \end{cases}$$

The assumption concerning the differentiability of the function $q(x)$ is not essential. It can be dropped upon approximating $q(x)$ by smooth functions and subsequently passing to the limit.

PROBLEMS

1. Check that the proof of Theorem 1.1.2 carries over with no modifications to the case of locally summable functions $q(x)$.

2. Show that the solution $K(x,t)$ of equation (1.2.18) is subject to the estimate

$$|K(x,t)| \leq \frac{1}{2} w(\frac{x+t}{2}) \exp \left\{2 \int_0^x w(\xi)d\xi\right\} , \qquad (1.2.35)$$

where

$$w(x) = \max_{0 \leq \eta \leq x} \left|\int_0^\eta q(t)dt\right| \qquad (1.2.35')$$

(if $q(x)$ changes sign, this estimate improves (1.2.20)).

<u>Hint.</u> Pass to equation (1.2.18') and find its solution as the sum $H_1(u,v) + H_2(u,v)$ of functions satisfying the equations

$$H_1(u,v) = \frac{1}{2} \int_0^u q(\xi)d\xi + \int_0^u \int_0^v q(\alpha + \beta)H_2(\alpha,\beta)d\beta d\alpha ,$$

and

$$H_2(u,v) = \int_0^u \int_0^v q(\alpha + \beta)H_1(\alpha,\beta)d\beta d\alpha .$$

Differentiate the first of these equations with respect to v and the second with respect to u to get the system of equations

$$A(u,v) = \int_0^u [Q(u + v) - Q(\alpha + v)]B(\alpha,v)d\alpha ,$$

$$B(u,v) = \frac{1}{2} Q(u)[Q(u+v) - Q(u)] + \int_0^v [Q(u+v) - Q(u+\beta)]A(u,\beta)d\beta ,$$

in which

$$A(u,v) = \frac{\partial}{\partial v} H_1(u,v) \quad , \quad B(u,v) = \frac{\partial}{\partial u} H_2(u,v) ,$$

and

$$Q(u) = \int_0^u q(t)dt .$$

Next use the method of successive approximations to obtain the bounds

$$|B(u,v)| \leqslant w(u)w(u+v)\text{ch} \left\{2 \int_0^{u+v} w(\xi)d\xi\right\} ,$$

and

$$|A(u,v)| \leqslant w(u)w(u+v)\text{sh} \left\{2 \int_0^{u+v} w(\xi)d\xi\right\} ,$$

which, upon returning to the functions $H_1(u,v)$, $H_2(u,v)$, yield an estimate equivalent to (1.2.35) for $H(u,v) = H_1(u,v) + H_2(u,v)$.

3. Show, on the basis of the preceding problem, that if the sequence $\int_0^x q_n(t)dt$ converges to $Q(x)$ uniformly on every bounded interval, then the corresponding sequence of kernels $K_n(x,y)$ of transformation operators converges to some function $K(x,y)$ uniformly on every bounded domain. Give examples of sequences of functions $q_n(x)$ such that

$$\lim_{n\to\infty} \inf_{-\infty<x<\infty} \int_x^{x+1} |q_n(t)|dt = \infty ,$$

but $\lim_{n\to\infty} e_0^{(n)}(\lambda,x) = e^{i\lambda x}$ uniformly in $\lambda \in (-\infty,\infty)$ and x belonging to any bounded interval, where $e_0^{(n)}(\lambda,x)$ is the solution of the equation $y''(x) - q_n(x)y(x) + \lambda^2 y(x) = 0$ with initial data $y(0) = 1$, $y''(0) = i\lambda$.

4. Let H be a separable Hilbert space, and let OH be the set of all bounded operators acting in H . The norm of the operator $f \in OH$ will be denoted by $|f|$. The operator Sturm-Liouville equation is by definition the equation

$$-y''(x) + q(x)y(x) = \lambda^2 y(x) , \tag{1.2.36}$$

where $q(x) \in \mathcal{O}H$ for every x and depends continuously on x. By the solution of the Cauchy problem for equation (1.2.36) we mean an operator-valued function $x \to y(x) \in \mathcal{O}H$, which satisfies this equation and the initial conditions $y(0) = A$, $y'(0) = B$, where $A,B \in \mathcal{O}H$. Generalize all the results of this section (including Problems 2 and 3) to the operator Sturm-Liouville equation.

5. The operator equation

$$By'(x) + \Omega(x)y(x) = \lambda y(x) , \tag{1.2.37}$$

(where the function $x \to \Omega(x) \in \mathcal{O}H$ is continuous, and the constant operator B belongs to $\mathcal{O}H$ is called the Dirac equation if $B^2 = I$ and $B\Omega(x) + \Omega(x)B \equiv 0$, where I is the identity operator. Show that the solution $e(\lambda,x)$ of equation (1.2.37) with initial data $e(\lambda,0) = I$ admits the representation

$$e(\lambda,x) = e^{-B\lambda x} + \int_{-x}^{x} K(x,t)e^{-B\lambda t}dt , \tag{1.2.38}$$

where $K(x,t) \in \mathcal{B}H$ and

$$1 + \int_{-x}^{x} |K(x,t)|dt \lesssim \exp \int_0^{|x|} |\Omega(t)|dt .$$

<u>Hint.</u> Upon setting $e(\lambda,x) = e^{-B\lambda x}u(\lambda,x)$ and observing that $e^{-B\lambda x}\Omega(x) \equiv \Omega(x)e^{B\lambda x}$, you obtain for $u(\lambda,x)$ the differential equation

$$u'(\lambda,x) = e^{2B\lambda x}B\Omega(x)u(\lambda,x) ,$$

which in turn is equivalent to the integral equation

$$u(\lambda,x) = I + \int_0^x e^{2B\lambda t}B\Omega(t)u(\lambda,t)dt .$$

If you now seek the solution of this equation in the form

$$u(\lambda,x) = I + \int_0^x e^{2B\lambda t}Q(x,t)dt ,$$

you get the equation

$$Q(x,t) = B\Omega(t) + B\int_0^{x-t} \Omega(t+\xi)Q(t+\xi,\xi)d\xi ,$$

for $Q(x,t)$, the solubility of which can be established by the method of successive approximation. Setting

$$K(x,t) = \frac{1}{4}\left[Q_+(x , \frac{x+t}{2}) + Q_-(x , \frac{x-t}{2})\right] ,$$

where $Q_+ = Q + BQB$, $Q_- = Q - BQB$, and using the fact that $BQ_+ + Q_+B = 0$ and $BQ_- - Q_-B = 0$, you get formula (1.2.38).

6. Since $e^{-B\lambda x} = I\cos\lambda x - B\sin\lambda x$, it follows from (1.2.38) that the solution of equation (1.2.37) with initial data $y(0) = A$ is representable in the form

$$y(x) = \left[\cos\lambda xI + \int_0^x K(x,t;0)\cos\lambda t dt\right] A -$$

$$- \left[\sin\lambda xI + \int_0^x K(x,t;\infty)\sin\lambda t dt\right] BA . \qquad (1.2.38')$$

Suppose the projector P is such that $BP + PB = B$. Let $\omega(\lambda,x;P)$ [resp. $\omega_0(\lambda,x;P)$] denote the solution of equation (1.2.37) [resp. (1.2.37) with $\Omega(x) \equiv 0$] with initial data $y(0) = P$;

$$\omega_0(\lambda,x;P) = \cos\lambda xP - \sin\lambda xBP .$$

Then it follows from formula (1.2.38') that

$$\omega(\lambda,x;P) = \omega_0(\lambda,x;P) + \int_0^x K_P(x,t)\omega_0(\lambda,t;P)dt , \qquad (1.2.39)$$

where $K_P(x,t) = K(x,t;0)P + K(x,t;\infty)(I - P)$. Solving equation (1.2.39) for $\omega_0(\lambda,x;P)$ you also get

$$\omega_0(\lambda,x;P) = \omega(\lambda,x;P) + \int_0^x L_P(x,t)\omega(\lambda,t;P)dt . \qquad (1.2.39')$$

Assuming that the operator-valued function $\Omega(x)$ is continuously differentiable, show that the kernel $L_P(x,t)$ satisfies the equation

$$B\frac{\partial}{\partial x} L_P(x,t) + \frac{\partial}{\partial t} L_P(x,t)B - L_P(x,t)\Omega(t) = 0 \qquad (1.2.40)$$

as well as the conditions

$$BL_P(x,x) - L_P(x,x)B = \Omega(x) , \tag{1.2.40'}$$

and

$$L_P(x,0)P = L_P(x,0) . \tag{1.2.40''}$$

7. Let $\omega(\lambda,x)$ and $\tilde{\omega}(\lambda,x)$ denote the solutions of the operator Sturm-Liouville equations

$$\omega'' - q(x)\omega + \lambda^2\omega = 0 \quad \text{and} \quad \omega'' - \tilde{\omega}q(x) + \lambda^2\tilde{\omega} = 0 \quad (0 \leqslant x < \infty) ,$$

with initial data $\omega(\lambda,0) = \tilde{\omega}(\lambda,0) = I$, $\omega_x(\lambda,0) = \tilde{\omega}_x(\lambda,0) = h$. From the existence of the transformation operators it follows that

$$\cos \lambda x I = \omega(\lambda,x) + \int_0^x L(x,t)\omega(\lambda,t)dt = \tilde{\omega}(\lambda,x) + \int_0^x \tilde{\omega}(\lambda,t)L(x,t)dt .$$

Let

$$f(x,y) = \begin{cases} L(x,y) + \int_0^y L(x,t)\tilde{L}(y,t)dt & (x \geqslant y) \\ \tilde{L}(y,x) + \int_0^x L(x,t)\tilde{L}(y,t)dt & (x \leqslant y) . \end{cases} \tag{1.2.41}$$

Show that

$$\left.\begin{aligned} f(x,y) &= \frac{1}{2} [L(x+y,0) + L(|x-y|,0)] , \\ f(x,y) &= \frac{1}{2} [\tilde{L}(x+y,0) + \tilde{L}(|x-y|,0)] . \end{aligned}\right\} \tag{1.2.42}$$

Hint. Consider first the case of continuously differentiable functions $q(x)$ and, using the analogs of equation (1.2.30), show that $f_{xx}(x,y) = f_{yy}(x,y)$ for $x \neq y$. This implies the validity of the first (second) equality in (1.2.42) for $0 < y < x$ (respectively, $0 < x < y$). Comparison of these equalities for $x = y$ yields the identity $L(x,0) = \tilde{L}(x,0)$ which shows that the formulas to be proved hold for all nonnegative values of x and y . In the general case you can approximate $q(x)$ by the continuously differentiable functions $q_\delta(x) = \frac{1}{\delta} \int_x^{x+\delta} q(t)dt$.

8. Show that the operator-valued function

$$u(x,y) = \frac{1}{2}\,[f(x+y) + f(|x-y|) + B\{f(x+y) - f(|x-y|)\}B] \tag{1.2.43}$$

is the unique solution of the operator equation

$$Bu_x(x,y) + u_y(x,y)B = 0 \quad (0 \leq y \leq x < \infty)$$

with smooth initial data $u(x,0) = f(x)$.

9. Along with the operator-valued functions $\omega(\lambda,x;P)$ and $\omega_0(\lambda,x,P)$ introduced in Problem 6, consider the operator-valued functions $\tilde{\omega}(\lambda,x;P)$ and $\tilde{\omega}_0(\lambda,x;P)$, which solve the equations

$$-\tilde{\omega}_x'B + \tilde{\omega}\Omega(x) = \lambda\tilde{\omega} \quad \text{and} \quad -\tilde{\omega}_{0x}'B = \lambda ,$$

with initial data $\tilde{\omega}(\lambda,0;P) = \tilde{\omega}_0(\lambda,0;P) = P$ (notice that $\tilde{\omega}_0(\lambda,x;P) = P\cos\lambda x + BP\sin\lambda x$). They are connected by the following analog of formula (1.2.39):

$$\tilde{\omega}_0(\lambda,x;P) = \tilde{\omega}(\lambda,x;P) + \int_0^x \tilde{\omega}(\lambda,t;P)\tilde{L}_P(x,t)\,dt . \tag{1.2.44}$$

Let

$$f(x,y) = \begin{cases} L_P(x,y) + \int_0^y L_P(x,t)\tilde{L}_P(y,t)\,dt & (x \geq y) , \\ \tilde{L}_P(x,y) + \int_0^x L_P(x,t)\tilde{L}_P(y,t)\,dt & (x \leq y) . \end{cases} \tag{1.2.45}$$

Show that the following equalities are valid for all nonnegative values of x , y :

$$f(x,y) = \frac{1}{2}\; L_P(x+y,0) + L_P(|x-y|,0) +$$

$$+\, B\, L_P(x+y,0) - L_P(|x-y|,0)\, B \;=$$

$$= \frac{1}{2}\,[\tilde{L}_P(x+y,0) + \tilde{L}_P(|x-y|,0) + B\{\tilde{L}_P(x+y,0) - \tilde{L}_P(|x-y|,0)\}B] , \tag{1.2.46}$$

and consequently observe that

$$L_P(x,0) + BL_P(x,0)B = \tilde{L}_P(x,0) + B\tilde{L}_P(x,0)B . \tag{1.2.47}$$

3. THE STURM-LIOUVILLE BOUNDARY VALUE PROBLEM ON A BOUNDED INTERVAL

Consider the boundary value problem generated on the interval $(0,\pi)$ by the Sturm-Liouville equation

$$-y'' + q(x)y = \lambda^2 y \tag{1.3.1}$$

and the boundary conditions

$$\Gamma_i(y) = a_{i1}y(0) + a_{i2}y'(0) + a_{i3}y(\pi) + a_{i4}y'(\pi) = 0 \ , \quad i = 1,2 \tag{1.3.2}$$

where $q(x)$ is a summable complex-valued function and the a_{ik} are arbitrary complex numbers.

The values of the parameter $\mu = \lambda^2$ for which this boundary value problem has a nonzero solution are called eigenvalues and the corresponding solutions are called eigenfunctions. In the following we shall let $c(\lambda,x)$ and $s(\lambda,x)$ denote the fundamental solutions of equation (1.3.1) which are defined by the initial data $c(\lambda,0) = s'(\lambda,0) = 1$ and $c'(\lambda,0) = s(\lambda,0) = 0$ (thus, in the notations of the preceding section, $c(\lambda,x) = \omega(\lambda,x;0)$ and $s(\lambda,x) = \omega(\lambda,x;\infty)$). Since the general solution $z(\lambda,x)$ of equation (1.3.1) is a linear combination of the functions $c(\lambda,x)$ and $s(\lambda,x)$:

$$z(\lambda,x) = A_1c(\lambda,x) + A_2s(\lambda,x) \ ,$$

we can write

$$\Gamma_i(z) = A_1[a_{i1} + a_{i3}c(\lambda,\pi) + a_{i4}c'(\lambda,\pi)] + $$
$$+ A_2[a_{i2} + a_{i3}s(\lambda,\pi) + a_{i4}s'(\lambda,\pi)] \ .$$

This shows that the boundary value problem (1.3.1), (1.3.2) has a nonzero solution if and only if the system of linear equations

$$A_1[a_{11} + a_{13}c(\lambda,\pi) + a_{14}c'(\lambda,\pi)] +$$
$$+ A_2[a_{12} + a_{13}s(\lambda,\pi) + a_{14}s'(\lambda,\pi)] = 0 \ ,$$
$$A_1[a_{21} + a_{23}c(\lambda,\pi) + a_{24}c'(\lambda,\pi)] +$$
$$+ A_2[a_{22} + a_{23}s(\lambda,\pi) + a_{24}s'(\lambda,\pi)] = 0$$

for the coefficients A_1, A_2 has a nonzero solution. Thus, the eigenvalues of the problem in question coincide with the roots of the characteristic function

$$\chi(\lambda) = \begin{vmatrix} a_{14} + a_{13}c(\lambda,\pi) + a_{14}c'(\lambda,\pi) & a_{12} + a_{13}s(\lambda,\pi) + a_{14}s'(\lambda,\pi) \\ a_{21} + a_{23}c(\lambda,\pi) + a_{24}c'(\lambda,\pi) & a_{22} + a_{23}s(\lambda,\pi) + a_{24}s'(\lambda,\pi) \end{vmatrix} . \tag{1.3.4}$$

Upon expanding this determinant and observing that the Wronskian $W(c,s) = c(\lambda,x)s'(\lambda,x) - c'(\lambda,x)s(\lambda,x)$ is identically equal to one, we find that

$$\chi(\lambda) = J_{12} + J_{34} + J_{13}s(\lambda,\pi) + J_{14}s'(\lambda,\pi) + J_{32}c(\lambda,\pi) + J_{42}c'(\lambda,\pi) , \tag{1.3.4'}$$

where $J_{\alpha\beta} = a_{1\alpha}a_{2\beta} - a_{2\alpha}a_{1\beta}$ is the determinant built of the α-th and β-th columns of the matrix of coefficients of the boundary conditions:

$$\begin{pmatrix} a_{11} & a_{12} & a_{13} & a_{14} \\ a_{21} & a_{22} & a_{23} & a_{24} \end{pmatrix} .$$

Letting $\chi_{ik} = \chi_{ik}(\lambda)$ denote the ik elements of the determinant (1.3.4), we construct the following solutions of equation (1.3.1):

$$\omega_i(\lambda,x) = \chi_{i2}c(\lambda,x) - \chi_{i4}s(\lambda,x) . \tag{1.3.5}$$

It follows from formula (1.3.3) that the equalities

$$\Gamma_1(\omega_1) = \Gamma_2(\omega_2) = 0 ,$$

$$\Gamma_1(\omega_2) = -\Gamma_2(\omega_1) = \chi_{11}(\lambda)\chi_{22}(\lambda) - \chi_{12}(\lambda)\chi_{21}(\lambda) = \chi(\lambda)$$

hold identically in λ .

An eigenvalue μ_n of the boundary valve problem (1.3.1), (1.3.2) is said to have multiplicity p if μ_n is a root of multiplicity p of the function $\chi(\sqrt{\mu})$.

Since

$$\frac{\partial^k}{\partial\mu^k}\Gamma_i(\omega_j) = \Gamma_i\left(\frac{\partial^k}{\partial\mu^k}\omega_j\right) \quad (\mu = \lambda^2) ,$$

the functions

$$\omega_{i,k}(x) = \frac{(-1)^k}{k!}\frac{\partial^k}{\partial\mu^k}\omega_i(\lambda,x) \quad (\mu = \lambda^2) \tag{1.3.6}$$

satisfy the two boundary conditions (1.3.2) for $\lambda^2 = \mu_n$ and $0 \leq k \leq p-1$.

The functions $\omega_{i,0}(x),\dots,\omega_{i,p-1}(x)$ $(i = 1,2)$ form a chain in which the first nonzero element, ω_{i,l_i}, is an eigenfunction, and its successors are the corresponding generalized eigenfunctions. Upon differentiating equation (1.3.1) k times with respect to $\mu = \lambda^2$, we conclude that the eigenfunctions and generalized eigenfunctions in the indicated chain satisfy the equations

$$-\omega''_{i,k}(x) + q(x)\omega_{i,k}(x) = \mu_n\omega_{i,k}(x) - \omega_{i,k-1}(x)$$

as well as the boundary conditions (1.3.2). To avoid misunderstandings, we emphasize that the two chains $\omega_{1,0}(x),\dots,\omega_{1,p-1}(x)$ and $\omega_{2,0}(x),\dots,\omega_{2,p-1}$ may in fact consist of the same functions. To us the important point is that, apart from eigenfunctions and generalized eigenfunctions, each chain contains only functions identically equal to zero.

A central position in the spectral theory of the boundary value problems (1.3.1), (1.3.2) is held by the theorem which asserts the completeness in $L_2(0,\pi)$ of the systems of eigenfunctions and generalized eigenfunctions. As is well known, in order that a system of vectors be complete in a Hilbert space, it is necessary and sufficient that the only vector orthogonal to all the vectors of this system be the null vector. Let M be the spectrum, i.e., the set of all eigenvalues μ_n of the boundary value problem (1.3.1), (1.3.2), and let P_n denote the multiplicity of μ_n. By the foregoing discussion, the functions

$$\frac{(-1)^k}{k!}\frac{\partial^k}{\partial\mu^k}\omega_i(\sqrt{\mu},x)\Big|_{\mu=\mu_n} \quad (0 \leq k \leq p_n-1\ ,\ \mu_n \in M\ ,\ i = 1,2)$$

either vanish identically, or are the eigenfunctions and generalized eigenfunctions of our boundary value problem. Hence, to prove the completeness of the system of eigenfunctions and generalized eigenfunctions in $L_2(0,\pi)$, it suffices to show that if $f(x) \in L_2(0,\pi)$ and

$$\int_0^\pi \frac{\partial^k}{\partial\mu^k}\omega_i(\sqrt{\mu},x)f(x)dx\Big|_{\mu=\mu_n} = 0 \tag{1.3.7}$$

for all $\mu_n \in M$, $k = 0,1,\dots,p_n-1$, $i = 1,2$, then $f(x) = 0$ a.e. The characteristic function $\chi(\sqrt{\mu})$ and the functions

$$\omega_i(\sqrt{\mu},f) = \int_0^x \omega_i(\sqrt{\mu},x)f(x)dx \quad (i = 1,2) \tag{1.3.7'}$$

are entire functions of μ. Equality (1.3.7) shows that every p_n-multiple root μ_n of the function $\chi(\sqrt{\mu})$ is also a root of at least the same multiplicity of both functions $\omega_i(\sqrt{\mu},f)$ $(i = 1,2)$. Consequently, equality (1.3.7) holds if and only if $\omega_i(\sqrt{\mu},f)[\chi(\sqrt{\mu})]^{-1}$ $(i = 1,2)$, and hence $\omega_i(\lambda,f)[\chi(\lambda)]^{-1}$ $(i = 1,2)$ too, are entire functions. Thus, to prove the completeness of the system of eigenfunctions and generalized eigenfunctions of the boundary value problem (1.3.1), (1.3.2), it suffices to show that the functions $\omega_i(\lambda,f)[\chi(\lambda)]^{-1}$ $(i = 1,2)$ can be entire if and only if $f(x) = 0$ a.e.

To this point we have imposed no restrictions whatsoever upon the boundary conditions (1.3.2). We now remark that for $q(x) \equiv 0$ the characteristic function $\chi_0(x)$ of the boundary value problem (1.3.1), (1.3.2) has the form

$$\chi_0(\lambda) = (J_{12} + J_{34}) + J_{13}\frac{\sin \lambda x}{\lambda} + (J_{14} + J_{32})\cos \lambda\pi - J_{42}\,\lambda \sin \lambda\pi$$

and in this, the simplest case, it makes sense to speak about the completeness of eigenfunctions and generalized eigenfunctions only for those boundary conditions for which $\chi_0(\lambda)$ is not a constant. This obviously can happen only in one of the following three cases:

$$\begin{aligned}
&1)\ J_{42} \neq 0\,,\\
&2)\ J_{42} = 0\,,\quad J_{14} + J_{32} \neq 0\,, \qquad\qquad (1.3.8)\\
&3)\ J_{42} = J_{14} + J_{32} = 0\,,\quad J_{13} \neq 0\,.
\end{aligned}$$

Boundary conditions satisfying one of these three conditions are called nondegenerate. Let us show that the system of eigenfunctions and generalized eigenfunctions of any boundary value problem (1.3.1), (1.3.2) with nondegenerate boundary conditions is complete in $L_2(0,\pi)$.

LEMMA 1.3.1. *If* $f(x) \in L_1(0,\pi)$, *then*

$$\lim_{|\lambda|\to\infty} e^{-|\operatorname{Im}\lambda\pi|}\int_0^\pi f(x)\cos\lambda x\,dx = \lim_{|\lambda|\to\infty} e^{-|\operatorname{Im}\lambda\pi|}\int_0^\pi f(x)\sin\lambda x\,dx = 0\,.$$

PROOF. The set $C^1(0,\pi)$ of continuously differentiable functions on $[0,\pi]$ is dense in $L_1(0,\pi)$, i.e., for every $f(x) \in L_1(0,\pi)$ and arbitrary $\varepsilon > 0$, one can find a function $g_\varepsilon(x) \in C^1(0,\pi)$ such that

$$\int_0^\pi |f(x) - g_\varepsilon(x)|dx < \varepsilon .$$

Now since

$$\int_0^\pi f(x)\cos \lambda x dx = \int_0^\pi \{f(x) - g_\varepsilon(x)\}\cos \lambda x dx + \int_0^\pi g_\varepsilon(x)\cos \lambda x dx =$$

$$= \int_0^\pi \{f(x) - g_\varepsilon(x)\}\cos \lambda x dx + \lambda^{-1}\left\{g_\varepsilon(\pi)\sin \lambda\pi - \int_0^\pi g_\varepsilon'(x)\sin \lambda x dx\right\}$$

and

$$|\cos \lambda x| \leq e^{|\mathrm{Im}\ \lambda\pi|} \quad , \quad |\sin \lambda x| \leq e^{|\mathrm{Im}\ \lambda x|}$$

for all $x \in [0,\pi]$, we have

$$\left|\int_0^\pi f(x)\cos \lambda x dx\right| \leq$$

$$\leq e^{|\mathrm{Im}\ \lambda\pi|}\left[\int_0^\pi |f(x) - g_\varepsilon(x)|dx + |\lambda|^{-1}\left\{|g_\varepsilon(\pi)| + \int_0^\pi |g_\varepsilon'(x)|dx\right\}\right] .$$

Therefore

$$\overline{\lim_{|\lambda|\to\infty}}\ e^{-|\mathrm{Im}\ \lambda\pi|}\left|\int_0^\pi f(x)\cos \lambda x dx\right| \leq \int_0^\pi |f(x) - g_\varepsilon(x)|dx < \varepsilon ,$$

and since $\varepsilon > 0$ is arbitrary, the lemma is proved.

□

Let $c_\pi(\lambda,x)$ and $s_\pi(\lambda,x)$ denote the solutions of equation with initial conditions $c_\pi(\lambda,\pi) = s_\pi'(\lambda,\pi) = 1$, and $c_\pi'(\lambda,\pi) = s_\pi(\lambda,\pi) = 0$.

COROLLARY 1. *If* $f(x) \in L_1(0,\pi)$, *then*

$$\lim_{|\lambda|\to\infty} e^{-|\mathrm{Im}\ \lambda\pi|}\int_0^\pi f(x)c(\lambda,x) = \lim_{|\lambda|\to\infty} e^{-|\mathrm{Im}\ \lambda\pi|}\int_0^\pi f(x)c_\pi(\lambda,x)dx = 0 ,$$

and

$$\lim_{|\lambda|\to\infty} e^{-|\mathrm{Im}\ \lambda\pi|}\int_0^\pi f(x)\lambda s(\lambda,x)dx = \lim_{|\lambda|\to\infty} e^{-|\mathrm{Im}\ \lambda\pi|}\int_0^\pi f(x)\lambda s_\pi(\lambda,x)dx = 0 .$$

PROOF. By formula (1.2.11'),

$$\lambda s_\pi(\lambda,x) = \sin \lambda(x - \pi) + \int_\pi^x K(x,t;\infty)\sin \lambda(t - \pi)dt$$

and hence

$$\int_0^\pi f(x)\lambda s_\pi(\lambda,x)dx = \int_0^\pi f(x)\left\{\sin \lambda(x - \pi) + \int_\pi^x K_\pi(x,t;\infty)\sin \lambda(t - \pi)dt\right\} dx =$$

$$= \int_0^\pi \sin \lambda t \left\{-f(\pi - t) + \int_0^{\pi-t} f(x)K_\pi(x,\pi-t;\infty)dx\right\} dt .$$

The function

$$\hat{f}(t) = -f(\pi - t) + \int_0^{\pi-t} f(x)K_\pi(x,\pi-t;\infty)dx ,$$

obviously belongs to $L_1(0,\pi)$, and now the last of the asserted equalities is seen to be a direct consequence of Lemma 1.3.1. The other three equalities are established in exactly the same way using formulas (1.2.10), (1.2.11), and (1.2.10').

□

COROLLARY 2. *If* $f(x) \in L_1(0,\pi)$, *then*

$$\lim_{|\lambda|\to\infty} e^{-|\text{Im } \lambda\pi|}\omega_i(\lambda,f) = 0 \quad (i = 1,2) .$$

PROOF. It follows from formula (1.3.6) and the definition of the functions $\chi_{i1}(\lambda)$ and $\chi_{i2}(\lambda)$ that

$$\omega_i(\lambda,x) = [a_{i2} + a_{i3}s(\lambda,\pi) + a_{i4}s'(\lambda,\pi)]c(\lambda,x) -$$

$$- [a_{i1} + a_{i3}c(\lambda,\pi) + a_{i4}c'(\lambda,\pi)]s(\lambda,x) =$$

$$= a_{i2}c(\lambda,x) - a_{i1}s(\lambda,x) - a_{i3}s_\pi(\lambda,x) + a_{i4}c_\pi(\lambda,x) , \tag{1.3.9}$$

where the functions

$$c_\pi(\lambda,x) = s'(\lambda,\pi)c(\lambda,x) - c'(\lambda,\pi)s(\lambda,x)$$

and

$$s_\pi(\lambda,x) = c(\lambda,\pi)s(\lambda,x) - s(\lambda,\pi)c(\lambda,x)$$

satisfy equation (1.3.1) and the initial conditions $c_\pi(\lambda,\pi) = s_\pi'(\lambda,\pi) = 1$ and $c_\pi'(\lambda,\pi) = s_\pi(\lambda,\pi) = 0$. Thus the assertion follows from the preceding corollary. □

LEMMA 1.3.2. *If the boundary conditions in the boundary value problem* (1.3.1), (1.3.2) *are nondegenerate, then there exists a constant* $C > 0$ *and a sequence of unboundedly expanding contours* K_n *on which*

$$|\chi(\lambda)| > |\lambda|^{-1}C\exp|\operatorname{Im}\lambda\pi| . \tag{1.3.10}$$

PROOF. Using representations (1.2.10), (1.2.11) and observing that the kernels $K(x,t;0)$ and $K(x,t;\infty)$ always have summable first order derivatives, we find that

$$\left.\begin{aligned}
c(\lambda,\pi) &= \cos\lambda\pi + \int_0^\pi K(\pi,t;0)\cos\lambda t\,dt = \\
&= \cos\lambda\pi + K(\pi,\pi;0)\frac{\sin\lambda\pi}{\lambda} - \int_0^\pi K'_t(\pi,t;0)\frac{\sin\lambda t}{\lambda}dt , \\
c'(\lambda,\pi) &= -\lambda\sin\lambda\pi + K(\pi,\pi;0)\cos\lambda\pi + \int_0^\pi K'_x(\pi,t;0)\cos\lambda t\,dt , \\
s(\lambda,\pi) &= \frac{\sin\lambda\pi}{\lambda} + \int_0^\pi K(\pi,t;\infty)\frac{\sin\lambda t}{\lambda}dt , \\
s'(\lambda,\pi) &= \cos\lambda\pi + K(\pi,\pi;\infty)\frac{\sin\lambda\pi}{\lambda} + \int_0^\pi K'_x(\pi,t;\infty)\frac{\sin\lambda t}{\lambda}dt ,
\end{aligned}\right\} \tag{1.3.11}$$

and

$$K(\pi,\pi;0) = K(\pi,\pi;\infty) = \frac{1}{2}\int_0^\pi q(t)dt .$$

It follows from these equalities, formula (1.3.4'), and Lemma 1.3.1 that, depending on which of the three cases (1.3.8) is considered, the characteristic function is expressible as

1) $\chi(\lambda) = J_{42}\{-\lambda\sin\lambda\pi\} + \lambda e^{|\operatorname{Im}\lambda\pi|}\varepsilon_1(\lambda)$,

2) $\chi(\lambda) = (J_{14} + J_{32})[\cos\lambda\pi + L_2] + e^{|\operatorname{Im}\lambda\pi|}\varepsilon_2(\lambda)$,

or

3) $\chi(\lambda) = J_{13}[\lambda^{-1}\sin\lambda\pi + L_3] + \lambda^{-1}e^{|\operatorname{Im}\lambda\pi|}\varepsilon_3(\lambda)$,

where

$$L_2 = (J_{12} + J_{34})(J_{14} + J_{32})^{-1} ,$$

$L_3 = (J_{12} + J_{34})J_{13}^{-1}$,

and

$$\lim_{|\lambda|\to\infty} \varepsilon_i(\lambda) = 0 \quad (i = 1,2,3) .$$

The functions $\chi^{(0)}(\lambda)$ given by

1) $\chi^{(0)}(\lambda) = -\lambda \sin \lambda\pi$;

2) $\chi^{(0)}(\lambda) = \cos \lambda\pi + L_2$;

3) $\chi^{(0)}(\lambda) = \lambda^{-1}\sin \lambda\pi + L_3$,

are called the principal parts of the corresponding characteristic functions. They essentially determine the behavior of $\chi(\lambda)$ as $|\lambda| \to \infty$.

We shall call the sequences of unboundedly expanding contours $K_n(1)$, $K_n(2)$, $K_n(3)$ admissible if for some $C > 0$,

$$|\sin z\pi| \geq C \exp|\text{Im } z\pi| \quad , \quad z \in K_n(1) ,$$

$$|\cos z\pi + L_2| \geq C \exp|\text{Im } z\pi| \quad , \quad z \in K_n(2) ,$$

and

$$|\sin z\pi + zL_3| \geq C \exp|\text{Im } z\pi| \quad , \quad z \in K_n(3) ,$$

respectively. Obviously, on an admissible sequence of contours, the characteristic function satisfies the inequalities

1) $|\chi(z)| > C_1|z|\exp|\text{Im } z\pi|$,

2) $|\chi(z)| > C_1 \exp|\text{Im } z\pi|$,

and

3) $|\chi(z)| > C_1|z|^{-1} \exp|\text{Im } z\pi|$,

respectively, for all sufficiently large n . Hence, to complete the proof of the lemma, it remains only to establish the existence of admissible contours.

We let Z_ρ (where ρ is an arbitrarily small positive number) denote the domain which is obtained by removing discs of radius ρ centered at the zeros of the function $\sin z\pi$, i.e., 0, ±1, ±2,..., from the complex plane.

Let us show that

$$\sup_{z\in Z_\rho} |\sin z\pi|^{-1} \exp|\text{Im } z\pi| = C_\rho < \infty .$$

Since the function $|\sin z\pi|\exp|\text{Im } z\pi|$ is even and periodic with period z, we have

$$C_\rho = \sup_{z\in Z_\rho} |\sin z\pi|^{-1} \exp|\text{Im } z\pi| = \sup_{z\in Z_\rho\cap\Pi^+} |\sin z\pi|\exp|\text{Im } z\pi| =$$

$$= \sup_{z\in Z_\rho\cap\Pi^+} |(\sin z\pi)^{-1} e^{-iz\pi}| ,$$

where Π^+ designates the half strip $0 < \text{Re } z < 2$, $\text{Im } z > 0$.

The function $(\sin z\pi)^{-1}e^{-iz\pi}$ is holomorphic in the domain $Z_\rho \cap \Pi^+$, it tends therein to $(-2i)^{-1}$ when $\text{Im } z \to +\infty$, and is continuous on the boundary of this domain. By the maximum modulus principle, the supremum of its modulus is finite. Consequently, $C_\rho < \infty$, and the inequality $|\sin z\pi| \geq C^{-1} \exp|\text{Im } z\pi|$ holds in Z_ρ, which shows that in case 1, one can take any sequence of expanding contours contained in the doamin Z_ρ as an admissible sequence $K_n(1)$.

It is proved in much the same way that in the second case one can take as an admissible sequence $K_n(2)$ any sequence of expanding contours, the distance of which to the zeros of the function $\cos z\pi + L_2$, i.e., to the set of points $\pm\theta_2$, $\pm(2+\theta_2)$, $\pm(2-\theta_2)$,..., with $\theta_2 = \pi^{-1}\arccos(-L_2)$, is bounded from below by some positive number ρ.

In the third case the same sequences as in the first case are obviously admissible for $L_3 = 0$. Now suppose that $L_3 = |L_3|e^{i\gamma\pi} \neq 0$, $-1 \leq \gamma \leq 1$. Then

$$|\sin z\pi + L_3 z| \geq \frac{1}{2} \exp|\text{Im } z\pi|\{1 - (2|zL_3| + 1)\exp(-|\text{Im } z\pi|)\} ,$$

for every z,

$$|\sin z\pi + L_3 z| \geq |zL_3| - 1$$

for $\text{Im } z = 0$ and, finally,

$$|\sin z\pi + L_3 z| = \frac{1}{2} |e^{i\alpha_n\pi}e^{-\text{Im } z\pi} - e^{-i\alpha_n\pi}e^{\text{Im } z\pi} + 2iz|L_3|e^{i\gamma\pi}| =$$

$$= \frac{1}{2} |- ie^{i(\alpha_n-\gamma)\pi-\text{Im } z\pi} + ie^{-i(\alpha_n+\gamma)\pi+\text{Im } z\pi} + 2z|L_3|| \geq$$

$$\geq \frac{1}{2} \text{Re}\{e^{-i\gamma(1+\beta(z))\pi-\text{Im } z\pi} + e^{-i\gamma(1-\beta(z))\pi+ \text{Im } z\pi} + 2 \frac{n}{|n|} zL_3\} \geq$$

$$\geqslant \frac{1}{2}\{e^{|\mathrm{Im}\, z\pi|} + (4|n| - 1)|L_3| - e^{-|\mathrm{Im}\, z|}\}$$

for $\beta(z) = \mathrm{sign}\ \mathrm{Im}\, z = \pm 1$ on the half lines $\mathrm{Re}\, z = \alpha_n = 2n + \frac{n}{|n|} - \gamma\beta(z)$. From these inequalities, it follows that on the contours $K_n(3)$ bounding the union of the rectangles

$|\mathrm{Re}\, z + \gamma| \leqslant 2n + \frac{1}{2}$, $0 \leqslant \mathrm{Im}\, z \leqslant 2n$, $|\mathrm{Re}\, z - \gamma| \leqslant 2n + \frac{1}{2}$,

$0 \geqslant \mathrm{Im}\, z \geqslant -2n$,

we have

$$|\sin z\pi + L_3 z| \geqslant \frac{1}{3}\exp|\mathrm{Im}\, z\pi|$$

for all sufficiently large n .

Therefore, sequences of admissible contours exist in all three cases and the lemma is proved. □

We next exhibit concrete sequences of admissible contours. Let $C(\ell)$ be the contour bounding the square $|\mathrm{Re}\, z| \leqslant \ell$, $|\mathrm{Im}\, z| \leqslant \ell$. Then by the foregoing discussion, the following are admissible contours:

In case 1), $C(n + \frac{1}{2})$;

In case 2), $C(n)$ if $\theta_2 \neq 0,1$, $C(2n + 1)$ if $\theta_2 = 0$, and $C(2n)$ if $\theta_2 = 1$;

In case 3), $C(n + \frac{1}{2})$ if $L_3 = 0$, and the sequence $K_n(3)$ constructed above if $L_3 \neq 0$.

Remark. The inequalities obtained in the course of proving the lemma show that the estimates

$$\left.\begin{aligned} &1)\ |\chi(z) - J_{42}\chi^{(0)}(z)| = o\,|\chi^{(0)}(z)| ,\\ &2)\ |\chi(z) - (J_{14} + J_{32})\chi^{(0)}(z)| = o\,|\chi^{(0)}(z)| ,\\ &3)\ |\chi(z) - J_{13}\chi^{(0)}(z)| = o\,|\chi^{(0)}(z)| \end{aligned}\right\} \tag{1.3.10'}$$

are valid on the boundaries of the domains

1) $\Pi_n(1) = \{z : |\mathrm{Re}\, z| < 2n + \frac{1}{2}\}$,

2) $\Pi_n(2) = \{z : |\mathrm{Re}\, z| < 2n\}$ if $\theta_2 \neq 0$,

and

$\tilde{\Pi}_n(2) = \{z : |\mathrm{Re}\, z| < 2n + 1\}$ if $\theta_2 = 0$.

3) $\Pi_n(1)$ if $L_3 = 0$,

and

$$\Pi_n(3) = \{z : |\mathrm{Re}\, z + \gamma| < 2n + \tfrac{1}{2} ,\ \mathrm{Im}\, z \geq 0\} \cap$$

$$\cap \{z : |\mathrm{Re}\, z - \gamma| < 2n + \tfrac{1}{2} ,\ \mathrm{Im}\, z \leq 0\} \text{ if } L_3 \neq 0 ,$$

respectively.

We now turn to the proof of the main result of this section.

THEOREM 1.3.1. *The system of eigenfunctions and generalized eigenfunctions of the boundary value problem* (1.3.1), (1.3.2) *with nondegenerate boundary conditions is complete in the space* $L_2(0,\pi)$.

PROOF. We showed above that in order to prove this theorem it suffices to show that the functions $\omega_i(\lambda,f)[\chi(\lambda)]^{-1}$ $(i = 1,2)$ are entire if and only if $f(x) = 0$ a.e. Suppose $f(x) \in L_2(0,\pi)$ is such that the functions $\omega_i(\lambda,f)[\chi(\lambda)]^{-1}$ $(i = 1,2)$ are entire. By Lemma 1.3.2, there exist a constant C and a sequence of unboundedly expanding contours K_n such that

$$|\omega_i(\lambda,f)[\chi(\lambda)]^{-1}| \leq C|\lambda||\omega_i(\lambda,f)|\exp(-|\mathrm{Im}\,\lambda\pi|) .$$

From this estimate and Corollary 2 it follows that

$$\lim_{n\to\infty} \max_{\lambda\in K_n} |\omega_i(\lambda,f)[\chi(\lambda)\lambda]^{-1}| = 0 \quad (i = 1,2) .$$

This shows that as $|\lambda| \to \infty$, the entire functions $\omega_i(\lambda,f)[\chi(\lambda)]^{-1}$ grow slower than $|\lambda|$, and hence they are identically equal to constants which we designate by f_i . Therefore, $\omega_i(\lambda,f) \equiv f_i\chi(\lambda)$ whence, in view of the definition (1.3.6) of the functions $\omega_i(\lambda,f)$,

$$\chi_{12}(\lambda)c(\lambda,f) - \chi_{11}(\lambda)s(\lambda,f) \equiv f_1\chi(\lambda) ,$$

$$\chi_{22}(\lambda)c(\lambda,f) - \chi_{21}(\lambda)s(\lambda,f) \equiv f_2\chi(\lambda) ,$$

where

$$c(\lambda,f) = \int_0^\pi f(x)c(\lambda,x)dx \quad , \quad s(\lambda,f) = \int_0^\pi f(x)s(\lambda,x)dx .$$

Consequently,

$$\{\chi_{11}(\lambda)\chi_{22}(\lambda) - \chi_{21}(\lambda)\chi_{12}(\lambda)\}s(\lambda,f) \equiv \chi(\lambda)\{f_2\chi_{12}(\lambda) - f_1\chi_{22}(\lambda)\}$$

and since $\chi_{11}(\lambda)\chi_{22}(\lambda) - \chi_{21}(\lambda)\chi_{12}(\lambda) \equiv \chi(\lambda)$, we get that

$$s(\lambda,f) \equiv f_2\chi_{12}(\lambda) - f_1\chi_{22}(\lambda) \equiv$$

$$\equiv (f_2a_{12} - f_1a_{22}) + (f_2a_{13} - f_1a_{22})s(\lambda,\pi) + (f_2a_{14} - f_1a_{24})s'(\lambda,\pi) .$$

Let us examine this identity for real values $\lambda \to \pm\infty$. Using Corollary 1 and formula (1.3.11), we can express it as

$$\lambda^{-1}\delta(\lambda) \equiv (f_2a_{12} - f_1a_{22}) + (f_2a_{13} - f_1a_{23})\lambda^{-1}(\sin\lambda\pi + \varepsilon_1(\lambda)) +$$

$$+ (f_2a_{14} - f_1a_{24})(\cos\lambda\pi + \varepsilon_2(\lambda)) ,$$

where the functions $\delta(\lambda)$, $\varepsilon_1(\lambda)$, and $\varepsilon_2(\lambda)$ tend to zero as $\lambda \to \pm\infty$. This is obviously possible if and only if $f_2a_{12} - f_1a_{22} = f_2a_{13} - f_1a_{23} = f_2a_{14} - f_1a_{24}$. It follows that $s(\lambda,f) \equiv 0$, which, using representation (1.2.11) of the solution $s(\lambda,x)$ in terms of the transformation operator, implies in turn that

$$\int_0^\pi f(x)\left\{\frac{\sin\lambda x}{\lambda} + \int_0^x K(x,t;\infty)\,\frac{\sin\lambda t}{\lambda}\,dt\right\} =$$

$$= \int_0^\pi \left\{f(t) + \int_t^\pi f(x)K(x,t;\infty)dx\right\} \frac{\sin\lambda t}{\lambda}\,dt \equiv 0 .$$

Hence, the Fourier sin-transform of the function

$$f(t) + \int_t^\pi f(x)K(x,t;\infty)dx \quad (0 \leqslant t \leqslant \pi)$$

vanishes identically, and therefore, by the uniqueness theorem for the Fourier transform,

$$f(t) + \int_t^\pi f(x)K(x,t;\infty)dx = 0$$

a.e. on the segment $[0,\pi]$. Since the homogeneous Volterra equation with continuous kernel $K(x,t;\infty)$ has only the trivial solution, we conclude that $f(x) = 0$ for a.e. $x \in [0,\pi]$, as asserted. □

Let us examine in more detail boundary value problems with separated boundary conditions

$$\Gamma_1(y) = hy(0) - y'(0) = 0 \ , \quad \Gamma_2(y) = h_1y(\pi) + y'(\pi) = 0 , \tag{1.3.12}$$

which are obtained by setting

$$a_{11} = h \ , \ a_{12} = -1 \ , \ a_{23} = h_1 \ , \ a_{24} = 1 \ ,$$
$$a_{13} = a_{14} = a_{21} = a_{23} = 0 \qquad (1.3.12')$$

in (1.3.2). Since in this case $J_{42} = 1$, the boundary condition (1.3.12) is nondegenerate, and hence the system of eigenfunctions and generalized eigenfunctions is complete in $L_2(0,\pi)$. Let us show that the eigenfunctions and generalized eigenfunctions of the boundary value problem (1.3.1), (1.3.12) form a basis in $L_2(0,\pi)$ and find the form of the expansion with respect to this basis. Upon substituting the values (1.3.12') into formulas (1.3.4') and (1.3.5), we see that in the present case

$$\chi(\lambda) = hh_1 s(\lambda,\pi) + hs'(\lambda,\pi) + h_1 c(\lambda,\pi) + c'(\lambda,\pi) =$$
$$= (hs'(\lambda,\pi) + c'(\lambda,\pi)) + h_1(hs(\lambda,\pi) + c(\lambda,\pi)) = \omega'(\lambda,\pi;h) + h_1\omega(\lambda,\pi;h) \ ,$$
$$\omega_1(\lambda,x) = -c(\lambda,x) - hs(\lambda,x) = -\omega(\lambda,x;h) \ , \qquad (1.3.13)$$
$$\omega_2(\lambda,x) = -h_1 s_\pi(\lambda,x) + c_\pi(\lambda,x) = \omega_\pi(\lambda,x;-h_1) \ ,$$

where $\omega(\lambda,x;h)$ and $\omega_\pi(\lambda,x;-h_1)$ designate the solutions of equation (1.3.1) with initial data $\omega(\lambda,0;h) = 1$, $\omega'(\lambda,0;h) = h$ and $\omega_\pi(\lambda,\pi;-h_1) = 1$, $\omega_\pi'(\lambda,\pi;-h_1) = -h_1$, respectively. It is readily established that the linear spans of the eigenfunctions and generalized eigenfunctions constructed from the solutions $\omega(\lambda,x,h)$ and $\omega_\pi(\lambda,x;-h_1)$ by means of formulas (1.3.6), coincide. It therefore suffices to consider the eigenfunctions and generalized eigenfunctions derived from the solution $\omega(\lambda,x;h)$.

LEMMA 1.3.3. *The boundary value problem* (1.3.1), (1.3.12) *may have only a finite number of multiple eigenvalues, and the eigenvalues of large modulus are of the form* $(n + \frac{a_n}{n})^2$, *where* $\sup|a_n| < \infty$.

PROOF. Since $\chi(\lambda)$ is an even entire function, its zeros (counted with multiplicities) can be arranged in a sequence $\ldots,-\lambda_n,-\lambda_{n-1},\ldots,-\lambda_1,-\lambda_0;\lambda_0,\lambda_1,\ldots,\lambda_{n-1},\lambda_n,\ldots$, where $|\omega_i| \leq |\lambda_{i+1}|$ and $\operatorname{Re}\lambda_i \geq 0$ for $i \geq 0$. Let us evaluate the number of zeros of $\chi(\lambda)$ in the strip $|\operatorname{Re}\lambda| \leq n + \frac{1}{2}$. On the boundary of this strip, $|\lambda \sin \pi\lambda| \geq \frac{1}{2}|\lambda|\exp|\operatorname{Im}\lambda\pi|$, and in view of (1.3.13) and (1.3.11), there exists a constant $C_2 > 0$ such that

$$|\chi(\lambda) + \lambda \sin \lambda\pi| < C_2 \exp|\operatorname{Im}\lambda\pi| \ . \qquad (1.3.14)$$

From these inequalities and the formula

$$\chi(\lambda) = -\lambda \sin \lambda\pi + (\chi(\lambda) + \lambda \sin \lambda\pi) , \tag{1.3.15}$$

we conclude, applying Rouché's theorem, that for $n > C_2$, the functions $\chi(\lambda)$ and $\lambda \sin \lambda\pi$ have the same number of zeros: $2n + 1$, in the strip $|\mathrm{Re}\,\lambda| \leqslant n + \frac{1}{2}$. Therefore, for large n, the zero λ_n is simple and lies in the strip $|\mathrm{Re}\,\lambda - n| < \frac{1}{2}$. Thus, the boundary value problem (1.3.1), (1.3.12) has at most a finite number of multiple eigenvalues.

Next, it is readily verified that for any $\rho < \frac{1}{2}$, the inequality $|\sin \lambda\pi| > C_\rho \exp|\mathrm{Im}\,\lambda\pi|$ holds for all values λ in the intersection of the strip $|\mathrm{Re}\,\lambda - n| < \frac{1}{2}$ with the set $|\lambda - n| \geqslant \rho$, where C is a constant which does not depend on n and ρ. Therefore, if $|n| > 2C^{-1}(C_2 + 1) + \frac{1}{2}$, then the inequality $|\lambda \sin \pi\lambda| > (C_2 + 1)\exp|\mathrm{Im}\,\lambda\pi|$ holds for any λ in the intersection of the strip $|\mathrm{Re}\,\lambda - n| \leqslant \frac{1}{2}$ with the exterior of the disc

$$|\lambda - n| \leqslant \frac{2(C_2 + 1)}{C(2|n| - 1)} . \tag{1.3.16}$$

In view of (1.3.14), this implies, for these values of λ, that $|\lambda \sin \lambda\pi| > |\chi(\lambda) + \lambda \sin \lambda\pi|$. From formula (1.3.15) and this inequality, it follows, by an application of Rouché's theorem, that the root λ_n lies in the disc (1.3.16), i.e., $\lambda_n = n + \frac{a_n}{n}$, where $|a_n| < \frac{C_2 + 1}{C} \frac{2|n|}{2|n| - 1}$. Since the eigenvalues are the square roots of the zeros λ_n, it follows that the eigenvalues of large enough modulus are of the form $(n + \frac{a_n}{n})^2$, where $\sup|a_n| < \infty$. This completes the proof of the lemma.

□

Remark. It follows from this lemma that the distinct eigenvalues of the boundary problem (1.3.1), (1.3.12) can be arranged in a sequence $\mu_{n_0}, \mu_{n_0+1}, \ldots, \mu_n, \ldots$, where $\mu_n = (n + \frac{a_n}{n})^2$, $\sup|a_n| < \infty$, $n_0 = \sum_{k=n_0}^{\infty} (p_k - 1)$, and p_k designates the multiplicity of the eigenvalue μ_k.

The function $\frac{(-1)^k}{k!} \frac{\partial^k}{\partial\mu^k} \omega(\sqrt{\mu},x;h)$, regarded as an element of $L_2(0,\pi)$, will be denoted by $\omega_k(\mu)$ (here μ is not necessarily an eigenvalue). We endow the space $L_2(0,\pi)$ with a pseudo-inner product $\langle f,g\rangle$ setting $\langle f,g\rangle = \int_0^\pi f(x)g(x)dx$. Let us calculate $\langle \omega_k(\mu),\omega_r(\nu)\rangle$. From the equations satisfied by the functions $\omega(\sqrt{\mu},x;h)$, we deduce the equality

$\omega''(\sqrt{\mu},x;h)\omega(\sqrt{\nu},x;h) - \omega(\sqrt{\mu},x;h)\omega''(\sqrt{\nu},x;h) = (\nu - \mu)\omega(\sqrt{\mu},x;h)\omega(\sqrt{\nu},x;h)$,

which, upon integration, yields

$(\nu - \mu)<\omega_0(\mu),\omega_0(\nu)> = \{\omega'(\sqrt{\mu},x;h)\omega(\sqrt{\nu},x,h) - \omega(\sqrt{\mu},x;h)\omega'(\sqrt{\nu},x;h)\}|_0^\pi =$
$= \omega'(\sqrt{\mu},\pi;h)\omega(\sqrt{\nu},\pi;h) - \omega(\sqrt{\mu},\pi;h)\omega'(\sqrt{\nu},\pi;h)$.

Recalling the definition of the characteristic function $\chi(\sqrt{\mu})$, we can reexpress the last equality as

$$(\nu - \mu)<\omega_0(\mu),\omega_0(\nu)> = \chi(\sqrt{\mu})\omega(\sqrt{\nu},\pi;h) - \chi(\sqrt{\nu})\omega(\sqrt{\mu},\pi;h) . \tag{1.3.17}$$

LEMMA 1.3.4. *If* μ_n *is an eigenvalue of multiplicity* p_n *of the boundary value problem* (1.3.1), (1.3.12), *then*

$<\omega_k(\mu_n),\omega_r(\mu_n) = a_{k+r} \quad (0 \leq k \leq p_n-1 \quad ; \quad 0 \leq r \leq p_n-1)$;

moreover, $a_{k+r} = 0$ *if* $k+r < p_{n-1}$ *and* $a_{p_n-1} \neq 0$.

If μ_m *is a different eigenvalue of multiplicity* p_m *of the same problem, then*

$<\omega_k(\mu_n),\omega_s(\mu_m)> = 0 \quad (0 \leq k \leq p_n-1 \quad , \quad 0 \leq s \leq p_m-1)$.

PROOF. By assumption, μ_n is a root of multiplicity p_n of the function $\chi(\sqrt{\mu})$. Hence, on setting $\mu = \mu_n$ in formula (1.3.17), we get

$(\nu - \mu_n) < \omega_0(\mu_n),\omega_0(\nu)> = -\chi(\sqrt{\nu})\omega(\sqrt{\mu_n},\pi;h)$.

Moreover, $\omega(\sqrt{\mu_n},\pi;h) \neq 0$ because the functions $\chi(\sqrt{\lambda}) = \omega''(\sqrt{\lambda},\pi;h) + h_1\omega(\sqrt{\lambda},\pi;h)$ and $\omega(\sqrt{\lambda},\pi;h)$ cannot vanish simultaneously. Therefore, the expansion of the right hand side of the last equality in a series of powers of $\nu - \mu_n$ has the form

$-\chi(\sqrt{\nu})\omega(\sqrt{\mu_n},\pi;h) = (\nu - \mu_n)^{p_n}(c_0 + (\nu - \mu_n)c_1 + \ldots)$,

whence

$<\omega_0(\mu_n),\omega_0(\nu)> = (\nu - \mu_n)^{p_n-1}(c_0 + (\nu - \mu_n)c_1 + \ldots)$,

where $c_0 \neq 0$. Differentiating here k times with respect to ν and recalling the definition of the functions $\omega_k(\mu_n)$, we get

$$<\omega_0(\mu_n),\omega_k(\mu_n)> = 0 \quad (0 \leq k < p_n-1) , \tag{1.3.18}$$

and

$$\langle\omega_0(\mu_n),\omega_{p_n-1}(\mu_n)\rangle = (-1)^{p_n-1} c_0 \neq 0 . \tag{1.3.19}$$

Now apply the operation $\frac{(-1)^{k+r+1}}{k!(r+1)!} \frac{\partial^{k+r+1}}{\partial\mu^k \partial\nu^{r+1}}$ $(0 \leq k \leq p_n-1 \ ; \ 0 \leq r+1 \leq p_{n-1})$ to both sides of formula (1.3.17) and set $\mu = \nu = \mu_n$ to get

$$\langle\omega_k(\mu_n),\omega_r(\mu_n)\rangle - \langle\omega_{k-1}(\mu_n),\omega_{r+1}(\mu_n)\rangle = 0 .$$

This shows that the quantity $\langle\omega_k(\mu_n),\omega_r(\mu_n)\rangle$ depends only on the sum $k+r$ of the indices k, r, and not upon each of them separately. Consequently,

$$\langle\omega_k(\mu_n),\omega_r(\mu_n)\rangle = a_{r+k} ,$$

and, as follows from (1.3.18) and (1.3.19), $a_{r+k} = 0$ for $r+k < p_n-1$, whereas $a_{r+k} \neq 0$ for $r+k = p_n-1$. Next, we divide both sides of formula (1.3.17) by $\nu-\mu$, apply the operation $\frac{(-1)^{k+s}}{k!s!} \frac{\partial^{k+s}}{\partial\mu^k \partial\nu^s}$ to the resulting equality and then set $\mu = \mu_n$, $\nu = \mu_m$. Since μ_n (μ_m) is a root of the function $\chi(\sqrt{z})$ of multiplicity p_n (respectively, p_m), the right hand side vanishes for $0 \leq k \leq p_n-1$, $0 \leq s \leq p_m-1$. Therefore

$$\langle\omega_k(\mu_n),\omega_s(\mu_m)\rangle = 0 ,$$

and the proof of the lemma is complete. □

Now let $\mu_{n_0},\mu_{n_1},\ldots,\mu_n = (n + \frac{a_n}{n})^2$ denote the distinct eigenvalues of the boundary value problem (1.3.1), (1.3.12) with respective multiplicities p_n. Suppose that the function $f(x) \in L_2(0,\pi)$ admits a series expansion in the eigenfunctions and generalized eigenfunctions of the problem which converges in the norm of $L_2(0,\pi)$:

$$f(x) = \underset{N\to\infty}{\text{l.i.m.}} \sum_{n=n_0}^{N} \left\{ \sum_{k=0}^{p_n-1} \omega_k(\mu_n,f)\omega_k(\mu_n) \right\} . \tag{1.3.20}$$

Then the coefficients of this series can be found using formulas similar to the classical Euler formulas for the ordinary Fourier coefficients. In fact, upon multiplying both sides of equality (1.3.20) by $\omega_s(\mu_m)$ and integrating over the interval $(0,\pi)$, we find that

$$\langle f,\omega_s(\mu_m)\rangle = \sum_{k=0}^{p_m-1} \omega_k(\mu_m,f)\langle\omega_k(\mu_m),\omega_s(\mu_m)\rangle , \tag{1.3.21}$$

(here we used the fact that, by the preceding lemma, $\langle \omega_k(\mu_n), \omega_s(\mu_m)\rangle = 0$ for $\mu_n \neq \mu_m$). Thus, in order to determine the coefficients of the series (1.3.20) we must solve the system of linear equations (1.3.21) with respect to $\omega_k(\mu_m, f)$, for each ω_m separately. Therefore, if μ_m is a simple eigenvalue (i.e., $p_m = 1$), then

$$\omega_0(\mu_m, f) = \frac{\langle f, \omega_0(\mu_m)\rangle}{\langle \omega_0(\mu_m), \omega_0(\mu_m)\rangle}, \tag{1.3.22}$$

whereas, if $p_m \neq 1$,

$$\omega_k(\mu_m, f) = \sum_{j=0}^{p_m - 1} b_{k,j} \langle f, \omega_j(\mu_m)\rangle , \tag{1.3.23}$$

where the $b_{k,j}$ are the entries of the inverse of the matrix with entries $a_{k,s} = \langle \omega_k(\mu_m), \omega_s(\mu_m)\rangle$.

Let us calculate the matrix $||b_{k,j}||$. By Lemma 1.3.4,

$$\left.\begin{aligned} &a_{k,s} = \langle \omega_k(\mu_m), \omega_s(\mu_m)\rangle = a_{k+s} , \\ &a_{k+s} = 0 \quad \text{for} \quad k + s < p_m - 1 , \quad a_{p_m - 1} \neq 0 . \end{aligned}\right\} \tag{1.3.24}$$

We show that

$$b_{k,j} = \begin{cases} b_{k+j} , & k + j \leq p_m - 1 \\ 0 , & k + j > p_m - 1 , \end{cases} \tag{1.3.25}$$

where the numbers $b_0, b_1, \ldots, b_{p_m - 1}$ are found from the system of equations

$$\left.\begin{aligned} &b_{p_m - 1} a_{p_m - 1} = 1 , \\ &b_{p_m - 1} a_{p_m} + b_{p_m - 2} a_{p_m - 1} = 0 , \\ &\cdot \\ &\cdot \\ &\cdot \\ &b_{p_m - 1} a_{2p_m - 2} + \ldots + b_0 a_{p_m - 1} = 0 , \end{aligned}\right\} \tag{1.3.26}$$

which obviously has a unique solution. In fact, for this choice of $b_{k,j}$, the entries $c_{k,j}$ in the product $||b_{k,j}|| \cdot ||a_{k,j}||$ of the matrices $||b_{k,j}||$ and $||a_{k,j}||$ are of the form

$$c_{k,j} = \sum_{i=0}^{p_m - 1} b_{k,i} a_{i,j} = \sum_{i=0}^{p_m - 1 - k} b_{k+i} a_{i+j} = \sum_{s=k}^{p_m - 1} b_s a_{s-k+j} ,$$

that is, by (1.3.24) and (1.3.25),

$$c_{k,j} = \begin{cases} 1 \ , & k = j \ , \\ 0 \ , & k \neq j \ , \end{cases}$$

as asserted.

It follows from formulas (1.3.23) and (1.3.25) that

$$\sum_{k=0}^{p_m-1} \omega_k(\mu_m,f)\omega_k(\mu_m) = \sum_{k=0}^{p_m-1} \omega_k(\mu_m) \sum_{j=0}^{p_m-1-k} b_{k+j} <f,\omega(\mu_m)> =$$

$$= \sum_{s=0}^{p_m-1} b_s \sum_{j=0}^{s} <f,\omega_j(\mu_m)>\omega_{s-j}(\mu_m) \ ,$$

whence, in view of the definition of the functions $\omega_k(\mu_m)$,

$$\sum_{k=0}^{p_m-1} \omega_k(\mu_m,f)\omega_k(\mu_m) =$$

$$= \sum_{s=0}^{p_m-1} \frac{(-1)^s b_s}{s!} \sum_{j=0}^{s} \frac{s!}{j!(s-j)!} \frac{\partial^j}{\partial\lambda^j} <f,\omega_0(\lambda)> \frac{\partial^{s-j}}{\partial\lambda^{s-j}} \omega_0(\lambda)\Big|_{\lambda=\mu_m}$$

or, equivalently,

$$\sum_{k=0}^{p_m-1} \omega_k(\mu_m,f)\omega_k(\mu_m) = \sum_{s=0}^{p_m-1} r_s(\mu_m) \frac{\partial^s}{\partial\lambda^s} \{<f,\omega_0(\lambda)>\omega_0(\lambda)\}\Big|_{\lambda=\mu_m},$$

where $r_s(\mu_m) = \frac{(-1)^s b_s}{s!}$ Therefore, if the function $f(x) \in L_2(0,\pi)$ admits a series expansion in terms of the eigenfunctions and generalized eigenfunctions of problem (1.3.1), (1.3.12), then this expansion has the form

$$f(x) = \sum_{n=n_0}^{\infty} \left\{ \sum_{s=0}^{p_n-1} r_s(\mu_n) \frac{\partial^s}{\partial\lambda^s} \{<f,\omega_0(\lambda)>\omega_0(\lambda)\}\Big|_{\lambda=\mu_n} \right\} , \tag{1.3.27}$$

where the coefficients $r_s(\mu_n)$ are independent of $f(x)$ and, moreover, $p_n = 1$, and $r_0(\mu_n) = \{<\omega_0(\mu_n),\omega_0(\mu_n)>\}^{-1}$ for all sufficiently large n . Conversely, if the series on the right hand side of (1.3.27) converges, then its sum $\varphi(x)$ satisfies the equalities

$$<\varphi,\omega_k(\mu_m)> = <f,\omega_k(\mu_m)> \quad (0 \leq k \leq p_m-1)$$

for all eigenvalues μ_m , i.e.,

$<\varphi-f,\omega_k(\mu_m)> = 0$.

Since the system of eigenfunctions and generalized eigenfunctions is complete, this yields $\varphi(x) = f(x)$. Hence, to prove that every function $f(x) \in L_2(0,\pi)$ admits a series expansion (1.3.27), it suffices to verify that the indicated series converges in the norm of $L_2(0,\pi)$. By the preceding discussion, the general term of this series is equal to

$$\frac{<f,\omega_0(\mu_n)>\omega_0(\mu_n)}{<\omega_0(\mu_n),\omega_0(\mu_n)>} ,$$

for all sufficiently large values of n , where $\mu_n = (n + \frac{a_n}{n})^2$, $\sup|a_n| < \infty$, and

$$\omega_0(\mu_n) = \cos(n + \frac{a_n}{n})x + \int_0^x K(x,t;h)\cos(n + \frac{a_n}{n})tdt .$$

Now since

$$\cos(n + \frac{a_n}{n})x = \cos nx - \frac{a_n x}{n} \sin nx + \frac{\beta_n(x)}{n^2} ,$$

where the functions $\beta_n(x)$ are continuous and the sequence $\max\limits_{0\leq x\leq\pi} |\beta_n(x)|$ is bounded, it follows that

$$\omega_0(\mu_n) = \cos nx - \frac{a_n x}{n} \sin nx + \frac{\beta_n(x)}{n^2} +$$

$$+ \int_0^x K(x,t;h) \left\{\cos nt - \frac{a_n t}{n} \sin nt + \frac{\beta_n(t)}{n^2}\right\} dt$$

or, equivalently, that

$$\omega_0(\mu_n) = \cos nx + \frac{K(x,x;h) - a_n x}{n} \sin nx + \gamma_n(x) , \tag{1.3.28}$$

where the functions

$$\gamma_n(x) = \frac{\beta_n(x)}{n^2} + \frac{1}{n^2}\int_0^x K(x,t;h)\beta_n(t)dt - \frac{1}{n}\int_0^x\{K_t'(x,t;h) + a_n K(x,t;h)\}\sin ntdt$$

are also continuous. Using the Bessel inequality for the Fourier sine-coefficients, the Cauchy-Bunyakovski inequality, and the boundedness of the sequences $|a_n|$ and $\max\limits_{0\leq x\leq\pi} |\beta_n(x)|$, we find that

$$\max_{0\leqslant x\leqslant\pi} \sum_{n=N}^{\infty} |\nu_n(x)| \leqslant \frac{C}{\sqrt{N}}, \tag{1.3.29}$$

where the constant C is dependent on N. From formulas (1.3.28) and (1.3.29) it follows that we can write

$$\langle f,\omega_0(\mu_n)\rangle = \langle f,\cos nx\rangle + \varepsilon_n^{(1)}$$

and

$$\{\langle\omega_0(\mu_n),\omega_0(\mu_n)\rangle\}^{-1} = \frac{2}{\pi} + \varepsilon_n^{(2)} ,$$

with

$$\sum_{n=N}^{\infty}\{|\varepsilon_n^{(1)}| + |\varepsilon_n^{(2)}|\} < \infty .$$

Therefore, for large values of n, the general term of series (1.3.27) is of the form

$$\frac{2}{\pi}\langle f,\cos nx\rangle \cos nx + \varepsilon_n(x)$$

in which the $\varepsilon_n(x)$ are continuous functions, with

$$\max_{0\leqslant x\leqslant\pi} \sum_{n=N}^{\infty} |\varepsilon_n(x)| < \frac{C'}{\sqrt{N}}, \tag{1.3.30}$$

where the constant C' is independent of N. Now the convergence of series (1.3.27) in the norm of $L_2(0,\pi)$ follows from the convergence of the series $\sum_{n=0}^{\infty}{}' \frac{2}{\pi}\langle f,\cos nx\rangle \cos nx$ in the same norm, which in turn is guaranteed by the Riesz-Fischer theorem. Moreover, it follows from the continuity of the functions $\varepsilon_n(x)$ and inequality (1.3.30) that the difference of the partial sums of the series (1.3.27) and $\sum_{n=0}^{\infty}{}' \frac{2}{\pi}\langle f,\cos nx\rangle \cos nx$ converges uniformly in x to a continuous function. The latter function must, however, vanish identically since the two indicated series converge in the L_2-norm to the same function $f(x)$. Hence, at any given point x, the series (1.3.27) converges to $f(x)$ if and only if the Fourier cosine series of f converges to $f(x)$; moreover, series (1.3.27) converges uniformly to $f(x)$ on some interval (α,β) if the Fourier cosine series of f converges uniformly on (α,β).

Summing up, we can state the following theorem.

THEOREM 1.3.2. *The system of eigenfunctions and generalized functions of the boundary value problem* (1.3.1), (1.3.12) *is complete in the space* $L_2(0,\pi)$ *and constitutes there a basis; every function* $f(x) \in L_2(0,\pi)$ *admits a unique series expansion* (1.3.27) *with respect to this system which converges in the norm of* $L_2(0,\pi)$. □

Series (1.3.27) is uniformly equiconvergent with Fourier cosine series of $f(x)$, i.e.,

$$\lim_{N\to\infty} \sup_{0\le x\le\pi} |\sigma_N(x) - s_N(x)| = 0 ,$$

where $\sigma_N(x)$ and $s_N(x)$ designate the partial sums of the series (1.3.27) and the Fourier cosine series, respectively.

PROBLEMS

1. Let $f(x) \in L_1(0,\pi)$ and let

$$F(\lambda) = \int_0^\pi f(x)\omega_1(\lambda,x;h_1)\omega_2(\lambda,x;h_2)dx ,$$

where $\omega_1(\lambda,x;h_1)$ and $\omega_2(\lambda,x;h_2)$ are the solutions of Sturm-Liouville equations with potentials $q_1(x)$ and $q_2(x)$, respectively. Prove that the identity $F(\lambda) \equiv 0$ holds if and only if $f(x) = 0$ a.e.

Hint. It follows from the existence of the transformation operators (1.2.10) that

$$2\omega_1(\lambda,x;h_1)\omega_2(\lambda,x;h_2) = 2\cos^2\lambda x + \int_0^x [K_1(x,t;h_1) + (K_2(x,t;h_2)] \times$$

$$\times 2\cos\lambda x\cos\lambda t dt + \int_0^x\int_0^x K_1(x,t;h_1)K_2(x,\xi;h_2)2\cos\lambda t\cos\lambda\xi dt d\xi$$

or, equivalently, that

$$2\omega_1(\lambda,x;h_1)\omega_2(\lambda,x;h_2) = 1 + \cos 2\lambda x + \int_0^x M(x,t)\cos 2\lambda t dt , \qquad (1.3.31)$$

where

$$M(x,t) = 2\{K_1(x,2t-x;h_1) + K_2(x,2t-x;h_2) + K_1(x,x-2t;h_1) +$$

$$+ K_2(x,x-2t;h_2)\} + 2\int_0^x \{K_1(x,2t-u;h_1) + K_1(x,2t+u;h_1)\}K_2(x,u;h_2)du$$

(here the functions $K_i(x,y;h)$ are set equal to zero for $y \notin [0,x]$). Consequently,

$$2F(\lambda) = \int_0^\pi f(x)dx + \int_0^\pi \left\{f(x) + \int_x^\pi M(\xi,x)f(\xi)d\xi\right\} \cos 2\lambda x dx$$

and so, if $F(\lambda) \equiv \text{const}$, then

$$\int_0^\pi \left\{f(x) + \int_x^\pi M(\xi,x)(\xi)d\xi\right\} \cos 2\lambda x dx \equiv \text{const} ,$$

from which in turn it follows, by the Riemann-Lebesgue lemma and the uniqueness theorem for the Fourier transform, that

$$f(x) + \int_x^\pi M(\xi,x)f(\xi)d\xi = 0 \quad \text{a.e.}$$

Since the kernel $M(\xi,x)$ is bounded, this Volterra integral equation has only the trivial solution, i.e., $f(x) = 0$ a.e.

2. Let $\{q(x),h,H\}$ denote the boundary value problem

$$-y'' + q(x)y = \lambda^2 y \quad (0 \leqslant x \leqslant \pi) ,$$

$$y'(0) - hy(0) = 0 \quad , \quad y'(\pi) + Hy(\pi) = 0 ,$$

and let $\{S\, q(x),h,H\}$ designate the set of its eigenvalues, multiplicities counted. Prove the following uniqueness theorem of G. Borg: if

$$S\{q\ (x),h_1,H_1\} = S\{q_2(x),h_2,H_2\}$$

and

$$S\{q_1(x),h_1,\tilde{H}_1\} = S\{q_2(x),h_2,\tilde{H}_2\} ,$$

and if $H_1 \neq \tilde{H}_1$, then $q_1(x) = q_2(x)$ a.e., $H_1 = H_2$ and $\tilde{H}_1 = \tilde{H}_2$.

Hint. According to (1.3.13)

$$\chi_i(\lambda,H) = \omega_i'(\lambda,\pi) + H\omega_i(\lambda,\pi) ,$$

where $\chi_i(\lambda,H)$ is the characteristic function of the boundary value problem $\{q_i(x),h_i,H\}$ and $\omega_i(\lambda,x)$ is the solution of the equation

$$-\omega_i'' + q_i(x)\omega_i = \lambda^2\omega_i$$

with initial conditions $\omega_i(\lambda,0) = 1$, $\omega_i'(\lambda,0) = h_i$. Since the eigenvalues are the squares of the roots of the characteristic function, the assumptions of the problem imply that the ratios

$$\chi_1(\lambda,H_1)[\chi_2(\lambda,H_2)]^{-1} \quad \text{and} \quad \chi_1(\lambda,\tilde{H}_1)[\chi_2(\lambda,\tilde{H}_2)]^{-1}$$

are entire functions. Furthermore, the principal parts of all the characteristic functions considered here are equal to $-\lambda \sin \lambda\pi$ and, according to (1.3.10'), there is a sequence of unboundedly expanding contours K_n such that

$$\lim_{n\to\infty} \chi_1(\lambda,H_1)[\chi_2(\lambda,H_2)]^{-1}\Big|_{\lambda\in K_n} = \lim_{n\to\infty} \chi_1(\lambda,\tilde{H}_1)[\chi_2(\lambda,\tilde{H}_2)]^{-1}\Big|_{\lambda\in K_n} = 1 .$$

Consequently,

$$\chi_1(\lambda,H_1) \equiv \chi_2(\lambda,H_2) \quad , \quad \chi_1(\lambda,\tilde{H}_1) \equiv \chi_2(\lambda,\tilde{H}_2) , \tag{1.3.34}$$

which, in view of (1.3.32), yields

$$(H_1 - \tilde{H}_1)\omega_1(\lambda,\pi) = \chi_1(\lambda,H_1) - \chi_1(\lambda,\tilde{H}_1) = \chi_2(\lambda,H_2) - \chi_2(\lambda,\tilde{H}_2) =$$

$$= (H_2 - \tilde{H}_2)\lambda_2(\lambda,\pi) ,$$

and since $\omega_i(\lambda,\pi) = \cos \lambda\pi + o(1)$ as $\lambda \to +\infty$, it follows that

$$H_1 - \tilde{H}_1 = H_2 - \tilde{H}_2 \quad , \quad \omega_1(\lambda,\pi) = \omega_2(\lambda,\pi) .$$

Upon multiplying the first of equations (1.3.33) by $\omega_2(\lambda,x)$ and the second by $\omega_1(\lambda,x)$, and subsequently integrating the difference of these products you get the equality

$$-\omega_1'(\lambda,\pi)\omega_2(\lambda,\pi) + \omega_1(\lambda,\pi)\omega_2'(\lambda,\pi) + h_1 - h_2 +$$

$$+ \int_0^\pi [q_1(x) - q_2(x)]\omega(\lambda,x)\omega_2(\lambda,x)dx = 0 ,$$

which, in view of (1.3.32), (1.3.34), and (1.3.35), can be written in the equivalent form

$$\int_0^\pi [q_1(x) - q_2(x)]\omega_1(\lambda,x)\omega_2(\lambda,x)dx + h_1 - h_2 = (\tilde{H}_1 - H_1)[\omega_1(\lambda,\pi)]^2 .$$

Next, it follows from formula (1.3.31) and the Riemann-Lebesgue lemma that as $\lambda \to +\infty$, the left hand side of this equality tends to

$\frac{1}{2}\int_0^\pi [q_1(x) - q_2(x)]dx + h_1 - h_2$, whereas the right hand side equals $(\tilde{H}_1 - H_1)[\cos^2 \lambda\pi + o(1)]$, which is possible only if $\tilde{H}_1 = H_1$. Therefore, $H_1 = \tilde{H}_1$, $H_2 = \tilde{H}_2$, and

$$\int_0^\pi [q_1(x) - q_2(x)]\omega_1(\lambda,x)\omega_2(\lambda,x)dx = h_2 - h_1 = \text{const} .$$

By the preceding problem, $q_1(x) = q_2(x)$ a.e. and $h_1 = h_2$.

3. Prove the analog of Theorem 1.3.1 for the boundary value problems generated by the Dirac equation

$$By'(x) + \Omega(x)y(x) = \lambda y(x) \quad (0 < x < \pi) \tag{1.3.36}$$

with boundary conditions

$$\Gamma(y) = A_0y(0) + A_1y(\pi) = 0 ,$$

where

$$B = \begin{pmatrix} 0 & 1 \\ -1 & 0 \end{pmatrix} , \quad \Omega(x) = \begin{pmatrix} p(x) & r(x) \\ r(x) & -p(x) \end{pmatrix} ,$$

$$A_0 = \begin{pmatrix} a_{11} & a_{12} \\ a_{21} & a_{22} \end{pmatrix} , \quad A_1 = \begin{pmatrix} a_{13} & a_{14} \\ a_{23} & a_{24} \end{pmatrix} ,$$

$p(x)$, $r(x)$ are summable complex-valued functions, and the a_{ik} are complex numbers.

Hint. The general solution of equation (1.3.36) is $e(\lambda,x)C$, where $e(\lambda,x)$ is the fundamental solution normalized by the condition $e(\lambda,0) = I$, and C is an arbitrary constant matrix. Hence, the eigenvalues of the problem in question coincide with the roots of the characteristic function $\chi(\lambda) = \operatorname{Det}(A_0 + A_1e(\lambda,\pi))$. Expanding this determinant, you get

$$\chi(\lambda) = J_{12} + J_{34} + J_{32}e_{11} + J_{13}e_{12} + J_{42}e_{21} + J_{14}e_{22} , \tag{1.3.37}$$

where $e_{ik} = e_{ik}(\lambda,\pi)$ is the ik entry of the matrix $e(\lambda,\pi)$, and $J_{\alpha\beta} = a_{1\alpha}a_{2\beta} - a_{2\alpha}a_{1\beta}$ (here you have to use the fact that $\operatorname{Det} e(\lambda,x) \equiv 1$).

Let $\tilde{\Gamma}$, $\tilde{A}_0$, $\tilde{A}_1$ be the matrices built from the cofactors of the matrix $\Gamma = A_0 + A_1e(\lambda,\pi), A_0, A_1$, and $\omega(\lambda,x) = e(\lambda,x)\tilde{\Gamma}$. You can readily check that

$$\omega(\lambda,x) = e(\lambda,x)\tilde{A}_0 + e_\pi(\lambda,x)\tilde{A}_1 , \quad A_0\omega(\lambda,0) + A_1\omega(\lambda,\pi) = \chi(\lambda)I , \tag{1.3.38}$$

where $e_\pi(\lambda,x) = e(\lambda,x)[e(\lambda,\pi)]^{-1}$. If λ_n is a root of multiplicity p of the function $\chi(\lambda)$, then the matrices

$$\omega_k(x) = \frac{1}{k!}\frac{\partial^k}{\partial\lambda^k}\omega(\lambda,x)\Big|_{\lambda=\lambda_n} \quad (k = 0,1,\dots,p-1)$$

satisfy the boundary conditions $A_0\omega_k(0) + A_1\omega_k(\pi)$ and the equations $B\omega_k' + \Omega(x)\omega_k' = \lambda_n\omega_k + \omega_{k-1}$, i.e., they form a chain made of a matrix eigenfunction $\omega_0(x)$ and generalized matrix eigenfunctions. Hence, to prove the completeness of the system of the set of matrix eigenfunctions and generalized eigenfunctions, it suffices to check that the matrix $[\chi(\lambda)]^{-1}\int_0^\pi f(x)\omega(x,\lambda)dx$ cannot be an entire function of λ unless the matrix function f vanishes a.e.

Using the transformation operators constructed in Problem 5, Section 2 of this chapter and Lemma 1.3.1, you can verify that

$$|\chi(\lambda) - \chi_0(\lambda)| = o(\exp|\mathrm{Im}\,\lambda\pi|) \quad \text{as} \quad |\lambda| \to \infty ,$$

where

$$\chi_0(\lambda) = J_{12} + J_{34} + (J_{14} + J_{32})\cos\lambda\pi + (J_{13} + J_{42})\sin\lambda\pi ,$$

is the characteristic function of the boundary value problem with $\Omega(x) \equiv 0$. If

$$(J_{14} + J_{32})^2 + (J_{13} + J_{42})^2 \neq 0 , \tag{1.3.39}$$

then there exists a sequence of unboundedly expanding contours on which $|\chi_0(\lambda)|$, and hence $|\chi(\lambda)|$ too, is larger than $C\exp|\mathrm{Im}\,\lambda\pi|$. Now using formula (1.3.8) and the transformation operators of Problem 5, Section 2, attached to the points 0 and π, you can prove the completeness of the set of matrix eigenfunctions and generalized eigenfunctions in much the same way as in Theorem 1.3.1. If, however, condition (1.3.39) is not satisfied, then completeness fails even for $\Omega(x) \equiv 0$.

4. Generalize Theorem 1.3.1 to the case in which the coefficients a_{ij} in the boundary conditions (1.3.2) are entire functions (in particular, polynomials) of λ.

4. ASYMPTOTIC FORMULAS FOR SOLUTIONS OF THE STURM-LIOUVILLE EQUATION

The proofs of the theorems given in the preceding section rested on rather crude estimates of the solutions $\omega(\lambda,x;h)$ of equation (1.3.1). In point of fact, the solutions of this equation admit asymptotic expansions in powers of λ^{-1} which become more precise as the number of derivatives that the function $q(x)$ has increases. Indeed, if the function $q(x)$ has n continuous derivatives, then by Theorem 1.2.2 the kernel $K(x,t)$, and hence $K(x,t;h)$ too, has a continuous derivative of order $n+1$. Therefore, one can integrate by parts $n+1$ times on the right hand side of formula (1.2.10) to obtain an expansion of the solution $\omega(\lambda,x;h)$ in powers of λ^{-1}, with a remainder on the order (λ^{-n-1}). It is however difficult to find the explicit form of the coefficients and the remainder of this expansion. It is considerably simpler to derive asymptotic expansions of this kind directly from equation (1.3.1).

We let $W_n^2[a,b]$ denote the Sobolev space of complex-valued functions defined on $[a,b]$ which have $n-1$ absolutely continuous derivatives, and a n-th order derivative which is square-summable over $[a,b]$. We consider a Sturm-Liouville operator $L = -\frac{d^2}{dx^2} + q(x)$ with $q(x) \in W_n^2[0,a]$ and seek solutions of the equation

$$L[y] = \lambda^2 y \quad (0 \leq x \leq a) \tag{1.4.1}$$

in the form

$$y = y(\lambda,x) = e^{i\lambda x}\left[u_0(x) + \frac{u_1(x)}{2i\lambda} + \dots + \frac{u_n(x)}{(2i\lambda)^n} + \frac{u_{n+1}(\lambda,x)}{(2i\lambda)^{n+1}}\right] . \tag{1.4.2}$$

LEMMA 1.4.1. *If* $q(x) \in W_2^n[0,a]$, *then equation* (1.4.1) *has a solution of the form* (1.4.2), *where*

$$u_0(x) = 1 \quad , \quad u_k(x) = \int_0^x L[u_{k-1}(\xi)]d\xi \ ,$$

and the function $u_{n+1}(\lambda,x)$ *and its derivatives admit the representations*

$$u_{n+1}(\lambda,x) = u_{n+1}(x) + \frac{1}{2i\lambda}\int_0^x q(t)u_{n+1}(t)dt -$$

$$- \int_0^x \{u'_{n+1}(x-\xi) + \frac{1}{2i\lambda} K_{n+1}^{(0)}(x,\xi)\ e^{-2i\lambda\xi}d\xi \ ,$$

and

$$u'_{n+1}(\lambda,x) = 2i\lambda \int_0^x \{u'_{n+1}(x-\xi) + \frac{1}{2i\lambda} K_{n+1}^{(1)}(x,\xi)\}e^{-2i\lambda\xi}d\xi \ ,$$

respectively, in which the kernels $K_{n+1}^{(0)}(x,\xi)$, $K_{n+1}^{(1)}(x,\xi)$ *and* $u'_{n+1}(x-\xi)$

are square-summable in the variable ξ *for every* $x \in [0,a]$.

PROOF. Substituting the right hand side of formula (1.4.2) into equation (1.4.1) and setting the coefficients of the powers $(2i\lambda)^{-k}$, $k = -1,0,1,\dots,n-1$, equal to zero, we get the system of equations

$$u_0'(x) = 0 \ , \quad u_k'(x) = L[u_{k-1}(x)] \quad (k = 1,2,\dots,n) \ ,$$

$$L[u_{n+1}(\lambda,x)] - 2i\lambda u_{n+1}'(\lambda,x) + 2i\lambda L[u_n(x)] = 0 \ . \tag{1.4.3}$$

It is convenient to write the last equation of this system as

$$L[e^{i\lambda x}u_{n+1}(\lambda,x)] - \lambda^2[e^{i\lambda x}u_{n+1}(\lambda,x)] = -2i\lambda e^{i\lambda x}L[u_n(x)] \ .$$

Thus, the function (1.4.2) is a solution of equation (1.4.1) if we put

$$u_0(x) = 1 \ , \quad u_k(x) = \int_0^x L[u_{k-1}(\xi)]d\xi \ (k = 1,2,\dots,n) \ , \tag{1.4.4}$$

$$u_{n+1}(x) = \int_0^x L[u_n(\xi)]d\xi \ , \quad u_{n+1}(\lambda,x) = e^{-i\lambda x}v_{n+1}(\lambda,x) \ ,$$

where $v_{n+1}(\lambda,x)$ is the solution of the equation

$$L[v_{n+1}] - \lambda^2 v_{n+1} = -2i\lambda e^{i\lambda x}u_{n+1}(x) \tag{1.4.3'}$$

with initial data $v_{n+1}(\lambda,0) = v_{n+1}'(\lambda,0) = 0$. From formulas (1.4.4) we deduce by induction that

$$u_k(x) + (-1)^k q^{(k-2)}(x) \in W_2^{n+3-k}[0,a] \tag{1.4.5}$$

whenever $q(x) \in W_2^n[0,a]$. Consequently, the functions $u_k(x)$ are correctly defined by formulas (1.4.4) for $k = 1,2,\dots,n+1$. Moreover, the function $v_{n+1}(x)$ has a square-summable derivative.

Let $c(\lambda,x)$ and $s(\lambda,x)$ denote the fundamental system of solutions of equation (1.4.1) with initial data $c(\lambda,0) = s'(\lambda,0) = 1$, $c'(\lambda,0) = s(\lambda,0) = 0$. Let

$$w(x,t,\lambda) = s(\lambda,x)c(\lambda,t) - s(\lambda,t)c(\lambda,x) \ .$$

The method of variation of constants leads to the following representation of the solution $v_{n+1}(\lambda,x)$ of equation (1.4.3'):

$$v_{n+1}(\lambda,x) = 2i\lambda \int_0^x w(x,\xi,\lambda)e^{i\lambda\xi}u_{n+1}'(\xi)d\xi = 2i\lambda \int_0^x w(x,x-t,\lambda)e^{i\lambda(x-t)}u_{n+1}'(x-t)dt \ ,$$

This in turn yields the formula

$$u_{n+1}(\lambda,x) = 2i\lambda \int_0^x w(x,x-t,\lambda)e^{-i\lambda t}u'_{n+1}(x - t)dt ,$$

Since the function $\omega(\lambda,t) = w(x,x-t,\lambda)$ satisfies the equation $\omega''_{tt} - q(x-t)\omega + \lambda^2\omega = 0$, the Corollary to Theorem 1.2.1 gives

$$\omega(\lambda,t) = w(x,x-t,\lambda) = \frac{\sin \lambda t}{\lambda} + \int_0^t K(x;t,\xi) \frac{\sin \lambda\xi}{\lambda} d\xi ,$$

where x serves as a parameter. Therefore

$$u_{n+1}(\lambda,x) = 2i \int_0^x \left\{\sin \lambda t + \int_0^t K(x;t,\xi)\sin \lambda\xi d\xi\right\} e^{-i\lambda t}u'_{n+1}(x - t)dt =$$

$$= u_{n+1}(x) + \int_0^x K_{n+1}(x,t)e^{-2i\lambda t}dt , \tag{1.4.6}$$

where

$$K_{n+1}(x,t) = -u'_{n+1}(x - t) + 2 \int_t^x \frac{\xi - 2t}{|\xi - 2t|} K(x;\xi,|\xi-2t|)u'_{n+1}(x - \xi)d\xi .$$

On the other hand, from (1.4.3') we derive the integral equation

$$v_{n+1}(\lambda,x) = \int_0^x \frac{\sin \lambda(x - 1)}{\lambda} \{2i\lambda e^{i\lambda t}u'_{n+1}(t) + q(t)v_{n+1}(\lambda,t)\}dt$$

and we also have that

$$u_{n+1}(\lambda,x) = \int_0^x \{1 - e^{-2i\lambda(x-t)}\} \left\{u'_{n+1}(t) + \frac{1}{2i\lambda} q(t)u_{n+1}(\lambda,t)\right\} dt . \tag{1.4.7}$$

Therefore, upon substituting here formula (1.4.6) for $u_{n+1}(\lambda,t)$, we get

$$u_{n+1}(\lambda,t) = \int_0^x \{1 - e^{-2i\lambda(x-t)}\} \{u'_{n+1}(t) + \frac{1}{2i\lambda} q(t)u_{n+1}(t)\} dt +$$

$$+ \frac{1}{2i\lambda} \int_0^x \{1 - e^{-2i\lambda(x-t)}\}q(t) \int_0^t K_{n+1}(t,\xi)e^{-2i\lambda\xi}d\xi dt ,$$

from which it follows, after some simple manipulations, that

$$u_{n+1}(\lambda,x) = u_{n+1}(x) + \frac{1}{2i\lambda} \int_0^x q(t)u_{n+1}(t)dt -$$

$$-\left\{ \int_0^x e^{-2i\lambda\xi}u'_{n+1}(x - \xi)d\xi + \frac{1}{2i\lambda} \int_0^x K^{(0)}_{n+1}(x,\xi)e^{-2i\lambda\xi}d\xi\right\} , \tag{1.4.8}$$

where

$$K_{n+1}^{(0)}(x,\xi) = q(x - \xi)u_{n+1}(x - \xi) - \int_{\xi}^{x} q(t)K_{n+1}(t,\xi)dt + $$
$$+ \int_{x-\xi}^{x} q(t)K_{n+1}(t,\xi - x + t)dt .$$

Differentiating equality (1.4.7) once with respect to x, we get

$$u'_{n+1}(\lambda,x) = 2i\lambda\left\{ \int_0^x e^{-2i\lambda(x-t)}u'_{n+1}(t)dt + \frac{1}{2i\lambda}\int_0^x e^{-2i\lambda(x-t)}q(t)u_{n+1}(\lambda,t)dt \right\} ,$$

whence, in view of (1.4.8)

$$u'_{n+1}(\lambda,x) = 2i\lambda \int_0^x u'_{n+1}(x - \xi)e^{-2i\lambda\xi}d\xi + \int_0^x K_{n+1}^{(1)}(x,\xi)e^{-2i\lambda\xi}d\xi , \tag{1.4.9}$$

where

$$K_{n+1}^{(1)}(x,\xi) = K_{n+1}^{(0)}(x,\xi) + \int_{\xi}^{x} q(t)K_{n+1}(t,\xi)dt .$$

The assertions of the lemma are now seen to be consequences of formulas (1.4.3), (1.4.8), and (1.4.9).

□

The recursion formulas (1.4.4) from which the functions $u_k(x)$ are found contain undesirable integrations. To get rid of them, one can proceed as follows: Setting

$$\sigma(\lambda,x) = \frac{d}{dx} \ln \left[1 + \frac{u_1(x)}{2i\lambda} + \dots + \frac{u_n(x)}{(2i\lambda)^n} + \frac{u_{n+1}(\lambda,x)}{(2i\lambda)^{n+1}}\right] , \tag{1.4.10}$$

to get the representation

$$y(\lambda,x) = \exp \left\{i\lambda x + \int_0^x \sigma(\lambda,t)dt\right\} , \tag{1.4.11}$$

for the solution of (1.4.2). Then from this we derive the equality

$$y'(\lambda,x) = \{i\lambda + \sigma(\lambda,x)\}y(\lambda,x) , \tag{1.4.11'}$$

and the equation

$$\sigma'(\lambda,x) + 2i\lambda\sigma(\lambda,x) + \sigma^2(\lambda,x) - q(x) = 0 . \tag{1.4.12}$$

for the function $\sigma(\lambda,x)$. To simplify these formulas, let us introduce the notations

$$P_n(\lambda,x) = 1 + \frac{u_1(x)}{2i\lambda} + \dots + \frac{u_n(x)}{(2i\lambda)^n} , \tag{1.4.13}$$

$$Q_n(\lambda,x) = P_n(\lambda,x) + \frac{u_{n+1}(\lambda,x)}{(2i\lambda)^{n+1}} \quad , \tag{1.4.13}$$

in which formulas (1.4.2) and (1.4.10) take the form

$$y(\lambda,x) = e^{i\lambda x}Q_n(\lambda,x) \ ,$$

and, respectively,

$$\sigma(\lambda,x) = \frac{P_n'(\lambda,x)}{P_n(\lambda,x)} + \frac{u_{n+1}'(\lambda,x)P_n(\lambda,x) - u_{n+1}(\lambda,x)P_n'(\lambda,x)}{(2i\lambda)^{n+1}P_n(\lambda,x)Q_n(\lambda,x)} \ .$$

The function $[P_n'(\lambda,x)\ P_n(\lambda,x)]^{-1}$ admits a series expansion in powers of $(2i\lambda)^{-1}$ in the neighborhood of infinity in the complex λ-plane. If we now single out the first n terms in this expansion, we get

$$\frac{P_n'(\lambda,x)}{P_n(\lambda,x)} = \sum_{k=1}^{n} \frac{\sigma_k(x)}{(2i\lambda)^n} + \frac{1}{(2i\lambda)^n} \sum_{k=1}^{\infty} \frac{\varphi_k(x)}{(2i\lambda)^k} \tag{1.4.14}$$

and

$$\sigma(\lambda,x) = \sum_{k=1}^{n} \frac{k(x)}{(2i\lambda)^k} + \frac{\sigma_n(\lambda,x)}{(2i\lambda)^n} \ , \tag{1.4.15}$$

where

$$\sigma_n(\lambda,x) = \sum_{k=1}^{\infty} \frac{\varphi_k(x)}{(2i\lambda)^k} + \frac{u_{n+1}'(\lambda,x)P_n(\lambda,x) - u_{n+1}(\lambda,x)P_n'(\lambda,x)}{2iP_n(\lambda,x)Q_n(\lambda,x)} \ . \tag{1.4.16}$$

Since $u_1(0) = u_2(0) = \ldots = u_n(0) = u_{n+1}(\lambda,0) = u_{n+1}'(\lambda,0)$, it follows that

$$\frac{P_n'(\lambda,0)}{P_n(\lambda,0)} = \sum_{k=1}^{n} \frac{u_k'(0)}{(2i\lambda)^k} = \sum_{k=1}^{n} \frac{\sigma_k(0)}{(2i\lambda)^k} + \frac{1}{(2i\lambda)^n} \sum_{k=1}^{\infty} \frac{\varphi_k(0)}{(2i\lambda)^k} \ ,$$

and hence that $u_k'(0) = \sigma_k(0)$ $(1 \leqslant k \leqslant n)$, $\varphi_k(0) = 0$ $(1 \leqslant k < \infty)$, and $\sigma_n(\lambda,0) = 0$.

We shall write $f(\lambda) = o(\lambda^m)$ $(O(\lambda^m))$ if in every strip $|\mathrm{Im}\ \lambda| \leqslant C < \infty$,

$$\overline{\lim_{|\lambda|\to\infty}} \sup_{|\mathrm{Im}\ \lambda| \leqslant C} |\lambda^{-m}f(\lambda)| = 0 \quad (\text{respectively}, < \infty\ ) \ .$$

Then, for example, it follows from formulas (1.4.8), (1.4.9), equation (1.4.3'), and the Riemann-Lebesgue lemma (on the convergence to zero at infinity of the Fourier transform of a summable function), that

$$u_{n+1}(\lambda,x) - u_{n+1}(x) = o(1) \quad , \quad u'_{n+1}(\lambda,x) = o(\lambda) \ , \tag{1.4.17}$$

and

$$(2i\lambda)^{-1}u''_{n+1}(\lambda,x) + u'_{n+1}(\lambda,x) - u'_{n+1}(x) = o(1) \ . \tag{1.4.18}$$

Substituting the right-hand side of formula (1.4.15) into equation (1.4.12) and using the relations $\sigma(\lambda,x) = o(1)$, $\sigma'(\lambda,x) = o(\lambda)$ (which are consequences of (1.4.16), (1.4.17), and (1.4.18)), we find that

$$\sigma_1(x) = q(x) \quad , \quad \sigma'_k(x) + \sigma_{k+1}(x) + \sum_{j=1}^{k-1} \sigma_{k-j}(x)\sigma_j(x) = 0 \quad (k = 1,2,\ldots,n-1)$$

and

$$\sigma'_n(\lambda,x) + 2i\lambda\sigma_n(\lambda,x) + \sigma'_n(x) + \sum_{j=1}^{n-1} \sigma_{n-j}(x)\sigma_j(x) =$$

$$= - \sum_{p=1}^{n} \sum_{j=p}^{n} \frac{\sigma_{n+p-j}(x)\sigma_j(x)}{(2i\lambda)^n} - \frac{\sigma_n^2(\lambda,x)}{(2i\lambda)^n} - 2\sigma_n(\lambda,x) \sum_{k=1}^{n} \frac{\sigma_k(x)}{(2i\lambda)^k} \ .$$

It thus follows that the functions $\sigma_k(x)$ are determined by the recursion relations

$$\sigma_1(x) = q(x) \quad , \quad \sigma_2(x) = -q'(x) \quad , \quad \sigma_3(x) = q''(x) - q(x)^2 \ ,$$

$$\sigma_{k+1}(x) = -\sigma'_k(x) - \sum_{j=1}^{k-1} \sigma_{k-j}(x)\sigma_j(x) \quad (k = 2,3,\ldots,n) \ ,$$

which contain no integrations; in particular,

$$\sigma_{k+1}(x) = (-1)^k q^{(k)}(x) + S_{k-2}(x) \ ,$$

where $S_{k-2}(x)$ is a polynomial in $q(x),q'(x),\ldots,q^{(k-2)}(x)$.

Furthermore, it follows from equality (1.4.15) that

$$\sigma'_n(\lambda,x) + 2i\lambda\sigma_n(\lambda,x) = \sigma_{n+1}(x) + o(1) \ ,$$

whereas formulas (1.4.16) and (1.4.17), (1.4.18) yield

$$\sigma'_n(\lambda,x) + 2i\lambda\sigma_n(\lambda,x) = \varphi_1(x) + u'_{n+1}(x) + o(1) \ .$$

Consequently,

$$u'_{n+1}(x) = \sigma_{n+1}(x) - \varphi_1(x)$$

and

$$\int_0^x u'_{n+1}(x-\xi)e^{-2i\lambda\xi}d\xi = \int_0^x \sigma_{n+1}(x-\xi)e^{-2i\lambda\xi}d\xi - \frac{\varphi_1(x)}{2i\lambda} +$$

$$+ \frac{1}{2i\lambda}\int_0^x \varphi_1'(x-\xi)e^{-2i\lambda\xi}d\xi ,$$

where we used the fact that $\varphi_1(x) \in W_2^2[0,a]$ and $\varphi_1(0) = 0$. This allows us to replace equalities (1.4.8) and (1.4.9) by

$$u_{n+1}(\lambda,x) = \int_0^x \{\sigma_{n+1}(\xi) - \varphi_1(\xi)\}d\xi + \frac{1}{2i\lambda}\left\{\varphi_1(x) + \int_0^x q(t)u_{n+1}(t)dt\right\} -$$

$$- \left\{ \int_0^x \sigma_{n+1}(x-\xi)e^{-2i\lambda\xi}d\xi + \frac{1}{2i\lambda}\int_0^x \tilde{K}_{n+1}^{(0)}(x,\xi)e^{-2i\lambda\xi}d\xi \right\} , \qquad (1.4.8')$$

and

$$u'_{n+1}(\lambda,x) = 2i\lambda \int_0^x \sigma_{n+1}(x-\xi)e^{-2i\lambda\xi}d\xi - \varphi_1(x) + \int_0^x \tilde{K}_{n+1}^{(1)}(x,\xi)e^{-2i\lambda\xi}d\xi , \quad (1.4.9')$$

in which the kernels $\tilde{K}_{n+1}^{(s)}(x,\xi) = K_{n+1}^{(s)}(x,\xi) + \varphi_1'(x-\xi)$ $(s = 0,1)$ are again square-summable in the variable ξ .

Finally, upon combining the equalities (1.4.16), (1.4.9') with the estimates (1.4.17), (1.4.18), we get

$$\sigma_n(\lambda,x) = \frac{2i\lambda \int_0^x \sigma_{n+1}(x-\xi)e^{-2i\lambda\xi}d\xi + \int_0^x \tilde{K}_{n+1}^{(1)}(x,\xi)e^{-2i\lambda\xi}d\xi}{2i\lambda Q_n(\lambda,x)} + o(\lambda^{-2}) . \quad (1.4.19)$$

Thus, the following lemma is also valid.

LEMMA 1.4.2. *Under the assumptions of Lemma 1.4.1:*

1) *the solution* (1.4.2) *admits the representation* (1.4.11), (1.4.15) ;

2) *the functions* $\sigma_k(t)$ $(1 \leq k \leq n+1)$ *are determined by the recursion formulas*

$$\sigma_1(t) = q(t) \quad , \quad \sigma_{k+1}(t) = -\sigma_k'(t) - \sum_{j=1}^{k-1} \sigma_{k-j}(t)\sigma_j(t) , \qquad (1.4.20)$$

which in turn imply that

$$\sigma_{k+1}(t) = (-1)^k q^{(k)}(t) + S_{k-2}(t) , \qquad (1.4.20')$$

where $S_{k-2}(t)$ *is a polynomial in* $q(t)$, $q'(t),\dots,q^{(k-2)}(t)$;

3) *the function* $\sigma_n(\lambda,x)$ *is given by formula* (1.4.16), *from which equality* (1.4.19), *which is useful for estimates, follows.*

□

We remark that by using formula (1.4.2) for $\lambda \neq 0$ we can obtain two solutions $y(\lambda,x)$ and $y(-\lambda,x)$ of equation (1.4.1). According to (1.4.11), the Wronskian

$$W[y(\lambda,x),y(-\lambda,x)] = y'(\lambda,x)y(-\lambda,x) - y(\lambda,x)y'(-\lambda,x)$$

of these solutions is of the form

$$W[y(\lambda,x),y(-\lambda,x)] = y(\lambda,x)y(-\lambda,x)[2i\lambda + \sigma(\lambda,x) - \sigma(-\lambda,x)] .$$

Since the Wronskian of two arbitrary solutions of the Sturm-Liouville equation does not depend on x, and since $y(\lambda,0) = y(-\lambda,0) = 1$,

$$y(\lambda,x)y(-\lambda,x)[2i\lambda + \sigma(\lambda,x) - \sigma(-\lambda,x)] = 2i\lambda + \sigma(\lambda,0) - \sigma(-\lambda,0) ,$$

whence

$$y(\lambda,x)y(-\lambda,x) = \frac{\omega(\lambda,0)}{\omega(\lambda,x)} \tag{1.4.21}$$

and

$$W[y(\lambda,x),y(-\lambda,x)] = \omega(\lambda,0) , \tag{1.4.22}$$

where

$$\omega(\lambda,x) = 2i\lambda + \sigma(\lambda,x) - \sigma(-\lambda,x) . \tag{1.4.23}$$

In particular, the solutions $y(\lambda,x)$ and $y(-\lambda,x)$ are linearly independent for all values of λ except for the roots of the polynomial $\lambda^n\omega(\lambda,0)$. It is readily checked that the previously constructed solutions $c(\lambda,x)$ and $s(\lambda,x)$ of equation (1.4.1) are expressible through $y(\lambda,x)$ and $y(-\lambda,x)$:

$$s(\lambda,x) = \frac{y(\lambda,x) - y(-\lambda,x)}{\omega(\lambda,0)} \tag{1.4.24}$$

and

$$c(\lambda,x) = \frac{y(\lambda,x)[i\lambda - \sigma(-\lambda,0)] + y(-\lambda,x)[i\lambda + \sigma(\lambda,0)]}{\omega(\lambda,0)} . \tag{1.4.24'}$$

In a number of problems, an important role is played by estimates of the functions $\sigma_n(\lambda,x)$, $u_{n+1}(\lambda,x)$, and their derivatives which are obtained from Lemma 1.3.1. For $\lambda = k + a + O(k^{-1})$, such estimates can be obtained by means of the following simple lemma.

LEMMA 1.4.3. *Suppose that the numerical sequence* $a_k = k + a + h_k$ *is such that* $h_k = O(k^{-1})$ *for* $k \to \pm\infty$. *Let* $f \in L_2(0,\pi)$ *and* $\tilde{f}(\lambda) = \int_0^\pi f(x)e^{-2i\lambda x}dx$.

Then

$$\tilde{f}(a_k) = \tilde{f}(k+a) + k^{-1}g(k)$$

and

$$\sum_{k=-\infty}^{\infty} |\tilde{f}(k+a)|^2 < \infty \quad , \quad \sum_{k=-\infty}^{\infty} |g(k)|^2 < \infty .$$

PROOF. From the equality

$$\tilde{f}(a_k) = \int_0^\pi f(x)e^{-2i(k+a)x}e^{-2ih_k x}dx = \int_0^\pi f(x)e^{-2i(k+a)x}[1 - 2ih_k x + O(h_k^2)]dx$$

it follows that

$$\tilde{f}(a_k) = \tilde{f}(k+a) + h_k\tilde{f}'(k+a) + O(h_k^2) = \tilde{f}(k+a) + k^{-1}g(k) ,$$

where

$$g(k) = kh_k\tilde{f}'(k+a) + k^{-1}O(k^2h_k^2) .$$

Since

$$\tilde{f}(k+a) = \int_0^\pi f(x)e^{-2iax}e^{-2ikx}dx ,$$

and

$$\tilde{f}'(k+a) = -2i \int_0^\pi f(x)xe^{-2iax}e^{-2ikx}dx$$

are the Fourier coefficients of the functions $f(x)e^{-2iax}$ and $-2if(x)xe^{-2iax}$, which belong to $L_2(0,\pi)$, Bessel's inequality implies that

$$\sum_{k=-\infty}^{\infty} |\tilde{f}(k + a)|^2 < \infty \quad , \quad \sum_{k=-\infty}^{\infty} |\tilde{f}'(k + a)|^2 < \infty ,$$

and hence also that $\sum_{k=-\infty}^{\infty} |g(k)|^2 < \infty$, because by assumption, $\sup|kh_k| < \infty$. □

LEMMA 1.4.4. *Suppose that* $q(x) \in W_2^n[0,\pi]$ *and the sequence* $a_k = k + a + h_k$ *is such that* $h_k = O(k^{-1})$. *Then*

$$\sigma_n(a_k,\pi) = \tilde{\sigma}_{n+1}(k+a) + k^{-1}g_1(k) ,$$

and

$$\int_0^\pi \sigma(a_k,x)dx = \sum_{j=1}^{n+2} c_j(2ia_k)^{-j} - (2ik)^{-n-1}\tilde{\sigma}_{n+1}(k+a) + k^{-n-2}g_2(k) ,$$

where

$$\tilde{\sigma}_{n+1}(\lambda) = \int_0^\pi \sigma_{n+1}(\pi - x)e^{-2i\lambda x}dx \ , \tag{1.4.25}$$

the coefficients $c_j = \int_0^\pi \sigma_j(x)dx$ $(j = 1,2,\dots,n,n+1)$ *do not depend upon* n *and* k *, and the coefficient* c_{n+2} *does not depend on* k *and vanishes for* $n = 0$. *Moreover,*

$$\sum_{k=-\infty}^{\infty} |g_1(k)|^2 < \infty \quad , \quad \sum_{k=-\infty}^{\infty} |g_2(k)|^2 < \infty \ .$$

PROOF. The assertion concerning the sequence $\sigma_n(a_k,\pi)$ itself is a straightforward consequence of the preceding lemma, formula (1.4.19), and the obvious estimates $Q_n(\lambda,\pi) = 1 + O(\lambda^{-1})$, and $a_k^{-1} = k^{-1} + O(k^{-2})$. To prove the assertion concerning the sequence $\int_0^\pi \sigma(a_k,x)dx$, we use equalities (1.4.10) and (1.4.14), or, more precisely, their consequences

$$\int_0^\pi \sigma(\lambda,x)dx = \ln\left[1 + \frac{u_1(\pi)}{2i\lambda} + \dots + \frac{u_n(\pi)}{(2i\lambda)^n} + \frac{u_{n+1}(\lambda,\pi)}{(2i\lambda)^{n+1}}\right] =$$

$$= \ln P_n(\lambda,\pi) + \ln[1 + \{(2i\lambda)^{n+1}P_n(\lambda,\pi)\}^{-1}u_{n+1}(\lambda,\pi)] \ , \tag{1.4.26}$$

and

$$\ln P_n(\lambda,\pi) = \sum_{j=1}^{n} c_j(2i\lambda)^{-j} + \sum_{j=1}^{\infty} b_j(2i\lambda)^{-n-j} \ ,$$

where

$$c_j = \int_\pi^\pi \sigma_j(x)dx \quad , \quad b_j = \int_0^\pi \varphi_j(x)dx \ .$$

Let us examine separately the two cases, $n = 0$ and $n \geqslant 1$.

In the first case, $P_0(\lambda,x) \equiv 1$, $\varphi_k(x) \equiv 0$, and

$$\int_0^\pi \sigma(\lambda,x)dx = \ln[1 + (2i\lambda)^{-1}u_1(\lambda,\pi)] \ .$$

Next, in view of the equalities $\sigma_1(x) = q(x)$,

$$u_1(x) = \int_0^x q(t)dt = \int_0^x \sigma_1(t)dt \ ,$$

and

$$\int_0^\pi q(t)u_1(t)dt = \frac{1}{2}\left[\int_0^\pi \sigma_1(x)dx\right]^2 = \frac{c_1^2}{2} \ ,$$

it follows from (1.4.8') that

$$u_1(\lambda,\pi) = c_1 + (4i\lambda)^{-1}c_1^2 - \tilde{\sigma}_1(\lambda) - (2i\lambda)^{-1}\int_0^\pi K_1^{(0)}(\pi,\xi)e^{-2i\lambda\xi}d\xi .$$

Furthermore, in view of Lemma 1.4.3, this yields

$$1 + (2ia_k)^{-1}u_1(a_k,\pi) = 1 + c_1(2ia_k)^{-1} +$$

$$+ \frac{1}{2}[c_1(2ia_k)^{-1}]^2 - (2ik)^{-1}\tilde{\sigma}_1(k+a) + k^{-2}\delta_1(k) =$$

$$= [1 + c_1(2ia_k)^{-1} + \frac{1}{2}\{c_1(2ia_k)^{-1}\}^2][1 - (2ik)^{-1}\tilde{\sigma}_1(k+a) + k^{-2}\delta_2(k)] ,$$

where the sequences $\delta_1(k)$ and $\delta_2(k)$ belong to ℓ_2 . Consequently,

$$\int_0^\pi \sigma(a_k,x)dx = \ln[1 + c_1(2ia_k)^{-1} + \frac{1}{2}\{c_1(2ia_k)^{-1}\}^2] +$$

$$+ \ln[1 - (2ik)^{-1}\tilde{\sigma}_1(k+a) + k^{-2}\delta_2(k)] =$$

$$= c_1(2ia_k)^{-1} - (2ik)^{-1}\tilde{\sigma}_1(k+a) + k^{-2}g_2(k) \quad , \quad \sum_{k=-\infty}^{\infty}|g_2(k)|^2 < \infty .$$

(Here we used the equalities $\ln(1+x) = x + O(x^2)$ and $\ln(1 + x + \frac{1}{2}x^2) = x + O(x^3)$, which hold for $x \to 0$.)

Now let $n \geqslant 1$. In this case,

$$P_n(\lambda,\pi) = 1 + (2i\lambda)^{-1}u_1(\pi) + O(\lambda^{-2})$$

and

$$\{(2i\lambda)^{n+1}P(\lambda,\pi)\}^{-1}u_{n+1}(\lambda,\pi) = (2i\lambda)^{-n-1}u_{n+1}(\lambda,\pi)[1 - (2i\lambda)^{-1}u_1(\pi)] + O(\lambda^{-n-3}) ,$$

which, upon using formula (1.4.8') and Lemma 1.4.3, implies that

$$\{(2ia_k)^{n+1}P_n(a_k,\pi)\}^{-1}u_{n+1}(a_k,\pi) = (2ia_k)^{-n-1}(c_{n+1} - b_1) +$$

$$+ (2ia_k)^{-n-2}d + (2ik)^{-n-1}\tilde{\sigma}_{n+1}(k+a) + (2ik)^{-n-2}\delta(k) ,$$

where $\sum_{k=-\infty}^{\infty}|\delta(k)|^2 < \infty$ and

$$d = \varphi_1(\pi) + \int_0^\pi q(t)u_{n+1}(t)dt - (c_{n+1} - b_1)u_1(\pi) .$$

Substituting this expression into formula (1.4.26), we get in the same way as in the previous case, after a number of elementary transformations,

$$\int_0^\pi \sigma(a_k,x)dx = \sum_{j=1}^{n+2} c_j(2ia_k)^{-j} - (2ik)^{-n-1}\tilde{\sigma}_{n+1}(k+a) + k^{-n-2}g_2(k) ,$$

where

$$\sum_{k=-\infty}^{\infty} |g_2(k)|^2 < \infty ,$$

and

$$c_{n+2} = b_2 + \varphi_1(\pi) + \int_0^\pi q(t)u_{n+1}(t)dt - (c_{n+1} - b_1)u_1(\pi) ,$$

as claimed.

PROBLEMS

1. Prove the following analog of Lemma 1.4.1: The Dirac equation (1.2.37) with $n \geqslant 0$ times continuously differentiable potential $\Omega(x)$ has solutions of the form

$$y_1(\lambda,x) = e^{i\lambda x}\{(I + iB)u_1(\lambda,x) + (I - iB)u_2(\lambda,x)\} ,$$

and

$$y_2(\lambda,x) = e^{-i\lambda x}\{(I - iB)u_1(-\lambda,x) - (I + iB)u_2(-\lambda,x)\} ,$$

where

$$u_1(\lambda,x) = I + \sum_{k=1}^{n} b_k(x)(2i\lambda)^{-k} + b_n(\lambda,x)(2i\ )^{-n} , \tag{1.4.27}$$

$$u_2(\lambda,x) = \sum_{k=1}^{n} a_k(x)(2i\lambda)^{-k} + a_n(\lambda,x)(2i\lambda)^{-n} , \tag{1.4.27'}$$

$$\left.\begin{aligned} a_1(x) &= i\Omega(x), \quad a_{k+1}(x) = -a_k'(x) + i\Omega(x)b_k(x) , \\ b_k(x) &= -i\int_0^x \Omega(t)a_k(t)dt , \end{aligned}\right\} \tag{1.4.27''}$$

$$a_n(\lambda,x) = \int_0^x A(x,t)^{-2i\lambda t}dt ,$$

and

$$b_n(\lambda,x) = -i\int_0^x \left\{\int_t^x \Omega(\xi)A(\xi,t)d\xi\right\} e^{-2i\lambda t}dt ;$$

moreover, $Bu_1(\lambda,x) - u_1(\lambda,x)B = 0$ and $Bu_2(\lambda,x) + u_2(\lambda,x)B = 0$.

Hint. It is a straightforward verification that the indicated operator-valued functions $y_1(\lambda,x)$ and $y_2(\lambda,x)$ satisfy equations (1.2.27) if $u_1(\lambda,x)$ and $u_2(\lambda,x)$ satisfy the system of equations

$$u_2'(\lambda,x) + 2i\lambda u_2(\lambda,x) - i\Omega(x)u_1(\lambda,x) = 0 \quad , \quad u_1'(\lambda,x) + i\Omega(x)u_2(\lambda,x) = 0 \; .$$

Now upon substituting the right-hand sides of formulas (1.4.27) and (1.4.27') into this system and using the recursion formulas (1.4.27"), you obtain the system of differential equations

$$b_n'(\lambda,x) = -i\Omega(x)a_n(\lambda,x) \quad , \quad a_n'(\lambda,x) + 2i\lambda a_n(\lambda,x) = a_{n+1}(x) + i\Omega(x)b_n(\lambda,x)$$

for $a_n(\lambda,x)$ and $b_n(\lambda,x)$, which in turn is equivalent to the equality

$$b_n(\lambda,x) = -i\int_0^x \Omega(t)a_n(\lambda,t)dt$$

and the integral equation

$$a_n(\lambda,x) = \int_0^x a_{n+1}(t)e^{-2i\lambda(x-t)}dt + \int_0^x e^{-2i\lambda(x-t)}\Omega(t)\int_0^t \Omega(\xi)a_n(\lambda,\xi)d\xi dt \; .$$

Setting $a_n(\lambda,x) = \int_0^x A(x,t)e^{-2i\lambda t}dt$, you get the equation

$$A(x,t) = a_{n+1}(x-t) + \int_0^t \Omega(x-u)\int_{t-u}^{x-u} \Omega(\xi)A(\xi,t-u)d\xi du \; ,$$

the solubility of which is checked by the method of successive approximations, which also provides estimates and properties of the kernel $A(x,t)$.

2. Prove the following analog of Lemma 1.4.2: if $n \geqslant 1$, then the operator-valued function $v(\lambda,x) = u_2(\lambda,x)[u_1(\lambda,x)]^{-1}$ admits a representation of the form

$$v(\lambda,x) = \sum_{k=1}^{n} \frac{u_k(x)}{(2i\lambda)^k} + \frac{v_n(\lambda,x)}{(2i\lambda)^n} \; ,$$

where

$$v_1(x) = i\Omega(x) \; ,$$

$$v_{k+1}(x) = -v_k'(x) - i\sum_{j=1}^{k-1} v_{k-j}(x)\Omega(x)v_j(x) \; ,$$

and

$$v_n(\lambda,x) = \int_0^x v_{n+1}(x-t)e^{2i\lambda t}dt + (2i\lambda)^{-1}\int_0^x B(x,t)e^{-4i\lambda t}dt + O(\lambda^{-2}) .$$

Hint. From the system of equations satisfied by the operator-valued functions $u_1(\lambda,x)$ and $u_2(\lambda,x)$ you get for $v(\lambda,x)$ the equation

$$v'(\lambda,x) + 2i\lambda v(\lambda,x) - iv(\lambda,x)\Omega(x)v(\lambda,x) - i\Omega(x) = 0 .$$

Now you have to use this equation in conjunction with formulas (1.4.27), (1.4.27'), just as in the proof of Lemma 1.4.2.

3. Show that, if the functions $p(r)$ and $r(x)$ belong to $W_2^n[0,\pi]$, then the Dirac equation (1.3.6) considered in Problem 3 of Section 3 of this chapter has a solution of the form

$$y(\lambda,x) = \begin{pmatrix} e^{i\lambda x}f^+(\lambda,x) & e^{-i\lambda x}g^-(-\lambda,x) \\ e^{i\lambda x}g^+(\lambda,x) & e^{-i\lambda x}f^-(-\lambda,x) \end{pmatrix} , \qquad (1.4.28)$$

where

$$f^{\pm}(\lambda,x) = u_1^{\pm}(\lambda,x) + u_2^{\mp}(\lambda,x) \quad , \quad g^{\pm}(\lambda,x) = -i\{u_1^{\pm}(\lambda,x) - u_2^{\mp}(\lambda,x)\} ,$$

$$u_1^{\pm}(\lambda,x) = 1 + \sum_{k=1}^{n} b_k^{\pm}(x)(2i\lambda)^{-k} + b_n^{\pm}(\lambda,x)(2i\lambda)^{-n} ,$$

$$u_2^{\pm}(\lambda,x) = \sum_{k=1}^{n} a_k^{\pm}(x)(2i\lambda)^{-k} + a_n^{\pm}(\lambda,x)(2i\lambda)^{-n} ,$$

$$a_1^{\pm}(x) = iq^{\pm}(x) \quad , \quad a_k^{\pm}(x) = -a_{k-1}^{\pm}(x)' + iq^{\pm}(x)b_k^{\mp}(x) ,$$

$$b_k^{\mp}(x) = -i\int_0^x q^{\mp}(t)a_k^{\pm}(t)dt ,$$

$$q^{\pm}(x) = p(r) \pm ir(x) ,$$

and

$$a_n^{\pm}(\lambda,x) = \int_0^x A^{\pm}(x,t)e^{-2i\lambda t}dt \quad , \quad b_n^{\mp}(\lambda,x) = -t\int_0^x \left\{\int_t^x q^{\mp}(\xi)A^{\pm}(\xi,t)d\xi\right\} e^{-2i\lambda t}dt .$$

Moreover, if $n \geq 1$, then

$$u_2^{\pm}(\lambda,x)[u_1^{\mp}(\lambda,x)]^{-1} = \sigma^{\pm}(\lambda,x) = \sum_{k=1}^{n}\sigma^{\pm}{}_k(x)(2i\lambda)^{-k} + \sigma_n^{\pm}(\lambda,x)(2i\lambda)^{-n} ,$$

$$\sigma_1^{\pm}(x) = iq^{\pm}(x) \quad , \quad \sigma_k^{\pm}(x) = -\sigma_{k-1}^{\pm}(x)' + iq^{\mp}(x) \sum_{j=1}^{k-2} \sigma_{k-j-1}^{\pm}(x)\sigma_j^{\pm}(x) \ ,$$

and

$$\sigma_n^{\pm}(\lambda,x) = \int_0^x \sigma_{n+1}^{\pm}(x-t)e^{-2i\lambda t}dt + (2i\lambda)^{-1} \int_0^x B^{\pm}(x,t)e^{-4i\lambda t}dt + o(\lambda^{-2}) \ .$$

Hint. Upon substituting (1.4.28) into equation (1.3.36), you get the following system of differential equations for the functions $u_1^{\pm}(\lambda,x)$ and $u_2^{\pm}(\lambda,x)$:

$$u_2^{\pm}(\lambda,x)' + 2i\lambda u_2^{\pm}(\lambda,x) - iq^{\pm}(x)u_1^{\mp}(\lambda,x) = 0 \ ,$$

$$u_1^{\mp}(\lambda,x)' + iq^{\mp}(x)u_2^{\pm}(\lambda,x) = 0 \ .$$

From this point, you can proceed much in the same way as in the proofs of Lemmas 1.4.1 and 1.4.2.

4. Express the entries $e_{ik}(\lambda,x)$ of the fundamental solution $e(\lambda,x)$ $(e(\lambda,0) = I)$ of equation (1.3.36) through the functions $u_1^{\pm}(\lambda,x)$ and $\sigma^{\pm}(\lambda,x)$.

Hint. From the identity

$$e(\lambda,x) = y(\lambda,x)[y(\lambda,0)]^{-1}$$

(the solution $y(\lambda,x)$ is defined by formula (1.4.28)) and the equalities

$$u_1^{\pm}(\lambda,0) = 1$$

$$f^{\pm}(\lambda,x) = u_1^{\pm}(\lambda,x)\{1 + \sigma^{\mp}(\lambda,x)\} \ ,$$

$$g^{\pm}(\lambda,x) = -iu_1^{\pm}(\lambda,x)\{1 - \sigma^{\mp}(\lambda,x)\} \ ,$$

and

$$\Delta = \mathrm{Det}\, y(\lambda,0) = 2\{1 + \sigma^{+}(-\lambda,0)\sigma^{-}(\lambda,0)\} \ ,$$

it follows that $e_{ik}(\lambda,x) = \hat{e}_{ik}\Delta^{-1}$, where

$$\hat{e}_{11} = e^{i\lambda x}u_1^{+}(\lambda,x)[1 + \sigma^{-}(\lambda,x)][1 + \sigma^{+}(-\lambda,0)] +$$

$$+ e^{-i\lambda x}u_1^{-}(-\lambda,x)[1 - \sigma^{+}(-\lambda,x)][1 - \sigma^{-}(\lambda,0)] \ ,$$

$$-i\hat{e}_{12} = e^{i\lambda x}u_1^{+}(\lambda,x)[1 + \sigma^{-}(\lambda,x)][1 - \sigma^{+}(-\lambda,0)] -$$

$$- e^{i\lambda x}u_1^-(\lambda,x)[1 - \sigma^+(-\lambda,x)][1 + \sigma^-(\lambda,0)] \; ,$$

$$i\hat{e}_{21} = -e^{i\lambda x}u_1^+(\lambda,x)[1 - \sigma^-(\lambda,x)][1 + \sigma^+(-\lambda,0)] +$$

$$+ e^{-i\lambda x}u_1^-(-\lambda,x)[1 + \sigma^+(-\lambda,x)][1 - \sigma^-(\lambda,0)] \; ,$$

$$\hat{e}_{22} = e^{i\lambda x}u_1^+(\lambda,x)[1 - \sigma^-(\lambda,x)][1 - \sigma^+(-\lambda,0)] +$$

$$+ e^{-i\lambda x}u_1^-(-\lambda,x)[1 + \sigma^+(-\lambda,x)][1 + \sigma^-(\lambda,0)] \; .$$

5. Suppose that the numerical sequence $a_k = k + a + h_k$ $(k = 0,\pm 1,\pm 2,\dots)$ satisfies the condition $\sup|h_k| = h < \infty$. Let $f(x) \in L_2(0,\pi)$ and $\tilde{f}(\lambda) = \int_0^\pi f(x)e^{-2i\lambda x}dx$. Show that the following equalities hold for every $n = 0,1,\dots$:

$$\tilde{f}(a_k) = \sum_{p=0}^{n} \tilde{f}^{(p)}(k+a)h_k^p(p!)^{-1} + h_k^{n+1}g_n(k) \; ,$$

where

$$\sum_{k=-\infty}^{\infty} |\tilde{f}^{(p)}(k+a)|^2 < \infty \quad (p = 0,1,\dots) \quad \text{and} \quad \sum_{k=-}^{\infty} |g_n(k)|^2 < \infty \; .$$

Hint. As in the proof of Lemma 1.4.3, you find that

$$\tilde{f}(a_k) = \sum_{p=0}^{n} \tilde{f}^{(p)}(k+a)h_k^p(p!)^{-1} + h_k^{n+1}g_n(k) \; ,$$

where

$$g_n(k) = \sum_{q=n+1}^{\infty} h_k^{q-n-1}(q!)^{-1} \int_0^\pi f(x)(-2ix)^q e^{-2iax}e^{-2ikx}dx \; .$$

By the Cauchy-Bunyakovski inequality

$$|g_n(k)|^2 \leq$$

$$\leq \left[\sum_{q=n+1}^{\infty} |h_k|^{2(q-n-1)}(q!)^{-1}\right]\left[\sum_{q=n+1}^{\infty} (q!)^{-1} \left|\int_0^\pi f(x)(-2ix)^q e^{-2iax}e^{-2ikx}dx\right|^2\right] \; ,$$

while the Parseval formula yields

$$\sum_{k=-\infty}^{\infty}\left|\int_0^\pi f(x)(-2ix)^q e^{-2iax}e^{-2ikx}dx\right|^2 = \pi \int_0^\pi |f(x)e^{-2iax}(-2ix)^q|^2dx \leq$$

$$\leq 2^{2q}\pi^{2q+1} \int_0^\pi |f(x)e^{-2iax}|^2 dx .$$

This implies that

$$\sum_{k=-\infty}^{\infty} |g_n(k)|^2 \leq$$

$$\leq \left[\sum_{q=n+1}^{\infty} h^{2(q-n-1)}(q!)^{-1}\right]\left[\sum_{q=n+1}^{\infty} (2\pi)^{2q}(q!)^{-1}\pi \int_0^\pi |f(x)e^{-2iax}|^2 dx\right] =$$

$$= \pi C_n(h) \int_0^\pi |f(x)e^{-2iax}|^2 dx < \infty .$$

5. ASYMPTOTIC FORMULAS FOR EIGENVALUES AND TRACE FORMULAS

We consider the boundary value problem (1.3.1), (1.3.2) with an arbitrary nondegenerate boundary condition and a complex-valued potential $q(x) \in W_2^n[0,\pi]$. Proceeding exactly as in the proof of Lemma 1.3.3, one can show, using Rouché's theorem and the estimates (1.3.10') derived in Lemma 1.3.2, that starting from some n , the characteristic function $\chi(z)$ and its principal part $\chi^{(0)}(z)$ have the same number of roots in the domain Π_n ; moreover, the distance between the roots of these functions lying in the domains $\Pi_{n+1} \setminus \Pi_n$ tends to zero as $n \to \infty$. The principal part of the characteristic function is: 1) $\chi^{(0)}(z) = -z \sin z\pi$; 2) $\chi^{(0)}(z) = \cos z\pi + L_2$; or 3) $\chi^{(0)}(z) = z^{-1}\sin z\pi + L_3$, depending upon which of the three cases (1.3.8) is considered, and its roots are:

1) $\pm 0, \pm 1, \pm 2, \ldots, \pm k, \ldots$;

2) $\pm\theta_2, \pm(2+\theta_2), \pm(2-\theta_2), \ldots, \pm(2k+\theta_2), \pm(2k-\theta_2), \ldots$, where $\pi\theta_2 = \arccos(L_2)$; or

3) $\pm 1, \pm 2, \ldots, \pm k, \ldots$, if $L_3 = 0$, and

$\pm(k + \frac{1}{2} + (-1)^k(i\pi)^{-1}\ln 2kL_3 + o(1))$, $k = 1,2,\ldots$, if $L_3 \neq 0$,

correspondingly. Therefore, the eigenvalues of the boundary value problems under consideration (i.e., the squares of the roots of the characteristic function) form the following sequences:

1) $(k + o(1))^2$;

2) $(2k \pm \theta_2 + o(1))^2$;

3) $(k + o(1))^2$ if $L_3 = 0$, and

$(k + \frac{1}{2} + (-1)^k(i\pi)^{-1}\ln 2kL_3 + o(1))^2$ if $L_3 \neq 0$.

The results of the previous section allow us to obtain asymptotic formulas for the eigenvalues that faithfully reflect the smoothness of the potential. To this end, we express the characteristic function through the solutions $y(\lambda,x)$, $y(-\lambda,x)$ of equations (1.3.1), which were introduced in Lemma 1.4.1. Upon replacing $c(z,x)$ and $s(z,x)$ in the right-hand side of equality (1.3.4') by their expressions (1.4.24) and (1.4.24') in terms of $y(z,x)$, $y(-z,x)$, we get

$$\chi(z) = 2C + A(z) + A(-z) \quad , \quad 2C = J_{12} + J_{34} , \tag{1.5.1}$$

where

$$A(z) = \frac{y'(z,\pi)[J_{14} - J_{42}y'(-z,0)] + y(z,\pi)[J_{13} - J_{32}y'(-z,0)]}{y'(z,0) - y'(-z,0)} . \tag{1.5.1'}$$

Next, using equalities (1.4.11') and (1.4.22), we rewrite the characteristic equation $\chi(z) = 0$ in the form

$$y(z,\pi)\,\frac{G(z)}{\omega(z,0)} - y(-z,\pi)\,\frac{G(-z)}{\omega(z,0)} + 2C = 0 ,$$

where

$$G(z) = [iz + \sigma(z,\pi)][J_{14} - J_{42}\{-iz + \sigma(-z,0)\}] +$$
$$+ J_{13} - J_{32}[-iz + \sigma(-z,0)] . \tag{1.5.2}$$

Upon multiplying this equation by $y(z,\pi)$ and observing that, by (1.4.21), $y(z,\pi)y(-z,\pi) = \omega(z,0)[\omega(z,\pi)]^{-1}$, we see that

$$y^2(z,\pi)\,\frac{G(z)}{\omega(z,0)} + 2Cy(z,\pi) - \frac{G(-z)}{\omega(z,\pi)} = 0 .$$

Hence

$$y(z,\pi) = \frac{\omega(z,0)}{G(z)}\left[-C \pm \sqrt{C^2 + \frac{G(z)G(-z)}{\omega(z,0)\omega(z,\pi)}}\right] .$$

Formulas (1.4.11), (1.4.12) allow us to put this equality into the more symmetric form

$$\exp\left\{iz\pi + \frac{1}{2}\int_0^\pi [\sigma(z,t) - \sigma(-z,t)]dt\right\} =$$
$$= \frac{\omega(z,0)}{G(z)}\sqrt{\frac{\omega(z,\pi)}{\omega(z,0)}}\left[-C \pm \sqrt{C^2 + \frac{G(z)G(-z)}{\omega(z,0)\omega(z,\pi)}}\right] . \tag{1.5.3}$$

In the following we shall restrict our discussion to boundary conditions which are either separated and of the form

$$y(0) = y(\pi) = 0 \ , \tag{1.5.4}$$

or

$$y(0) = y'(\pi) = 0 \ , \tag{1.5.5}$$

or periodic and anti-periodic:

$$y(0) - y(\pi) = y'(0) - y'(\pi) = 0 \tag{1.5.6}$$

and

$$y(0) + y(\pi) = y'(0) + y(\pi) \ , \tag{1.5.7}$$

respectively.

The asymptotic formulas for any other kind of nondegenerate boundary conditions can be derived in much the same way. From the crude asymptotic formulas it follows that the square roots of the eigenvalues of the boundary value problems in question form the following sequences:

$$\pm\sqrt{\lambda_1},\ \pm\sqrt{\lambda_2},\ \ldots,\ \pm\sqrt{\lambda_k},\ \ldots\ ;\quad \sqrt{\lambda_k} = k + \theta(k) \ , \tag{1.5.4'}$$

$$\pm\sqrt{\nu_1},\ \pm\sqrt{\nu_2},\ \ldots,\ \pm\sqrt{\nu_k},\ \ldots\ ;\quad \sqrt{\nu_k} = k - \frac{1}{2} + \delta(k) \ , \tag{1.5.5'}$$

$$\pm\sqrt{\mu_0},\ \pm\sqrt{\mu_2^-},\ \pm\sqrt{\mu_2^+},\ \ldots,\ \pm\sqrt{\mu_{2k}^-},\ \pm\sqrt{\mu_{2k}^+},\ \ldots\ ;$$

$$\sqrt{\mu_{2k}^{\pm}} = 2k + \varepsilon^{\pm}(2k) \ , \tag{1.5.6'}$$

and

$$\pm\sqrt{\mu_1^-},\ \pm\sqrt{\mu_1^+},\ \ldots,\ \pm\sqrt{\mu_{2k+1}^-},\ \pm\sqrt{\mu_{2k+1}^+},\ \ldots\ ;$$

$$\sqrt{\mu_{2k+1}^{\pm}} = 2k + 1 + \varepsilon^{\pm}(2k + 1) \ , \tag{1.5.7'}$$

respectively, where the quantities $\theta(k)$, $\delta(k)$, $\varepsilon^{\pm}(2k)$, and $\varepsilon^{\pm}(2k+1)$ tend to zero as $k \to \infty$ (the eigenvalues of the periodic (anti-periodic) problem are labelled by even (respectively, odd) n numbers to ensure that the asymptotic equality $\sqrt{\mu_k^{\pm}} = k + o(1)$ holds for $k \to \infty$).

Upon computing the constant C and the function $G(z)$ by means of formulas (1.5.1) and (1.5.2), it may be verified that for the present choice of boundary value problems,

$$C = 0 \quad , \quad G(z) = 1 \ , \tag{1.5.4''}$$

$$C = 0 \quad , \quad G(z) = iz + \sigma(z,\pi) \ , \tag{1.5.5''}$$

$$C = 1 \quad , \quad G(z) = -\omega(z,0)[1 + \Delta(z)] \ , \tag{1.5.6''}$$

and

$$C = 1 \quad , \quad G(z) = \omega(z,0)[1 + \Delta(z)] \ , \tag{1.5.7''}$$

where

$$\Delta(z) = \frac{\sigma(z,\pi) - \sigma(z,0)}{\omega(a,0)} \ . \tag{1.5.8}$$

Next, upon inserting these expressions for C and $G(z)$ in the equation (1.5.3) satisfied by the square roots of the eigenvalues, we obtain the following equations for the quantities $\theta(k)$, $\delta(k)$, and $\varepsilon^{\pm}(k)$:

$$i\theta(k)\pi = -\frac{1}{2}\int_0^\pi [\sigma(z,t) - \sigma(-z,t)]dt\Big|_{z=h+\theta(k)} \ , \tag{1.5.9}$$

$$i\delta(k)\pi = \left\{-\frac{1}{2}\int_0^\pi [\sigma(z,t) - \sigma(-z,t)]dt + \right.$$

$$\left. + \frac{1}{2}\ln\frac{1 + (-iz)^{-1}\sigma(-z,\pi)}{1 + (iz)^{-1}\sigma(z,\pi)}\right\}\Bigg|_{z=k - \frac{1}{2} + \delta(k)} \ , \tag{1.5.10}$$

and

$$i\varepsilon^{\pm}(k)\pi + \frac{1}{2}\int_0^\pi [\sigma(z,t) - \sigma(-z,t)]dt\Big|_{z=k+\varepsilon^{\pm}(k)} =$$

$$= \left\{\frac{1}{2}\ln(1 + \Delta(z) + \Delta(-z)) - \ln(1 + \Delta(z)) + \right.$$

$$\left. + \ln\left[1 \pm i\sqrt{\frac{\Delta(z)\Delta(-z)}{1 + \Delta(z) + \Delta(-z)}}\right]\right\}\Bigg|_{z=k+\varepsilon^{\pm}(k)} \ . \tag{1.5.11}$$

To avoid any misunderstanding, we remark, first, that these equations are satisfied starting from some sufficiently large value of k , secondly, that here we chose the branch of $\ln(1+w)$ that vanishes for $w = 0$, and, thirdly, that in equations (1.5.11), the even (odd) values of k correspond to periodic (respectively, anti-periodic) boundary conditions.

It follows from formulas (1.4.26), (1.4.15), (1.5.8), and (1.4.22), that the right-hand sides of equations (1.5.9)-(1.5.11) decay on the order of k^{-1} as $k \to \infty$. Hence, the estimates $\theta(k) = O(k^{-1})$, $\theta(k) = O(k^{-1})$, $\varepsilon^{\pm}(k) = O(k^{-1})$ hold as $k \to \infty$, and allow us to apply Lemma 1.4.4.

THEOREM 1.5.1. *If* $q(x) \in W_2^n[0,\pi]$ *, then square roots of the eigenvalues* λ_k *and* ν_k *of the boundary value problems* (1.3.1), (1.5.4), *and* (1.3.1), (1.5.5), *respectively, can be expressed as*

$$\sqrt{\lambda_k} = k + \sum_{1\leqslant 2j+1\leqslant n+2} a_{2j}(2k)^{-2j-1} + \frac{1}{2}(-1)^n s_n(2k)(2k)^{-n-1} + k^{-n-2}\alpha_k ,$$

and

$$\sqrt{\nu_k} = k - \frac{1}{2} + \sum_{1\leqslant 2j+1\leqslant n+2} b_{2j+1}(2k-1)^{-2j-1} -$$

$$- \frac{1}{2}(-1)^n s_n(2k-1)(2k-1)^{-n-1} + k^{-n-2}\beta_k ,$$

where the numbers a_{2j+1} , b_{2j+1} *do not depend on* k ,

$$a_1 = b_1 = \frac{1}{\pi}\int_0^\pi q(t)dt ,$$

$$s_n(p) = \frac{2}{\pi}\int_0^\pi q^{(n)}(t)\sin[pt - \frac{\pi}{2}(n+1)]dt ,$$

and

$$\sum_{k=1}^\infty |\alpha_k|^2 < \infty , \quad \sum_{k=1}^\infty |\beta_k|^2 < \infty .$$

PROOF. According to Lemma 1.4.4,

$$\int_0^\pi [\sigma(z,t) - \sigma(-z,t)]dt\Big|_{z=k+a+h_k} =$$

$$= 2\sum_{1\leqslant 2j+1\leqslant n+2} c_{2j+1}(2iz)^{-2j-1}\Big|_{x=k+a+h_k} -$$

$$- (2ik)^{-n-1}[\tilde\sigma_{n+1}(k+a) - (-1)^{n+1}\tilde\sigma_{n+1}(-k-a)] + k^{-n-2}\nu_k , \qquad (1.5.12)$$

and $\sum_{k=1}^\infty |\gamma_k|^2 < \infty$ if $h_k = O(k^{-1})$ as $k \to \infty$. Using this equality and introducing for convenience the function

$$F_1(w) = \frac{1}{\pi}\sum_{1\leqslant 2j+1\leqslant n+2} (-1)^j c_{2j+1}w^{2j+1} , \qquad (1.5.13)$$

we can rewrite rewrite equation (1.5.9) in the form

$$\theta(k) - F_1(w)\Big|_{w=(2k+2\theta(k))^{-1}} = \frac{\tilde{\sigma}_{n+1}(k) - (-1)^{n+1}\tilde{\sigma}_{n+1}(-k)}{2\pi i(2ik)^{n+1}} + \frac{\alpha_k^{(1)}}{k^{n+2}}, \tag{1.5.14}$$

where $\sum_{k=1}^{\infty} |\alpha_k^{(1)}|^2 < \infty$. Furthermore, it follows from the equalities

$$\ln[1 + (iz)^{-1}\sigma(z,\pi)] =$$

$$= \ln\left[1 + 2\sum_{j=1}^{n} \sigma_j(\pi)(2iz)^{-j-1} + 2\sigma_n(z,\pi)(2iz)^{-n-1}\right] =$$

$$= \ln\left[1 + 2\sum_{j=1}^{n} \sigma_j(\pi)(2iz)^{-j-1}\right][1 + 2\sigma_n(z,\pi)(2iz)^{-n-1}\{1 + O(z^{-2})\}] =$$

$$= \ln\left[1 + 2\sum_{j=1}^{n} \sigma_j(\pi)(2iz)^{-j-1}\right] +$$

$$+ 2\sigma_n(z,\pi)(2iz)^{-n-1}[1 - \sigma_n(z,\pi)(2iz)^{-n-1}] + O(z^{-n-3})$$

and Lemma 1.4.4, that

$$\frac{1}{2\pi i} \ln \frac{1 + (-iz)^{-1}\sigma(-z,\pi)}{1 + (iz)^{-1}\sigma(z,\pi)}\Bigg|_{z = k - \frac{1}{2} + \delta(k)} =$$

$$= \frac{1}{2\pi i} \ln \frac{1 + 2\sum_{j=1}^{n} \sigma_j(\pi)(-2iz)^{-j-1}}{1 + 2\sum_{j=1}^{n} \sigma_j(\pi)(2iz)^{-j-1}}\Bigg|_{z = k - \frac{1}{2} + \delta(k)} -$$

$$- 2\,\frac{\tilde{\sigma}_{n+1}(k - \frac{1}{2}) - (-1)^{n+1}\tilde{\sigma}_{n+1}(-k + \frac{1}{2})}{2\pi i[i(2k - 1)]^{n+1}} + \frac{\beta_k^{(0)}}{k^{n+2}},$$

where $\sum_{k=1}^{\infty} |\beta_k^{(0)}|^2 < \infty$. This equality and formulas (1.5.12), (1.5.13) allow us to reexpress equation (1.5.10) in the form

$$\delta(k) - F_2(w)\Big|_{w=[2k-1+2\delta(k)]^{-1}} =$$

$$= \frac{\tilde{\sigma}_{n+1}(k - \frac{1}{2}) - (-1)^{n+1}\tilde{\sigma}_{n+1}(-k + \frac{1}{2})}{2\pi i[i(2k - 1)]^{n+1}} + \frac{\beta_k^{(1)}}{k^{n+2}}. \tag{1.5.15}$$

Here $\sum_{k=1}^{\infty} |\beta_k^{(1)}|^2 < \infty$ and the function

$$F_2(w) = F_1(w) + \frac{1}{2i} \ln \frac{1 + 2\sum_{j=1}^{n} \sigma_j(\pi)(-iw)^{j+1}}{1 + 2\sum_{j=1}^{n} \sigma_j(\pi)(iw)^{j+1}}$$

is holomorphic in the neighborhood of zero, is odd, and

$$F_2'(0) = F_1'(0) = \frac{c_1}{\pi} = \frac{1}{\pi}\int_0^{\pi} q(t)dt .$$

Since equation (1.5.15) can be investigated in much the same way as equation (1.5.14), we will focus on the latter. According to known theorems on functions which are defined implicitly by analytic equations, one can find a number $r > 0$ such that in the disc $|y| < r$ there is a unique analytic function $\theta_1(y)$ (with $\theta_1(0) = 0$) which satisfies the equation

$$\theta_1 - F_1\left(\frac{y}{1 + 2y\theta_1}\right) = 0 . \tag{1.5.16}$$

Since the function $F_1(w)$ is odd, the analytic function $-\theta_1(-y)$ also satisfies this equation and, in view of the uniqueness of the solution, $\theta_1(y) = -\theta_1(-y)$. Hence, the function $\theta_1(y)$ is odd: $\theta_1(y) = \sum_{j=0}^{\infty} a_{2j+1}y^{2j+1}$ ($|y| < r$); moreover, $a_1 = \theta_1'(0) = F_1'(0) = \frac{1}{\pi}\int_0^{\pi} q(t)dt$. Now, upon setting $y = (2k)^{-1}$ ($(2k)^{-1} < r$) in equation (1.5.16) and subsequently substracting it from (1.5.14), we obtain the equality

$$\theta(k) - \theta_1([2k]^{-1}) - F_1([2k + 2\theta(k)]^{-1}) + F_1([2k + \theta_1([2k]^{-1})]^{-1}) =$$

$$= \frac{\tilde{\sigma}_{n+1}(k) - (-1)^{n+1}\tilde{\sigma}_{n+1}(-k)}{2\pi i(2ik)^{n+1}} + \frac{\alpha_k^{(1)}}{k^{n+2}} ,$$

from which we see that

$$\theta(k) - \theta_1([2k]^{-1}) = \left\{\frac{\tilde{\sigma}_{n+1}(k) - (-1)^{n+1}\tilde{\sigma}_{n+1}(-k)}{2\pi i(2ik)^{n+1}} + \frac{\alpha_k^{(1)}}{k^{n+2}}\right\}(1 + O(k^{-2})) , \tag{1.5.17}$$

because

$$F_1([2k + 2\theta(k)]^{-1}) - F_1([2k + \theta_1([2k]^{-1})]^{-1}) = \{\theta(k) - \theta_1([2k]^{-1})\}O(k^{-2}) .$$

Furthermore, according to (1.4.25) and (1.4.20'),

$$\tilde{\sigma}_{n+1}(k) = \int_0^\pi \sigma_{n+1}(\pi - t)e^{-2ikt}dt = \int_0^\pi \sigma_{n+1}(x)e^{2ikx}dx =$$

$$= (-1)^n \int_0^\pi q^{(n)}(x)e^{2ikx}dx + \int_0^\pi S_{n-2}(x)e^{2ikx}dx =$$

$$= (-1)^n \int_0^\pi q^{(n)}(x)e^{2ikx}dx + \frac{S_{n-2}(\pi) - S_{n-2}(0)}{2ik} - \frac{1}{2ik}\int_0^\pi S'_{n-2}(x)e^{2ikx}dx =$$

$$= (-1)^n \int_0^\pi q^{(n)}(x)e^{2ikx}dx + \frac{S_{n-2}(\pi) - S_{n-2}(0)}{2ik} - \frac{p(k)}{k}, \qquad (1.5.18)$$

where $\sum_{k=-\infty}^{\infty} |p(k)|^2 < \infty$, since the function $S_{n-2}(x)$, being a polynomial in $q(x), q'(x), \dots, q^{(n-2)}(x)$, has a square-summable derivative. Consequently,

$$\tilde{\sigma}_{n+1}(k) - (-1)^{n+1}\tilde{\sigma}_{n+1}(-k) =$$

$$= (-1)^n \int_0^\pi q^{(n)}(x)[e^{2ikx} - (-1)^{n+1}e^{-2ikx}]dx +$$

$$+ \frac{S_{n-2}(\pi) - S_{n-2}(0)}{2ik}(1 + (-1)^{n+1}) - \frac{p(k) + (-1)^{n+1}p(-k)}{k} =$$

$$= (i)^{n+2}(-1)^n \pi s_n(2k) + \frac{S_{n-2}(\pi) - S_{n-2}(0)}{2ik}(1 + (-1)^{n+1}) + \frac{p^{(1)}(k)}{k}$$

where $\sum_{k=1}^{\infty} |p^{(1)}(k)|^2 < \infty$ and

$$s_n(2k) = \frac{2}{\pi}\int_0^\pi q^{(n)}(x)\sin[2kx - \tfrac{\pi}{2}(n+1)]dx .$$

Substituting this expression into equality (1.5.17), we get

$$\theta(k) - \theta_1([2k]^{-1}) = \frac{(-1)^n}{2}\frac{s_n(2k)}{(2k)^{n+1}} + \frac{S_{n-2}(\pi) - S_{n-2}(0)}{2\pi i(2ik)^{n+2}}(1 + (-1)^{n+1}) + \frac{\alpha_k^{(2)}}{k^{n+2}},$$

where $\sum_{k=1}^{\infty} |\alpha_k^{(2)}|^2 < \infty$. Upon observing that $\theta_1([2k]^{-1}) = \sum_{1 \le 2j+1 \le n+2} a_{2j+1}(2k)^{-2j-1} + O(k^{-n-3})$, this yields the final formula

$$\sqrt{\lambda_k} = k + \theta(k) = k + \sum_{1 \le 2j+1 \le n+2} a_{2j+1}(2k)^{-2j-1} +$$

$$+ \hat{a}_{n+2}(2k)^{-n-2} + \frac{(-1)^n}{2}\frac{s_n(2k)}{(2k)^{n+1}} + \frac{\alpha_k}{k^{n+2}} ,$$

in which $\sum_{k=1}^{\infty} |\alpha_k|^2 < \infty$, $a_{n+2} = 0$ if $n+2$ is even, and $a_{n+2} =$ $= \pi^{-1}(i)^{-n-3}[S_{n-2}(\pi) - S_{n-2}(0)]$, if $n+2$ is odd.

The asymptotic formula for $\sqrt{\nu_k}$ is deduced from equality (1.5.15) in a similar fashion. □

COROLLARY. *In order that the complex-valued function* $q(x) \in L_2(0,\pi)$ *belong to the space* $W_2^n[0,\pi]$, *it is necessary and sufficient that the eigenvalues* λ_k *and* ν_k *of the boundary value problems* (1.3.1), (1.5.4) *and* (1.3.1), (1.5.5), *respectively, satisfy the asymptotic equalities*

$$\sqrt{\lambda_k} = k + \sum_{1 \le 2j+1 \le n+2} \tilde{a}_{2j+1}(2k)^{-2j-1} + \tilde{\alpha}_k k^{-n-1} ,$$

and

$$\sqrt{\nu_k} = k - \frac{1}{2} + \sum_{1 \le 2j+1 \le n+2} \tilde{b}_{2j+1}(2k-1)^{-2j-1} + \tilde{\beta}_k k^{-n-1} ,$$

where $\sum_{k=1}^{\infty} \{|\tilde{\alpha}_k|^2 + |\tilde{\beta}_k|^2 < \infty$.

PROOF. The necessity of these conditions is an immediate consequence of Theorem 1.5.1. To prove the sufficiency, we proceed by reduction ad absurdum. Suppose that $q(x) \in W_2^m[0,\pi]$, but $q(x) \notin W_2^{m+1}[0,\pi]$, where $m < n$. Then, by Theorem 1.5.1,

$$\sqrt{\lambda_k} = k + \sum_{1 \le 2j+1 \le m+2} a_{2j+1}(2k)^{-2j-1} + \frac{(-1)^m}{2}\frac{s_m(2k)}{(2k)^{m+1}} + \frac{\alpha_k}{k^{m+2}} ,$$

and

$$\sqrt{\nu_k} = k - \frac{1}{2} + \sum_{1 \le 2j+1 \le m+2} b_{2j+1}(2k-1)^{-2j-1} - \frac{(-1)^m}{2}\frac{s_m(2k-1)}{(2k-1)^{m+1}}\frac{\beta_k}{k^{m+2}} ,$$

where $\sum_{k=1}^{\infty} \{|\alpha_k|^2 + |\beta_k|^2\} < \infty$. Comparing these equalities with the given ones, we see that $a_{2j+1} = \tilde{a}_{2j+1}$, $b_{2j+1} = \tilde{b}_{2j+1}$ for $2j + 1 \le m + 1$, and hence that

$$(-1)^m s_m(p) = \begin{cases} p^{-1}\varepsilon_p , & m = 2l , \\ p^{-1}[\tilde{a}_{m+2}(1 + (-1)^p) - \tilde{b}_{m+2}(1 - (-1)^p) + \delta_p] , & m = 2l - 1 , \end{cases}$$

with $\sum_{p=1}^{\infty} |\varepsilon_p|^2 < \infty$ and $\sum_{p=1}^{\infty} |\delta_p|^2 < \infty$. Now observe that the numbers $s_m(p)$ are the Fourier coefficients of the function $q^{(m)}(x)$ with respect to the orthogonal system $\sin[px - \frac{\pi}{2}(m+1)]$. It follows that for m even (odd), the function $(-1)^m q^{(m)}(x)$ admits the Fourier series expansion

$$\text{const} + (-1)^{l+1} \sum_{p=1}^{\infty} \frac{\varepsilon_p}{p} \cos px ,$$

(respectively,

$$(-1)^l \left\{\tilde{a}_{m+2} \sum_{p=1}^{\infty} \frac{\sin 2px}{p} - 2\tilde{b}_{m+2} \sum_{p=1}^{\infty} \frac{\sin(2p-1)x}{2p-1} + \sum_{p=1}^{\infty} \frac{\delta_p}{p} \sin px\right\}).$$

Since

$$\sum_{p=1}^{\infty} \frac{\sin 2px}{p} = \frac{\pi}{2} - x \quad \text{and} \quad \sum_{p=1}^{\infty} \frac{\sin(2p-1)x}{2p-1} = \frac{\pi}{4}$$

for $0 < x < \pi$, $\sum_{p=1}^{\infty} |\varepsilon_p|^2 < \infty$ and $\sum_{p=1}^{\infty} |\delta_p|^2 < \infty$, these expansions show that in both cases the function $q^{(m)}(x)$ has a square summable derivative. Therefore, $q(x) \in W_2^{m+1}[0,\pi]$, which is a contradiction. □

We shall write $\tilde{W}_2^n[0,\pi]$ to denote the subspace of $W_2^n[0,\pi]$ consisting of the functions $f(x) \in W_2^n[0,\pi]$, which satisfies the periodic boundary conditions $f^{(k)}(0) = f^{(k)}(\pi)$, $k = 0,1,2,\ldots,n-1$. Notice that $W_2^0[0,\pi] = \tilde{W}_2^0[0,\pi] = L_2(0,\pi)$.

THEOREM 1.5.2. *If* $q(x) \in W_2^n[0,\pi]$ *and* $\operatorname{Im} q(x) = 0$, *then the square roots of the eigenvalues* $\mu_{2k}^{\pm}$ *and* $\mu_{2k+1}^{\pm}$ *of the periodic and anti-periodic boundary value problems* (1.3.1), (1.5.6) *and* (1.3.1), (1.5.7), *respectively, satisfy the equalities*

$$\sqrt{\mu_k^{\pm}} = k + \sum_{1 \leq 2j+1 \leq n+2} a_{2j+1}(2k)^{-2j-1} \pm |e_n(2k)|(2k)^{-n-1} + \gamma_k^{\pm} k^{-n-2} ,$$

where a_{2j+1} *are the same numbers as in* Theorem 1.5.1,

$$e_n(p) = \frac{1}{\pi}\int_0^{\pi} q^{(n)}(x)e^{-ipx}dx ,$$

and

$$\sum_{k=0}^{\infty} |\gamma_k^{\pm}|^2 < \infty .$$

PROOF. From the fact that the potential $q(x)$ belongs to $\tilde{W}_2^n[0,\pi]$, it follows that $\sigma_k(0) = \sigma_k(\pi)$ $(1 \leqslant k \leqslant n)$, because the functions are polynomials in $q(x),\ldots,q^{(k-1)}(x)$. Consequently,

$$\sigma(z,\pi) - \sigma(z,0) = \sum_{k=1}^{n} \frac{\sigma_k(\pi) - \sigma_k(0)}{(2iz)^k} + \frac{\sigma_n(z,\pi)}{(2iz)^n} = \frac{\sigma_n(z,\pi)}{(2iz)^n} ,$$

and, in view of (1.5.8) and (1.4.23),

$$\Delta(z) = \frac{\sigma_n(z,\pi)}{(2iz)^{n+1}}\left[1 + \frac{\sigma(z,0) - \sigma(-z,0)}{2iz}\right]^{-1} = \frac{\sigma_n(z,\pi)}{(2iz)^{n+1}}\,[1 + O(z^{-2})] .$$

Using this formula, the estimate $\varepsilon^{\pm}(k) = O(k^{-1})$, the expansions $\ln(1+x) = x - \frac{x^2}{2} + O(x^3)$, $\sqrt{1+x} = 1 + \frac{1}{2}x + O(x^2)$ $(x \to 0)$, and Lemma 1.4.4, one can check that the right-hand side of equation (1.5.11) is equal to

$$-\frac{\tilde{\sigma}_{n+1}(k) - (-1)^{n+1}\tilde{\sigma}_{n+1}(-k)}{2(2ik)^{n+1}} \pm i\left.\frac{\sqrt{\sigma_n(z,\pi)\sigma_n(-z,\pi)}}{(2k)^{n+1}}\right|_{z=k+\varepsilon^{\pm}(k)} + \frac{\gamma_k^{(0)\pm}}{k^{n+2}} ,$$

where $\sum_{k=1}^{\infty} |\gamma_k^{(0)\pm}|^2 < \infty$. On the other hand, the equalities (1.5.12) and (1.5.18) show that the left-hand side of equation (1.5.11) is equal to

$$i\pi\varepsilon^{\pm}(k) - i\pi F_1(w)\Big|_{w=[2k+2\varepsilon^{\pm}(k)]^{-1}} - \frac{\tilde{\sigma}_{n+1}(k) - (-1)^{n+1}\tilde{\sigma}_{n+1}(-k)}{2(2ik)^{n+1}} + \frac{\gamma_k^{(1)\pm}}{k^{n+2}} ,$$

where $\sum_{k=0}^{\infty} |\gamma_k^{(1)\pm}|^2 < \infty$. It follows that the numbers $\varepsilon^{\pm}(k)$ satisfy the equations

$$\varepsilon^{\pm}(k) - F_1(w)\Big|_{w=[2k+2\varepsilon^{\pm}(k)]^{-1}} =$$

$$= \pm \left. \frac{\sqrt{\sigma_n(z,\pi)\sigma_n(-z,\pi)}}{\pi(2k)^{n+1}} \right|_{z=k+\varepsilon^\pm(k)} + \frac{\gamma_{(k)}^{(0)\pm} - \gamma_k^{(1)\pm}}{k^{n+2}} ,$$

which differ from equations (1.5.14) only by the right-hand side. From this we conclude, just as in the proof of Theorem 1.5.1, that $\varepsilon^\pm(k)$ satisfy the equalities

$$\varepsilon^\pm(k) = \sum_{1\leqslant 2j+1\leqslant n+2} a_{2j+1}(2k)^{-2j-1} \pm \left. \frac{\sqrt{\sigma_n(z,\pi)\sigma_n(-z,\pi)}}{\pi(2k)^{n+1}} \right|_{z=k+\varepsilon^\pm(k)} + \frac{\gamma_k^{(2)\pm}}{k^{n+2}} , \tag{1.5.19}$$

where $\sum_{k=1}^{\infty} |\gamma_k^{(2)\pm}|^2 < \infty$.

To this point we have not used the assumption that the potential $q(x)$ is real. We now remark that the periodic and anti-periodic boundary value problems with a real potential are self-adjoint, their eigenvalues $\mu_k^\pm = [k + \varepsilon^\pm(k)]^2$ are real, and $\sigma_n(z,\pi) = \overline{\sigma_n(-z,\pi)}$ for real values of z . Consequently, for large values of k ,

$$\left. \sqrt{\sigma_n(z,\pi)\sigma_n(-z,\pi)} \right|_{z=k+\varepsilon^\pm(k)} = |\sigma_n(k + \varepsilon^\pm(k),\pi)| ,$$

whence, in view of Lemma 1.4.4 and equality (1.5.18),

$$\left. \sqrt{\sigma_n(z,\pi)\sigma_n(-z,\pi)} \right|_{z=h+\varepsilon^\pm(k)} = \left| \int_0^\pi q^{(n)}(x)e^{-2ikx}dx \right| + \frac{\gamma_k^{(3)\pm}}{k} ,$$

where $\sum_{k=1}^{\infty} |\gamma_k^{(3)\pm}|^2 < \infty$. (Here we used the equality $S_{n-2}(\pi) - S_{n-2}(0) = 0$, which in turn follows from the fact that $q(x) \in \tilde{W}_2^n[0,\pi]$.) Substituting this expression into the right-hand side of equality (1.5.19), we finally obtain

$$\sqrt{\mu_k^\pm} = k + \varepsilon^\pm(k) = k + \sum_{1\leqslant 2j+1\leqslant n+2} a_{2j+1}(2k)^{-2j-1} \pm |e_n(2k)|(2k)^{-n-1} + \gamma_k^\pm k^{-n-2} ,$$

where $\sum_{k=0}^{\infty} |\gamma_k^\pm|^2 < \infty$ and

$$e_n(2k) = \frac{1}{\pi} \int_0^\pi q^{(n)}(x)e^{-2ikx}dx ,$$

as asserted. □

COROLLARY. *In order for the real-valued function* $q(x) \in L_2(0,\pi)$ *to belong to the space* $\tilde{W}_2^n[0,\pi]$ *, it is necessary and sufficient that*

$$\sum_{k=1}^{\infty} k^{2(n+1)} |\sqrt{\mu_k^+} - \sqrt{\mu_k^-}|^2 < \infty , \qquad (1.5.20)$$

where $\mu_k^{\pm}$ *are the eigenvalues of the periodic and anti-periodic boundary value problems generated by equation* (1.3.1).

PROOF. The necessity of condition (1.5.20) is an abvious consequence of Theorem 1.5.2. To prove its sufficiency, we proceed by reductio ad absurdum. Suppose that $q(x) \in \tilde{W}_2^m[0,\pi]$, but $q(x) \notin \tilde{W}_2^{m+1}[0,\pi]$, where $m < n$. Then, by Theorem 1.5.2,

$$\sqrt{\mu_k^+} - \sqrt{\mu_k^-} = \frac{2|e_m(2k)|}{(2k)^{m+1}} + \frac{\gamma_k^+ - \gamma_k^-}{k^{m+2}} ,$$

or, equivalently,

$$4k|e_m(2k)| = 2^{m+2}(\gamma_k^- - \gamma_k^+) + (\sqrt{\mu_k^+} - \sqrt{\mu_k^-})(2k)^{n+1}(2k)^{m+1-n} ,$$

where, by assumption, $m + 1 - n \leq 0$. Hence, condition (1.5.20) implies that $\sum_{k=1}^{\infty} |2ke_m(2k)|^2 < \infty$, and, since the numbers $e_m(2k)$ are the Fourier coefficients of the function $q^{(m)}(x)$ with respect to the orthogonal system e^{2ikx} , it follows that $q^{(m)}(0) = q^{(m)}(\pi)$, and that the function $q^{(m)}(x)$ has a square summable derivative. Consequently, $q(x) \in \tilde{W}_2^{m+1}[0,\pi]$, contrary to assumption.

□

We remark in conclusion that in the case where the potential $q(x)$ is infinitely differentiable, the eigenvalues of the boundary value problems under consideration admit asymptotic series expansions as $k \to \infty$. For example, it follows from Theorem 1.5.1 that in this case the square roots of the eigenvalues of the boundary value problems (1.3.1), (1.5.4) and (1.3.1), (1.5.5) admit the asymptotic expansions

$$\sqrt{\lambda_k} \simeq k + \sum_{j=0}^{\infty} a_{2j+1}(2k)^{-2j-1} ,$$

and

$$\sqrt{\nu_k} \simeq k - \frac{1}{2} + \sum_{j=0}^{\infty} b_{2j+1}(2k - 1)^{-2j-1} ,$$

respectively. Asymptotic expansions of this kind can be obtained for arbitrary nondegenerate boundary conditions proceeding from equation (1.5.3), and the analogs of the auxiliary equations (1.5.16) that follow from (1.5.3). In the general case, the eigenvalues of the boundary value problem are partitioned into two sequences, and for each of these sequences one obtains its own asymptotic series. The reason for this is that in equation (1.5.3), the square root appears with two signs, to each of which there corresponds an auxiliary equation of the type (1.5.16).

It follows from the crude asymptotic formulas given in the beginning of this section that, starting with some value $k = k_0$, each domain $\Pi_k \setminus \Pi_{k-1}$ contains four roots of the characteristic function, which are situated symmetrically with respect to zero, since the characteristic function is even. We denote the pair of these roots with positive real part by α_k^-, α_k^+ ($\mathrm{Re}\,\alpha_k^- \leq \mathrm{Re}\,\alpha_k^+$) and the corresponding eigenvalues by $\mu_k^- = (\alpha_k^-)^2$ and $\mu_k^+ = (\alpha_k^+)^2$. We label all the remaining eigenvalues so that the inequalities $0 \leq \mathrm{Re}\sqrt{\mu_k^-} \leq \mathrm{Re}\sqrt{\mu_k^+} \leq \mathrm{Re}\sqrt{\mu_{k+1}^-}$ hold also for $k < k_0$. The eigenvalues, this way labelled, form a sequence that will be denoted by μ. From the crude asymptotic formulas we see that the enumeration starts either with 0 or with 1, depending on the type of boundary conditions.

Without going into the proofs, which essentially repeat the first part of the proof of Theorem 1.5.1, we give the asymptotic series for $\sqrt{\mu_k^-}$ and $\sqrt{\mu_k^+}$:

1) $J_{42} \neq 0$, $\mu = \{\mu_0^+, \mu_1^-, \mu_1^+, \mu_2^-, \mu_2^+, \dots\}$,

$$\sqrt{\mu_k^-} \simeq (2k-1) + \sum_{j=0}^{\infty} c_{2j+1}^-(2k-1)^{-2j-1},$$

$$\sqrt{\mu_k^+} \simeq 2k + \sum_{j=0}^{\infty} c_{2j+1}^+(2k)^{-2j-1};$$

2) $J_{42} = 0$, $J_{14} + J_{32} \neq 0$, $L_2 = (J_{12} + J_{34})(J_{14} + J_{32})^{-1}$,

$\theta_2 = \pi^{-1} \arccos(-L_2)$.

If $\theta_2 \neq 0$, then

$\mu = \{\mu_1^-, \mu_1^+, \mu_2^-, \mu_2^+, \dots\}$,

$$\sqrt{\mu_k^-} \simeq 2k - 2 + \theta_2 + \sum_{j=0}^{\infty} d_{2j+1}(2k - 2 + \theta_2)^{-2j-1} + \sum_{j=1}^{\infty} d_{2j}(2k - 2 + \theta_2)^{-2j} ,$$

$$\sqrt{\mu_k^+} \simeq 2k - \theta_2 + \sum_{j=0}^{\infty} d_{2j+1}(2k - \theta_2)^{-2j-1} - \sum_{j=1}^{\infty} d_{2j}(2k - \theta_2)^{-2j} .$$

If $\theta_2 = 0$, then

$\mu = \{\mu_0^+, \mu_1^-, \mu_1^+, \mu_2^-, \mu_2^+, \dots\}$,

$$\sqrt{\mu_k^-} \simeq 2k + \sum_{j=0}^{\infty} l_{2j+1}(2k)^{-2j-1} + \sum_{j=1}^{\infty} l_{2j}(2k)^{-2j} ,$$

$$\sqrt{\mu_j^+} \simeq 2k + \sum_{j=0}^{\infty} l_{2j+1}(2k)^{-2j-1} - \sum_{j=1}^{\infty} l_{2j}(2k)^{-2j} ;$$

3°) $J_{42} = 0$, $J_{14} + J_{32} = J_{12} + J_{34} = 0$, $J_{13} \neq 0$,

$\mu = \{\mu_1^-, \mu_1^+, \mu_2^-, \mu_2^+, \dots\}$,

$$\sqrt{\mu_k^-} \simeq (2k - 1) + \sum_{j=0}^{\infty} p_{2j+1}(2k - 1)^{-2j-1} ,$$

$$\sqrt{\mu_k^+} \simeq 2k + \sum_{j=0}^{\infty} p_{2j+1}(2k)^{-2j-1} .$$

The coefficients of these asymptotic series may be expressed in terms of the potential $q(x)$ and the determinants $J_{\alpha\beta}$ which are built from the columns of the matrix of boundary conditions. The actual computation of these coefficients is conveniently carried out by means of recursion formulas. If the potential $q(x)$ has only a finite number n of derivatives, then in the series only the terms decaying for $k \to \infty$ no faster than k^{-n} are preserved, and the remainder has order $o(k^{-n})$ and in general does not admit an asymptotic expansion. On raising these asymptotic series to even powers $2m$, we obtain the following equalities:

$$1)\ (\mu_k^{\pm})^m = M_{2m}^{\pm}(k) + o(k^{-2}) ,$$

$$2)\ (\mu_k^{\pm})^m = M_{2m}^{\pm}(k) \pm \frac{\nu_{2m}}{2k} + o(k^{-2}) , \tag{1.5.21}$$

3°) $(\mu_k^{\pm})^m = M_{2m}(k) + O(k^{-2})$, (1.5.21)

where $M_{2m}^{\pm}(x)$ are polynomials of degree $2m$. These equalities show that

$$(\mu_k^{+})^m - M_{2m}^{+}(k) + (\mu_k^{-})^m - M_{2m}^{-}(k) = O(k^{-2}) ,$$

and hence that the series

$$s_m = \sum_k{}' \, [(\mu_k^{+})^m - M_{2m}^{+}(k) + (\mu_k^{-})^m - M_{2m}^{-}(k)]$$

converges absolutely; here the prime indicates that the term with $k = 0$ is missing (is equal to $(\mu_0^{+})^m - M_{2m}^{m}(0)$) if the first index in the sequence is 1 (respectively, 0). The sum s_m of such a series is called the regularized trace (of order m), and the formulas which express the regularized trace s_m in terms of the potential $q(x)$ and the determinants $J_{\alpha\beta}$ are known as trace formulas. Since the eigenvalues are the squares of the roots of the characteristic function $\chi(z) = \chi(-z)$, we have by the residue theorem,

$$2 \sum_{k\leqslant l} [(\mu_k^{+})^m + (\mu_k^{-})^m] = \frac{1}{2\pi i} \oint_{K_\ell} z^{2m} d \ln \chi(z) , \tag{1.5.22}$$

where K_ℓ is the admissible sequence of contours for the given boundary value problem, introduced in Lemma 1.3.2 (i.e., in cases 1 and 3°, $K_\ell = C(2\ell + \frac{1}{2})$, and in case 2, $K_\ell = C(2\ell + 1)$ if $\theta_2 = 0$, and $K_\ell = C(2\ell)$ if $\theta_2 \neq 0$).

Let $q(x) \in W_2^n[0,\pi]$. We introduce the auxiliary function

$$y_n(z,x) = e^{izx} P_n(z,x) = e^{izx} \left[1 + \frac{u_1(x)}{2iz} + \dots + \frac{u_n(x)}{(2iz)^n}\right] .$$

Let $\chi_n(z)$ and $A_n(z)$ be the functions obtained on replacing $y(z,x)$ and $y(-z,x)$ by $y_n(z,x)$ and $y_n(-z,x)$ in formulas (1.5.1) and (1.5.1'), respectively:

$$\chi_n(z) = 2C + A_n(z) + A_n(-z) \quad , \quad 2C = J_{12} + J_{34} , \tag{1.5.23}$$

and

$$A_n(z) = e^{iz\pi}\{2iz + P_n'(z,0) - P_n'(-z,0)\}^{-1}\{[izP_n(z,\pi) + P_n'(z,\pi)] \times$$

$$\times [J_{14} - J_{42}\{-iz + P_n'(-z,0)\}] + P_n(z,\pi)[J_{13} - J_{32}(-iz + P_n'(-z,0))]\} . \tag{1.5.24}$$

LEMMA 1.5.2. *If* $2m \leqslant n$, *then for all nondegenerate boundary conditions and potentials* $q(x) \in W_2^n[0,\pi]$, *the following equality holds:*

$$\lim_{\ell\to\infty} \oint_{K_\ell} z^{2m} d\{\ln \chi(z) - \ln \chi_n(z)\} = 0 ,$$

where K_ℓ *is any of the sequences of contours which are admissible for the given boundary value problem.*

PROOF. According to Lemma 1.4.1,

$$y(z,\pi) - y_n(z,\pi) = e^{iz\pi} u_{n+1}(z,\pi)(2iz)^{-n-1} ,$$

$$y'(z,\pi) - y_n'(z,\pi) = e^{iz\pi}\{u_{n+1}'(z,\pi)(2iz)^{-1} + 0\ 5u_n(z,\pi)\}(2iz)^{-n} ,$$

and, in addition,

$$y'(z,0) = y_n'(z,0) = iz + P_n'(z,0) = iz + \sum_{k=0}^{n} \sigma_k(0)(2iz)^{-k} ,$$

because $u_{n+1}(z,0) = u_{n+1}'(z,0) = 0$. From these equalities and formulas (1.4.8), (1.4.9), and Lemma 1.3.1, it follows that as $|z| \to \infty$,

$$\chi_n(z) - \chi(z) = e^{|\mathrm{Im}\, z\pi|}(2iz)^{-n-1}\{zJ_{42}o(1) + (J_{32} + J_{14})O(1) + o(1)\},$$

Upon using the estimates for $|\chi(z)|$ on the contours K_ℓ derived in Lemma 1.3.2, this further implies that

$$\frac{\chi_n(z) - \chi(z)}{\chi(z)} = o(z^{-n})$$

uniformly on K_ℓ , starting with some ℓ . Consequently, the estimate

$$\ln \frac{\chi_n(z)}{\chi(z)} = \ln \left\{1 + \frac{\chi_n(z) - \chi(z)}{\chi(z)}\right\} = o(z^{-n})$$

on the contours K_ℓ , also holds uniformly, and the function $\ln(\chi_n(z)[\chi(z)]^{-1})$ is single-valued and holomorphic on them. It follows that

$$\oint_{K_\ell} z^{2m} d\{\ln \chi(z) - \ln \chi_n(z)\} = -\oint_{K_\ell} z^{2m} d \ln \frac{\chi_n(z)}{\chi(z)} =$$

$$= 2m \oint_{K_\ell} z^{2m-1} \ln \frac{\chi_n(z)}{\chi(z)} dz = 2m \oint_{K_\ell} z^{2m-1-n} o(1) dz .$$

For $2m - n \leqslant 0$, this yields the desired equality because the modulus of

z^{2m-1-n} is bounded by $C_1\ell^{-1}$ on the contour K_ℓ, and the length of the contour does not exceed $C_2\ell$. □

We need also the following simple lemma.

LEMMA 1.5.2. *Let* $F(z) = \sum_{\alpha=1}^{N} g_\alpha(z)P_\alpha(z)$, *where* $g_\alpha(z)$ *are meromorphic functions which are periodic with period* 2 *and the* $P_\alpha(z)$ *are polynomials of degree at most* m. *Then the residues of the function* $F(z)$ *at the sequence of points* $2k+\theta$ $(k = 0,\pm 1,\dots)$ *are equal to* $R_\theta(2k+\theta)$, *where* $R_\theta(z)$ *is a polynomial of degree at most* m.

PROOF. Since $g_\alpha(z+2k) \equiv g_\alpha(z)$, we have

$$\operatorname{Res} F(z)\big|_{z=2k+\theta} = \operatorname{Res} F(w+2k+\theta)\big|_{w=0} =$$

$$= \operatorname{Res} \sum_{\alpha=1}^{N} g_\alpha(w+2k+\theta)P_\alpha(w+2k+\theta)\big|_{w=0} =$$

$$= \operatorname{Res} \sum_{\alpha=1}^{N} \left\{ g_\alpha(w+\theta) \sum_{j=0}^{m} \frac{P_\alpha^{(j)}(2k+\theta)}{j!} w^j \right\} \Bigg|_{w=0} = R_\theta(2k+\theta) ,$$

where $R_\theta(z)_m$ is a polynomial of degree at most m, and

$$R_\theta(z) = \sum_{j=0}^{m} \frac{P_\alpha^{(j)}(z)}{j!} \left\{ \sum_{\alpha=1}^{N} \operatorname{Res} g_\alpha(w+\theta)w^j\big|_{w=0} \right\} ,$$

as asserted. □

To derive the trace formulas, it is convenient to isolate the factors which determine the behavior of the functions $A_n(z)$ as $z \to \infty$ in expression (1.5.24). To this end we take the function $P_n(z,\pi)$ out of parentheses and then use the identity (1.4.14). After some simple manipulations we get

$$A_n(z) = e^{iz\pi}B_n(z) \quad , \quad B_n(z) = \frac{P_n(z,\pi)H_n(z)}{2Q_n(z)} (1 + z^{-n-1}r_n(z)) , \tag{1.5.25}$$

where

$$P_n(z,\pi) = 1 + \sum_{j=1}^{n} u_j(\pi)(2iz)^{-j} , \tag{1.5.25'}$$

$$Q_n(z) = 1 + 2 \sum_{1 \le 2j+1 \le n} \sigma_{2j+1}(0)(2iz)^{-2j-2} , \tag{1.5.25'}$$

$$H_n(z) = izJ_{42} + (J_{14} + J_{32}) + (iz)^{-1}J_{13} +$$

$$+ J_{42}\left[\sum_{j=1}^{n} \{\sigma_j(\pi) - (-1)^j\sigma_j(0)\}(2iz)^{-j} + \sum_{j=2}^{n-1} \gamma_j(2iz)^{-j-1}\right] +$$

$$+ (iz)^{-1}\left[J_{14}\sum_{j=1}^{n} \sigma_j(\pi)(2iz)^{-j} - J_{32}\sum_{j=1}^{n} \sigma_j(0)(-2iz)^{-j}\right] , \tag{1.5.25''}$$

$$\gamma_j = 2\sum_{l=1}^{j-1} (-1)^l\sigma_l(0)\sigma_{j-l}(\pi) ,$$

and the function $r_n(z) = \sum_{j=0}^{\infty} r_n(j)z^{-j}$ is holomorphic and bounded in the neighborhood of infinity.

THEOREM 1.5.3. *If* $q(x) \in W_2^n[0,\pi]$ *and* $2m \le n$, *then the following trace formulas are valid:*

1) $J_{42} \ne 0$; $2s_m = -M_{2m}^+(0) + \sum\{p_i^{2m} - q_j^{2m} + h_l^{2m}\}$;

2) $J_{42} = 0$, $J_{14} + J_{32} \ne 0$; $2s_m = \sum\{p_i^{2m} - q_j^{2m} + h_l^{2m}\}$;

3°) $J_{42} = J_{14} + J_{32} = J_{12} + J_{34} = 0$, $J_{13} \ne 0$;

$2s_m = M_{2m}^+(0) + \sum\{p_i^{2m} - q_j^{2m} + h_l^{2m}\}$,

where p_i , q_j , *and* h_l *are the roots of the polynomials* $z^nP_n(z,\pi)$, $z^{n+1}Q_n(z)$, *and* $z^{n+1}H_n(z)$, *respectively.*

PROOF. It follows from formulas (1.5.25), (1.5.25'), (1.5.25") that the function $B_n(z)$ is equal to

$$1)\ \frac{izJ_{42}}{2}(1 + b_n(z)) ,$$

$$2)\ \frac{(J_{14} + J_{32})}{2}(1 + b_n(z)) , \tag{1.5.26}$$

$$3°)\ \frac{J_{13}}{2iz}(1 + b_n(z)) , \tag{1.5.26}$$

in cases 1, 2, and 3, respectively, where the functions $b_n(z) = \sum_{j=1}^{\infty} b_n(j)z^{-j}$ are holomorphic in a neighborhood of infinity. We put

$$\rho(z) = \sqrt{(1 + b_n(z))(1 + b_n(-z))} \quad \text{and} \quad \gamma(z) = \frac{1}{2\pi i} \ln \frac{1 + b_n(z)}{1 + b_n(-z)} .$$

Then

$$1 + b_n(z) = \rho(z)e^{i\pi\gamma(z)} . \tag{1.5.27}$$

Since the function $\rho(z)$ is even, the function $\gamma(z)$ is odd, and both are holomorphic in a neighborhood of infinity; we have

$$\rho(z) = 1 + \sum_{j=1}^{\infty} \rho_{2j}z^{-2j} \quad \text{and} \quad \gamma(z) = \sum_{j=0}^{\infty} \gamma_{2j+1}z^{-2j-1} .$$

Comparing equalities (1.5.23), (1.5.25), and (1.5.26), (1.5.27), we get

$$1)\ \chi_n(z) = -zJ_{42}\rho(z)\left[1 + \frac{\delta_1(z)}{\sin w\pi}\right] \sin w\pi ,$$

$$2)\ \chi_n(z) = (J_{14} + J_{42})\rho(z)\left[1 + \frac{\delta_2(z)}{\cos w\pi + L_2}\right][\cos w\pi + L_2] , \tag{1.5.28}$$

$$3°)\ \chi_n(z) = z^{-1}J_{13}\rho(z)\sin w\pi ,$$

in which the functions

$$w = w(z) = z + \gamma(z) = z + \sum_{j=0}^{\infty} \gamma_{2j+1}z^{-2j-1} , \tag{1.5.29}$$

$$\delta_1(z) = \frac{-2C}{z\rho(z)J_{42}} = \sum_{j=0}^{\infty} \delta_{2j+1}^{(1)}z^{-2j-1} , \tag{1.5.30}$$

$$\delta_2(z) = \frac{2C}{J_{14} + J_{32}}\,\frac{1 - \rho(z)}{\rho(z)} = \sum_{j=1}^{\infty} \delta_{2j}^{(2)}z^{-2j} \tag{1.5.30'}$$

are holomorphic in a neighborhood of infinity.

An important property of the function $w(z)$ is that it is univalent in a neighborhood of infinity, as follows from the inequality

$$|w(z_1) - w(z_2)| \geq |z_1 - z_2| - |\gamma(z_1) - \gamma(z_2)| \geq$$

$$\geq |z_1 - z_2| \left\{1 - \sum_{j=0}^{\infty} |\gamma_{2j+1}|R^{2-j-2}(2j + 1)\right\} \geq \frac{|z_1 - z_2|}{2} ,$$

which holds in the domain $|z| > R$ for R large enough. The fact that the function $w(z)$ is univalent and odd guarantees the existence of a holomorphic even inverse function

$$z(w) = w + \sum_{j=0}^{\infty} g_{2j+1} w^{-2j-1} . \qquad (1.5.31)$$

We give a detailed proof of trace formulas only for the case 1. Let $K_\ell = C(\ell + \frac{1}{2})$ be a sequence of contours which is admissible for case 1. From equality (1.5.28) it follows that

$$\frac{1}{2\pi i} \oint_{K_\ell} z^{2m} d \ln \chi_n(z) = \frac{1}{2\pi i} \oint_{K_\ell} z^{2m} d \ln z\rho(z) + \frac{1}{2\pi i} \oint_{K_\ell} z^{2m} d \ln \sin w\pi +$$

$$+ \frac{1}{2\pi i} \oint_{K_\ell} z^{2m} d \ln\left(1 + \frac{\delta_1(z)}{\sin w\pi}\right) = J_1 + J_2 + J_3 . \qquad (1.5.32)$$

We examine separately each of the integrals J_1, J_2, and J_3. Since

$$B_n(z) = \frac{izJ_{42}}{2} (1 + b_n(z)) = \frac{izJ_{42}}{2} \rho(z) e^{i\pi\gamma(z)} ,$$

in the case in question we have, according to (1.5.25),

$$d \ln z\rho(z) = d \ln B_n(z) - i\pi\gamma'(z)dz = d \ln z^n P_n(z,\pi) + d \ln z^{n+1} H_n(z) -$$

$$- d \ln z^{n+1} Q_n(z) - d \ln z^n - i\pi\gamma'(z)dz + d \ln(1 + z^{-n-1} r_n(z)) ,$$

and hence

$$J_1 = \frac{1}{2\pi i} \oint_{K_\ell} z^{2m} d[\ln z^n P_n(z,\pi) + \ln z^{n+1} H_n(z) - \ln z^n Q_n(z)] - \frac{n}{2\pi i} \oint_{K_\ell} z^{2m-1} dz -$$

$$- \frac{1}{2} \oint_{K_\ell} z^{2m} \gamma'(z)dz + \frac{1}{2\pi i} \oint_{K_\ell} \frac{z^{2m-n-2}[zr_n'(z) - (n + 1)r_n(z)]}{1 + z^{-n-1} r_n(z)} dz .$$

It is clear that for all sufficiently large values of ℓ, the first term is equal to $\sum\{p_i^{2m} + h_j^{2m} - q_s^{2m}\}$, where p_i, h_j, and q_s are the roots of the polynomials $z^n P_n(z,\pi)$, $z^{n+1} H_n(z)$, and $z^{n+1} Q_n(z)$, respectively. The next two terms vanish, the first in view of the holomorphy of the function z^{2m-1}, and the second in view of the symmetry of the contours $K_\ell = C(\ell + \frac{1}{2})$ with respect to zero and the evenness of the function $z^{2m}\gamma'(z)$. Finally, since $r_n(z) = O(1)$ and $r_n'(z) = O(z^{-2})$ for $2m \leq n$, it is readily established that the integrand in the last term is $O(\ell^{-2})$ on the contour K_ℓ, whereas the length of K_ℓ is $O(\ell^{-1})$; consequently, this term tends to zero as ℓ^{-1}

when $\ell \to \infty$. We conclude that

$$J_1 = \sum\{p_i^{2m} + h_j^{2m} - q_s^{2m}\} + O(\ell^{-1}) \tag{1.5.32'}$$

as $\ell \to \infty$.

For large values of ℓ, the contour K_ℓ lies in the domain of univalence of the function $w(z)$, which permits us to change the integration variable to w in the second integral J_2 (see (1.5.32)). This yields

$$J_2 = \frac{1}{2\pi i} \oint_{K'_\ell} [z(w)]^{2m} d \ln \sin w\pi ,$$

where the contour K'_ℓ is the image of K_ℓ under the mapping $w(z) = z + \gamma(z)$ which tends to the identity mapping as $z \to \infty$. Consequently, the integrand has no singularities in the domain bounded by the contours K'_ℓ and K_ℓ, and hence we can replace integration along K'_ℓ by integration along the contour K_ℓ without affecting the value of J_2. Furthermore, it follows from equality (1.5.31) that $[z(w)]^{2m} = A_{2m}(w) + O(w^{-2})$ as $w \to \infty$, where $A_{2m}(w)$ is an even polynomial of degree $2m$. Therefore,

$$J_2 = \frac{1}{2\pi i} \oint_{K_\ell} [z(w)]^{2m} d \ln \sin w\pi = \frac{1}{2\pi i} \oint_{K_\ell} [z(w)]^{2m} d \ln \sin w\pi =$$

$$= \frac{1}{2\pi i} \oint_{K_\ell} A_{2m}(w) d \ln \sin w\pi + \oint_{K_\ell} O(w^{-2}) \frac{\cos w\pi}{\sin w\pi} dw .$$

Using the residue theorem and observing that on the contours $K_\ell = C(\ell + \frac{1}{2})$ the function $\operatorname{cotan} w\pi$ is uniformly bounded, we get

$$J_2 = \sum_{k=-\ell}^{\ell} A_{2m}(k) + O(\ell^{-1}) \tag{1.5.32''}$$

as $\ell \to \infty$.

Finally, we turn to the third integral, J_3. From equalities (1.5.29) and (1.5.30) it follows that for large values of ℓ,

$$\left| \frac{\delta_1(z)}{\sin w} \right| \leq \frac{C_1}{|z|} e^{-|\operatorname{Im} w\pi|} \leq \frac{C_2}{|z|} e^{-|\operatorname{Im} z\pi|}$$

on the contours $K_\ell = C(\ell + \frac{1}{2})$, which guarantees that the function $\ln(1 + \delta_1(z)[\sin w\pi]^{-1})$ is single-valued and holomorphic in a neighborhood of these contours. Consequently,

$$J_3 = \frac{1}{2\pi i}\oint_{K_\ell} z^{2m} d \ln\left(1 + \frac{\delta_1(z)}{\sin w\pi}\right) = \frac{-2m}{2\pi i}\oint_{K_\ell} z^{2m-1}\ln\left(1 + \frac{\delta_1(z)}{\sin w\pi}\right) dz =$$

$$= \frac{1}{2\pi i}\oint_{K_\ell}\left\{\sum_{p=1}^{2m-1}\frac{2mz^{2m-1}[\delta_1(p)]^p(-1)^p}{p[\sin w\pi]^p} + O(|z|^{-1}e^{-|\mathrm{Im}\, z\pi|})\right\} dz =$$

$$= \frac{1}{2\pi i}\oint_{K_\ell}\sum_{p=1}^{2m-1}\frac{2mz^{2m-1}[\delta_1(z)]^p(-1)^p}{p[\sin w\pi]^p} dz + O(\ell^{-1})$$

since

$$\oint_{K_\ell} O(|z|^{-1}e^{-|\mathrm{Im}\, z\pi|})dz = O(\ell^{-1})$$

as $\ell \to \infty$. Changing the integration variable to w and subsequently deforming the contour K_ℓ' into K_ℓ (which, as we showed above, is permitted for large ℓ), we get

$$J_3 = \frac{1}{2\pi i}\oint_{K_\ell}\sum_{p=1}^{2m-1}\frac{2m[z(w)]^{2m-1}[\delta_1(z(w))]^p(-1)^p z'(w)}{p[\sin w\pi]^p} dw + O(\ell^{-1}) .$$

Further, it follows from (1.5.30) and (1.5.31) that

$$2m[z(w)]^{2m-1}[\delta_1(w)]^p z'(w)p^{-1}(-1)^p = T_{2m-p-1}(w) + O(w^{-1})$$

as $w \to \infty$, where $T_{2m-p-1}(w)$ is a polynomial of degree $2m - p - 1$. By Lemma 1.5.2,

$$\mathrm{Res}\sum_{p=1}^{2m-1} T_{2m-1-p}(w)[\sin w\pi]^{-p}\Big|_{w=2k+0} = R_0(2k) ,$$

and

$$\mathrm{Res}\sum_{p=1}^{2m-1} T_{2m-1-p}(w)[\sin w\ ]^{-p}\Big|_{w=2k-1} = R_{-1}(2k - 1) ,$$

where $R_0(w)$ and $R_{-1}(w)$ are polynomials of degree $\leqslant 2m - 2$. Therefore,

$$J_3 = \frac{1}{2\pi i}\oint_{K_\ell}\sum_{p=1}^{2m-1}\{T_{2m-p-1}(w) + O(w^{-1})\}[\sin w\pi]^{-p}dw + O(\ell^{-1}) =$$

$$= \sum_{|2k|\leqslant\ell} R_0(2k) + \sum_{|2k-1|\leqslant\ell} R_{-1}(2k - 1) + O(\ell^{-1}) , \tag{1.5.32''}$$

since for $p \geqslant 1$ and $\ell \to \infty$,

$$\oint_{K_\ell} O(w^{-1})[\sin w\pi]^{-p}dw = O(\ell^{-1}) .$$

On combining equalities (1.5.32), (1.5.32'), (1.5.32"), (1.5.32'''), Lemma 1.5.1, and formula (1.5.22), we get

$$2\{(\mu_0^+)^m + (\mu_1^+)^m + \dots + (\mu_{\ell_1}^+)^m + (\mu_1^-)^m + \dots + (\mu_{\ell_2}^-)^m\} =$$

$$= 2\left\{\sum_{k=0}^{\ell_1} A_{2m}^+(k) + \sum_{k=1}^{\ell_2} A_{2m}^-(k)\right\} + A_{2m}^+(0) + \sum\{p_i^{2m} + h_j^{2m} - q_s^{2m}\} + o(1) , \quad (1.5.33)$$

where $\ell_1 = [\frac{\ell}{2}]$, $\ell_2 = [\frac{\ell+1}{2}]$, and

$$A_{2m}^+(w) = A_{2m}(2w) + \frac{1}{2}[R_0(2w) + R_0(-2w)] ,$$

$$A_{2m}^-(w) = A_{2m}(2w - 1) + \frac{1}{2}[R_{-1}(2w - 1) + R_{-1}(-2w + 1)] .$$

Setting $\ell = 2s + 1$ and $\ell = 2s$ (respectively, $\ell = 2s$, $\ell = 2s - 1$) in these equalities, and substracting the results one from another, we get

$$(\mu_{s+1}^-)^m = A_{2m}^-(s + 1) + O(1) ,$$

$$(\text{respectively, } (\mu_s^+)^m = A_{2m}^+(s) + O(1))$$

as $s \to \infty$. Comparing the relations obtained with formulas (1.5.21), we see that $A_{2m}^-(w) \equiv M_{2m}^-(w)$, $A_{2m}^+(w) \equiv M_{2m}^+(w)$, whence, in view of (1.5.33),

$$2\sum_{k=0}^{\ell_1} [(\mu_k^+)^m - M_{2m}^+(k)] + 2\sum_{k=1}^{\ell_2} [(\mu_k^-)^m - M_{2m}^-(k)] =$$

$$= -M_{2m}^+(0) + \sum\{p_i^{2m} + h_j^{2m} - q_s^{2m}\} + o(1) ,$$

as $\ell \to \infty$, which is equivalent to the asserted trace formula for case 1.

The proof in cases 3 and 2 for $\theta \neq 0, 1$, is carried out in much the same way. For $\theta_2 = 0$ (respectively, $\theta_2 = 1$), the last part of the proof must be slightly modified. The point is that for these values of θ , the role of the function $\sin w\pi$ is played by the function $\cos w\pi - 1$ (respectively, $\cos w\pi + 1$), the roots of which are $\dots,-4,-2,0,0,2,4,\dots$ (respectively, $\dots,-3,-1,1,3,\dots$), and, according to Lemma 1.5.2, the right-hand side of equality (1.5.32''') takes the form

$$\sum_{k=-\ell}^{\ell} R(2k) + O(\ell^{-1})$$

$$(\text{respectively, } \sum_{k=-\ell}^{\ell-1} R(2k+1) + O(\ell^{-1})),$$

while the right-hand side of equality (1.5.33) becomes

$$2 \sum_{k=0}^{\ell} A(k) - A(0) + \sum \{p_i^{2m} + h_j^{2m} - q_s^{2m}\} + o(1)$$

$$(\text{respectively, } 2 \sum_{k=1}^{\ell} A(k) + \sum \{p_i^{2m} + h_j^{2m} - q_s^{2m}\} + o(1)),$$

where

$$A(k) = 2A_{2m}(2k) + \frac{1}{2} [R(2k) + R(-2k)]$$

$$(\text{respectively, } A(k) = 2A_{2m}(2k-1) + \frac{1}{2} [R(2k-1) + R(-2k+1)]).$$

Further, for these values of θ, the neighboring admissible contours are $C(\ell + \frac{1}{2})$, $C(2(\ell-1) + \frac{1}{2})$ (respectively, $C(2\ell)$, $C(2\ell - 2)$), which allows to prove only the identity $M_{2m}^{+}(k) + M_{2m}^{-}(k) \equiv A(k)$. For $\theta_0 = 1$, this identity is obviously sufficient to prove the trace formula, whereas for $\theta_0 = 0$, it only implies that

$$(\mu_0^{+})^m - M_{2m}^{+}(0) + \sum_{k=1}^{\infty} \{(\mu_k^{+})^m - M_{2m}^{+}(k) + (\mu_k^{-})^m - M_{2m}^{-}(k)\} =$$

$$= \frac{1}{2} [M_{2m}^{-}(0) - M_{2m}^{+}(0)] + \frac{1}{2} \sum \{p_i^{2m} + h_j^{2m} - q_s^{2m}\},$$

and we still have to check the equality $M_{2m}^{+}(0) = M_{2m}^{-}(0)$. For $\theta_2 = 0$, the latter is a corollary of the asymptotic expansions of $\sqrt{\mu_k^{+}}$ and $\sqrt{\mu_k^{-}}$, from which it follows, as is readily verified, that the coefficients of the even (odd) powers of w in the polynomials $M_{2m}^{+}(w)$ and $M_{2m}^{-}(w)$ are equal (respectively, differ only by sign). □

<u>Remark</u>. To compute the sum $\alpha_\ell = \sum_{i=1}^{n} x_i^{\ell}$ of powers of the roots of the polynomial $z^n + a_1 z^{n-1} + \ldots + a_n$, one can use Newton's formulas

$$\alpha_\ell + a_1\alpha_{\ell-1} + a_2\alpha_{\ell-2} + \ldots + a_{\ell-1}\alpha_1 + \ell a_\ell = 0 \quad (\ell \leqslant n),$$

from which it follows, in particular, that the sums α_ℓ are polynomials in

$a_1, a_2, \ldots, a_\ell$ with integral coefficients:

$$\alpha_1 = -a_1 \ , \quad \alpha_2 = -2a_2 + a_1^2 \ , \quad \alpha_3 = -3a_3 + 3a_1a_2 - a_1^3, \ldots \ .$$

Consequently, the right-hand sides of the trace formulas are polynomials in $u_1(\pi), \ldots, u_n(\pi)$, $q^{(k)}(0)$, $q^{(k)}(\pi)$ $(k = 0,1,\ldots,n)$.

COROLLARY. *The eigenvalues* λ_k , $\mu_{2k}^{\pm}$, *and* $\mu_{2k-1}^{\pm}$ *of the boundary value problems generated by equation* (1.3.1) *with a potential* $q(x) \in \tilde{W}_2^n[0,\pi]$ *and the boundary conditions* (1.5.4), (1.5.6), *and* (1.5.7), *respectively, satisfy the relation*

$$(\mu_0^+)^m + \sum_{k=1}^{\infty} \{(\mu_k^+)^m + (\mu_k^-)^m - 2(\lambda_k)^m\} = \sum q_i^{2m} \ , \tag{1.5.34}$$

for $m = 1,2,\ldots,[\frac{n}{2}]$, *where* q_i *are the roots of the polynomial*

$$z^{n+1}Q_n(z) = z^{n+1}\left[1 + 2 \sum_{1\leqslant 2j+1\leqslant n} \sigma_{2j+1}(0)(2iz)^{-2j-2}\right] \ .$$

PROOF. As we remarked above, it follows from the fact that the potential $q(x)$ belongs to $\tilde{W}_2^n[0,\pi]$ that $\sigma_j(0) = \sigma_j(\pi)$ for $j = 1,2,\ldots,n$. This in turn implies that, for the boundary conditions (1.5.4), (1.5.6), and (1.5.7), the functions $H_n(z)$ given by formula (1.5.25") are equal to $(iz)^{-1}$, $-2\left(1 + 2 \sum_{1\leqslant 2j+1\leqslant n} \sigma_{2j+1}(0)(2iz)^{-2j-2}\right)$, and $2\left(1 + 2 \sum_{1\leqslant 2j+1\leqslant n} \sigma_{2j+1}(0)(2iz)^{-2j-2}\right)$, respectively. Moreover, it follows from Theorems 1.5.1 and 1.5.2 that the polynomial parts of $(\lambda_k)^m$ and $(\mu_k^{\pm})^m$ are identical for $m = 1,2,\ldots,[\frac{n}{2}]$. Therefore, to prove equalities (1.5.34), it suffices, by Theorem 1.5.3, to substract twice the regularized trace of the boundary value problem (1.5.4) from the sum of the regularized traces of the boundary value problems with boundary conditions (1.5.6) and (1.5.7). In conclusion, we remark that, in view of Newton's formulas and equalities (1.5.34), the quantities

$$\Delta_m = (\mu_0^+)^m + \sum_{k=1}^{\infty} [(\mu_k^+)^m + (\mu_k^-)^m - 2(\lambda_k)^m] \tag{1.5.35}$$

satisfy the recursion formulas

$$\Delta_m = -2\left[\sum_{j=0}^{m-2} \sigma_{2j+1}(0)(-4)^{-j-1}\Delta_{m-j-1} + 2m\sigma_{2m-1}(0)(-4)^{-m}\right] \ . \tag{1.5.35'}$$

□

PROBLEMS

1. Derive the asymptotic formula for the eigenvalues μ_k of the boundary value problem $\{q(x),h,H\}$ generated by equation (1.3.1) with the boundary conditions $y'(0) - hy(0) = 0$, $y'(\pi) + Hy(\pi) = 0$, and a potential $q(x) \in W_2^n[0,\pi]$.

Hint. Proceeding as in the proof of Theorem 1.5.1, but using the expansion $\tilde{f}(a_k) = \tilde{f}(k+a) + h_k\tilde{f}'(k+a) + h_k^2 g_2(k)$, which was obatined in Problem 5, § 5 of this chapter in place of Lemma 1.4.3, you find that

$$\sqrt{\mu_k} = k + \sum_{1\leqslant 2j+1\leqslant n+3} d_{2j+1}(2k)^{-2j-1} +$$

$$+ \frac{(-1)^{n-1}}{2}[s_n(2k) + \tilde{s}_n(2k)k^{-1}](2k)^{-n-1} + \delta_k k^{-n-2} + \varepsilon_k(h,H)k^{-n-3} ,$$

where d_{2j+1} are numbers that do not depend on k ,

$$d_1 = \frac{2h + 2H + \int_0^\pi q(t)dt}{\pi} ,$$

$$s_n(p) = \frac{2}{\pi}\int_0^\pi q^{(n)}(t)\sin\{pt - \frac{\pi}{2}(n+1)\}dt ,$$

$$\tilde{s}_n(p) = \frac{2}{\pi}\int_0^\pi q^{(n)}(t)[2h - d_1 t]\sin\{pt - \frac{\pi}{2}(n+2)\}dt ,$$

δ_k does not depend on h , H , and

$$\sum_{k=1}^\infty |\delta_k|^2 < \infty \quad , \quad \sum_{k=1}^\infty |\varepsilon_k(h,H)|^2 < \infty .$$

2. Using the results of the preceding problem, show that in order for the potential $q(x)$ to belong to $W_n^2[0,\pi]$ it is necessary and sufficient sufficient that the eigenvalues $\mu_k(1)$ and $\mu_k(2)$ of the boundary value problems $\{q(x),h,H_1\}$ and $\{q(x),h,H_2\}$ $(H_1 \neq H_2)$ satisfy the asymptotic equalities

$$\sqrt{\mu_k(1)} = k + \sum_{1\leqslant 2j+1\leqslant n+2} d_{2j+1}(2k)^{-2j-1} + \beta_k k^{-n-1} ,$$

$$\sqrt{\mu_k(1)} - \sqrt{\mu_k(2)} = \sum_{1\leqslant 2j+1\leqslant n+3} c_{2j+1}(2k)^{-2j-1} + \nu_k k^{-n-2} ,$$

where the numbers d_{2j+1} and c_{2j+1} do not depend on k, and

$$\sum_{k=1}^{\infty} |\beta_k|^2 < \infty \ , \quad \sum_{k=1}^{\infty} |\gamma_k|^2 < \infty \ .$$

3. Derive the asymptotic formulas for the eigenvalues of the periodic ($y(0) = y(\pi)$) and anti-periodic ($y(0) = -y(\pi)$) boundary value problems generated by the Dirac equation (1.3.36), in which the matrix $\Omega(x)$ has real-valued entries $p(x), r(x) \in W_2^n[0,\pi]$ $(n \geq 1)$.

Hint. According to (1.3.37), the characteristic functions of the indicated boundary value problems have the form $\chi(\lambda) = 2 \mp \{e_{11}(\lambda,\pi) + e_{22}(\lambda,\pi)\}$. The fact that the matrix $\Omega(x)$ is real guarantees that these problems are self-adjoint. Consequently, the eigenvalues $\ldots < \mu_{2k}^- \leq \mu_{2k}^+ < \ldots$ and $\ldots < \mu_{2k+1}^- \leq \mu_{2k+1}^+ < \ldots$ of the periodic and anti-periodic problems respectively, are real. Furthermore, the reality of the functions $p(x)$ and $r(x)$ implies the equality $q^-(x) = \overline{q^+(x)}$, from which in turn it follows that for real values of λ,

$$u_1^+(\lambda,x) = \overline{u_1^-(-\lambda,x)} = u(\lambda,x) \ ,$$

$$u_2^-(\lambda,x) = \overline{-u_2^+(-\lambda,x)} \ ,$$

and

$$\sigma^-(\lambda,x) = \overline{-\sigma^+(-\lambda,x)} = \sigma(\lambda,x) \ .$$

From this you find, upon using the formulas obtained in Problem 4, Section 4 of this chapter, that

$$e_{11}(\lambda,\pi) + e_{22}(\lambda,\pi) = \frac{2\ \mathrm{Re}\{e^{i\lambda\pi}u(\lambda,\pi)[1 - \sigma(\lambda,\pi)\overline{\sigma(\lambda,0)}]\}}{1 - \sigma(\lambda,0)\overline{\sigma(\lambda,0)}}$$

and

$$\chi(\lambda) = 2 \mp \{e^{i\lambda\pi}u(\lambda,\pi)[1 - \Delta(\lambda)\overline{\sigma(\lambda,0)}] + e^{-i\lambda\pi}\,\overline{u(\lambda,\pi)}[1 - \overline{\Delta(\lambda)}\sigma(\lambda,0)]\} \ .$$

Consequently, $\mu_k^{\pm}$ are the roots of the equation

$$e^{i\lambda\pi}u(\lambda,\pi)[1 - \Delta(\lambda)\overline{\sigma(\lambda,0)}] + e^{-i\lambda\pi}\overline{u(\lambda,\pi)}[1 - \overline{\Delta(\lambda)}\sigma(\lambda,0)] - (-1)^k 2 = 0 \ ,$$

where $\Delta(\lambda) = [\sigma(\lambda,\pi) - \sigma(\lambda,0)][1 - \sigma(\lambda,0)\overline{\sigma(\lambda,0)}]^{-1}$. Multiplying this equation by

$$\varphi(\lambda) = e^{i\lambda\pi}u(\lambda,\pi) = \exp i\left\{\lambda\pi - \int_0^\pi q(\xi)\sigma(\lambda,\xi)d\xi\right\},$$

and using the equality

$$u(\lambda,\pi)\overline{u(\lambda,\pi)} = [1 - \sigma(\lambda,0)\overline{\sigma(\lambda,0)}][1 - \sigma(\lambda,\pi)\overline{\sigma(\lambda,\pi)}]^{-1},$$

you obtain a quadratic equation for $\varphi(\lambda)$, from which you deduce that

$$\varphi(\lambda) = \exp i\left\{\lambda\pi - \int_0^\pi q(\xi)\sigma(\lambda,\xi)d\xi\right\} =$$
$$= (-1)^k[1 \pm i|\Delta(\lambda)u(\lambda,\pi)|][1 - \Delta(\lambda)\overline{\sigma(\lambda,0)}]^{-1}.$$

Using this equation and proceeding as in the proof of Theorem 1.5.2, you obtain the formula

$$\mu_k^\pm = k + \sum_{j=1}^{n+1} c_j(2k)^{-j} \pm |e_n(k)|(2k)^{-n} + \varepsilon_k^\pm k^{-n-1},$$

where the numbers c_j do not depend on k, $\sum_{k=-\infty}^{\infty}|\varepsilon_k|^2 < \infty$, $e_n(k) = \frac{1}{\pi}\int_0^\pi q^{(n)}(t)e^{-2ikt}dt$, and $q(t) = p(t) + ir(t)$. (For $n = 0$, the $\mu_k^\pm$ are expressible as $\mu_k^\pm = k + \varepsilon_k^\pm$, where $\sum_{k=-\infty}^{\infty}|\varepsilon_k^\pm|^2 < \infty$.)

4. Find the explicit expressions for the polynomials $M_2^\pm(k)$ and for the regularized trace s_1 of order one.

<u>Answer:</u>

1) $(J_{42} \neq 0)$

$$M_2^\pm(k) = (2k - \tfrac{1}{2} \pm \tfrac{1}{2})^2 + \frac{1}{\pi}\left\{c_1 + 2\,\frac{(J_{14} + J_{32}) \pm (J_{12} + J_{34})}{J_{42}}\right\},$$

$$s_1 = -\frac{1}{2}M_2^+(0) + \frac{q(0) + q(\pi)}{4} - \frac{1}{2}\left(\frac{J_{14} + J_{32}}{J_{42}}\right)^2 + \frac{J_{13}}{J_{42}};$$

2) ($J_{42} = 0$, $J_{14} + J_{32} \neq 0$, $\theta_2 = \arccos[-(J_{12} + J_{31})(J_{14} + J_{32})^{-1}]$, $0 < \operatorname{Re}\theta_2 \leqslant 1$)

$$M_2^\pm(k) = [2k - 1 \pm (1 - \theta_2)]^2 + \frac{1}{\pi}\left\{c_1 + \frac{2J_{13}}{J_{14} + J_{32}}\right\},$$

$$s_1 = \frac{J_{14} - J_{32}}{4(J_{14} + J_{32})}[q(\pi) - q(0)] - \frac{J_{13}^2}{2(J_{14} + J_{32})^2};$$

3°) $(J_{42} = J_{14} + J_{32} = J_{12} + J_{34} = 0 \quad , \quad J_{13} \neq 0)$

$$M_2^{\pm}(k) = \left[2k - \frac{1}{2} \pm \frac{1}{2}\right]^2 + \frac{1}{\pi}\left\{c_1 + \frac{J_{14}}{J_{13}}[q(\pi) - q(0)]\right\} ,$$

$$s_1 = \frac{1}{2} M_2^+(0) - \frac{q(0) + q(\pi)}{4} + \frac{J_{14}^2}{8J_{13}^2}[q(\pi) - q(0)] - \frac{J_{14}}{4J_{13}}[q'(\pi) + q'(0)] .$$

(Here $c_1 = \int_0^\pi q(t)dt$.)

5. In theorem 1.5.3, the m-th order trace formulas were proved for potentials which possessed at least 2m square-summable derivatives. Actually, these formulas are valid for less smooth potentials. For example, the first-order trace formula for the boundary value problem (1.3.1), (1.5.4) is valid for potentials with Fourier series which converges at the point $x = 0$. Prove the last assertion.

Hint. It follows from formula (1.4.24) and Lemma 1.4.1 that for $n = 0$, the characteristic function $\chi(z) = s(z,\lambda)$ of the problem in question can be expressed as

$$\chi(z) = (2iz)^{-1}[y(z,\pi) - y(-z,\pi)] ,$$

where

$$y(z,\pi) = e^{iz\pi}[1 + (2iz)^{-1}u_1(z,\pi)] ,$$

$$u_1(z,\pi) = u_1(\pi) + (2iz)^{-1}\int_0^\pi q(t)u_1(t)dt - \int_0^\pi u_1'(\pi - \xi)e^{-2iz\xi}d\xi -$$

$$- (2iz)^{-1}\int_0^\pi K_1^{(1)}(\pi,\xi)e^{-2iz\xi}d\xi ,$$

and

$$u_1(x) = \int_0^x q(t)dt \quad , \quad \int_0^\pi q(t)u_1(t)dt = \frac{1}{2}u_1(\pi)^2 \quad , \quad u_1'(x) = q(x) .$$

From this, upon using Lemma 1.3.1 and denoting $\tilde{g}(\lambda) = \int_0^\pi g(x)e^{-i\lambda x}dx$ for brevity, you readily obtain that

$$\chi(z) = \frac{\sin z\pi}{z}[1 - (2z)^{-1}u_1(\pi)\mathrm{ctg}\, z\pi - (8z^2)^{-1}u_1(\pi)^2 +$$

$$+ (4z \sin z\pi)^{-1}\{e^{-iz\pi}\tilde{q}(-2z) + e^{iz\pi}\tilde{q}(2z)\} + (z^2 \sin z\pi)^{-1}e^{|\mathrm{Im}\, z\pi|}o(1)] =$$

$$= \frac{\sin z\pi}{z} [1 + r(z)]$$

for $|z| \to \infty$. Since the functions $|\mathrm{cotan}\, z\pi|$ and $|\sin z\pi|^{-1}\exp|\mathrm{Im}\, z\pi|$ are bounded by a constant on the contours $K_n = C(n + \frac{1}{2})$ which does not depend on n, the maximum of the modulus of $r(z)$ on these contours tends to zero as $n \to \infty$. Hence, if λ_k are the eigenvalues of the boundary value problem (1.3.1), (1.5.4), you have

$$2 \sum_{k=1}^{n} \lambda_k = \frac{1}{2\pi i} \oint_{K_n} z^2 d \ln \chi(z) = \frac{1}{2\pi i} \Big\{ \oint_{K_n} z^2 d \ln \frac{\sin z\pi}{z} +$$

$$+ \oint_{K_n} z^2 d \ln(1 + r(z)) \Big\} = 2 \sum_{k=1}^{n} k^2 - \frac{1}{2\pi i} \oint_{K_n} 2z \ln(1 + r(z))dz =$$

$$= 2 \sum_{k=1}^{n} k^2 - \frac{1}{2\pi i} \oint_{K_n} 2z[-(2z)^{-1}u_1(\pi) \,\mathrm{cotan}\, z\pi - (8z^2)^{-1}u_1^2(\pi) +$$

$$+ (4z \sin z\pi)^{-1}\{e^{-iz\pi}\tilde{q}(-2z) + e^{iz\pi}\tilde{q}(2z)\} - (8z^2)^{-1}u_1^2(\pi) \,\mathrm{cotan}^2 z\pi + o(z^{-2})]dz ,$$

whence

$$2 \sum_{k=1}^{n} \lambda_k = 2 \sum_{k=1}^{n} k^2 + \frac{u_1(\pi)}{\pi} (2n + 1) - \sum_{k=-n}^{n} \frac{\tilde{q}(2k)}{\pi} + o(1) ,$$

since $1 + \mathrm{cotan}^2 z\pi \to 0$ as $|\mathrm{Im}\, z| \to \infty$ or, equivalently,

$$\sum_{k=1}^{n} [\lambda_k - k^2 - \pi^{-1}u_1(\pi)] = \frac{u_1(\pi)}{2\pi} - \frac{1}{2} \sum_{k=-n}^{n} \frac{\tilde{q}(2k)}{\pi} + o(1) .$$

Therefore,

$$\lim_{n\to\infty} \Big\{ \sum_{k=1}^{n} [\lambda_k - k^2 - \pi^{-1}u_1(\pi)] + \frac{1}{2} \sum_{k=-n}^{n} a_k \Big\} = \frac{u_i(\pi)}{2\pi} ,$$

where $a_k = \pi^{-1}\tilde{q}(2k)$ are the Fourier coefficients of the function $q(x)$. If the Fourier series of $q(x)$ converges at the point $x = 0$, then, upon denoting its sum by $z^{-1}[q(0+) + q(\pi - 0)]$ for brevity, you obtain the following expression for the first-order regularized trace:

$$\sum_{k=1}^{\infty} [\lambda_k - k^2 - \pi^{-1}u_1(\pi)] = \frac{u_1(\pi)}{2\pi} - \frac{q(0+) + q(\pi - 0)}{4} .$$

6. Let $\lambda_k(1)$, $\lambda_k(2)$, $\lambda_k(3)$, and $\lambda_k(4)$ be the eigenvalues of the boundary value problems generated on the interval $(0,\pi)$ by the Dirac equation (1.3.36) and the boundary conditions with the following matrices of coefficients:

$$1)\begin{pmatrix} 1 & 0 & 0 & 0 \\ 0 & 0 & 1 & 0 \end{pmatrix},\ 2)\begin{pmatrix} 0 & 1 & 0 & 0 \\ 0 & 0 & 0 & 1 \end{pmatrix},\ 3)\begin{pmatrix} 1 & i & 0 & 0 \\ 0 & 0 & 1 & i \end{pmatrix},\ 4)\begin{pmatrix} 1 & -i & 0 & 0 \\ 0 & 0 & 1 & -i \end{pmatrix}$$

(see Section 3 of this chapter). Further, let $\ldots,\lambda_{2k}^-,\lambda_{2k}^+,\ldots$ and $\ldots,\lambda_{2k+1}^-,\lambda_{2k+1}^+,\ldots$ be the eigenvalues of the periodic and anti-periodic boundary value problem, respectively, generated by the same equation. Suppose that the entries $p(x)$ and $r(x)$ of the matrix $\Omega(x)$ belong to $\tilde{W}_2^n[0,\pi]$. Show that in this case the following equalities hold for $1 \leqslant m \leqslant n$:

$$\sum_{k=-\infty}^{\infty} \{(\lambda_k(1))^m - (\lambda_k(2))^m\} = [\lambda^n + C^-(\lambda)]_m + [(-\lambda)^n - C^+(-\lambda)]_m +$$

$$+ [\lambda^n - C^-(\lambda)]_m + [(-\lambda)^n + C^+(-\lambda)]_m ,$$

$$\sum_{k=-\infty}^{\infty} \{(\lambda_k(1))^m - (\lambda_k(3))^m\} = [\lambda^n + C^-(\lambda)]_m + [(-\lambda)^n - C^+(-\lambda)]_m - [C^+(-\lambda)]_m ,$$

$$\sum_{k=-\infty}^{\infty} \{(\lambda_k(1))^m - (\lambda_k(4))^m\} = [\lambda^n + C^-(\lambda)]_m + [(-\lambda)^n - C^+(-\lambda)]_m - [C^-(\lambda)]_m ,$$

and

$$\frac{1}{2}\sum_{k=-\infty}^{\infty} \{2(\lambda_k(1))^m - (\lambda_k^-)^m - (\lambda_k^+)^m\} = [\lambda^n + C^-(\lambda)]_m +$$

$$+ [(-\lambda)^n - C^+(-\lambda)]_m - [(-1)^n\lambda^{2n} + C^-(\lambda)C^+(-\lambda)]_m ,$$

where $[Q(\lambda)]_m$ designates the sum of the m-th powers of the roots of the polynomial $Q(\lambda)$, and

$$C^{\pm}(\lambda) = \lambda^n\sigma^{\mp}(\lambda,0) = \sum_{k=1}^{n} \lambda^{n-k}\sigma_k^{\mp}(0)(2i)^{-k} .$$

Hint. According to (1.3.37), the characteristic functions of the problems under consideration may be expressed through $e_{ik}(\lambda,\pi)$ as follows:

$$\chi_1(\lambda) = e_{12}(\lambda,\pi) \ , \quad \chi_2(\lambda) = -e_{21}(\lambda,\pi) \ ,$$

$$\chi_3(\lambda) = e_{12}(\lambda,\pi) + e_{21}(\lambda,\pi) - i\{e_{11}(\lambda,\pi) - e_{22}(\lambda,\pi)\} ,$$

$$\chi_4(\lambda) = e_{12}(\lambda,\pi) + e_{21}(\lambda,\pi) + i\{e_{11}(\lambda,\pi) - e_{22}(\lambda,\pi)\} ,$$

$$\chi_p(\lambda) = 2 - \{e_{11}(\lambda,\pi) + e_{22}(\lambda,\pi)\} \quad , \quad \chi_a(\lambda) = 2 + \{e_{11}(\lambda,\pi) + e_{22}(\lambda,\pi)\} .$$

Since in the present case $\sigma_k^{\pm}(\pi) = \sigma_k^{\pm}(0)$ $(k = 1,2,\ldots,n)$, it follows from the results obtained in Problems 3, 4, of Section 4 of this chapter and Lemma 1.3.1, that

$$e^{\pm i\lambda\pi}u_1^{\pm}(\lambda,\pi)\sigma^{\mp}(\lambda,\pi) = e^{\pm i\lambda\pi}u_1^{\pm}(\lambda,\pi)\sigma^{\mp}(\lambda,0) + o(\lambda^{-n}e^{|\text{Im }\lambda\pi|}) ,$$

and

$$\chi_1(\lambda) = \frac{i(1+\sigma^-(\lambda,0))(1-\sigma^+(-\lambda,0))}{2(1+\sigma^-(\lambda,0)\sigma^+(-\lambda,0))} F(\lambda) + o(\lambda^{-n}e^{|\text{Im }\lambda\pi|}) ,$$

$$\chi_2(\lambda) = -\frac{i(1-\sigma^-(\lambda,0))(1+\sigma^+(-\lambda,0))}{2(1+\sigma^-(\lambda,0)\sigma^+(-\lambda,0))} F(\lambda) + o(\lambda^{-n}e^{|\text{Im }\lambda\pi|}) ,$$

$$\chi_3(\lambda) = -\frac{2i\sigma^+(-\lambda,0)}{1+\sigma^-(\lambda,0)\sigma^+(-\lambda,0)} F(\lambda) + o(\lambda^{-n}e^{|\text{Im }\lambda\pi|}) ,$$

$$\chi_4(\lambda) = \frac{2i\sigma^-(\lambda,0)}{1+\sigma^-(\lambda,0)\sigma^+(-\lambda,0)} F(\lambda) + o(\lambda^{-n}e^{|\text{Im }\lambda\pi|}) ,$$

$$\chi_p(\lambda) = 2 - G(\lambda) + o(\lambda^{-n}e^{|\text{Im }\lambda\pi|}) \quad , \quad \chi_a(\lambda) = 2 + G(\lambda) + o(\lambda^{-n}e^{|\text{Im }\lambda\pi|}) ,$$

as $|\lambda| \to \infty$, where

$$F(\lambda) = e^{i\lambda\pi}u_1^+(\lambda,\pi) - e^{-i\lambda\pi}u_1^-(-\lambda,\pi) \quad , \quad G(\lambda) = e^{i\lambda\pi}u_1^+(\lambda,\pi) + e^{-i\lambda\pi}u_1^-(-\lambda,\pi) .$$

Moreover,

$$u_1^+(\lambda,\pi)u_1^-(-\lambda,\pi) = 1 + o(\lambda^{-n}e^{2|\text{Im }\lambda\pi|}) ,$$

so that

$$\chi_a(\lambda)\chi_p(\lambda) = -F^2(\lambda) + o(\lambda^{-n}e^{2|\text{Im }\lambda\pi|}) .$$

Now notice that for large N,

$$\sum_{k=-N}^{N} \{2(\lambda_k(1))^m - (\lambda_k^-)^m - (\lambda_k^+)^m\} = \frac{1}{2\pi i}\oint_{K_N} \lambda^m d\ \ln \frac{(\chi_1(\lambda))^2}{\chi_p(\lambda)\chi_a(\lambda)} ,$$

where K_N are the contours which bound the squares $|\text{Re }\lambda| \leqslant N + \frac{1}{2}$, $|\text{Im }\lambda| \leqslant N + \frac{1}{2}$. On the contour K_N

$$|F(\lambda)| \geq Ce^{|\operatorname{Im} \lambda\pi|} ,$$

where the constant C does not depend on N. Consequently, for $\lambda \in K_N$ and $N \to \infty$,

$$\frac{(\chi_1(\lambda))^2}{\chi_p(\lambda)\chi_a(\lambda)} = \left[\frac{(1 + \sigma^-(\lambda,0))(1 - \sigma^+(-\lambda,0))}{2(1 + \sigma^-(\lambda,0)\sigma^+(-\lambda,0)} \right]^2 (1 + o(N^{-n})) ,$$

while

$$\frac{1}{2\pi i} \oint_{K_N} \lambda^m d \ln \frac{(\chi_1(\lambda))^2}{\chi_p(\lambda)\chi_a(\lambda)} =$$

$$= \frac{2}{2\pi i} \oint_{K_N} \lambda^m d \ln \frac{(1 + \sigma^-(\lambda,0))(1 - \sigma^+(-\lambda,0))}{1 + \sigma^-(\lambda,0)\sigma^+(-\lambda,0)} + o(N^{m-n}) .$$

From this you immediately obtain the last of the equalities to be proved. The other three equalities are proved in much the same way. (In verifying the equalities which involve the eigenvalues of the third and fourth boundary value problems, you have to assume that $q^{\pm}(0) \neq 0$ and $m < n$.) For $m = 1$ the right-hand sides of the four equalities become

1) $-2p(0)$; 2) $-p(0) - i[\ln q^+(x)]'_{x=0}$;

3) $-p(0) + i[\ln q^-(x)]'_{x=0}$; 4) $-p(0)$.

CHAPTER 2

THE STURM-LIOUVILLE BOUNDARY VALUE PROBLEM ON THE HALF LINE

1. SOME INFORMATION ON DISTRIBUTIONS

The function $f(x)$, defined on the half line $0 \leqslant x < \infty$, is said to have compact support if it vanishes outside some bounded segment. We denote the set of all square-summable functions with compact support by K^2, and the set of all functions $f \in K^2$, such that $f(x) = 0$ for $x > \sigma$, by $K^2(\sigma)$. The Fourier cosine transform of a function $f(x)$ will be denoted by

$$C(\lambda,f) = \int_0^\infty f(x)\cos \lambda x dx , \tag{2.1.1}$$

and $CK^2(\sigma)$ will designate the set of Fourier cosine transforms of elements of $K^2(\sigma)$. It is not hard to give an intrinsic characterization of $CK^2(\sigma)$: it coincides with the set of all even entire functions $g(\lambda)$ which are square-summable on the real line and satisfy the inequality

$$|g(\lambda)| \leqslant C \exp\{\sigma|\text{Im } \lambda|\} \tag{2.1.2}$$

for all complex values of λ, where C is a constant that depends on g.

In fact, if $g(\lambda) \in CK^2(\sigma)$, then by definition $g(\lambda) = C(\lambda,f)$, where $f(x) \in K^2(\sigma)$. It follows immediately from (2.1.1) that $g(\lambda) = C(\lambda,f)$ is an even entire function which satisfies the inequality (2.1.2) with $C = \int_0^\sigma |f(x)|dx$, and, by the Plancherel theorem, $|g(\lambda)|^2$ is summable on the real line. Conversely, if the function $g(\lambda)$ satisfies the indicated conditions, then by Jordan's lemma,

$$f_\varepsilon(x) = \frac{1}{2\pi} \int_{-\infty}^{\infty} g(\lambda) \frac{\sin \varepsilon\lambda}{\varepsilon\lambda} e^{i\lambda x} d = 0$$

for $|x| > \sigma + \varepsilon$. Hence

$$g(\lambda) \frac{\sin \varepsilon\lambda}{\varepsilon\lambda} \int_{-\sigma-\varepsilon}^{\sigma+\varepsilon} f_\varepsilon(x) e^{-i\lambda x} dx = \int_0^{\sigma+\varepsilon} 2f_\varepsilon(x) \cos \lambda x dx ,$$

and since $g(\lambda) \frac{\sin \varepsilon\lambda}{\varepsilon\lambda}$ converges to $g(\lambda)$ in the $L_2(-\infty,\infty)$ norm as $\varepsilon \to 0$, it follows that the limit $\lim_{\varepsilon\to 0} 2f_\varepsilon(x) = f(x)$ exists in $L_2(0,\infty)$, and clearly $f(x) = 0$ for $x > \sigma$. Therefore, $f(x) \in K^2(\sigma)$ and $g(\lambda) = \int_0^\sigma f(x) \cos \lambda x dx$, i.e., $g(\lambda) = C(\lambda,f) \in CK^2(\sigma)$.

We remark that, according to the Paley-Wiener theorem, condition (2.1.2) can be weakened to

$$\overline{\lim_{r\to\infty}}\, r^{-1} \ln\{ \max_{|z|=r} |g(z)|\} = \sigma_g \leq \sigma .$$

The entire functions satisfying this condition are called entire functions of exponential type σ_g . Thus, $CK^2(\sigma)$ is precisely the set of even entire functions of exponential type at most σ , which are square-summable on the real line. (Incidentally, we shall make no use of this weaker version of condition (2.1.2) in the sequel.) Let $Z(\sigma)$ denote the normed linear space of even entire functions $f(\lambda)$ which are summable on the real line and satisfy inequality (2.1.2), with the usual operations of addition and multiplication by complex numbers equipped with the norm $||f|| = \int_{-\infty}^{\infty} |f(\lambda)| d\lambda$. The spaces $Z(\sigma)$ are obviously subspaces of $L_1(-\infty,\infty)$, and $Z(\sigma) \subset Z(\sigma')$ if $\sigma < \sigma'$. Let $Z = \cup_\sigma Z(\sigma)$. Then clearly Z is a linear manifold in the space $L_1(-\infty,\infty)$. We define convergence in Z as follows: the sequence $f_n(x)$ converges to $f(\lambda)$ if there is a σ such that all the functions $f_n(\lambda)$ belong to $Z(\sigma)$ and converge to $f(\lambda)$ in the metric of this space. We take the linear manifold Z with this notion of convergence as the space of test functions.

DEFINITION 2.1.1. *The space of test functions* Z *is the set of all even entire functions* $f(\lambda)$ *which are summable on the real line and satisfy inequality* (2.1.2) (*in which the constants* σ *and* C *depend on* $f(\lambda)$), *with the usual operations of addition and multiplication by complex numbers. The sequence*

$f_n(\lambda) \in Z$ *converges to* $f(\lambda)$ *if*

$$\lim_{n\to\infty} \int_{-\infty}^{\infty} |f_n(\lambda) - f(\lambda)| d\lambda = 0 ,$$

and the types σ_n *of the functions* $f_n(\lambda)$ *are bounded:* $\sup \sigma_n < \infty$.

We list a number of properties of the space Z . Let CK^2 designate the set of Fourier cosine transforms of the functions from K^2 , i.e., $CK^2 = \cup_\sigma CK^2(\sigma)$. Then $CK^2 \supset Z$. In fact, if $f(\lambda) \in Z$, then $f(\lambda) \in Z(\sigma)$ for some σ , and hence $\sup_{-\infty<\lambda<\infty} |f(\lambda)| < \infty$. But then

$$\int_{-\infty}^{\infty} |f(\lambda)|^2 d\lambda \leq \sup_{-\infty<\lambda<\infty} |f(\lambda)| \int_{-\infty}^{\infty} |f(\lambda)| d\lambda < \infty ,$$

whence $f(\lambda) \in CK^2(\sigma) \subset CK^2$. It is also clear that the product of any two functions from CK^2 belongs to Z and that the set of such products is dense in Z , because if $f(\lambda) \in Z$, then $f(\lambda) \in CK^2$, and the product $f_\varepsilon(\lambda) = f(\lambda) \dfrac{\sin \varepsilon\lambda}{\varepsilon\lambda} \to f(\lambda)$ in Z as $\varepsilon \to 0$. Multiplication of elements in the space Z of test functions by arbitrary even entire functions $\varphi(\lambda)$ which satisfy the inequality (1.2.1) is permitted because every such product belongs to Z . Accordingly, we shall refer to the functions $\varphi(\lambda)$ as multipliers in the space Z .

DEFINITION 2.1.2. *The additive, homogeneous, and continuous functionals* $R[f(\lambda)] = (f(\lambda),R)$, *defined on the space* Z *of test functions, are called distributions (generalized functions). The set of all distributions will be denoted by* Z' .

From this definition and the definition of convergence in the space Z , it follows that the distributions in Z' are precisely the additive homogeneous functionals on Z whose restrictions to the $Z(\sigma)$ are linear functionals on these normed spaces.

The sequence $R_n \in Z'$ of distributions converges to $R \in Z'$ if

$$\lim_{n\to\infty} (f(\lambda),R_n) = (f(\lambda),R) \tag{2.1.3}$$

for all test functions $f(\lambda) \in Z$. Notice that here the requirement that $R \in Z'$ is not essential, because if the limit on the left-hand side of the equality exists for every $f(\lambda) \in Z$, then the restriction of the sequence R_n

to any space $Z(\sigma)$ converges weakly, and therefore, by a well-known theorem of Banach, the limit is a linear functional on $Z(\sigma)$. Consequently, the functional defined by the left-hand side of (2.1.3) is a distribution, i.e., the functional R belongs automatically to Z'.

We call the distribution $R \in Z'$ regular if it is given by the formula

$$(f(\lambda),R) = \int_0^\infty f(\lambda)R(\lambda)d\lambda ,$$

where $R(\lambda)$ is an arbitrary bounded measurable function on the half line $0 \leq \lambda < \infty$. That is to say, R is regular if it defines a functional which is continuous in the metric of the space $L_1(-\infty,\infty)$. Distributions $R \in Z'$ can be multiplied by the multipliers $\varphi(\lambda)$ of the space Z: $(f(\lambda),R\cdot\varphi(\lambda)) = (f(\lambda)\varphi(\lambda),R)$. If the function $A(x)$ is summable on the half line $0 \leq x < \infty$, then its Fourier cosine transform $C(\lambda,A)$ is a bounded continuous function. Hence, it can be identified with the regular distribution $C(A)$ acting according to the rule

$$(f(\lambda),C(A)) = \int_0^\infty f(\lambda)C(\lambda,A)d\lambda .$$

Using the definition of the Fourier cosine transform, we can reexpress this as

$$(f(\lambda),C(A)) = \int_0^\infty \left[\int_0^\infty f(\lambda)\cos \lambda x dx \right] A(x)dx . \tag{2.1.4}$$

Notice that the right-hand side of this formula is meaningful for every locally summable (i.e., summable on every bounded segment $[0,\sigma]$) function $A(x)$, since $f(\lambda) \in Z \subset CK^2$, and hence $\int_0^\infty f(\lambda)\cos \lambda x d\lambda$ is a continuous function with compact support. It is readily verified that the local summability of the function $A(x)$ is actually sufficient for the functional on Z defined by the right-hand side of formula (2.1.4) to be continuous, i.e., a distribution. This permits us to give the following definition of the Fourier cosine transform for arbitrary locally summable functions.

DEFINITION 2.1.3. *The Fourier cosine transform of the locally summable function* $A(x)$ *is the distribution* $C(A) \in Z'$, *which acts in the space of test functions* Z *by formula* (2.1.4).

We list without proofs a number of easily verified properties of the

Fourier cosine transform. We shall write $R \sim B(x)$ if the distribution $R \in Z'$ is the Fourier cosine transform of the locally summable function $B(x)$.

1. If the sequence of locally summable functions $A_n(x)$ converges in the mean on every bounded interval of the half line to the function $A(x)$, then $\lim_{n\to\infty} C(A_n) = C(A)$ in the sense of convergence of distributions. In particular, one always has

$$C(A) = \lim_{n\to\infty} \int_0^n A(x)\cos \lambda x dx ,$$

where the regular functions $\int_0^n A(x)\cos \lambda x dx$ converge to $C(A)$ in the sense of convergence of distributions.

2. If $C(A) \sim A(x)$, then

$$\cos \lambda a C(A) \sim \frac{1}{2}\{A(|x+a|) + A(|x-a|)\} .$$

3. If $C(A) \sim A(x)$ and $\varphi(\lambda) = \int_0^\infty g(\xi)\cos \lambda\xi d\xi$, where $g(\xi)$ is a summable function with compact support, then

$$\varphi(\lambda)C(A) \sim \frac{1}{2}\int_0^\infty g(\xi)[A(x+\xi) + A(|x-\xi|)]d\xi .$$

4. If $C(A) \sim A(x)$, then

$$\frac{2}{\pi}\left(\frac{1 - \cos\ x}{\lambda^2}, C(A)\right) = \int_0^x (x-t)A(t)dt .$$

A distribution $R \in Z'$ is said to be positive if $(f(\lambda),R) \geq 0$ for every test function $f(\lambda) \in Z$ which satisfies the inequality $f(\sqrt{\mu}) \geq 0$ $(-\infty < \mu < \infty)$.

THEOREM 2.1.1. *For every positive distribution* $R \in Z'$ *there is a nondecreasing function* $\rho(\mu)$ $(-\infty < \mu < \infty)$ *such that*

$$(f(\lambda)g(\lambda),R) = \int_{-\infty}^\infty f(\sqrt{\mu})g(\sqrt{\mu})d\rho(\mu)$$

for all functions $f(\lambda),g(\lambda) \in CK^2$. *The Stieltjes integral on the right-hand side of this formula converges absolutely.*

The proof of this theorem rests on the method for extending positive functionals due to M. Riesz. For this reason we first give an account of this method.

Consider an arbitrary linear manifold $\mathcal{U}$ of continuous real-valued functions $f(x)$ $(-\infty < x < \infty)$ in which for every real number r, one can find a nonnegative function $r(x)$ such that $r(x) \geq 1$ for all $x \leq r$. A functional R given on this manifold is said to be positive if $R[f(x)] \geq 0$ for every nonnegative function $f(x) \in \mathcal{U}$. We say that the function $f(x) \in \mathcal{U}$ admits a majorant if there is a nonnegative function $g(x) \in \mathcal{U}$ such that

$$\lim_{|x|\to\infty} \frac{|f(x)|}{g(x)} = 0 .$$

THEOREM (Riesz). *Let* R *be an additive, homogeneous, and positive functional on* $\mathcal{U}$. *Then there exists a nondecreasing function* $\rho(t)$ $(-\infty < t < \infty)$ *such that*

$$R[f(x)] = \int_{-\infty}^{\infty} f(t)d\rho(t)$$

for every function $f(x)$ *admitting a majorant.*

PROOF. We arrange the rational numbers in a sequence $r_1, r_2, \ldots$ and introduce the functions

$$h(x;r_k) = \begin{cases} 1 , & x \leq r_k , \\ 0 , & x > r_k . \end{cases}$$

Let M (respectively N) denote the set of all functions $f(x) \in \mathcal{U}$ (respectively $g(x) \in \mathcal{U}$) such that

$f(x) \leq h(x;r_1)$ (respectively $h(x;r_1) \leq g(x)$) .

The set M is not empty, since the function $f(x) \equiv 0$ always belongs to M; the set N is nonempty by the assumption on $\mathcal{U}$. Since the functional R is positive,

$$0 \leq \sup_{f(x)\in M} R[f(x)] \leq \inf_{g(x)\in N} R[g(x)] < \infty .$$

Consequently, there is a nonnegative number γ_1 such that

$$\sup_{f(x)\in M} R[f(x)] \leqslant \gamma_1 \leqslant \inf_{g(x)\in N} R[g(x)] .$$

Now extend the functional R to the linear manifold $\mathcal{U}_1$ of all functions of the form $f(x) + \alpha h(x;r_1)$, where $f(x) \in \mathcal{U}$ and α is an arbitrary real number, by the formula

$$R[f(x) + \alpha h(x;r_1)] = R[f(x)] + \alpha\gamma_1 .$$

It is readily checked that the functional R defined by this formula is also additive, homogeneous, and positive on the entire set $\mathcal{U}_1$.

This construction can obviously be repeated for the set $\mathcal{U}_1$ and the function $h(x;r_2)$. We thus extend the functional R to the set $\mathcal{U}_2$ of all functions of the form $f(x) + \alpha_1 h(x;r_1) + \alpha_2 h(x;r_2)$ $(f(x) \in \mathcal{U}$, α_1 , α_2 arbitrary real numbers). Continuing in this manner, we extend the functional R to the set of all functions of the form $f(x) + \sum_{k=1}^{n} \alpha_k h(x;r_k)$ $(n = 1,2,\ldots)$, and the result is also an additive homogeneous positive functional.

Now we set

$$\rho(t) = \sup_{r_k\leqslant t} R[h(x;r_k)] . \tag{2.1.5}$$

If $r_k > r_{k'}$, then $h(x;r_k) \geqslant h(x;r_{k'}) \geqslant 0$ $(-\infty < x < \infty)$, from which it follows, in view of the positivity of the functional R , that the function $\rho(t)$ does not decrease, and that at the rational points r_m it coincides with the value of the functional R on $h(x;r_m)$: $\rho(r_m) = R[h(x;r_m)]$.

Suppose the function $f(x) \in \mathcal{U}$ has a majorant $g(x) \in \mathcal{U}$ such that $g(x) \geqslant 0$ and

$$\lim_{|x|\to\infty} \frac{|f(x)|}{g(x)} = 0 . \tag{2.1.6}$$

Pick arbitrary positive integers n , N , and construct the piecewise constant function

$$\varphi(x) = \sum_{k=-N+1}^{N} f(\tfrac{nk}{N})[h(x;\tfrac{nk}{N}) - h(x;\tfrac{n(k-1)}{N})] , \tag{2.1.7}$$

which vanishes in the exterior of the semi-interval $(-n,n]$, and which is equal to $f(\frac{nk}{N})$ on the semi-intervals $(\frac{n(k-1)}{N} , \frac{nk}{N}]$ $(k = -N+1,-N+2,\ldots,0,\ldots,N-1,N)$. Hence, for $x \notin (-n,n]$,

$$|f(x) - \varphi(x)| = |f(x)| \leq \delta_n g(x) , \tag{2.1.8}$$

where $\delta_n = \sup\limits_{|x|>n} \dfrac{|f(x)|}{g(x)}$, while for $x \in (-n,n]$,

$$|f(x) - \varphi(x)| \leq \omega(\tfrac{n}{N}) = \omega(\tfrac{n}{N})h(x;n) , \tag{2.1.9}$$

where $\omega(h) = \sup\limits_{|x|\leq n} \sup\limits_{|t|\leq h} |f(x+t) - f(x)|$. Moreover, by (2.1.6), $\lim\limits_{n\to\infty} \delta_n = 0$, and in view of the continuity of the function $f(x)$, $\lim\limits_{h\to 0} \omega(h) = 0$. From inequalities (2.1.8) and (2.1.9) it follows that for all values of x ,

$$-\omega(\tfrac{n}{N})h(x;n) - \delta_n g(x) \leq f(x) - \varphi(x) \leq \omega(\tfrac{n}{N})h(x;n) + \delta_n g(x) .$$

In view of the positivity of the functional R , this implies the inequality

$$|R[f(x)] - R[\varphi(x)]| \leq \omega(\tfrac{n}{N})R[h(x;n)] + \delta_n R[g(x)] .$$

By definitions (2.1.5) and (2.1.7) of functions $\rho(t)$ and $\varphi(x)$,

$$R[\varphi(x)] = \sum_{k=-N+1}^{N} f(\tfrac{nk}{N})[\rho(\tfrac{nk}{N}) - \rho(\tfrac{n(k-1)}{N})] ,$$

and hence

$$|R[f(x)] - \sum_{k=-N+1}^{N} f(\tfrac{nk}{N})[\rho(\tfrac{nk}{N}) - \rho(\tfrac{n(k-1)}{N})]| \leq \omega(\tfrac{n}{N})R[h(x;n)] + \delta_n R[g(x)] .$$

Letting first $N \to \infty$, and then $n \to \infty$, we get

$$R[f(x)] - \lim_{n\to\infty} \int_{-n}^{n} f(x)d\rho(x) = 0 ,$$

i.e.,

$$R[f(x)] = \int_{-\infty}^{\infty} f(x)d\rho(x) ,$$

as asserted. □

We can now find the form of the positive distributions $R \in Z'$. In fact, let us take for U the set of all real-valued functions of the variable μ $(-\infty < \mu < \infty)$ of the form $f(\mu) = \hat{f}(\sqrt{\mu})$, where $\hat{f}(\lambda)$ is an arbitrary function from Z which assumes real values for real and purely imaginary values of λ . The set U meets the conditions under which Riesz's theorem was proved, because the function $2\left(\frac{\sin \varepsilon\sqrt{\mu}}{\varepsilon\sqrt{\mu}}\right)^2$ belongs to U and $2\left(\frac{\sin \varepsilon\sqrt{\mu}}{\varepsilon\sqrt{\mu}}\right)^2 \geq 1$ for all $\mu \leq r$, provided only that ε is sufficiently small. A positive distribution $R \in Z'$ defines an additive, homogeneous, positive functional $R[f(\mu)] = (\hat{f}(\lambda),R)$ on this set. Hence, Riesz's theorem provides a nonnegative function $\rho(\mu)$ such that

$$(\hat{f}(\lambda),R) = R[f(\mu)] = \int_{-\infty}^{\infty} f(\mu)d\rho(\mu) = \int_{-\infty}^{\infty} \hat{f}(\sqrt{\mu})d\rho(\mu)$$

for all functions in U admitting a majorant.

Let $\hat{f}(\lambda) \in Z$ and $\hat{f}(\sqrt{\mu}) \geq 0$ $(-\infty < \mu < \infty)$. Then the function $\hat{f}(\lambda)(\frac{\sin \lambda h}{\lambda h})^8$ also belongs to Z, and the function $\hat{f}(\sqrt{\mu})\left(\frac{\sin \sqrt{\mu}\, h}{\sqrt{\mu}\, h}\right)$ belongs to U and is majorized there by $\hat{f}(\sqrt{\mu})\left(\frac{\sin \sqrt{\mu}\, h}{\sqrt{\mu}\, h}\right)^8(1 + \mu^2)$. Hence

$$(\hat{f}(\lambda)(\frac{\sin \lambda h}{\lambda h})^8,R) = \int_{-\infty}^{\infty} \hat{f}(\sqrt{\mu})\left(\frac{\sin \sqrt{\mu}\, h}{\sqrt{\mu}\, h}\right)^8 d\rho(\mu) .$$

Letting $h \to 0$ and observing that

$$\lim_{h\to 0} \hat{f}(\lambda)(\frac{\sin \lambda h}{\lambda h})^8 = \hat{f}(\lambda)$$

in the sense of convergence in Z, we get

$$(\hat{f}(\lambda),R) = \int_{-\infty}^{\infty} \hat{f}(\sqrt{\mu})d\rho(\mu)$$

for every $\hat{f}(\lambda) \in Z$ such that $\hat{f}(\sqrt{\mu}) \geq 0$ $(-\infty < \mu < \infty)$.

The function $f(\lambda)g(\lambda)$, where $f(\lambda),g(\lambda) \in CK^2$ and assume real values for real and purely imaginary values of λ, is expressible as

$$f(\lambda)g(\lambda) = \frac{1}{4}\{[f(\lambda) + g(\lambda)]^2 - [f(\lambda) - g(\lambda)]^2\} .$$

Consequently,

$$(f(\lambda)g(\lambda),R) = \frac{1}{4}\int_{-\infty}^{\infty} [f(\sqrt{\mu}) + g(\sqrt{\mu})]^2 d\rho(\mu) - \frac{1}{4}\int_{-\infty}^{\infty} [f(\sqrt{\mu}) - g(\sqrt{\mu})]^2 d\rho(\mu) =$$

$$= \int_{-\infty}^{\infty} f(\sqrt{\mu})g(\sqrt{\mu})d\rho(\mu) ,$$

which proves the theorem for functions $f(\lambda)$ and $g(\lambda)$ which assume real values for real and purely imaginary values of λ. The general case is readily reduced to the one just considered.

To test the positivity of distributions $R \in Z'$, one can resort to the following lemma.

LEMMA 2.1.1. *The distribution* $R \in Z'$ *is positive if and only if the inequality* $(f(\lambda)\overline{f(\bar{\lambda})},R) \geq 0$ *holds for all functions* $f(\lambda) \in CK^2$.

PROOF. Let $g(\lambda) \in Z$, and suppose that $g(\sqrt{\mu}) \geq 0$ for all $\mu \in (-\infty,\infty)$. Then there are constants C and σ such that

$$|g(\lambda)| \leq C \exp\{\sigma|\operatorname{Im} \lambda|\} . \tag{2.1.10}$$

Consider the auxiliary function

$$g_\varepsilon(\lambda) = g(\lambda)\left(\frac{\sin \varepsilon\lambda}{\varepsilon\lambda}\right)^8 + \left(\frac{\varepsilon \sin \sigma_1\lambda}{\lambda}\right)^2 ,$$

where $\varepsilon > 0$ is arbitrary and $\sigma_1 = \frac{1}{2}\sigma + 4\varepsilon$. It is readily verified that $g_\varepsilon(0) > 0$, that $g_\varepsilon(\sqrt{\mu}) \geq 0$ for all $\mu \in (-\infty,\infty)$, and that $\lim\limits_{\varepsilon\to 0} g_\varepsilon(\lambda) =$ $= g(\lambda)$ in the sense of convergence in Z. Consequently, $(g(\lambda),R) =$ $= \lim\limits_{\varepsilon\to 0} (g_\varepsilon(\lambda),R)$, and therefore, in order to prove the lemma it suffices to check that for every $\varepsilon > 0$ the function $g_\varepsilon(\lambda)$ can be written as a product $f_\varepsilon(\lambda)\overline{f_\varepsilon(\bar{\lambda})}$, with factors belonging to CK^2. In view of the inequality (2.1.10),

$$\left|g(\lambda)\left(\frac{\sin \varepsilon\lambda}{\varepsilon\lambda}\right)^8\right| \leq C|\varepsilon\lambda|^{-8} \exp(2\sigma_1|\operatorname{Im} \lambda|) . \tag{2.1.11}$$

Upon using the self-evident equality

$$\left|\frac{\varepsilon \sin \sigma_1\lambda}{\lambda}\right|^2 = \frac{\varepsilon^2}{4\lambda^2} e^{2\sigma_1|\operatorname{Im} \lambda|}\left|1 - e^{2i\sigma_1|\operatorname{Re} \lambda|-2\sigma_1|\operatorname{Im} \lambda|}\right|^2 \tag{2.1.12}$$

and Rouché's theorem, this implies that for every sufficiently large positive integer n_1, function $g_\varepsilon(\lambda)$ has exactly $4n_1$ zeros in the strip $|\operatorname{Re} \lambda| \leq \frac{\pi}{\sigma_1}(n_1 + \frac{1}{2})$, whereas for integers $n > n_1$, it has exactly two zeros which both lie at a distance no greater than An^{-3} from the point $\frac{\pi n}{\sigma_1}$ in every strip $\left|\operatorname{Re} \lambda - \frac{\pi n}{\sigma_1}\right| \leq \frac{\pi}{2\sigma_1}$, where the constant A does not depend on n. Since the function $g_\varepsilon(\lambda)$ is even and nonnegative for real and purely imaginary λ , this leads to the conclusion that the zeros of $g_\varepsilon(\lambda)$ can be arranged in the following sequence:

$$\alpha_1, \bar{\alpha}_1, -\alpha_1, -\bar{\alpha}_1 \ ; \ \alpha_2, \bar{\alpha}_2, -\alpha_2, -\bar{\alpha}_2 \ ; \ \dots ,$$

where all $\alpha_n \neq 0$,

$$\alpha_n = \frac{n\pi}{\sigma_1} + \delta_n \quad \text{and} \quad \sup_n |n^3 \delta_n| = A < \infty . \tag{2.1.13}$$

We consider the functions

$$f_\varepsilon(\lambda) = \sqrt{g_\varepsilon(0)} \prod_{k=1}^{\infty} \left(1 - \frac{\lambda^2}{\alpha_k^2}\right) \quad \text{and} \quad \overline{f_\varepsilon(\bar{\lambda})} = \sqrt{g_\varepsilon(0)} \prod_{k=1}^{\infty} \left(1 - \frac{\lambda^2}{\bar{\alpha}_k^2}\right) ,$$

and we show that

$$g_\varepsilon(\lambda) = f_\varepsilon(\lambda)\overline{f_\varepsilon(\bar{\lambda})} . \tag{2.1.14}$$

From the definition of $f_\varepsilon(\lambda)$ and $\overline{f_\varepsilon(\bar{\lambda})}$, it follows that the function

$$\frac{f_\varepsilon(\lambda)\overline{f_\varepsilon(\bar{\lambda})}}{g_\varepsilon(\lambda)} \tag{2.1.15}$$

is entire and is equal to one at $\lambda = 0$. Hence, to prove the identity (2.1.14) it suffices, in view of Liouville's theorem, to establish the existence of a sequence of unboundedly expanding closed contours on which the modulus of the function (2.1.15) stays bounded. We claim that the boundaries L_n of the squares

$$|\mathrm{Re}\ \lambda| \leq \frac{\pi}{\sigma_1}\left(n + \frac{1}{2}\right) \quad , \quad |\mathrm{Im}\ \lambda| \leq \frac{\pi}{\sigma_1}\left(n + \frac{1}{2}\right) \quad (n = 1,2,\dots)$$

form such a sequence. According to (2.1.11) and (2.1.12), we have

$$|g_\varepsilon(\lambda)| > \frac{\varepsilon^2}{8|\lambda^2|} \exp\{2\sigma_1 |\mathrm{Im}\ \lambda|\} \tag{2.1.16}$$

on the contours L_n for sufficiently large n . On the other hand,

$$\left|\lambda^2 - \frac{k^2\pi^2}{\sigma_1^2}\right| \geq \frac{\pi^2}{2\sigma_1^2}(|n|+|k|) \geq \frac{\pi^2|k|}{2\sigma_1^2}$$

for $\lambda \in L_n$, and hence

$$\left| \frac{\sigma_1 \lambda f_\varepsilon(\lambda)}{\sin \sigma_1 \lambda} \right| = \sqrt{g_\varepsilon(0)} \prod_{k=1}^{\infty} \left| \frac{1 - \dfrac{\lambda^2}{\alpha_k^2}}{1 - \dfrac{\lambda^2 \sigma_1^2}{k^2 \pi^2}} \right| =$$

$$= \sqrt{g_\varepsilon(0)} \prod_{k=1}^{\infty} \left| \frac{k\pi}{\alpha_k \sigma_1} \right|^2 \left| 1 + \frac{\sigma_1^2 \alpha_k^2 - k^2\pi^2}{k^2\pi^2 - \sigma_1^2 \lambda^2} \right| \leqslant$$

$$\leqslant \sqrt{g_\varepsilon(0)} \prod_{k=1}^{\infty} \left| \frac{k\pi}{\alpha_k \sigma_1} \right|^2 \left(1 + \frac{2|\delta_k|\sigma_1(2k\pi + \sigma_1|\delta_k|)}{\pi^2 k} \right) = M ,$$

where, by (2.1.13), $M < \infty$. This shows that the following inequalities hold on the contours L_n :

$$|f_\varepsilon(\lambda)| \leqslant M|\sigma_1\lambda|^{-1}|\sin \sigma_1\lambda| \leqslant M|\sigma_1\lambda|^{-1} \exp\{\sigma_1|\operatorname{Im}\lambda|\} , \tag{2.1.17}$$

and

$$|\overline{f(\overline{\lambda})}| \leqslant M|\sigma_1\lambda|^{-1} \sin \sigma_1\overline{\lambda} \leqslant M|\sigma_1\lambda|^{-1} \exp\{\sigma_1|\operatorname{Im}\lambda|\} , \tag{2.1.18}$$

from which in turn it follows, in view of (2.1.16), that, for sufficiently large n ,

$$\max_{\lambda \in L_n} \left| \frac{f_\varepsilon(\lambda)\overline{f_\varepsilon(\overline{\lambda})}}{g_\varepsilon(\lambda)} \right| \leqslant 8M^2(\sigma_1\varepsilon)^{-2} .$$

We have thus established identity (2.1.14). To show that the condition of the lemma is sufficient, it remains to check that the functions $f_\varepsilon(\lambda)$ and $\overline{f_\varepsilon(\lambda)}$ belong to CK^2 . Considering (2.1.14) for real values of λ , we see that $f_\varepsilon(\lambda)$ and $\overline{f_\varepsilon(\lambda)}$ are square summable on the real line. Therefore, since they are even entire functions which satisfy inequalities (2.1.17), (2.1.18), they belong to the set $CK^2(\sigma_1) \subset CK^2$, as asserted. The necessity of the condition of the lemma is obvious. □

<u>Remark</u>. It is not hard to establish the existence of the limit $\lim_{\varepsilon \to 0} f_\varepsilon(\lambda) = f(\lambda) \in CK^2$. Therefore, every function $g(\lambda) \in Z$ which satisfies the condition $g(\sqrt{\mu}) \geqslant 0$ $(-\infty < \mu < \infty)$ can be expressed in the form $g(\lambda) = f(\lambda)\overline{f(\overline{\lambda})}$, where $f(\lambda) \in CK^2$.

PROBLEMS

1. Suppose that the distribution $R \in Z'$ assumes nonnegative values: $(g(\lambda),R) \geq 0$ for all test functions $g(\lambda) \in Z$ which satisfy the condition

$$g(\lambda) \geq 0 \quad (-\infty < \lambda < \infty) . \tag{2.1.19}$$

Show that in this case there is a nondecreasing function $\rho(\mu)$ such that

$$(f_1(\lambda)f_2(\lambda),R) = \int_0^\infty f_1(\sqrt{\mu})f_2(\sqrt{\mu})d\rho(\mu)$$

for all functions $f_1(\lambda),f_2(\lambda) \in CK^2$.

2. Show that every function $g(\lambda) \in Z$ which satisfies (2.1.19) is expressible as

$$g(\lambda) = g_1(\lambda) + \lambda^2 g_2(\lambda) ,$$

where $g\ (\lambda) \in Z$ and $g_i(\sqrt{\mu}) \geq 0$ for all $\mu \in (-\infty,\infty)$ $(i = 1,2)$.

3. Show that the distribution $R \in Z'$ satisfies the condition of Problem 1 whenever

$$(f(\lambda)\overline{f(\overline{\lambda})},R) \geq 0 ,$$

and

$$(\lambda^2 f(\lambda)\overline{f(\overline{\lambda})},R) \geq 0$$

for all functions $f(\lambda) \in CK^2$ such that $\lambda^2 f(\lambda) \in CK^2$ also.

4. Let OH denote the space of bounded linear operators acting on the separable Hilbert space H , and let $OH[0,\infty)$ designate the set of all continuous operator-valued functions $f(x)$ with compact support $(f(x) \in OH$ for all $x \in [0,\infty))$. The inversion formulas and the Parseval identity are valid for the Fourier cosine and sine transforms of functions $f(x) \in OH[0,\infty)$ defined by

$$C(\lambda,f) = \int_0^\infty f(x)\cos \lambda x dx \quad , \quad S(\lambda,f) = \int_0^\infty f(x)\sin \lambda x dx .$$

Define the Fourier ω_0- and ω_0-transforms by the rules

$$\omega_0(\lambda,f;P) = \int_0^\infty f(x)\omega_0(\lambda,x;P)dx$$

and

$$\tilde{\omega}_0(\lambda,f;P) = \int_0^\infty f(x)\tilde{\omega}_0(\lambda,x;P)dx$$

where

$$\omega_0(\lambda,x;P) = \cos \lambda xP - \sin \lambda xBP$$

and

$$\tilde{\omega}_0(\lambda,x;P) = P \cos \lambda x + BP \sin \lambda x ,$$

respectively (see Problems 6 and 9 in Section 2, Chapter 1). Check the Parseval identity

$$\int_0^\infty f(x)g(x)dx = \frac{1}{\pi} \int_{-\infty}^\infty \omega_0(\lambda,f;P)\tilde{\omega}_0(\lambda,g;P)d\lambda$$

and the inversion formulas

$$f(x) = \frac{1}{\pi} \int_{-\infty}^\infty \omega_0(\lambda,f;P)\tilde{\omega}_0(\lambda,x;P)d\lambda = \frac{1}{\pi} \int_{-\infty}^\infty \omega_0(\lambda,x;P)\tilde{\omega}_0(\lambda,f;P)dx .$$

We list the main notations and definitions used in handling operator boundary-value problems.

$K^2(-\infty,\infty)$ and $\tilde{K}^2(-\infty,\infty)$ designate the set of square-summable functions with compact support and the set of their Fourier transforms, respectively.

$Z(-\infty,\infty)$ is the space of test functions, consisting of entire functions $f(\lambda)$ of exponential type which are summable on the real axis; convergence in $Z(-\infty,\infty)$ is defined as follows: $\lim_{n\to\infty} f_n(\lambda) = f(\lambda)$ if the types σ_n of functions $f_n(\lambda)$ form a bounded set and $\lim_{n\to\infty} \int_{-\infty}^\infty |f_n(\lambda) - f(\lambda)|d\lambda = 0$.

The set of distributions on $Z(-\infty,\infty)$ is denoted by $Z'(-\infty,\infty)$. We let $OHZ'(-\infty,\infty)$ denote the set of operator-valued distributions, which by definition are homogeneous, additive, and continuous mappings $f(\lambda) \to (f(\lambda),R)$ of the space $Z(-\infty,\infty)$ into OH . The set OHZ' is defined in a similar manner.

The Fourier cosine and sine transforms of an arbitrary continuous operator-valued function $L(x)$ belong to $OHZ'(-\infty,\infty)$ and are defined by the formulas

$$(f(\lambda),C(L)) = \frac{1}{2} \int_{-\infty}^\infty [L(x) + L(-x)] \int_{-\infty}^\infty f(\lambda)e^{i\lambda x}\, d\lambda dx$$

and

$$(f(\lambda),S(L)) = \frac{1}{2}\int_{-\infty}^{\infty} [L(x) - L(-x)] \int_{-\infty}^{\infty} f(\lambda)e^{i\lambda x}d\lambda dx .$$

The operator-valued distribution $C(L)$ belongs also to the set OHZ' and acts according to the rule

$$(C(\lambda,f),C(L)) = \int_{0}^{\infty} [L(x) + L(-x)] \int_{0}^{\infty} C(\lambda,f)\cos \lambda x d\lambda dx .$$

It is often convenient to use the language of coordinates. Let $e_1,e_2,\ldots$ be an orthonormal basis of H. The matrix of the operator A in this basis will be denoted by $[A_{ij}]$. Similarly, the matrix of the operator-valued distribution $R \in OHZ'$ (or $R \in OHZ'(-\infty,\infty)$) will be denoted by $[R_{ij}]$; its entries are, by definition, distributions $R_{ij} \in Z'$ (respectively, $R_{ij} \in Z'(-\infty,\infty)$), which act by the formulas

$$(f(\lambda),R_{ij}) = (f(\lambda),R)_{ij} .$$

We let (A,R) and (A,R,B) denote the operators with the matrices $[c_{ij}]$ and $[d_{ij}]$, where

$$c_{ij} = \sum_{\alpha=1}^{\infty}(A_{i\alpha},R_{\alpha j})$$

and

$$d_{ij} = \sum_{\alpha=1}^{\infty}\sum_{\beta=1}^{\infty}(A_{i\alpha}B_{\beta j},R_{\alpha\beta}) ,$$

respectively; here it is assumed that the entries $A_{i\alpha}$ in the first formula, and the entries $A_{i\alpha}B_{\beta j}$ in the second formula, belong to the space of test functions over which the distribution R is defined.

5. The operator-valued distribution $R \in OHZ'(-\infty,\infty)$ is said to be positive if the operator $(f(\lambda),R)$ is nonnegative for every function $f(\lambda) \in Z(-\infty,\infty)$ such that $f(\lambda) \geq 0$ for all $\lambda \in (-\infty,\infty)$. Show that for every positive distribution $R \in OHZ'(-\infty,\infty)$ there is an operator-valued measure $\rho(\lambda)$ such that

$$(f(\lambda)g(\lambda),R) = \int_{-\infty}^{\infty} f(\lambda)g(\lambda)d\rho(\lambda)\cdot$$

for all $f(\lambda),g(\lambda) \in \tilde{K}^2(-\infty,\infty)$. Here $\rho(\lambda) \in OH$ for every $\lambda \in (-\infty,\infty)$, and the operators $\rho(\lambda') - \rho(\lambda)$ are nonnegative for $\lambda' > \lambda$.

Hint. Indexing the set of rational numbers as $r_1, r_2, \ldots$, we set

$$h(\lambda,r_k) = \begin{cases} 1 - (1+e^{\lambda})^{-1} , & \lambda \leqslant r_k , \\ 0 , & \lambda > r_k . \end{cases}$$

An important step in the proof is to check the following simple assertion: if the function $\chi(\lambda)$ is continuous at the point r_k and

$$\frac{c_1}{1+\lambda^2} > \chi(\lambda) > h(\lambda,r_k) + \frac{c_2}{1+\lambda^2} \quad (c_1,c_2 > 0) ,$$

then there is a function $f(\lambda) \in Z(-\infty,\infty)$ such that $\chi(\lambda) > f(\lambda) > h(\lambda,r_k)$.

Let M (N) denote the set of all functions $f \in Z(-\infty,\infty)$ (respectively $g \in Z(-\infty,\infty)$) such that

$f(\lambda) \leqslant h(\lambda,r_1)$ (respectively $h(\lambda,r_1) \leqslant g(\lambda)$) .

Let $a_1,a_2,\ldots$ be a countable dense subset of the unit sphere of the space H , let

$$\gamma_k = \inf_{g\in N} ((g(\lambda),R)a_k,a_k) ,$$

and let $g_{k,m} = g_{k,m}(\lambda) \in N$ be functions such that

$$\gamma_k > ((g_{k,m},R)a_k,a_k) - \frac{1}{2m} .$$

Then for sufficiently small $\varepsilon > 0$, the functions

$$g'_{k,m} = [g_{k,m}(\lambda) + \varepsilon^2] \frac{\cosh 1 - \cos \varepsilon\lambda}{(1+\varepsilon^2\lambda^2)(\cosh 1 - 1)}$$

satisfy the inequalities

$$\gamma_k > ((g'_{k,m},R)a_k,a_k) - \frac{1}{m} .$$

By the assertion formulated above, there is a function $f_N(\lambda) \in Z(-\infty,\infty)$ such that

$$\min_{1\leqslant k\leqslant n} g'_{k,N}(\lambda) > f_N(\lambda) > h(\lambda,r_1) ,$$

and hence

$$\gamma_k > ((f_N,R)a_k,a_k) - \frac{1}{N} \quad (k = 1,2,\ldots,N) .$$

Since the operators (f_N,R) are bounded and the set $\{a_k\}$ is dense in the unit sphere, the sequence of quadratic forms $((f_N,R)a,a)$ converges as $N \to \infty$ to a nonnegative quadratic form associated with some nonnegative operator γ ; moreover,

$$((g(\lambda),R)a,a) \geq (\gamma a,a) \geq ((f(\lambda),R)a,a) \quad (f \in M \ , \ g \in N) .$$

This guarantees that the operator-valued distribution R can be extended to a positive distribution on the set of functions of the form $f(\lambda) + c_1 h(\lambda,r_1)$: just set $(f(\lambda) + c_1 h(\lambda,r_1),R) = (f(\lambda),R) + c_1\gamma$. From here on, the proof continues just as in the scalar case.

6. Suppose that $R \in OHZ'$ is such that the operators $(f(\lambda),R)$ are nonnegative for every function $f(\lambda) \in Z$ with $f(\lambda) \geq 0$ for all $\lambda \in (-\infty,\infty)$. Show that there is an operator-valued measure $\rho(\mu)$ such that

$$(f(\lambda)g(\lambda),R) = \int_0^\infty f(\sqrt{\mu})g(\sqrt{\mu})d\rho(\mu)$$

for all functions $f(\lambda),g(\lambda) \in CK^2$, where $\rho(\mu) \in OH$ for all $\mu \in [0,\infty)$, and the operators $\rho(\mu') - \rho(\mu)$ are nonnegative if $\mu' > \mu$.

Hint. See the preceding problem.

2. DISTRIBUTION-VALUED SPECTRAL FUNCTIONS

Consider the boundary value problem generated on the half line $0 \leq x < \infty$ by the differential equation

$$-y''(x) + q(x)y(x) = \lambda^2 y(x) \qquad (2.2.1)$$

and the boundary condition

$$y'(0) - hy(0) = 0 , \qquad (2.2.2)$$

where $q(x)$ is an arbitrary complex-valued function and h is an arbitrary complex number. Since h will be fixed throughout this section, we shall omit it, for the sake of brevity, from the notation introduced in Section 2 of Chapter 2. Thus, instead of $\omega(\lambda,x;h)$ (the solution of equation (2.2.1) with initial data $\omega(\lambda,0;h) = 1$, $\omega'(\lambda,0;h) = h$), we shall write $\omega(\lambda,x)$, $K(x,t)$ and $L(x,t)$ and similarly, instead of $K(x,t;h)$ and $L(x,t;h)$, and so on.

For an arbitrary function $f(x) \in K^2$, the Fourier cosine transform $C(\lambda,f)$ and the Fourier ω-transform $\omega(\lambda,f)$ are defined by the formulas

$$C(\lambda,f) = \int_0^\infty f(x)\cos \lambda x dx$$

and

$$\omega(\lambda,f) = \int_0^\infty f(x)\omega(\lambda,x)dx .$$

From the existence of transformation operators, it follows that

$$\int_0^\infty f(x)\omega(\lambda,x)dx = \int_0^\infty \left[f(x) + \int_x^\infty f(\xi)K(\xi,x)d\xi\right] \cos \lambda x dx , \tag{2.2.3}$$

and

$$\int_0^\infty g(x)\cos \lambda x dx = \int_0^\infty \left[g(x) + \int_x^\infty g(\xi)L(\xi,x)d\xi\right] \omega(\lambda,x)dx , \tag{2.2.4}$$

where the integrals are actually taken over bounded intervals, since the functions $f(x)$ and $g(x)$ have compact support. Consequently, the following relationships hold between the Fourier ω-transform and the Fourier cosines transform of the functions $f(x),g(x) \in K^2(\sigma)$:

$$\omega(\lambda,f) = C(\lambda,\hat{f}) \quad , \quad C(\lambda,g) = \omega(\lambda,\check{g}) ,$$

where the functions $\hat{f}(x)$ and $\check{g}(x)$, which are defined by the formulas

$$\hat{f}(x) = f(x) + \int_x^\infty f(\xi)K(\xi,x)d\xi$$

and

$$\check{g}(x) = g(x) + \int_x^\infty g(\xi)L(\xi,x)d\xi$$

also belong to $K^2(\sigma)$. Hence, the set $CK^2(\sigma)$ coincides with both the set of Fourier cosine transforms of the functions in $K^2(\sigma)$, and the set of Fourier ω-transforms of the functions in $K^2(\sigma)$.

Let $f(x)$ and $g(x)$ be arbitrary elements of K^2 , and let $\omega(\lambda,f)$ and $\omega(\lambda,g)$ be their Fourier ω-transforms. To each such pair of functions $\omega(\lambda,f)$, $\omega(\lambda,g)$, we assign the number

$$R[\omega(\lambda,f),\omega(\lambda,g)] = \int_0^\infty f(x)g(x)dx .$$

This formula defines a functional of two variables, $\omega(\lambda,f)$ and $\omega(\lambda,g)$, each of which runs through CK^2 . Actually, this functional depends only on the product $\omega(\lambda,f)\omega(\lambda,g)$. Moreover, as we shall presently prove, there is a distribution $R \in Z'$ such that

$$(\omega(\lambda,f)\omega(\lambda,g),R) = R[\omega(\lambda,f),\omega(\lambda,g)] = \int_0^\infty f(x)g(x)dx .$$

First we sketch the main steps of the proof. Suppose that we succeeded in constructing a sequence of summable functions $R_n^\sigma(\lambda)$ on the half line $0 \leq \lambda < \infty$, such that the sequence

$$U_n^\sigma(x,y) = \int_0^\infty R_n^\sigma(\lambda)\omega(\lambda,x)\omega(\lambda,y)d\lambda$$

converges to Dirac's δ-function $\delta(x-y)$ in the domain $0 < x < \sigma$, $0 < y < \sigma$, as $n \to \infty$. Then, upon multiplying both sides of this equality by $f(x)g(y)$, where $f(x)$ and $g(x)$ are arbitrary elements of $CK^2(\sigma)$, and integrating with respect to both variables, we get

$$\int_0^\sigma\int_0^\sigma U_n^\sigma(x,y)f(x)g(y)dxdy = \int_0^\infty R_n^\sigma(\lambda)\omega(\lambda,f)\omega(\lambda,g)d\lambda .$$

Hence

$$\lim_{n\to\infty}\int_0^\sigma R_n^\sigma(\lambda)\omega(\lambda,f)\omega(\lambda,g)d\lambda = \lim_{n\to\infty}\int_0^\sigma\int_0^\sigma U_n^\sigma(x,y)f(x)g(y)dxdy =$$

$$= \int_0^\sigma\int_0^\sigma \delta(x-y)f(x)g(y)dxdy = \int_0^\infty f(x)g(x)dx ,$$

i.e.,

$$R[\omega(\lambda,f),\omega(\lambda,g)] = \int_0^\infty f(x)g(x)dx = \lim_{n\to\infty}\int_0^\infty R_n^\sigma(\lambda)\omega(\lambda,f)\omega(\lambda,g)d\lambda .$$

The $R_n^\sigma(\lambda)$ are obviously regular distributions which belong to Z', and if $\lim_{n\to\infty} R_n^\sigma(\lambda)$ exists in the sense of convergence of distributions, then, upon denoting it by R^σ, we get

$$(\omega(\lambda,f)\omega(\lambda,g),R^\sigma) = \int_0^\infty f(x)g(x)dx$$

for all elements $f(x),g(x)$ of $K^2(\sigma)$. Finally, if $\lim_{\sigma\to\infty} R^\sigma = R \in Z'$ exists, then the equality

$$(\omega(\lambda,f)\omega(\lambda,g),R) = \int_0^\infty f(x)g(x)dx$$

holds for all $f(x),g(x) \in K^2$, as asserted.

Now let us show how to construct the desired sequence $R_n^\sigma(\lambda)$. First of all, we remark that, in particular, we must have

$$U_n^\sigma(x,0) = \int_0^\infty R_n^\sigma(\lambda)\omega(\lambda,x)d\lambda = \delta_n(x) \to \delta(x) \quad (0 < x < \sigma) .$$

Applying here the transformation operator which takes $\omega(\lambda,x)$ into $\cos \lambda x$, we get

$$\int_0^\infty R_n^\sigma(\lambda)\cos \lambda x d\lambda = \delta_n(x) + \int_0^x L(x,t)\delta_n(t)dt \quad (0 < x < \sigma) .$$

This shows that the function $\frac{\pi}{2} R_n^\sigma(\lambda)$ must be the Fourier cosine transform of a function that coincides on the interval $0 < x < \sigma$ with $\delta_n(x) + \int_0^x L(x,t)\delta_n(t)dt$, where $\delta_n(x) \to \delta(x)$ as $n \to \infty$. Accordingly, we proceed as follows: we choose two sufficiently smooth functions $\delta_n(x)$ and $\gamma_\sigma(x)$ subject to the conditions

$$\int_0^\infty \delta_n(x)dx = 1 ,$$

$$\delta_n(x) = 0 \quad \text{for} \quad x = 0 \quad \text{and} \quad x \geq \frac{1}{n} ,$$

$$\delta_n(x) \geq 0 \quad \text{for} \quad 0 < x < \frac{1}{n} ,$$

$$\gamma_\sigma(x) = 1 \quad \text{for} \quad 0 \leq x \leq 2\sigma ,$$

$$\gamma_\sigma(x) = 0 \quad \text{for} \quad x \geq 2\sigma+1 ,$$

and then set

$$\frac{\pi}{2} R_n^\sigma(\lambda) = \int_0^\infty \left[\delta_n(x) + \int_0^x L(x,t)\delta_n(t)dt\right]\gamma_\sigma(x)\cos \lambda x dx . \tag{2.2.5}$$

Since the function $\left[\delta_n(x) + \int_0^x L(x,t)\delta_n(t)dt\right]\gamma_\sigma(x)$ has compact support and is continuously differentiable, the function $R_n^\sigma(\lambda)$ is bounded and summable on the half line $0 \leq \lambda < \infty$. Hence, the integral

$$\int_0^\infty R_n^\sigma(\lambda)\cos \lambda x d\lambda = \left[\delta_n(x) + \int_0^x L(x,t)\delta_n(t)dt\right]\gamma_\sigma(x)$$

converges absolutely. Now applying the transformation operator which takes $\cos \lambda x$ into $\omega(\lambda,x)$ to both sides of the last equality, and recalling that $\gamma_\sigma(x) = 1$ for $0 \leq x \leq 2\sigma$, we get

$$\int_0^\infty R_n^\sigma(\lambda)\omega(\lambda,x)d\lambda = \delta_n(x) \quad (0 \leq x \leq 2\sigma) .$$

Next, let

$$U_n^\sigma(x,y) = \int_0^\infty R_n^\sigma(\lambda)\omega(\lambda,x)\omega(\lambda,y)d\lambda \ . \tag{2.2.6}$$

The functions $\int_0^N R_n^\sigma(\lambda)\omega(\lambda,x)\omega(\lambda,y)d\lambda$ are twice continuously differentiable and satisfy the differential equation

$$v''_{xx} - q(x)v = v''_{yy} - q(y)v$$

as well as the initial conditions

$$v(x,0) = \int_0^N R_n^\sigma(\lambda)\omega(\lambda,x)d\lambda = \delta_n^N(x) \ ,$$

$$v'_y(x,0) = h\int_0^N R_n^\sigma(\lambda)\omega(\lambda,x)d\lambda = h\delta_n^N(x)$$

$(0 \leq x \leq 2\sigma)$; moreover, $\lim_{N\to\infty} \delta_n^N(x) = \delta(x)$. Hence, by Riemann's formula (1.1.7)

$$\int_0^N R_n(\lambda)\omega(\lambda,x)\omega(\lambda,y)d\lambda = \frac{\delta_n^N(x+y) + \delta_n^N(x-y)}{2} + \int_{x-y}^{x+y} W(x,y,t)\delta_n^N(t)dt$$

for $0 \leqslant y \leqslant x \leqslant \sigma$, where $W(x,y,t)$ is a continuous function. Letting $N \to \infty$, we get

$$U_n^\sigma(x,y) = \frac{1}{2}\,[\delta_n(x+y) - \delta_n(x-y)] + \int_{x-y}^{x+y} W(x,y,t)\delta_n(t)dt$$

for $0 \leq y \leq x \leq \sigma$. From the definition (2.2.6) of the functions $U_n^\sigma(x,y)$ it follows that $U_n^\sigma(x,y) = U_n^\sigma(y,x)$. Hence, we can write

$$U_n^\sigma(x,y) = \frac{1}{2}\,[\delta_n(x+y) + \delta_n(|x-y|)] + \theta_n(x,y)$$

in the whole domain $0 \leq x \leq \sigma$, $0 \leq y \leq \sigma$, where the function $\theta_n(x,y)$ is symmetric in the variables x , y , and is defined for $0 \leq y \leq x \leq \sigma$ by

$$\theta_n(x,y) = \int_{x-y}^{x+y} W(x,y,t)\delta_n(t)dt \ .$$

The function $W(x,y,t)$ is bounded in every bounded domain. Consequently, there is a constant $C(\sigma)$, depending only on σ , such that

$$|W(x,y,t)| \leq C(\sigma) \quad (0 \leq y \leq x \leq \sigma \ , \ 0 \leq t \leq 2\sigma) \ ,$$

and hence

$$|\theta_n(x,y)| \leqslant C(\sigma) \int_0^\infty \delta_n(t)dt = C(\sigma) \quad (0 \leqslant y \leqslant x \leqslant \sigma) .$$

Furthermore, $\theta_n(x,y) = 0$ for $|x - y| > \frac{1}{n}$, because $\delta_n(t) = 0$ for $t > \frac{1}{n}$. From these estimates it follows that, if the functions $f(x)$ and $g(x)$ belong to $K^2(\sigma)$, then

$$\int_0^\infty \int_0^\infty U_n^\sigma(x,y)f(x)g(y)dxdy =$$

$$= \int_0^\infty \int_0^\infty \{\frac{1}{2} [\delta_n(x+y) + \delta_n(|x-y|)] + \theta_n(x,y)\}f(x)g(y)dxdy ;$$

moreover,

$$\left| \int_0^\infty \int_0^\infty \theta_n(x,y)f(x)g(y)dxdy \right| \leqslant C(\sigma) \iint_{D_n} |f(x)g(y)|dxdy ,$$

where $D_n = \{(x,y) : |x-y| \leqslant \frac{1}{n}, 0 \leqslant x \leqslant \sigma , 0 \leqslant y \leqslant \sigma\}$. Since the measure of the domain D_n tends to zero as $n \to \infty$ and the function $|f(x)g(y)|$ is summable,

$$\lim_{n\to\infty} \left| \int_0^\infty \int_0^\infty \theta_n(x,y)f(x)g(y)dxdy \right| = 0 ,$$

whence

$$\lim_{n\to\infty} \int_0^\infty R_n^\sigma(\lambda)\omega(\lambda,f)\omega(\lambda,g)d\lambda =$$

$$= \lim_{n\to\infty} \int_0^\infty \int_0^\infty U_n^\sigma(x,y)f(x)g(y)dxdy =$$

$$= \lim_{n\to\infty} \int_0^\infty \int_0^\infty \frac{1}{2} [\delta_n(x+y) + \delta_n(|x-y|)]f(x)g(y)dxdy =$$

$$= \int_0^\infty f(x)g(x)dx ,$$

i.e.,

$$\int_0^\infty f(x)g(x)dx = \lim_{n\to\infty} \int_0^\infty R_n^\sigma(\lambda)\omega(\lambda,f)\omega(\lambda,g)d\lambda \tag{2.2.7}$$

for all functions $f(x),g(x) \in K^2(\sigma)$.

It follows from the definition (2.2.5) of $R_n^\sigma(\lambda)$ that

$$\lim_{n\to\infty} R_n^\sigma(\lambda) = \frac{2}{\pi}(1 + C(\lambda,\gamma_\sigma L)) = R^\sigma ,$$

where $C(\lambda,\gamma_\sigma L)$ is the Fourier cosine transform of the function $\gamma_\sigma(x)L(x,0)$. Since $\gamma_\sigma(x)L(x,0)$ tends uniformly to $L(x,0)$ on every bounded interval as $\sigma \to \infty$, we deduce from property 1 of the Fourier transform (see Section 1) that $C(\lambda,\gamma_\sigma L)$ tends (in the sense of distributions) to the Fourier cosine transform $C(L)$ of the function $L(x,0)$ as $\sigma \to \infty$. Thus,

$$\lim_{\sigma\to\infty} \{\lim_{n\to\infty} R_n^\sigma(\lambda)\} = \frac{2}{\pi}(1 + C(L)) = R \in Z' ,$$

in which both limits exist in the sense of convergence of distributions. From this it follows, in view of (2.2.7), that the equality

$$\int_0^\infty f(x)g(x)dx = (\omega(\lambda,f)\omega(\lambda,g),R)$$

actually holds for all functions $f(x),g(x) \in K^2$.

We have thus proved the following basic result:

THEOREM 2.2.1. *To the boundary value problem* (2.2.1), (2.2.2) *there corresponds a distribution* $R \in Z'$ *such that*

$$\int_0^\infty f(x)g(x)dx = (\omega(\lambda,f)\omega(\lambda,g),R) , \tag{2.2.8}$$

where $f(x),g(x)$ *are arbitrary elements of* $L_2[0,\infty)$ *and* $\omega(\lambda,f),\omega(\lambda,g)$ *designate their Fourier* ω*-transforms.*

The distribution R *is connected with the kernel* $L(x,t)$ *of the transformation operator taking* $\omega(\lambda,x)$ *into* $\cos \lambda x$ *by the formula*

$$R = \frac{2}{\pi}(1 + C(L)) , \tag{2.2.9}$$

where $C(L)$ *is the Fourier cosine transform of the function* $L(x,0)$. □

Equality (2.2.8) is an analog of the Parseval identity. For this reason we shall call the distribution $R \in Z'$ a generalized (distribution) spectral function of the boundary value problem (2.2.1), (2.2.2).

We now show how to deduce the analog of the eigenfunction expansion for the boundary value problem (2.2.1), (2.2.2) from Theorem 2.2.1. According to (2.2.8),

$$\frac{1}{\delta}\int_x^{x+\delta} f(t)dt = \left(\omega(\lambda,f)\,\frac{1}{\delta}\int_x^{x+\delta}\omega(\lambda,t)dt,R\right), \tag{2.2.10}$$

for all $f(x) \in K^2$. If the Fourier ω-transform $\omega(\lambda,f)$ of the function $f(x)$ belongs to Z (i.e., if it is summable on the real line), then the product $\omega(\lambda,f)\,\frac{1}{\delta}\int_x^{x+\delta}\omega(\lambda,t)dt$ converges in Z to the function $\omega(\lambda,f)\omega(\lambda,x)$ as $\delta \to 0$. Hence,

$$\lim_{\delta\to 0}\left(\omega(\lambda,f)\,\frac{1}{\delta}\int_x^{x+\delta}\omega(\lambda,t)dt,R\right) = (\omega(\lambda,f)\omega(\lambda,x),R). \tag{2.2.11}$$

On the other hand, according to (2.2.3), $\omega(\lambda,f)$ is the Fourier cosine transform of the function $f(x) + \int_x^\infty f(\xi)K(\xi,x)d\xi$, and hence

$$\frac{2}{\pi}\int_0^\infty \omega(\lambda,f)\cos\lambda x d\lambda = f(x) + \int_x^\infty f(\xi)K(\xi,x)d\xi .$$

From this equality and the summability of the function $\omega(\lambda,f)$ on the real line it follows that the function $f(x) + \int_x^\infty f(\xi)K(\xi,x)d\xi$, and hence $f(x)$, are continuous. Consequently,

$$\lim_{\delta\to 0}\frac{1}{\delta}\int_x^{x+\delta} f(t)dt = f(x)$$

whence, in view of (2.2.10) and (2.2.11),

$$f(x) = (\omega(\lambda,f)\omega(\lambda,x),R). \tag{2.2.12}$$

COROLLARY. *Formula* (2.2.12) *holds for all functions* $f(x) \in K^2$ *whose Fourier* ω-*transform* $\omega(\lambda,f)$ *belongs to the space* Z (*i.e., is summable on the real line*).

□

The most important and most studied boundary value problem is (2.2.1), (2.2.) with a real valued function $q(x)$ and a real h. From the viewpoint of the theory of operators in Hilbert spaces, this is the case in which the boundary value problem is generated by a symmetric operator that admits self-adjoint extensions (which are not necessarily unique). Such problems will be referred to as symmetric.

In the symmetric case the function $\omega(\lambda,x)$ has the property that $\omega(\overline{\lambda},x) = \overline{\omega(\lambda,x)}$. It follows that if $f(\lambda) \in CK^2$ is the Fourier ω-transform of the function $g(x)$, then

$$\overline{f(\overline{\lambda})} = \int_0^\infty \overline{g(x)\omega(\overline{\lambda},x)}dx = \int_0^\infty \overline{g(x)}\omega(\lambda,x)dx$$

is the Fourier ω-transform of the function $\overline{g(x)}$, and hence, by (2.2.8),

$$(f(\lambda)\overline{f(\overline{\lambda})},R) = \int_0^\infty g(x)\overline{g(x)}dx \geqslant 0$$

for every function $f(\lambda) \in K^2$. Using Lemma 2.1.1 and Theorem 2.1.1, we conclude that there is a nondecreasing function $\rho(\mu)$ $(-\infty < \mu < \infty)$ such that

$$(f(\lambda)g(\lambda),R) = \int_{-\infty}^\infty f(\sqrt{\mu})g(\sqrt{\mu})d\rho(\mu) .$$

We thus proved the following result:

THEOREM 2.2.2. *If the boundary value problem* (2.2.1), (2.2.2) *is symmetric* (*i.e.*, $q(x)$ *and* h *are real*), *then there exists a nondecreasing function* $\rho(\mu)$ $(-\infty < \mu < \infty)$ *such that*

$$\int_0^\infty f(x)g(x)dx = \int_{-\infty}^\infty \omega(\sqrt{\mu},f)\omega(\sqrt{\mu},g)d\rho(\mu) \tag{2.2.13}$$

for all functions $f(x),g(x) \in K^2$.

□

For every function $f(x) \in L_2[0,\infty)$ we put

$$\omega_n(\lambda,f) = \int_0^n f(x)\omega(\lambda,x)dx .$$

By the theorem we have just proved,

$$\int_{-\infty}^\infty |\omega_n(\sqrt{\mu},f) - \omega_m(\sqrt{\mu},f)|^2 d\rho(\mu) = \int_n^m |f(x)|^2 dx ,$$

which shows that the functions $\omega_n(\sqrt{\mu},f)$ converge in the metric of the space $L_{2,\rho}(-\infty,\infty)$ to a function $\omega(\sqrt{\mu},f)$; moreover,

$$\int_{-\infty}^\infty |\omega(\sqrt{\mu},f)|^2 d\rho(\mu) = \int_0^\infty |f(x)|^2 dx .$$

Hence, formula (2.2.13) is valid for all functions $f(x),g(x) \in L_2[0,\infty)$.

Now let

$$f_N(x) = \int_{-N}^N \omega(\sqrt{\mu},f)\omega(\sqrt{\mu},x)d\rho(\mu) .$$

Then for every $g(x) \in K^2$,

$$\left| \int_0^\infty \{f_N(x) - f(x)\}g(x)dx \right| =$$

$$= \left| \int_{-N}^{N} \omega(\sqrt{\mu},f)\omega(\sqrt{\mu},g)d\rho(\mu) - \int_{-\infty}^{\infty} \omega(\sqrt{\mu},f)\omega(\sqrt{\mu},g)d\rho(\mu) \right| =$$

$$= \left| \int_{|\mu|>N} \omega(\sqrt{\mu},f)\omega(\sqrt{\mu},g)d\rho(\mu) \right| \leq$$

$$\leq \left[\int_{-\infty}^{\infty} |\omega(\sqrt{\mu},g)|^2 d\rho(\mu) \cdot \int_{|\mu|>N} |\omega(\sqrt{\mu},f)|^2 d\rho(\mu) \right]^{\frac{1}{2}} .$$

Setting here

$$g(x) = \begin{cases} \overline{f_N(x) - f(x)} , & 0 \leq x \leq n , \\ 0 , & x > n , \end{cases}$$

we thus see that

$$\int_0^n |f_N(x) - f(x)|^2 dx \leq \left[\int_0^n |f_N(x) - f(x)|^2 dx \cdot \int_{|\mu|>N} |\omega(\sqrt{\mu},f)|^2 d\rho(\mu) \right]^{\frac{1}{2}},$$

or, equivalently, that

$$\int_0^h |f_N(x) - f(x)|^2 dx \leq \int_{|\mu|>N} |\omega(\sqrt{\mu},f)|^2 d\rho(\mu) .$$

Letting $n \to \infty$, we get

$$\int_0^\infty |f_N(x) - f(x)|^2 dx \leq \int_{|\mu|>N} |\omega(\sqrt{\mu},f)|^2 d\rho(\mu) .$$

Therefore, $f_N(x) \in L_2[0,\infty)$, and

$$\lim_{N\to\infty} \int_0^\infty |f_N(x) - f(x)|^2 dx = 0 .$$

From this we obtain the following classical theorem of Weyl.

THEOREM 2.2.3. *To every symmetric boundary value problem* (2.2.1), (2.2.2) *there corresponds at least one nondecreasing function* $\rho(\mu)$ $(-\infty < \mu < \infty)$ *such that the following inversion formulas*

$$\omega(\sqrt{\mu},f) = \int_0^\infty f(x)\omega(\sqrt{\mu},x)dx$$

and

$$f(x) = \int_{-\infty}^\infty \omega(\sqrt{\mu},f)\omega(\sqrt{\mu},x)d\rho(\mu)$$

(where the integrals converge in the metrics of the spaces $L_{2,\rho}(-\infty,\infty)$ *and* $L_2[0,\infty)$ *, respectively) and the Parseval equality*

$$\int_0^\infty f(x)\overline{g(x)}dx = \int_{-\infty}^\infty \omega(\sqrt{\mu},f)\overline{\omega(\sqrt{\mu},g)}d\rho(\mu)$$

hold for every function $f(x) \in L_2[0,\infty)$.

□

PROBLEMS

1. Prove the existence of a distribution spectral function for the boundary value problem

$$-y'' + q(x)y = \lambda^2 y \quad (0 \leqslant x < \infty) \quad , \quad y(0) = 0 . \tag{2.2.14}$$

Hint. Instead of (2.2.6), consider the sequence

$$U_n^\sigma(x,y) = \int_0^\infty f^2 R_n^\sigma(\lambda)\omega(\lambda,x;\infty)\omega(\lambda,y;\infty)d\lambda ,$$

in which the functions $R_n^\sigma(\lambda)$ are determined from the condition

$$\int_0^\infty \lambda^2 R_n^\sigma(\lambda) \frac{\sin \lambda x}{\lambda} d\lambda = - \frac{d}{dx}\left\{\gamma_\sigma(x) \left[\delta_n(x) + \int_0^x d\xi \int_0^\infty L(\xi,t;\infty)\delta_n'(t)dt\right]\right\} .$$

When $n \to \infty$, $\sigma \to \infty$, the regular distributions R_n^σ converge in the space Z' and

$$\lim_{\sigma\to\infty} \{\lim_{n\to\infty} R_n^\sigma\} = R = \frac{2}{\pi} [1 + C(M)] ,$$

where $C(M)$ is the Fourier cosine transform of the function

$$M(x) = - \int_0^x L_t'(u,0;\infty)du .$$

Moreover,

$$\int_0^\infty f_1(x)f_2(x)dx = (\lambda^2\omega(\lambda,f_1,\infty)\omega(\lambda,f_2,\infty),R) = (\omega(\lambda,f_1,\infty)\omega(\lambda,f_2,\infty),\lambda^2R)$$

for all $f_1(x),f_2(x) \in K^2$, where

$$\omega(\lambda,f;\infty) = \int_0^\infty f(x)\omega(\lambda,x;\infty)dx .$$

Hence, λ^2R is a distribution spectral function of the boundary value problem (2.2.14).

2. Generalize Theorem 2.2.3 to symmetric boundary value problems with the boundary condition $y(0) = 0$.

3. Show that for every symmetric boundary-value problem (2.2.1), (2.2.2) such that $16q(x) \geq 9(|h| - h)^2$, the distribution spectral functions R are generated by measures $d\rho(\mu)$ which are supported on the positive half line:

$$\omega(\lambda,f;h)\omega(\lambda,g;h),R) = \int_0^\infty \omega(\sqrt{\mu},f;h)\omega(\sqrt{\mu},g;h)d\rho(\mu) .$$

Hint. A straightforward verification will convince you that if $f(\lambda) \in Z$ and $\lambda^2f(\lambda) \in Z$, then the function $y(x) = (f(\lambda)\omega(\lambda,x;h),R)$ has compact support, is twice continuously differentiable, and that $y'(0) - hy(0) = 0$, and

$$-y''(x) + q(x)y(x) = (\lambda^2f(\lambda)\omega(\lambda,x;h),R) .$$

Consequently,

$$(\lambda^2f(\lambda)\overline{f(\overline{\lambda})},R) = \int_0^\infty \{-y''(x) + q(x)y(x)\}\overline{y(x)}dx =$$

$$= h|y(0)|^2 + \int_0^\infty |y'(x)|^2dx + \int_0^\infty q(x)|y(x)|^2dx \geq$$

$$\geq h|y(0)|^2 + \int_0^\infty |y'(x)|^2dx + \frac{9}{16}(|h|-h)^2\int_0^\infty |y(x)|^2dx ,$$

so that $(\lambda^2f(\lambda)\overline{f(\overline{\lambda})},R) \geq 0$ if $h \geq 0$. If $h < 0$, then

$$(\lambda^2f(\lambda)\overline{f(\overline{\lambda})},R) \geq -|h||y(0)|^2 + \int_0^\infty |y'(x)|^2dx + \frac{9}{4}|h|^2\int_0^\infty |y(x)|^2dx ,$$

which in turn implies, in view of the inequality

$$|y(x)| \geq |y(0)| - \int_0^x |y'(t)|dt \geq |y(0)| - \left[\int_0^\infty |y'(t)|^2 dt \right]^{\frac{1}{2}} \sqrt{x} ,$$

that in this case,

$$(\lambda^2 f(\lambda)\overline{f(\overline{\lambda})},R) \geq -|h||y(0)|^2 + \int_0^\infty |y'(x)|^2 dx + \frac{|hy(0)|^2 |y(0)|^2}{4 \int_0^\infty |y'(x)|^2 dx} \geq 0 .$$

For the rest, you can use the results of Problems 1 and 3 in Section 1.

4. Show that the Sturm-Liouville boundary value problem on the full line

$$-y'' + q(x)y = \lambda^2 y \quad (-\infty < x < \infty)$$

has a distribution spectral matrix

$$R = \begin{bmatrix} R_{11} & R_{12} \\ R_{21} & R_{22} \end{bmatrix} \quad (R_{11}, R_{12}, R_{21} \in Z' \quad ; \quad R_{22} = \lambda^2 R'_{22} \quad , \quad R'_{22} \in Z') ,$$

such that for any pair of functions $f_1(x), f_2(x) \in K(-\infty,\infty)$,

$$\int_{-\infty}^\infty f_1(x) f_2(x) dx = \vec{\omega}(\lambda,f_1) R \vec{\omega}(\lambda,f_2) ;$$

in the right-hand side of this equality, the vector $\vec{\omega}(\lambda,f) =$ $= \{\omega(\lambda,f;0), \omega(\lambda,f;\infty)\}$ is regarded as a single-row matrix, and hence

$$\vec{\omega}(\lambda,f_1) R \vec{\omega}(\lambda,f_2) =$$

$$= (\omega(\lambda,f_1;0)\omega(\lambda,f_2;0),R_{11}) + (\omega(\lambda,f_1;0)\omega(\lambda,f_2;\infty),R_{12}) +$$

$$+ (\omega(\lambda,f_1;\infty)\omega(\lambda,f_2;0),R_{21}) + (\omega(\lambda,f_1;\infty)\omega(\lambda,f_2;\infty),R_{22}) .$$

The distributions R_{jk} are connected with the kernel $L(x,t)$ of the transformation operator taking $e_0(\lambda,x)$ into $e^{i\lambda x}$ through the formulas

$$R_{11} = \frac{1}{2\pi}(1 + C(L)) , \quad R_{12} = \frac{1}{2\pi} C(L'_t) ,$$

$$R_{21} = \frac{1}{2\pi} C(L'_x) , \quad R'_{22} = \frac{1}{2\pi}(1 + C(M)) ,$$

where

$L = L(x,0)$, $L'_x = L'_x(x,0)$, $L'_t = L'_t(x,0)$,

and

$$M = -\int_0^x L'_t(\xi,0)d\xi .$$

5. Show that the operator Sturm-Liouville problem has a distribution spectral matrix $R = [R_{ik}]$ with entries $R_{ik} \in Z'$, such that

$$\int_0^\infty f(x)g(x)dx = [\omega(\lambda,f;h)R\tilde{\omega}(\lambda,g;h)] ,$$

where $f(x),g(x) \in OH(0,\infty)$ and, by definition,

$$[F(\lambda)RG(\lambda)] = \left[\sum_{j,\ell=1}^n (F_{ij}(\lambda)G_{\ell k}(\lambda),R_{j\ell})\right] .$$

The spectral matrix R is connected with the kernel $L(x,t)$ $(\tilde{L}(x,t))$ of the transformation operator taking $\omega(\lambda,x;h)$ into $I\cos\lambda x$ through the formula

$$R = \frac{2}{\pi}\{I + C(L)\} = \frac{2}{\pi}\{I + C(\tilde{L})\} .$$

Hint. See the next problem.

6. Show that the operator Dirac problem has a distribution spectral matrix $R = [R_{ik}]$, such that $PRP = R$, $R_{ik} \in Z'(-\infty,\infty)$, and

$$\int_0^\infty f(x)g(x)dx = [\omega(\lambda,f;P)PRP\tilde{\omega}(\lambda,g;P)] .$$

Moreover,

$$R = \frac{1}{\pi}(P + \tilde{\omega}_0(L;P)) = \frac{1}{\pi}(P + \omega_0(\tilde{L};P)) .$$

Hint. It follows from the existence of the transformation operators (1.2.39') that

$$\omega(\lambda,f;P) = \omega_0(\lambda,F;P) \quad , \quad \tilde{\omega}(\lambda,g;P) = \tilde{\omega}_0(\lambda,G;P)$$

if

$$f(x) = F(x) + \int_x^\infty F(t)L_p(t,x)dt \quad \text{and} \quad g(x) = G(x) + \int_x^\infty \tilde{L}_p(t,x)dt .$$

Furthermore,

$$\int_0^\infty f(x)g(x)dx = \int_0^\infty F(x)G(x)dx +$$

$$+ \int_0^\infty \int_0^\infty F(x) \left[L_p(x,y) + \tilde{L}_p(x,y) + \int_0^\infty L_p(x,\xi)\tilde{L}_p(y,\xi)d\xi \right] G(y)dxdy ,$$

and according to the results of Problem 9, Section 2, Chapter 1,

$$\int_0^\infty f(x)g(x)dx = \int_0^\infty F(x)G(x)dx + \int_0^\infty \int_0^\infty F(x)f(x,y)G(y)dxdy ,$$

where the operator-valued function $f(x,y)$ is defined by (1.2.46). It is readily verified that if the operator-valued functions with compact support $\varphi(x)$ and $\tilde{\varphi}(x)$ satisfy the conditions

$$\varphi(x)P = \varphi(x) \quad , \quad P\tilde{\varphi}(x) = \tilde{\varphi}(x) ,$$

and

$$\varphi(x) + B\varphi(x) = \tilde{\varphi}(x) + B\tilde{\varphi}(x)B ,$$

then $P\tilde{\omega}_0(\lambda,\varphi;P) = \omega_0(\lambda,\tilde{\varphi};P)P$. Formula (1.2.47) shows that all these conditions are satisfied by the functions $\varphi_\sigma(x) = \gamma_\sigma(x)L_p(x,0)$, $\tilde{\varphi}_\sigma(x) = \gamma_\sigma(x)\tilde{L}_p(x,0)$. Hence, the operator-valued function

$$f_\sigma(x,y) = \frac{1}{\pi} \int_{-\infty}^\infty \omega_0(\lambda,x;P)\tilde{\omega}_0(\lambda,\varphi_\sigma;P)\tilde{\omega}_0(\lambda,y;P)d\lambda =$$

$$= \frac{1}{\pi} \int_{-\infty}^\infty \omega_0(\lambda,x;P)\omega(\lambda,\varphi_\sigma;P)\tilde{\omega}_0(\lambda,y;P)d\lambda$$

satisfies the equation

$$Bu_x + u_yB = 0$$

and the conditions

$$u(x,0) = \gamma_\sigma(x)L_p(x,0) \quad , \quad u(0,y) = \gamma_\sigma(y)\tilde{L}_p(y,0)$$

for all positive values of x and y (see Problem 4, Section 1). Hence, $f_\sigma(x,y) \equiv f(x,y)$ in the square $0 \leqslant x \leqslant \sigma$, $0 \leqslant y \leqslant \sigma$, and

$$\int_0^\infty f(x)g(x)dx = \frac{1}{\pi} \int_{-\infty}^\infty \omega_0(\lambda,F;P)\{I + \tilde{\omega}_0(\lambda,\varphi_\sigma;P)\}\tilde{\omega}_0(\lambda,G;P)d\lambda =$$

$$= \frac{1}{\pi} \int_{-\infty}^\infty \omega(\lambda,f;P)\{P + \omega_0(\lambda,\tilde{\varphi}_\sigma;P)\}\tilde{\omega}(\lambda,g;P)d\lambda$$

if $f(x) = g(x) = 0$ for $x > \sigma$. Letting $\sigma \to \infty$, you get the needed result.

7. Let h be a bounded self-adjoint operator acting in the Hilbert space H, and let h_1 be the infimum of the values of the quadratic form (hf,f) on the unit sphere $(f,f) = 1$ of H. Show that the distribution spectral matrices R of any symmetric $(q(x) = q(x)^*$, $h = h^*)$ operator Sturm-Liouville boundary value problem for which $16(q(x)f,f) - 9(|h_1|-h_1)^2(f,f) \geq 0$ satisfy the inequalities

$$(f(\lambda)IRI\overline{f(\bar{\lambda})} \geq 0 \quad , \quad (\lambda^2 g(\lambda)IRI\overline{g(\bar{\lambda})}) \geq 0$$

(of course, you have to assume that the functions $f(\lambda)$, $g(\lambda)$, and $\lambda^2 g(\lambda)$ belong to CK^2).

8. Show that the distribution spectral matrices R of the symmetric operator Sturm-Liouville boundary value problems are generated by operator-valued measures $d\rho(\mu)$:

$$[\omega(\lambda,f)\tilde{R}\omega(\lambda,g)] = \int_0^\infty f(x)g(x)dx = \int_{-\infty}^\infty \omega(\sqrt{\mu},f)d\rho(\mu)\tilde{\omega}(\sqrt{\mu},g) ,$$

where $\rho(\mu) = \rho(\mu)^* \in OH$ for every $\mu \in (-\infty,\infty)$, and the operator $\rho(\mu') - \rho(\mu)$ is nonnegative for $\mu' > \mu$.

<u>Hint</u>. First suppose that the conditions of the previous problem are satisfied. Using Problems 1, 3, and 6 of Section 1, you can prove that in this case

$$\int_0^\infty f(x)g(x)dx = \int_0^\infty \omega(\sqrt{\mu},f)d\rho(\mu)\tilde{\omega}(\sqrt{\mu},g) .$$

The change of variable $\mu = \mu' + a^2$ in this equality shows that

$$\int_0^\infty f(x)g(x)dx = \int_{-a^2}^\infty \omega(\sqrt{\mu},f)d\rho(\mu)\tilde{\omega}(\sqrt{\mu},g)$$

provided that

$$16(q(x)f,f) - 9(|h_1| - h_1)^2(f,f) \geq -a^2(f,f) . \qquad (2.2.15)$$

In the general case you can approximate $q(x)$ by an operator-valued function which satisfies conditions (2.2.15), and then subsequently let $a \to \infty$.

9. Show that for every symmetric $(q(x)$ real$)$ Sturm-Liouville boundary value problem on the full line, each spectral matrix R is generated by a matrix-valued measure $d\rho(\mu)$:

$$\int_{-\infty}^{\infty} f(x)g(x)dx = \int_{-\infty}^{\infty} \vec{\omega}(\sqrt{\mu},f)d\rho(\mu)\vec{\tilde{\omega}}(\sqrt{\mu},g)$$

(the matrices $\rho(\mu') - \rho(\mu)$ are nonnegative for $\mu' > \mu$).

10. Show that for every symmetric ($\Omega(x) = \Omega(x)^*$, $B^* = -B$) operator Dirac boundary value problem, each distribution spectral function R is generated by an operator-valued measure $d\rho(\mu)$:

$$\int_{0}^{\infty} f(x)g(x)dx = \int_{-\infty}^{\infty} \omega(\lambda,f;P)d\rho(\lambda)\tilde{\omega}(\lambda,g;P)$$

(the operators $\rho(\lambda') - \rho(\lambda)$ are nonnegative for $\lambda' > \lambda$, and $P\rho(\lambda)P = \rho(\lambda)$).

Hint. If $F(\lambda) \in Z(-\infty,\infty)$ and $F(\lambda) \geqslant 0$, then $F(\lambda) = G(\lambda)\overline{G(\lambda)}$, where

$$G(\lambda) = \int_{0}^{\infty} G(x)e^{-i\lambda x}dx = C(\lambda,G) - iS(\lambda,G) ,$$

and

$$\overline{G(\lambda)} = \int_{0}^{\infty} G(x)e^{-i\lambda x}dx = C(\lambda,\overline{G}) + iS(\lambda,\overline{G}) .$$

Hence,

$$(F(\lambda),R) = (C(\lambda,G)C(\lambda,\overline{G}),R) - i(S(\lambda,G)C(\lambda,\overline{G}),R) +$$
$$+ i(S(\lambda,\overline{G})C(\lambda,G),R) + (S(\lambda,G)S(\lambda,\overline{G}),R) .$$

On the other hand, the operator

$$A = (\omega_0(\lambda,G;P)R\tilde{\omega}_0(\lambda,\overline{G};R)) = \int_{0}^{\infty} g(x)g(x)^*dx$$

is nonnegative. From the equalities

$$A = (\omega_0(\lambda,G;P)R\tilde{\omega}_0(\lambda,\overline{G};P)) =$$
$$= (\{C(\lambda,G)P - S(\lambda,G)BP\}R\{PC(\lambda,\overline{G}) + PBS(\lambda,\overline{G})\}) =$$
$$= (C(\lambda,G)C(\lambda,\overline{G}),R) - B(S(\lambda,G)C(\lambda,\overline{G}),R) +$$
$$+ (S(\lambda,\overline{G})C(\lambda,G),R)B + B(S(\lambda,G)S(\lambda,\overline{G}),R)B$$

you find, upon observing that $PRP = R$ and $PBP = 0$, that

$$PAP = (C(\lambda,G)C(\lambda,\overline{G}),R) ,$$

$(I - P)AP = -B(S(\lambda,G)C(\lambda,\overline{G}),R)$,

$PA(I - P) = (S(\lambda,\overline{G})C(\lambda,G)R)$,

and

$(I - P)A(I - P) = B(S(\lambda,G)S(\lambda,\overline{G}),R)B$.

This implies further that

$(F(\lambda),R) = PAP - iB(I - P)AP - iPA(I - P)B + B(I - P)A(I - P)B =$

$= P(A - iBA + iAB + BAB)P = P(I - iB)A(I - iB)P$.

Therefore the operator $(F(\lambda),R)$ is nonnegative whenever $F(\lambda) \geqslant 0$. Now you can use Problem 5 of Section 1.

3. THE INVERSE PROBLEM

The questions considered in the previous section belong to the class of inverse problems of spectral analysis: in them, for a given boundary value problem, we have looked for a spectral function yielding expansion formulas. The term "inverse problems of spectral analysis" is used for those problems in which, given some spectral data, it is required to retrieve properties of the original operator, or the operator itself. The problem considered in this section is that of retrieving a boundary value problem of the form (2.2.1), (2.2.2) from its spectral function. First we shall establish the conditions that the distribution spectral function R of such a boundary value problem must satisfy.

LEMMA 2.3.1. *The spectral function* R *of the boundary value problem* (2.2.1), (2.2.2) *enjoys the following properties:*

1) if $f(\lambda) \in CK^2(\sigma)$ *and* $(f(\lambda)y(\lambda),R) = 0$ *for all* $y(\lambda) \in CK^2(\sigma)$, *then* $f(\lambda) \equiv 0$;

2) the function

$$\Phi(x) = \left(\frac{1 - \cos \lambda x}{\lambda^2} ,R \right) \quad (0 < x < \infty)$$

is thrice continuously differentiable, $\Phi'(0+) = 1$, *and* $\Phi''(0+) = -h$.

PROOF. Let $\hat{f}(x),\hat{y}(x) \in K^2(\ )$, and let $f(\lambda)$ and $y(\lambda)$ denote their Fourier ω-transforms. Then

$$(f(\lambda)y(\lambda),R) = \int_0^\infty \hat{f}(x)\hat{y}(x)dx ,$$

and since $\hat{y}(x)$ exhausts $K^2(\sigma)$ as $y(\lambda)$ runs through $CK^2(\sigma)$, it follows that the equality $(f(\lambda)y(\lambda),R) = 0$ can hold for all $y(x) \in CK^2(\sigma)$ if and only if $\hat{f}(x)$, and hence $f(\lambda)$, vanish identically. This proves assertion 1.

Next recall that $\lambda^{-2}(1 - \cos \lambda x)$ is the Fourier cosine transform of the function

$$\varphi(t) = \begin{cases} x-t , & 0 \leqslant t \leqslant x , \\ 0 , & t \geqslant x . \end{cases} \tag{2.3.1}$$

Therefore,

$$\Phi(x) = \left(\frac{1 - \cos \lambda x}{\lambda^2} , R \right) = \left(\frac{1 - \cos \lambda x}{\lambda^2} , \frac{2}{\pi} \right) + \left(\frac{1 - \cos \lambda x}{\lambda^2} , R - \frac{2}{\pi} \right) =$$

$$= \varphi(0) + \frac{2}{\pi}\left(\frac{1 - \cos \lambda x}{\lambda^2} , C(L)\right) ,$$

whence, in view of (2.1.4) and (2.3.1),

$$\Phi(x) = x + \int_0^x (x-t)L(t,0)dt \quad (0 \leqslant x < \infty) . \tag{2.3.2}$$

Since the function $L(t,0)$ is continuously differentiable, $\Phi(x)$ is thrice continuously differentiable; moreover,

$$\Phi'(0+) = 1 + \lim_{x\to 0} \int_0^x L(t,0)dt = 1 ,$$

and

$$\Phi''(0+) = \lim_{x\to 0} L(x,0) = L(0,0) = -h .$$

If the function $q(x)$ has $n \geqslant 0$ continuous derivatives, then $L(x,0)$ has $n+1$ continuous derivatives, and hence $\Phi(x)$ has $n+3$ continuous derivatives. □

We now derive the linear integral equation which is satisfied for each fixed point x by the kernel $K(x,y) = K(x,y;h)$ of the transformation operator.

The function $\omega(\lambda,x)$ is an even entire function of exponential type (in λ) which is bounded on whole real line for every fixed value of x .

Thus, $\omega(\lambda,x)$ is a multiplier in the space Z, and one can multiply any distribution from Z' by it. Consider the product $(R - \frac{2}{\pi})\omega(\lambda,x)$, where R is the spectral function of problem (2.2.1), (2.2.2). It follows from the representation

$$\omega(\lambda,x) = \cos \lambda x + \int_0^x K(x,t)\cos \lambda t dt$$

that

$$(R - \frac{2}{\pi})\omega(\lambda,x) = (R - \frac{2}{\pi})\cos \lambda x + (R - \frac{2}{\pi}) \int_0^x K(x,t)\cos \lambda t dt$$

and hence, upon recalling that $R - \frac{2}{\pi}$ is the Fourier cosine transform of the function $\frac{2}{\pi} L(y,0)$, and invoking properties 2 and 3 of the Fourier cosine transform (see Section 1), that

$$(R - \frac{2}{\pi})\omega(\lambda,x) \sim \frac{1}{\pi} \{L(x+y,0) + L(|x-y|,0\} +$$
$$+ \frac{1}{\pi} \int_0^x K(x,t)\{L(t+y,0) + L(|t-y|,0)\}dt .$$

Every function $F(\lambda) \in Z$ is the Fourier cosine transform of some continuous function $f(x)$ with compact support:

$$F(\lambda) = \int_0^\infty f(x)\cos \lambda x \quad , \quad f(x) = \frac{2}{\pi} \int_0^\infty F(\lambda)\cos \lambda x d\lambda . \tag{2.3.4}$$

Thus, upon expressing $\cos \lambda x$ through $\omega(\lambda,x)$ by means of the transformation operator $\mathbb{I} + \mathbb{L}$, we get

$$F(\lambda) = \int_0^\infty [f(x) + \int_x^\infty f(y)L(y,x)dy]\omega(\lambda,x)dx .$$

Hence, $F(\lambda)$ is also the Fourier ω-transform of the function

$$f(x) + \int_x^\infty f(y)L(y,x)dy ,$$

and according to the Corollary to Theorem 2.2.1,

$$(F(\lambda),R\omega(\lambda,x)) = (F(\lambda)\omega(\lambda,x),R) = f(x) + \int_x^\infty f(y)L(y,x)dy .$$

Moreover,

$$(F(\lambda), \frac{2}{\pi}\omega(\lambda,x)) = (F(\lambda)\omega(\lambda,x), \frac{2}{\pi}) =$$

$$= \frac{2}{\pi}\int_0^\infty F(\lambda)\left[\cos\lambda x + \int_0^x K(x,y)\cos\lambda y dy\right] d\lambda =$$

$$= f(x) + \int_0^x K(x,y)f(y)dy .$$

Hence,

$$(F(\lambda),(R - \frac{2}{\pi})\omega(\lambda,x)) = \int_x^\infty f(y)L(y,x)dy - \int_0^x K(x,y)f(y)dy =$$

$$= \int_0^\infty f(y)\{L(y,x) - K(x,y)\}dy \quad ,$$

from which in turn it follows, in view of (2.3.4) and the definition of the Fourier cosine transform of summable functions, that

$$(R - \frac{2}{\pi})\omega(\lambda,x) \sim \frac{2}{\pi}\{L(y,x) - K(x,y)\} . \tag{2.3.5}$$

Comparing formulas (2.3.3) and (2.3.5), we get the identity

$$L(y,x) - K(x,y) = \frac{1}{2}\{L(x+y,0) + L(|x-y|,0)\} +$$

$$+ \frac{1}{2}\int_0^x K(x,t)\{L(t+y,0) + L(|t-y|,0)\}dt ,$$

from which we derive the following integral equation for the kernel $K(x,y)$ on the triangle $0 \leqslant y \leqslant x$:

$$f(x,y) + K(x,y) + \int_0^x K(x,y)f(t,y)dt = 0 , \tag{2.3.6}$$

where

$$f(x,y) = \frac{1}{2}\{L(x+y,0) + L(|x-y|,0)\} . \tag{2.3.7}$$

On the other hand, by formula (2.3.2),

$$f(x,y) = \frac{1}{2}\{\Phi''(x+y) + \Phi''(|x-y|)\} , \tag{2.3.8}$$

i.e., the kernel and the free term of the integral equation (2.3.6) are both expressible directly through the spectral function R of the boundary value problem in question. Hence, upon solving equation (2.3.6) (we shall prove below that for every fixed x it has a unique solution), we recover the kernel $K(x,y)$, and together with it, the boundary value problem, from its spectral function. The uniqueness of the solution to equation (2.3.6) guarantees that there is only one boundary value problem associated with the given spectral function.

LEMMA 2.3.2. *Suppose the distribution* $R \in Z'$ *enjoys properties* 1 *and* 2 *of Lemma* 2.3.1. *Set*

$$\Phi(x) = \left(\frac{1 - \cos \lambda x}{\lambda^2} , R \right)$$

and

$$f(x,y) = \frac{1}{2} \{\Phi''(x+y) + \Phi''(|x-y|)\} ,$$

and write the integral equation (2.3.6) *for the unknown function* $K(x,y)$. *For every value of* $x \geq 0$, *this integral equation has a unique solution* $K(x,y)$. *This solution is continuous and has, with respect to both variables, as many continuous derivatives as the function* $\Phi''(x)$ *has with respect to the variable* x *on* $[0,\infty)$.

PROOF. According to the Fredholm alternative, the solvability of equation (2.3.6) for every $x = a$ will be established if we show that the corresponding homogeneous equation

$$g(y) + \int_0^a g(t)f(t,y)dt = 0 \quad (0 \leq y \leq a) \tag{2.3.9}$$

has only the trivial solution. We pick an arbitrary function $g(y) \in K^2(a)$ and calculate the Fourier cosine transform of the locally summable function

$$a(y) = g(y) + \int_0^\infty g(t)f(t,y)dt \quad (0 \leq y \leq \infty) .$$

From formulas (2.3.8) and properties 3 and 4 of the Fourier cosine transform (see Section 1) it follows that

$$C(a) = C(\lambda,g) + C(\lambda,g)(\tfrac{\pi}{2} - R - \Phi'(0+)) = \frac{\pi}{2} C(\lambda,g)R .$$

Hence, the equality

$$\int_0^\infty a(y)z(y)dy = \frac{2}{\pi} (C(\lambda,z), \frac{\pi}{2} C(\lambda,g)R) = (C(\lambda,g)C(\lambda,a),R)$$

holds for every function $z(y) \in K^2(a)$. If $g(y)$ satisfies the homogeneous equation (2.3.9), then the left hand side of this equality vanishes for every function $z(y) \in K^2(a)$, and hence $(C(\lambda,g)u(\lambda),R) = 0$ for every function $u(\lambda) \in CK^2(a)$. In view of property 1 (which by assumption is enjoyed by the distribution R), this shows that the function $C(\lambda,g)$ vanishes identically, and hence that $g(y) \equiv 0$. Thus, the homogeneous equation (2.3.9) has only the trivial solution, and the nonhomogeneous equation (2.3.6) has, for each fixed x , a unique solution $K(x,y)$.

Next we investigate the smoothness of this solution. The change of variables $y = xy'$, $t = xt'$ transforms (2.3.6) into the equation

$$f(x,xy') + K(x,xy') + \int_0^1 K(x,xt')f(xt',xy')x\,dt' = 0 , \qquad (2.3.10)$$

which can be written in the form

$$(\mathbb{I} + \mathbb{F}(x))K + f(x) = 0 ,$$

where the operators $\mathbb{F}(x)$, parametrized by x , act in the fixed space $C[0,1]$ and $f(x)$ is an element of $C[0,1]$ which depends continuously on the parameter x . The unique solvability of equation (2.3.6) established above guarantees the existence of the inverse operators $(\mathbb{I} + \mathbb{F}(x))^{-1}$ for all values of x . The kernel

$$\frac{x}{2}\{\Phi''(x(t'+y')) + \Phi''(x|t'-y'|)\}$$

of the integral operator $\mathbb{F}(x)$ depends continuously on x , and has $n+1$ continuous derivatives with respect to x if the function $\Phi(x)$ has $n+3$ continuous derivatives. Consequently, the inverse operator $(\mathbb{I} + \mathbb{F}(x))^{-1}$ depends continuously on x and has a derivative of order $n+1$ with respect to x . Since the function $f(x,xy')$ also has a continuous derivative of order $n+1$ with respect to x , it follows that the solution $K(x,xy')$ of equation (2.3.10) has a continuous derivative or order $n+1$ with respect to x . Substituting the expression for the function f in terms of Φ'' into (2.3.10), we get

$$K(x,xy') = \frac{1}{2}\{\Phi''(x(1+y')) + \Phi''(x(1-y'))\} + \frac{x}{2}\int_0^1 K(x,xt')\Phi''(x(t'+y'))dt' +$$

$$+ \frac{x}{2}\left\{\int_0^{y'} K(x,xt')\Phi''(x(y'-t'))dt' + \int_{y'}^1 K(x,xt')\Phi''(x(t'-y'))dt'\right\} . \qquad (2.3.11)$$

This formula shows that the function $K(x,xy')$ will have a continuous derivative of order $n+1$ with respect to the variable y' if the function $\Phi(x)$ has $n+3$ continuous derivatives. The lemma is proved. □

Remark. The preceding proof shows that equation (2.3.6) and all the functions that appear in it can be differentiated $n+1$ times with respect to x if the function $\Phi(x)$ has $n+3$ derivatives. This equation can also be differentiated $n+1$ times with respect to y, even though the function $f(t,y)$ may not have a derivative of this order with respect to y at the point $y = t$. To differentiate equation (2.3.6) correctly with respect to y, we must write it in a form analogous to (2.3.11):

$$f(x,y) + K(x,y) + \frac{1}{2}\int_0^x K(x,t)\Phi''(t+y)dt +$$

$$+ \frac{1}{2}\int_0^y K(x,t)\Phi''(y-t)dt + \frac{1}{2}\int_y^x K(x,t)\Phi''(t-y)dt = 0 .$$

In particular, for the first two derivatives, we get the equalities

$$f_y(x,y) + K_y(x,y) + \int_0^x K(x,t)f_y(t,y)dt = 0 \tag{2.3.12}$$

and

$$f_{yy}(x,y) + K_{yy}(x,y) + \int_0^x K(x,t)f_{yy}(t,y)dt + K(x,y)\Phi'''(0+) = 0 . \tag{2.3.13}$$

LEMMA 2.3.3. *Suppose the function* $\Phi(x)$ $(0 \leq x < \infty)$ *is four times continuously differentiable, set*

$$f(x,y) = \frac{1}{2}\{\Phi''(x+y) + \Phi''(|x-y|)\} , \tag{2.3.14}$$

and suppose that the homogeneous equation (2.3.9) *has only the trivial solution for every choice of* $a \geq 0$. *Then the solution* $K(x,y)$ *of the nonhomogeneous equation* (2.3.6) *satisfies the partial differential equation*

$$\frac{\partial^2}{\partial x^2}K(x,y) - q(x)K(x,y) = \frac{\partial^2}{\partial y^2}K(x,y) \quad (0 \leq y \leq x < \infty) ; \tag{2.3.15}$$

moreover,

$$q(x) = 2\frac{d}{dx}K(x,x) \quad , \quad K_y(x,0) = 0 . \tag{2.3.15'}$$

PROOF. By the foregoing discussion, the conditions of the lemma guarantee the existence and continuity of the partial derivatives $K_{xx}(x,y)$ and $K_{yy}(x,y)$. Furthermore, it follows from definition (2.3.14) that

$$f_{xx}(x,y) = f_{yy}(x,y) \quad (x \neq y) ,$$

$$f_x(0,y) = f_y(x,0) = 0 . \tag{2.3.16}$$

Differentiating equation (2.3.6) once with respect to y, we get equality (2.3.12), from which in turn we obtain, upon setting $y = 0$ and using (2.3.16),

$$K_y(x,y)\big|_{y=0} = 0 . \tag{2.3.17}$$

Differentiating equation (2.3.6) with respect to y twice, we get (2.3.13). But

$$\int_0^x K(x,t)f_{yy}(t,y)dt = \int_0^x K(x,t)f_{tt}(t,y)dt =$$

$$= \int_0^y K(x,t)f_{tt}(t,y)dt + \int_y^x K(x,t)f_{tt}(t,y)dt =$$

$$= K(x,t)f_t(t,y)\Big|_0^y - K_t(x,t)f(t,y)\Big|_0^y + \int_0^y K_{tt}(x,t)f(t,y)dt +$$

$$+ K(x,t)f_t(t,y)\Big|_y^x - K_t(x,t)f(t,y)\Big|_y^x + \int_y^x K_{tt}(x,t)f(t,y)dt ,$$

whence, in view of (2.3.14), (2.3.16), and (2.3.17),

$$\int_0^x K(x,t)f_{yy}(t,y)dt = -K(x,y)\Phi'''(0+) + K(x,x)f_x(x,y) -$$

$$- K_t(x,t)f(t,y)\Big|_{t=x} + \int_0^x K_{tt}(x,t)f(t,y)dt .$$

Therefore, equality (2.3.13) can be reexpressed as

$$f_{yy}(x,y) + K_{yy}(x,y) + \int_0^x K_{tt}(x,t)f(t,y)dt +$$

$$+ K(x,x)f_x(x,y) - K_t(x,t)f(t,y)\Big|_{t=x} = 0 . \tag{2.3.13'}$$

Next, we differentiate equation (2.3.6) twice with respect to x :

$$f_{xx}(x,y) + K_{xx}(x,y) + \frac{d}{dx}\{K(x,x)f(x,y)\} +$$

$$+ K_x(x,t)f(t,y)\Big|_{t=x} + \int_0^x K_{xx}(x,t)f(t,y)dt = 0$$

and substract (2.3.13') from this equality, to get

$$f_{xx}(x,y) - f_{yy}(x,y) + K_{xx}(x,y) - K_{yy}(x,y) + f(x,y)\frac{d}{dx} K(x,x) +$$

$$+ \{K_x(x,t) + K_t(x,t)\}f(t,y)\Big|_{t=x} + \int_0^x \{K_{xx}(x,t) - K_{tt}(x,t)\}f(t,y)dt = 0.$$

Since

$$f_{xx}(x,y) = f_{yy}(x,y)$$

and

$$\{K_x(x,t) + K_t(x,t)\}\Big|_{t=x} = \frac{d}{dx} K(x,x) ,$$

the last equality yields

$$K_{xx}(x,y) - K_{yy}(x,y) + q(x)f(x,y) + \int_0^x \{K_{xx}(x,t) - K_{tt}(x,t)\}f(t,y)dt = 0 ,$$

where $q(x) = 2\frac{d}{dx} K(x,x)$.

Finally, upon substracting equation (2.3.6), multiplied by $q(x)$, from this equality, we get

$$\{K_{xx}(x,y) - q(x)K(x,y) - K_{yy}(x,y)\} +$$

$$+ \int_0^x \{K_{xx}(x,t) - q(x)K(x,t) - K_{tt}(x,t)\}f(t,y)dt = 0 ,$$

whence

$$K_{xx}(x,y) - q(x)K(x,y) - K_{yy}(x,y) = 0$$

because the homogeneous equation (2.3.9) has, by assumption, only the null solution. The lemma is proved. □

THEOREM 2.3.1. *In order that the distribution* $R \in Z'$ *be the spectral function of a boundary value problem* (2.2.1), (2.2.2), *it is necessary and sufficient that the following conditions be satisfied:*

1) for every $\sigma > 0$ *, there is no nonzero function* $f(\lambda) \in CK^2(\sigma)$ *such that* $(f(\lambda)u(\lambda),R) = 0$ *for all* $u(\lambda) \in CK^2(\sigma)$ *;*

2) the function

$$\Phi(x) = \left(\frac{1 - \cos \lambda x}{\lambda^2} , R \right) \quad (0 < x < \infty)$$

is thrice continuously differentiable and $\Phi'(0+) = 1$.

If this is the case, then the function $q(x)$ *in equation (2.2.1) has as many continuous derivatives as the function* $\Phi'''(x)$ $(0 \leq x < \infty)$ *has.*

PROOF. The necessity of conditions 1 and 2 was proved in Lemma 2.3.1. To show that these conditions are also sufficient, we construct the integral equation (2.3.6) in which $f(x,y)$ is defined by formula (2.3.8). By Lemma 2.3.2, this equation has a unique solution $K(x,y)$ for every $x \geq 0$, and this solution is continuously differentiable in both arguments. Hence, the function

$$q(x) = 2 \frac{d}{dx} K(x,x) \tag{2.3.18}$$

exists and is continuous. We now consider the boundary value problem (2.2.1), (2.2.2), in which $q(x)$ is given by (2.3.18) and $h = K(0,0)$. We shall presently show that the given distribution R is a spectral function of this boundary value problem.

First of all, we must verify that the solution $K(x,y)$ of equation (2.3.6) is the kernel of the transformation operator of problem (2.2.1), (2.2.2), i.e., that the functions

$$\omega(\lambda,x) = \cos \lambda x + \int_0^x K(x,t)\cos \lambda t dt \tag{2.3.19}$$

satisfy equation (2.2.1) and the initial conditions

$$\omega(\lambda,0) = 1 \quad , \quad \omega'(\lambda,0) = h = K(0,0) \ . \tag{2.3.20}$$

We assume first that the function $\Phi(x)$ has four continuous derivatives. In Lemma 2.3.2 we proved that in this case, $K(x,y)$ satisfies the paartial differential equation (2.3.15) as well as conditions (2.3.15'). Hence,

$$\omega''(\lambda,x) - q(x)\omega(\lambda,x) + \lambda^2\omega(\lambda,x) =$$

$$= -\lambda^2 \cos \lambda x + \{K(x,x)\cos \lambda x\}' + K_x(x,t)\cos \lambda t\big|_{t=x} +$$

$$+ \int_0^x \{K_{xx}(x,t) - q(x)K(x,t)\}\cos \lambda t dt - q(x)\cos \lambda x +$$

$$+ \lambda^2 \cos \lambda x + \lambda^2 \int_0^x K(x,t)\cos \lambda t dt ,$$

whence, upon integrating the last term by parts twice,

$$\omega''(\lambda,x) - q(x)\omega(\lambda,x) + \lambda^2\omega(\lambda,x) =$$

$$= \int_0^x \{K_{xx}(x,t) - q(x)K(x,t) - K_{tt}(x,t)\}\cos \lambda t dt +$$

$$+ \{\frac{d}{dx} K(x,x) + [K_x(x,t) + K_t(x,t)]\Big|_{t=x} - q(x)\}\cos \lambda x - K_t(x,t)\Big|_{t=0} .$$

In view of (2.3.15) and (2.3.15'), this yields

$$\omega''(\lambda,x) - q(x)\omega(\lambda,x) + \lambda^2\omega(\lambda,x) = 0 .$$

It is readily checked that the function $\omega(\lambda,x)$ also satisfies the initial conditions (2.3.20).

Now let the function $\Phi(x)$ have only three continuous derivatives. Then the functions

$$\Phi_\delta(x) = \frac{1}{\delta} \int_x^{x+\delta} \Phi(t)dt$$

have four continuous derivatives. As $\delta \to 0$, the function $\Phi_\delta(x)$ and its first three derivatives tend to $\Phi(x)$, and its first three derivatives, respectively, uniformly on every bounded interval. Consequently, for sufficiently small δ , the equation

$$f_\delta(x,y) + K_\delta(x,y) + \int_0^x K_\delta(x,t)f_\delta(t,y)dt = 0 \quad (0 \leq y \leq x) ,$$

with

$$f_\delta(x,y) = \frac{1}{2} \{\Phi''_\delta(x+y) + \Phi''_\delta(|x-y|)\} ,$$

has a unique solution, and

$$\lim_{\delta\to 0} K_\delta(x,y) = K(x,y) \quad , \quad \lim_{\delta\to 0} \frac{d}{dx} K_\delta(x,x) = \frac{d}{dx} K(x,x)$$

uniformly on every bounded domain of variation of the variables x , y .

Moreover, by the preceding analysis, the function

$$\omega_\delta(\lambda,x) = \cos \lambda x + \int_0^x K_\delta(x,t)\cos \lambda t dt$$

satisfies the equation

$$y'' - q_\delta(x)y + \lambda^2 y = 0 \quad , \quad q_\delta(x) = 2 \frac{d}{dx} K_\delta(x,x)$$

and the initial conditions

$$\omega_\delta(\lambda,0) = 1 \quad , \quad \omega'_\delta(\lambda,0) = K_\delta(0,0) \ .$$

Letting $\delta \to 0$ in these formulas, we see that the function (2.3.19) satisfies equation (2.2.1), as well as the initial conditions (2.3.20). Thus, in this case, too, the solution $K(x,y)$ of equation (2.3.6) is the kernel of the transformation operator for the boundary value problem (2.2.1), (2.2.2).

Let R_0 denote the spectral function of the boundary value problem just constructed, and set

$$\Phi_0(x) = \left(\frac{1 - \cos \lambda x}{\lambda^2} , R \right) ,$$

and

$$f_0(x,y) = \frac{1}{2} \{\Phi_0''(x+y) + \Phi_0''(|x+y|)\} \ .$$

As we showed above, the kernel $K(x,y)$ of the transformation operator for this problem must satisfy the equation

$$f_0(x,y) + K(x,y) + \int_0^x K(x,t)f_0(t,y)dt = 0 \quad (0 \leq y \leq x) \ .$$

Substracting this equation from (2.3.6) and subsequently setting $y = 0$, we get

$$f(x,0) - f_0(x,0) + \int_0^x K(x,t)\{f(t,0) - f_0(t,0)\}dt = 0 \ .$$

Since the homogeneous Volterra equation

$$\varphi(x) + \int_0^x K(x,t)\varphi(t) = 0$$

obviously has only the null solution, it follows that

$f(x,0) - f_0(x,0) \equiv 0$.

Therefore, $\Phi''(x) = \Phi_0''(x)$, and since, in addition, $\Phi(0) = \Phi_0(0)$, $\Phi'(0+) = \Phi_0'(0+) = 1$, we have $\Phi(x) = \Phi_0$, and hence $R = R_0$.

Finally, we remark that, as was established in Lemma 2.3.2, the function $K(x,y)$ has as many continuous derivatives as $\Phi''(x)$. Hence, $q(x)$ has the same number of continuous derivatives as $\Phi'''(x)$.

The theorem is proved. □

PROBLEMS

1. Show that the distribution

$$R = \frac{2}{\pi} + \frac{i}{2n!}\,\delta^{(2n)}(\lambda) ,$$

acting in the space Z by the formula

$$(F(\lambda),R) = \frac{2}{\pi}\int_0^\infty F(\lambda)d\lambda + \frac{i}{2n!}\,\frac{d^{2n}}{d\lambda^{2n}}\,F(\lambda)\Big|_{\lambda=0} ,$$

is the spectral function for a boundary value problem

$$y'' - q(x)y + \lambda^2 y = 0 \quad (0 \le x < \infty) \quad , \quad y'(0) = 0 .$$

Show that in this problem the function $\omega_0(x) = \omega(\lambda,x;0)\big|_{\lambda=0}$ and the adjoined functions $\omega_k(x)$ $(k = 1,2,\ldots)$, which are defined as the solutions of the equations

$$\omega_k''(x) - q(x)\omega_k(x) = \omega_{k-1}(x) \quad , \quad \omega_k(0) = \omega_k'(0) = 0 ,$$

belong to $L_2[0,\infty)$. Show further that in the present case, the Parseval identity can be written in the form

$$\int_0^\infty f(x)g(x)dx = \frac{2}{\pi}\int_0^\infty \omega(\lambda,f)\omega(\lambda,g)d\lambda + (-1)^{n}i\sum_{k=0}^{n}\omega_{n-k}(f)\omega_k(g) ,$$

where

$$\omega_k(f) = \int_0^\infty f(x)\omega_k(x)dx \quad , \quad \omega_k(g) = \int_0^\infty g(x)\omega_k(x)dx ,$$

and that it can be extended to all functions $f(x),g(x) \in L_2[0,\infty)$.

2. Consider two Sturm-Liouville boundary value problems

$$-y'' + q_j(x)y = \lambda^2 y \quad , \quad y_j'(0) - h_j y_j(0) = 0 \quad (0 < x < \infty) \; ,$$

and the corresponding transformation operators. The operator $\mathbb{I} + \mathbb{L}_1$ transforms the solution $\omega_1(\lambda,x;h_1)$ of the first boundary value problem into $\cos \lambda x$, whereas $\mathbb{I} + \mathbb{K}_2$ transforms $\cos \lambda x$ into the solution $\omega_2(\lambda,x;h_2)$ of the second boundary value problem. Hence, the operator $\mathbb{I} + \mathbb{K}_{2,1} = (\mathbb{I} + \mathbb{K}_2)(\mathbb{I} + \mathbb{L}_1)$ transforms $\omega_1(\lambda,x;h)$ into $\omega_2(\lambda,x;h)$; moreover, it is also a Volterra integral operator. Show that the kernel $K_{2,1}(x,y)$ of the operator $\mathbb{K}_{2,1}$ satisfies the equation

$$f(x,y) + K_{2,1}(x,y) + \int_0^x K_{2,1}(x,t)f(t,y)dt = 0 \quad (0 \leqslant y \leqslant x) \; , \tag{2.3.21}$$

which is analogous to (2.3.6). Here,

$$f(x,y) = F_{x,y}(x,y) \quad , \quad F(x,y) = \left(\int_0^x \omega_1(\lambda,t)dt \int_0^y \omega_1(\lambda,t)dt, R_2 - R_1 \right) ,$$

and the R_j $(j=1,2)$ are distribution spectral functions of the boundary value problems under consideration.

Hint. From the formula

$$\omega_1(\lambda,x) = \omega_2(\lambda,x) + \int_0^x K_{1,2}(x,t)\omega_2(\lambda,t)dt$$

it follows that $\omega_1(\lambda,f) = \omega_2(\lambda,\hat{f})$, where

$$\hat{f}(x) = f(x) + \int_x^\infty f(t)K_{1,2}(t,x)dt \; .$$

Consequently,

$$(\omega_1(\lambda,f)\omega_1(\lambda,g), R_2 - R_1) =$$

$$= (\omega_2(\lambda,\hat{f})\omega_2(\lambda,\hat{g}), R_2) - (\omega_1(\lambda,f)\omega_1(\lambda,g)R_1) =$$

$$= \int_0^\infty \{\hat{f}(x)\hat{g}(x) - f(x)g(x)\}dx =$$

$$= \int_0 \int_0 f(x) \left\{K_{1,2}(x,y) + K_{1,2}(y,x) + \int_0^x K_{1,2}(x,t)K_{1,2}(y,t)dt\right\} g(y)dxdy ,$$

and

$$F(x,y) = \int_0^x \int_0^y \left\{K_{1,2}(\xi,\eta) + K_{1,2}(\eta,\xi) + \int_0^\xi K_{1,2}(\xi,t)K_{1,2}(\eta,t)dt\right\} d\eta d\xi .$$

Hence

$$f(x,y) = K_{1,2}(x,y) + K_{1,2}(y,x) + \int_0^x K_{1,2}(x,t)K_{1,2}(y,t)dt .$$

This yields the identity

$$\mathbb{I} + \mathbb{F} = (\mathbb{I} + \mathbb{K}_{1,2})(\mathbb{I} + \mathbb{K}_{1,2}^+)$$

for the corresponding integral operators, where $\mathbb{K}_{1,2}^+$ designates the transpose of $\mathbb{K}_{1,2}$. Since $\mathbb{I} + \mathbb{K}_{2,1} = (\mathbb{I} + \mathbb{K}_{1,2})^{-1}$,

$$(\mathbb{I} + \mathbb{K}_{2,1})(\mathbb{I} + \mathbb{F}) = \mathbb{I} + \mathbb{K}_{1,2}^+$$

or, in terms of the kernels,

$$K_{2,1}(x,y) + \int_0^x K_{2,1}(x,t)f(t,y)dt + f(x,y) = K_{1,2}(y,x) ,$$

which yields (2.3.21) for $y < x$.

3. Let R_1 be a distribution spectral function of the boundary value problem

$$-y'' + q_1(x)y = \lambda^2 y , \quad y(0)h_1 - y'(0) = 0 , \tag{2.3.22}$$

and let $\omega_1(\lambda,x) = \omega_1(\lambda,x;h_1)$ be the corresponding solution of equation (2.3.22). Show that the distribution $R_2 = R_1 + c\delta(\lambda-\mu)$ is a distribution spectral function of a boundary value problem of the same form if the function

$$1 + c \int_0^x \omega_1(\mu,t)^2 dt$$

has no zeros on the positive half line $0 \le x < \infty$, and find this spectral function.

Hint. Let $\omega_2(\lambda,x)$ be the solution of the sought-for boundary value problem

$$-y'' + q_2(x)y = \lambda^2 y \quad , \quad y(0)h_2 - y' = 0 \ ,$$

and let $K_{2,1}(x,y)$ designate the kernel of the transformation operator that takes $\omega_1(\lambda,x)$ into $\omega_2(\lambda,x)$. To find this kernel, you can use equation (2.3.21), which in the present case is degenerate:

$$c\omega_1(\mu,x)\omega_1(\mu,y) + K_{2,1}(x,y) + c\int_0^x K_{2,1}(x,t)\omega_1(\mu,t)\omega_1(\mu,y)dt = 0 \ ,$$

and admits the explicit solution:

$$K_{2,1}(x,y) = -\frac{c\omega_1(\mu,x)\omega_1(\mu,y)}{1 + c\int_0^x \omega_1(\mu,t)^2 dt} \ .$$

Hence

$$\left.\begin{aligned} &\omega_2(\lambda,x) = \omega_1(\lambda,x) - \frac{c\omega_1(\mu,x)}{1 + c\int_0^x \omega_1(\mu,t)^2 dt}\int_0^x \omega_1(\mu,t)\omega_1(\lambda,t)dt \ , \\ &h_2 = \omega_2'(\lambda,0) = h_1 - c \ , \\ &\text{and} \\ &q_2(x) = q_1(x) - 2c\left\{\frac{\omega_1(\mu,x)^2}{1 + c\int_0^x \omega_1(\mu,t)^2 dt}\right\}' \end{aligned}\right\} \quad (2.3.23)$$

4. Using the Wronskian

$$W\{u(x),v(x)\} = u'(x)v(x) - u(x)v'(x) \ ,$$

you can rewrite formula (2.3.23) for $\omega_2(\lambda,x)$ in the form

$$\omega_2(\lambda,x) = \omega_1(\lambda,x) + \frac{c\omega_2(\mu,x)W\{\omega_1(\mu,x),\omega_1(\lambda,x)\}}{\mu^2 - \lambda^2} \ ,$$

which suggests interesting generalizations.

Suppose that on the bounded or unbounded interval (a,b) you are given two Sturm-Liouville equations

$$-y_j'' + q_j(x)y_j = \lambda^2 y_j \quad (j = 1,2)$$

in which the functions $q_j(x)$ are assumed to be continuous only in the interior points of this interval. Let $y_j(x)$ denote a fixed solution of the j-th equation for $\lambda = \mu$, and let $\varphi_1(\lambda,x)$ denote an arbitrary solution of the first equation. Show that the functions

$$\left.\begin{aligned} r(\lambda,x) &= (\mu^2 - \lambda^2)^{-1} y_2(x) W\{y_1(x),\varphi_1(\lambda,x)\}, \\ &\text{and} \\ \psi(\lambda,x) &= \varphi_1(\lambda,x) + r(\lambda,x) \end{aligned}\right\} \tag{2.3.24}$$

satisfy the equations

$$-r''(\lambda,x) + q_2(x)r(\lambda,x) = \lambda^2 r(\lambda,x) - 2\varphi_1(\lambda,x)\{y_2(x)y_1(x)\}'$$

and

$$-\psi''(\lambda,x) + q_2(x)\psi(\lambda,x) = \lambda^2\psi(\lambda,x) +$$

$$+ \frac{\varphi_1(\lambda,x)}{y_2(x)y_1(x)} \{[y_2(x)y_1(x)]^2 + W\{y_2(x),y_1(x)\}\}' ,$$

respectively. In particular, if the solution $y_1(x)$ has no zeros in the interval (a,b), then the functions

$$\varphi_2(\lambda,x) = \frac{W\{y_1(x),\varphi_1(\lambda,x)\}}{y_1(x)(\mu^2 - \lambda^2)} \tag{2.3.25}$$

satisfy the equation

$$-\varphi_2(\lambda,x)'' + q_2(x)\varphi_2(\lambda,x) = \lambda^2\varphi_2(\lambda,x) ,$$

where

$$q_2(x) = y_1(x)\left(\frac{1}{y_1(x)}\right)'' + \mu^2 = q_1(x) - 2\frac{d}{dx}\left\{\frac{y_1'(x)}{y_1(x)}\right\} .$$

Suppose the function $[y_1(x)]^2$ has a primitive $J_1(x)$ which does not vanish for $x \in (a,b)$. Then the function

$$\varphi_2(\lambda,x) = \varphi_1(\lambda,x) + \frac{y_1(x)}{J_1(x)} \frac{W\{y_1(x),\varphi_1(\lambda,x)\}}{\mu^2 - \lambda^2} \tag{2.3.26}$$

satisfies the equation

$$-\varphi_2(\lambda,x)'' + q_2(x)\varphi_2(\lambda,x) = \lambda^2\varphi_2(\lambda,x)$$

in which

$$q_2(x) = q_1(x) - 2\frac{d}{dx}\left\{\frac{J_1'(x)}{J_1(x)}\right\}.$$

Thus, formulas (2.3.25) and (2.3.26) actually define certain transformation operators. Find their inverse operators.

5. Let $K^2(a,b)$ denote the set of square summable functions on the interval (a,b) with compact support, and let $K_1^2(a,b)$ designate the subset of differentiable functions whose derivative also belongs to $K^2(a,b)$. Also, let Z be a space of test functions which contains all the products $\varphi_1(\lambda,f_1)\varphi_1(\lambda,f_2)$ with $f_j(x) \in K^2(a,b)$, where $\varphi_1(\lambda,f) = \int_a^b f(x)\varphi_1(\lambda,x)dx$ for $f \in K^2(a,b)$. Suppose further that R_1 is a distribution on $Z(a,b)$, such that

$$(\varphi_1(\lambda,f_1)\varphi_1(\lambda,f_2),R_1) = \int_a^b f_1(x)f_2(x)dx .$$

A. Show that for every pair of functions $f(x) \in K^2(a,b)$ and $g(x) \in K_1^2(a,b)$, such that $\int_a^b f(x)y_2(x)dx = 0$, the following identity holds:

$$(r(\lambda,f)r(\lambda,g)(\mu^2 - \lambda^2),R_1) = -\int_a^b f(x)g(x)[y_2(x)y_1(x)]^2dx ,$$

where

$$r(\lambda,f) = \int_a^b f(x)r(\lambda,x)dx ,$$

and the function $r(\lambda,x)$ is defined by formula (2.3.24). Verify, in particular, that

$$(\varphi_2(\lambda,x)\varphi_2(\lambda,g)(\mu^2 - \lambda^2),R_1) = -\int_a^b f(x)g(x)dx$$

if the function $r(\lambda,x) = \varphi_2(\lambda,x)$ is defined by formula (2.3.25).

B. Show that, if the function $\varphi_2(\lambda,x)$ is defined by formula (2.3.26), then the identity

$$(\varphi_2(\lambda,f_1)\varphi_2(\lambda,f_2),R_1) = \int_a^b f_1(x)f_2(x)dx$$

holds for all functions $f_j \in K^2(a,b)$ which satisfy the constraint

$$\int_a^b f_j(x)\,\frac{y_1(x)}{J_1(x)}\,dx = 0\ .$$

<u>Hint</u>. Suppose that the conditions of problem A are fulfilled. Then

$$r(\lambda,f) = \frac{1}{\mu^2-\lambda^2}\int_a^b f(x)y_2(x)W\{y_1(x),\varphi_1(\lambda,x)\}dx =$$

$$= \frac{1}{\mu^2-\lambda^2}\int_a^b W\{y_1(x),\varphi_1(\lambda,x)\}d\left\{\int_a^x f(t)y_2(t)dt\right\} =$$

$$= \int_a^b y_1(x)\varphi_1(\lambda,x)\left\{\int_a^x f(t)y_2(t)dt\right\} = \varphi_1\left(\lambda, y_1\int_0^x f(t)y_2(t)dt\right),$$

and

$$r(\lambda,g)(\mu^2-\lambda^2) = \int_a^b g(x)y_2(x)\{y_1'(x)\varphi_1(\lambda,x) - y_1(x)\varphi_1'(\lambda,x)\}dx =$$

$$= \int_a^b \{g(x)y_2(x)y_1'(x) + [g(x)y_2(x)y_1(x)]'\}\varphi_1(\lambda,x)dx =$$

$$= \int_a^b \frac{[g(x)y_2(x)y_1^2(x)]'}{y_1(x)}\varphi_1(\lambda,x)dx = \varphi_1\left(\lambda, \frac{[gy_1^2y_2]'}{y_1}\right).$$

Hence

$$(r(\lambda,f)r(\lambda,g)(\mu^2-\lambda^2),R_1) =$$

$$= \left(\varphi_1\left(\lambda, y_1\int_0^x f(t)y_2(t)dt\right)\varphi_1\left(\lambda, \frac{gy_1^2y_2{}'}{y_1}\right), R_1\right) =$$

$$= \int_a^b [gy_1^2y_2]'\left\{\int_0^x f(t)y_2(t)dt\right\}dx = -\int_a^b f(x)g(x)[y_1(x)y_2(x)]^2dx\ .$$

You can solve problem B in much the same way, by taking advantage of the fact that

$$\varphi_2(\lambda,f) = \varphi_1\left(\lambda, f + y_1\int_0^x f(t)y_2(t)dt\right).$$

6. The results discussed above may be used to extend various theorems to boundary value problems of the form

$$-y'' + \{q(x) + \ell(\ell+1)x^{-2}\}y = \lambda^2 y \quad , \quad y(0) = 0 \; , \tag{2.3.27}$$

where $\ell \geqslant 0$ is an integer and function $q(x)$ is continuous for $0 \leqslant x < a$. To this end you must eliminate the singularity $\ell(\ell+1)x^{-2}$ by applying successively the operators (2.3.25) and (2.3.26).

A. Prove the existence of distribution spectral functions for problem (2.3.27) (with $a = \infty$) and generalize Theorem 2.3.1.

B. Let μ_k and ν_k denote the eigenvalues of the boundary value problems

$$-y'' + \{q(x) + 2(\sin^2 x)^{-1}\}y = \mu_k y \quad , \quad y(0) = y(\pi) = 0$$

and

$$-y'' + \{q(x) + (2\sin^2 \tfrac{x}{2})^{-1}\}y = \nu_k y \quad , \quad y(0) = y'(\pi) = 0 \; ,$$

respectively, where the functions $q(x)$ are differentiable (and hence continuous) on the segment $[0,\pi]$. Verify the identities

$$\sum_{k=1}^{\infty} \{\mu_k - (k+1)^2\} = -\frac{3}{4}\{q(0) + q(\pi)\}$$

and

$$\sum_{k=1}^{\infty} \{\nu_k - k^2\} = -q(0) + \frac{1}{4}\{q(0) + q(\pi)\} \; .$$

4. THE ASYMPTOTIC FORMULA FOR THE SPECTRAL FUNCTIONS OF SYMMETRIC BOUNDARY VALUE PROBLEMS AND THE EQUICONVERGENCE THEOREM

According to Theorem 2.2.2, the distribution spectral functions of symmetric boundary value problems (h and $q(x)$ real) are generated by measures, i.e., to each spectral function $R \in Z'$ there corresponds a nondecreasing function $\rho(\mu)$ $(-\infty < \mu < \infty)$, not necessarily unique, such that

$$(f(\lambda)g(\lambda),R) = \int_{-\infty}^{\infty} f(\sqrt{\mu})g(\sqrt{\mu})d\rho(\mu)$$

for all functions $f(\lambda), g(\lambda) \in CK^2$.

In this section we are concerned only with symmetric boundary value problems, and we shall use the term spectral function for $\rho(\mu)$. The first question we approach is that of the behavior of the spectral function $\rho(\mu)$ as $\mu \to \infty$. To this end it is convenient to use formula $R = \frac{2}{\pi}(1 + C(L))$ (Theorem 2.2.1) and its consequence

$$\left(\frac{1 - \cos \lambda x}{\lambda^2} , R \right) = x + \int_0^x (x-t)L(t,0;h)dt \quad (0 \leqslant x < \infty) ,$$

which was obtained in the proof of Lemma 2.3.1 (see formula (2.3.2)). Since

$$\frac{1 - \cos \lambda x}{\lambda^2} = 2 \frac{\sin \frac{\lambda x}{2}}{\lambda} \quad \frac{\sin \frac{\lambda x}{2}}{\lambda}$$

is clearly the product of two functions from CK^2 , the last equality yields the formula

$$\int_{-\infty}^{\infty} \frac{1 - \cos \sqrt{\mu}\, x}{\mu} d\rho(\mu) = x + \int_0^x (x-t)L(t,0;h)dt \quad , \quad x \in [0,\infty) , \tag{2.4.1}$$

for symmetric boundary value problems. From it, we can extract information on the behavior of $\rho(\mu)$ as $\mu \to -\infty$. In fact, since the function $\rho(\mu)$ is nondecreasing, and $\mu^{-1}(1 - \cos \sqrt{\mu}\, x) \geqslant 0$, we have

$$0 \leqslant \int_{-\infty}^{0+} \frac{\cosh \sqrt{\mu}\, x - 1}{|\mu|} d\rho(\mu) = \int_{-\infty}^{0+} \frac{1 - \cos \sqrt{\mu}\, x}{\mu} d\rho(\mu) \leqslant$$

$$\leqslant x + \int_0^x (x-t)L(t,0;h)dt$$

for all $x \in [0,\infty)$. From this inequality we conclude, upon observing that for every n ,

$$\lim_{\mu \to -\infty} \frac{|\mu|^n \cosh \sqrt{\mu}\, x}{\cosh \sqrt{\mu}\,(x+1) - 1} = 0 ,$$

that the integral

$$L_1(x) = \int_{-\infty}^{0+} \cosh \sqrt{\mu}\, x \, d\rho(\mu) = \int_{-\infty}^{0+} \cos \sqrt{\mu}\, x \, d\rho(\mu) \tag{2.4.2}$$

converges for every value of x , and that the function $L_1(x)$ is infinitely differentiable. Hence, the spectral function $\rho(\mu)$ of a symmetric boundary value problem always has a finite limit $\rho(-\infty)$ as $\mu \to -\infty$. Moreover,

$$\lim_{\mu\to-\infty} \exp \sqrt{|\mu|}\, x\{\rho(\mu) - \rho(-\infty)\} = 0 \tag{2.4.3}$$

for all $x \in [0,\infty)$, since otherwise the integral (2.4.2) cannot converge for all values of x.

To investigate the behavior of the spectral function for $\mu \to +\infty$, we rewrite equality (2.4.1) in the form

$$\int_{0+}^{\infty} \frac{1 - \cos\sqrt{\mu}\, x}{\mu}\, d\rho(\mu) = x + \int_0^x (x-t)\{L(t,0;h) - L_1(t)\}dt\ , \tag{2.4.4}$$

which is obtained by using the obvious identity

$$\int_{-\infty}^{0+} \frac{1 - \cos\sqrt{\mu}\, x}{\mu}\, d\rho(\mu) = \int_0^x (x-t) \int_{-\infty}^{0+} \cos\sqrt{\mu}\, t\, d\rho(\mu)dt = \int_0^x (x-t)L_1(t)dt\ .$$

Let $f(x)$ be an arbitrary infinitely differentiable even function with compact support. Upon multiplying both sides of equality (2.4.4) by $f''(x)$ and integrating over the half line $[0,\infty)$, we get

$$\int_{0+}^{\infty} C(\sqrt{\mu},f)d\rho(\mu) = f(0) + \int_0^{\infty} f(x)\{L(x,0;h) - L_1(x)\}dx\ .$$

Next, upon setting

$$\left.\begin{aligned} &\sigma_1(\lambda) = \frac{\lambda}{\pi}\ , \quad \sigma_2(\lambda) = \frac{2\lambda}{|\lambda|}\{\rho(\lambda^2) - \rho(0+)\}\ , \\ &M(x) = L(|x|,0;h) - L_1(x)\ , \quad \tilde{f}(\lambda) = \int_{-\infty}^{\infty} f(x)e^{-i\lambda x}dx\ , \end{aligned}\right\} \tag{2.4.5}$$

we rewrite the last equality in the form

$$\int_{-\infty}^{\infty} \tilde{f}(\lambda)d\sigma_2(\lambda) = \int_{-\infty}^{\infty} \tilde{f}(\lambda)d\sigma_1(\lambda) + \int_{-\infty}^{\infty} f(x)M(x)dx\ , \tag{2.4.6}$$

and remark that, since $M(x)$ is even and $\sigma_1(x)$ and $\sigma_2(x)$ are odd, it remains valid for arbitrary (and not only even) infinitely differentiable functions $f(x)$ with compact support.

We note that the function $M(x)$ is continuous and has bounded variation in every neighborhood of zero (and is even absolutely continuous); this follows from the differentiability of the functions $L(x,0;h)$ and

$L_1(x)$ on $[0,\infty)$. Moreover,

$$M(0) = -h - \rho(0+) + \rho(-\infty) , \tag{2.4.7}$$

because $L(0,0;h) = -h$ and $L_1(0) = \rho(0+) - \rho(-\infty)$. Now, if the function $M(x)$ would be summable on the real line and formula (2.4.6) would hold for all functions $f(x)$ which are square-summable on the real line, then, upon taking

$$f(x) = \frac{\sin Nx}{\pi x} , \quad \tilde{f}(\lambda) = \begin{cases} 1 , & |\lambda| < N , \\ 0 , & |\lambda| > N , \end{cases}$$

we would obtain the equality

$$\sigma_2(N) - \sigma_2(-N) = \sigma_1(N) - \sigma_1(-N) + \frac{1}{\pi} \int_{-\infty}^{\infty} \frac{\sin Nx}{x} M(x)dx .$$

But this in turn implies, upon returning to the function $\rho(\mu)$ and letting $N \to \infty$, that

$$\lim_{N\to\infty} \{\rho(N^2) - \rho(0+) - \frac{2}{\pi} N\} = -h - \rho(0+) + \rho(-\infty) ,$$

or equivalently, that

$$\rho(\mu) = \frac{2}{\pi} \sqrt{\mu} - h + \rho(-\infty) + o(1) , \tag{2.4.8}$$

as $\mu \to \infty$. In point of fact, all these operations are of course illegitimate. Nevertheless, the final conclusion is valid. To derive it rigorously, we must first prove a Tauberian-type theorem which will play an essential role in the following.

Suppose that the nondecreasing and left continuous functions $\sigma_1(\lambda)$ and $\sigma_2(\lambda)$ satisfy the following conditions:

I. *The following identity holds on the set of infinitely differentiable functions* $f(x)$ *which vanish outside the interval* $(-b,b)$ *:*

$$\int_{-\infty}^{\infty} \tilde{f}(\lambda)d\sigma_2(\lambda) = \int_{-\infty}^{\infty} \tilde{f}(\lambda)d\sigma_1(\lambda) + \int_{-\infty}^{\infty} f(x)G(x)dx , \tag{2.4.9}$$

where $G(x)$ *is a function which is defined and square-summable on* $(-b,b)$ *: and* $\tilde{f}(x)$ *designates the Fourier transform of* $f(x)$ *:*

$$\tilde{f}(\lambda) = \int_{-\infty}^{\infty} f(x)e^{-i\lambda x}dx .$$

II. *One of the functions, say,* $\sigma_1(\lambda)$, *has the property that*

$$\lim_{|T|\to\infty} \int_{-\infty}^{\infty} \frac{d\sigma_1(\lambda+T)}{1+\lambda^2} < \infty .$$

Under these assumptions, we shall presently find estimates for the difference $\sigma_2(\lambda) - \sigma_1(\lambda)$ for large values of $|\lambda|$. These estimates will depend upon the differentiability properties of the function $G(x)$.

First of all, we remark that condition II guarantees that the quantity

$$\sigma_1^*(\frac{1}{p}) = \overline{\lim_{|T|\to\infty}} \int_{-\infty}^{\infty} \frac{d\sigma_1(\lambda+T)}{1+p^2\lambda^2}$$

remains bounded for all values of $p > 0$ and does not increase when p is increased.

LEMMA 2.4.1. *The function* $\sigma_2(\lambda)$ *also satisfies condition* II, *and for every* $p \in (0,b)$

$$\sigma_1^*(\frac{1}{p}) \leqslant 5\sigma_1^*(\frac{1}{p}) . \tag{2.4.10}$$

PROOF. Let

$$\tilde{g}(\lambda) = \frac{\cosh 1 - \cos\lambda}{1+\lambda^2} , \quad g(x) = \frac{1}{2\pi} \int_{-\infty}^{\infty} \tilde{g}(\lambda)e^{i\lambda x}d\lambda . \tag{2.4.11}$$

It is readily verified that the function $g(x)$ vanishes for $|x| \geqslant 1$. Hence, the function $g(p(\lambda-T))$ is the Fourier transform of the function $p^{-1}g(p^{-1}x)e^{iTx}$, which vanishes for $|x| \geqslant p$, and

$$\tilde{g}_\varepsilon(\lambda) = \tilde{g}(p(\lambda-T)) \prod_{k=1}^{n} \left(\frac{\sin 2^{-k-1}\varepsilon\lambda}{2^{-k-1}\varepsilon\lambda} \right)^2$$

is the Fourier transform of an infinitely differentiable function $g_\varepsilon(x)$, which vanishes for $|x| \geqslant p+\varepsilon$.

It follows that if $p+\varepsilon \leqslant b$, then we are allowed to take $f(x) = g_\varepsilon(x)$ (and $\tilde{f}(\lambda) = \tilde{g}_\varepsilon(\lambda)$) in identity (2.4.9). In the resulting identity we now let $\varepsilon \to 0$. Since $\lim_{\varepsilon\to 0} \tilde{g}_\varepsilon(\lambda) = \tilde{g}(p(\lambda-T))$ and $\lim_{\varepsilon\to 0} \tilde{g}_\varepsilon(x) = p^{-1}g(p^{-1}x)$, we get, in view of conditions I and II,

$$\lim_{\varepsilon\to 0} \int_{-\infty}^{\infty} \tilde{g}_\varepsilon(\lambda)d\sigma_2(\lambda) = \int_{-\infty}^{\infty} \tilde{g}(p(\lambda-T))d\sigma_1(\lambda) + p^{-1} \int_{-\infty}^{\infty} g(p^{-1}x)e^{iTx}G(x)dx ,$$

which, upon using the well-known theorem on passing to the limit under the integral sign, yields the equality

$$\int_{-\infty}^{\infty} \tilde{g}(p(\lambda-T))d\sigma_2(\lambda) = \int_{-\infty}^{\infty} \tilde{g}(p(\lambda-T))d\sigma_1(\lambda) + p^{-1} \int_{-\infty}^{\infty} g(p^{-1}x)e^{iTx}G(x)dx . \quad (2.4.12)$$

From the definition of $g(\lambda)$, it follows that

$$\frac{\cosh 1 + 1}{1+\lambda^2} \geqslant \tilde{g}(\lambda) \geqslant \frac{\cosh 1 - 1}{1+\lambda^2} .$$

From these inequalities and (2.4.12) we get, after a change of variable,

$$\int_{-\infty}^{\infty} \frac{d\sigma_2(\lambda+T)}{1+p^2\lambda^2} \leqslant \frac{\cosh 1 + 1}{\cosh 1 - 1} \int_{-\infty}^{\infty} \frac{d\sigma_1(\lambda+T)}{1+p^2\lambda^2} + \frac{p^{-1}}{\cosh 1 - 1} \int_{-\infty}^{\infty} g(p^{-1}x)e^{iTx}G(x)dx .$$

Letting $T \to \infty$ and recalling that by the Riemann-Lebesgue theorem, the Fourier coefficients of a summable function tend to zero, we conclude that

$$\sigma_2^*(\tfrac{1}{p}) \leqslant \frac{\cosh 1 + 1}{\cosh 1 - 1} \sigma_1^*(\tfrac{1}{p}) \leqslant 5\sigma_1^*(\tfrac{1}{p}) ,$$

as claimed.

□

Remark. It follows from the proof of Lemma 2.4.1 that identity (2.4.9) can be extended by continuity to all functions $f(x)$ that vanish for $|x| > b$, and whose Fourier transform $\tilde{f}(\lambda)$ satisfies the condition

$$\sup_{-\infty<\lambda<\infty} (1+\lambda^2)|\tilde{f}(\lambda)| < \infty .$$

Now let $\varphi(x)$ be any even function with a bounded second derivative such that $\varphi(0) = 1$ and $\varphi(x) = 0$ for $|x| \geqslant 1$. An example is

$$\varphi(x) = \begin{cases} (1-x^2)^2 , & |x| \leqslant 1 , \\ 0 , & |x| \geqslant 1 . \end{cases} \quad (2.4.13)$$

We set

$$f(x) = \frac{\sin Nx}{\pi x} \varphi(p^{-1}x)e^{iTx} , \quad (2.4.14)$$

and calculate the Fourier transform of $f(x)$:

$$\tilde{f}(\lambda) = \frac{1}{\pi}\int_{-\infty}^{\infty}\frac{\sin Nx}{x}\varphi(p^{-1}x)e^{-i(\lambda-T)x}dx =$$

$$= \frac{1}{\pi}\int_{-\infty}^{\infty}\frac{\sin Nx}{x}e^{-i(\lambda-T)x}dx + \frac{1}{\pi}\int_{-\infty}^{\infty}\frac{\varphi(p^{-1}x)-1}{x}\sin Nx\, e^{-i(\lambda-T)x}dx =$$

$$= D(\frac{\lambda - T}{N}) + \frac{1}{2\pi i}\int_{-\infty}^{\infty}\frac{\varphi(t)-1}{t}\{e^{-i(\lambda-T-N)pt} - e^{-i(\lambda-T+N)pt}\}dt ,$$

where $D(\mu)$ is the Dirichlet discontinuous multiplier:

$$D(\mu) = \begin{cases} 1 , & |\mu| < 1 , \\ \frac{1}{2} , & |\mu| = 1 , \\ 0 , & |\mu| > 1 . \end{cases}$$

In view of the evenness of the function $\varphi(t)$, this in turn implies that

$$\tilde{f}(\lambda) = D(\frac{\lambda - T}{N}) + S_\varphi(p(\lambda - T + N)) - S_\varphi(p(\lambda - T - N)) , \tag{2.4.15}$$

where we put

$$S_\varphi(\mu) = \frac{1}{\pi}\int_{-\infty}^{\infty}\frac{\varphi(t)-1}{t}\sin \mu t dt .$$

Since

$$(1 + \mu^2)S_\varphi(\mu) = \frac{1}{\pi}\int_{-\infty}^{\infty}\{\frac{\varphi(t)-1}{t} - [\frac{\varphi(t)-1}{t}]''\}\sin \mu t dt ,$$

we have that

$$|S_\varphi(\mu)| \leqslant \frac{C_\varphi}{1+\mu^2} , \tag{2.4.16}$$

where

$$C_\varphi = \sup_{0\leqslant\mu<\infty}\left|\frac{1}{\pi}\int_{-\infty}^{\infty}\{\frac{\varphi(t)-1}{t} - [\frac{\varphi(t)-1}{t}]''\}\sin \mu t dt\right| .$$

In particular, if $\varphi(x)$ is given by formula (2.4.13), then

$$C_\varphi = \sup_{0\leqslant\mu<\infty}\frac{1}{\pi}\left|-\int_0^1\{t(2-t^2) - 6t\}\sin \mu t dt - \int_1^{\infty}\frac{\sin \mu t}{t} - 2\int_1^{\infty}\frac{\sin \mu t}{t^2}dt\right| < 2 ,$$

so that in this case,

$$|S_\varphi(\mu)| \leqslant \frac{2}{1+\mu^2} \; . \tag{2.4.16'}$$

We let $\tilde{G}_b(\lambda)$ denote the Fourier transform of the function $\varphi(b^{-1}x)G(x)$ and define the function $\tau(\lambda)$ by the formula

$$\tau(\lambda) = \sigma_2(\lambda) - \sigma_1(\lambda) - \frac{1}{2\pi}\int_{-\lambda}^{0} \tilde{G}_b(\xi)d\xi \tag{2.4.17}$$

at the points of continuity of the right-hand side, and by

$$\tau(\lambda) = \frac{\tau(\lambda+0) + \tau(\lambda-0)}{2}$$

at the points of discontinuity.

LEMMA 2.4.2. *If* $A_n > B_n$ *and* $\lim_{n\to\infty} |A_n| = \lim_{n\to\infty} |B_n| = \infty$, *then*

$$\varlimsup_{n\to\infty} |\tau(A_n) - \tau(B_n)| \leqslant 24\sigma_1^*(\tfrac{1}{b}) \; .$$

PROOF. We set $p = b$ in formula (2.4.14). Then the function $f(x)$ defined by this formula vanishes for $|x| \geqslant b$ and, in view of formulas (2.4.15) and (2.4.16), its Fourier transform $\tilde{f}(\lambda)$ satisfies the inequality $\sup_{\infty<\lambda<\infty} (1 + \lambda^2)|\tilde{f}(\lambda)| < \infty$. Consequently, identity (2.4.9) holds for $f(x)$ and $\tilde{f}(\lambda)$:

$$\int_{-\infty}^{\infty} \tilde{f}(\lambda)d\sigma_2(\lambda) = \int_{-\infty}^{\infty} \tilde{f}(\lambda)d\sigma_1(\lambda) + \frac{1}{\pi}\int_{-\infty}^{\infty} \varphi(b^{-1}x)e^{iTx}\frac{\sin Nx}{x}G(x)dx \; .$$

Upon using formula (2.4.15) and the definition of $\tau(\lambda)$, this can be reexpressed in the form

$$\tau(T+N) - \tau(T-N) = -\sum_{j=1,2} (-1)^j \int_{-\infty}^{\infty} S_\varphi(b\lambda)d\{\sigma_j(\lambda+T-N) - \sigma_j(\lambda+T+N)\} \; . \tag{2.4.18}$$

Now we choose T and N so that $T + N = A_n$, $T - N = B_n$. Then on the basis of inequality (2.4.16) and Lemma 2.4.1, we get

$$\varlimsup_{n\to\infty} |\tau(A_n) - \tau(B_n)| \leqslant 2\varlimsup_{n\to\infty}\sum_{j=1,2}\int_{-\infty}^{\infty}\frac{d\{\sigma_j(\lambda+A_n) + \sigma_j(\lambda+B_n)\}}{1+b^2\lambda^2} \leqslant$$

$$\leqslant 4\{\sigma_1^*(\tfrac{1}{b}) + \sigma_2^*(\tfrac{1}{b})\} \leqslant 24\sigma_1^*(\tfrac{1}{b}) \; ,$$

as asserted.

□

We now remark that the function $\tau(\lambda)$ is real-valued. In fact, from its definition (formula (2.4.17)), it follows that $\operatorname{Im}\tau(0) = 0$, and since the right-hand side of formula (2.4.18) is real, this further yields the reality of $\tau(\lambda)$. Hence, we can introduce the meaningful quantities $\overline{\lim}_{|\lambda|\to\infty} \tau(\lambda) = M$ and $\underline{\lim}_{|\lambda|\to\infty} \tau(\lambda) = m$, which by Lemma 2.4.2 are finite and tisfy the inequality $M - m \leqslant 24\sigma_1^*(\frac{1}{b})$.

This shows that all the limit values of the function $\tau(\lambda)$ for $|\lambda| \to \infty$ lie on the segment $[m,M]$, the length of which does not exceed $24\sigma_1^*(\frac{1}{b})$. Thus, upon setting $C = \frac{1}{2}(M+m)$, we get

$$\overline{\lim_{|\lambda|\to\infty}} |\tau(\lambda) - C| \leqslant 12\sigma_1^*(\tfrac{1}{b}) . \tag{2.4.19}$$

The next result is a straightforward corollary of this inequality.

THEOREM 2.4.1. *Suppose that the functions* $\sigma_2(\lambda)$ *and* $\sigma_1(\lambda)$ *satisfy conditions* I, II, *and that* $\tilde{G}_1(\lambda)$ *is the Fourier transform of an arbitrary function* $G_1(x) \in L_1(-\infty,\infty)$, *which coincides with* $G(x)$ *in some neighborhood of zero. Then there is a constant* C_1 *such that*

$$\overline{\lim_{|\lambda|\to\infty}} \left|\sigma_2(\lambda) - \sigma_1(\lambda) - \frac{1}{\pi}\int_{-\infty}^{0} \tilde{G}_1(\mu) - C_1\right| \leqslant 12\sigma_1^*(\tfrac{1}{b}) .$$

PROOF. If $G_1(x)$ coincides with $G(x)$ in some neighborhood of zero, then the function $\frac{G(x)\varphi(b^{-1}x) - G_1(x)}{ix}$ is also summable over the real line, and by the Riemann-Lebesgue theorem (which asserts that the Fourier transform of a summable function tends to zero at infinity),

$$\lim_{|\lambda|\to\infty} \frac{1}{2\pi}\int_{-\lambda}^{0} \{\tilde{G}_b(\mu) - \tilde{G}_1(\mu)\}d\mu =$$

$$= \lim_{|\lambda|\to\infty} \frac{1}{2\pi}\int_{-\infty}^{\infty} \frac{G(x)\varphi(b^{-1}x) - G_1(x)}{ix}(e^{i\lambda x} - 1)dx =$$

$$= -\frac{1}{2\pi}\int_{-\infty}^{\infty} \frac{G(x)\varphi(b^{-1}x) - G_1(x)}{ix}dx .$$

From this it follows, by definition (2.4.17) and inequality (2.4.19), that

$$\overline{\lim_{|\lambda|\to\infty}} \left|\sigma_2(\lambda) - \sigma_1(\lambda) - \frac{1}{2\pi}\int_{-\lambda}^{0} \tilde{G}(\mu)d\mu + C_1\right| \leq 12\sigma_1^*(\tfrac{1}{b}) ,$$

where

$$C_1 = C - \frac{1}{2\pi}\int_{-\infty}^{\infty} \frac{G(x)\varphi(b^{-1}x) - G_1(x)}{ix} dx .$$

The theorem is proved. □

COROLLARY. *Suppose that the functions* $\sigma_2(\lambda)$ *and* $\sigma_1(\lambda)$ *satisfy conditions* I, II, *and the function* $G(x)$ *satisfies conditions in the neighborhood of zero which guarantee the convergence of its Fourier series at the point* $x = 0$ *to* $G(x)$ *(for example,* $G(x)$ *is continuous at* $x = 0$ *and has bounded variation in a neighborhood of this point). Then*

$$\lim_{\lambda\to+\infty} |\sigma_2(\lambda) - \sigma_2(-\lambda) - \sigma_1(\lambda) + \sigma_1(-\lambda) - G(0)| \leq 24\sigma_1^*(\tfrac{1}{b}) . \tag{2.4.20}$$

PROOF. According to Theorem 2.4.1,

$$\overline{\lim_{\lambda\to+\infty}} \left|\sigma_2(\lambda) - \sigma_2(-\lambda) - \sigma_1(\lambda) + \sigma_1(-\lambda) - \frac{1}{2\pi}\int_{-\lambda}^{\lambda} \tilde{G}_1(\mu)d\mu\right| \leq 24\sigma_1^*(\tfrac{1}{b}) ,$$

where for $\tilde{G}_1(\mu)$ one can take the Fourier transform of a compactly supported function $G_1(x)$ which coincides with $G(x)$ in a neighborhood of zero. Under our assumptions on $G(x)$, the localization principle for Fourier integrals yields

$$\lim_{\lambda\to+\infty} \frac{1}{2\pi}\int_{-\lambda}^{\lambda} \tilde{G}_1(\mu)d\mu = G_1(0) = G(0) ,$$

as desired. □

We are now ready for the rigorous proof of the asymptotic formula (2.4.8). In fact, equality (2.4.6) shows that the functions $\sigma_1(\lambda)$ and $\sigma_2(\lambda)$ defined by formulas (2.4.5) satisfy conditions I, II for every value of b. Moreover, the function $G(x) = M(x)$ is continuous and has bounded variation in a neighborhood of zero, and

$$\sigma_1^*(\tfrac{1}{b}) = \overline{\lim_{T\to\infty}} \int_{-\infty}^{\infty} \frac{d\sigma_1(\lambda+T)}{1 + b^2\lambda^2} = \frac{1}{\pi}\int_{-\infty}^{\infty} \frac{d\lambda}{1 + b^2\lambda^2} = \frac{1}{b} .$$

Thus, all the conditions under which we proved inequality (2.4.20) are fulfilled. Recalling formulas (2.4.5) and (2.4.7), we find that in the present case,

$$\sigma_2(\lambda) - \sigma_2(-\lambda) = \rho(\lambda^2) - \rho(0+) ,$$

$$\sigma_1(\lambda) - \sigma_1(-\lambda) = \frac{2\pi}{\lambda} ,$$

$$G(0) = M(0) = -h - \rho(0+) + \rho(-\infty) ,$$

while inequality (2.4.20) takes the form

$$\overline{\lim_{\lambda\to+\infty}} \left|\rho(\lambda^2) - \frac{2\lambda}{\pi} + h - \rho(-\infty)\right| \leq \frac{24}{b} ;$$

we note that here $b > 0$ may be taken arbitrarily large. Therefore,

$$\lim_{\lambda\to+\infty} \left|\rho(\lambda^2) - \frac{2\lambda}{\pi} + h - \rho(-\infty)\right| = 0 ,$$

which is equivalent to formula (2.4.8).

We have thus proved the following result:

THEOREM 2.4.2. *The spectral functions* $\rho(\mu)$ *of a symmetric boundary value problem* (2.2.1), (2.2.2), *must satisfy the following conditions:*

$$\lim_{\mu\to-\infty} e^{\sqrt{|\mu|}x} (\rho(\mu) - \rho(-\infty)) = 0 \quad (0 \leq x < \infty)$$

and

$$\lim_{\mu\to+\infty} \{\rho(\mu) - \rho(-\infty) - \frac{2}{\pi}\sqrt{\mu}\} = -h .$$

□

According to Theorem 2.2.3, every function $f(x) \in L_2[0,\infty)$ admits an integral expansion with respect to the eigenfunctions of a symmetric boundary value problem, which converges in the norm of $L_2[0,\infty)$. We next prove a theorem asserting that such an expansion and the Fourier cosine integral expansion of the same function $f(x)$ are equiconvergent. But first we need a lemma.

LEMMA 2.4.3. *If* $\rho(\mu)$ *is a spectral function of a symmetric boundary value problem and* $F(\mu) \in L_{2,\rho}(-\infty,\infty)$, *then for every* $b > 0$,

$$\lim_{|N|\to\infty} \int_{0+}^{\infty} \frac{|F(\mu^2)|}{1 + b^2(\mu-N)^2}\, d\rho(\mu^2) = 0 .$$

PROOF. It obviously suffices to prove this equality for $N \to +\infty$. Let $I_1(N)$ and $I_2(N)$ denote the segment $[\frac{N}{6}, \frac{3N}{2}]$ and its complement in the half line $[0,\infty)$, respectively. Upon writing the integral of interest as a sum of two integrals, one over $I_1(N)$ and one over $I_2(N)$, and applying the Cauchy-Bunyakovskii inequality to each of these, we get

$$\int_{0+}^{\infty} \frac{|F(\mu^2)|}{1 + b^2(\mu-N)^2}\, d\rho(\mu^2) \leq$$

$$\leq \sum_{k=1,2} \left\{ \int_{I_k(N)} |F(\mu^2)|^2 d\rho(\mu^2) \right\}^{\frac{1}{2}} \left\{ \int_{I_k(N)} \frac{d\rho(\mu^2)}{[1 + b^2(\mu-N)^2]^2} \right\}^{\frac{1}{2}} .$$

Since for $\mu \in I_2(N)$,

$$1 + b^2(\mu-N)^2 > 1 + (\frac{bN}{2})^2 ,$$

we have a fortiori

$$\int_{0+}^{\infty} \frac{|F(\mu^2)|}{1 + b^2(\mu-N)^2}\, d\rho(\mu^2) \leq \left\{ \int_{0+}^{\infty} \frac{d\rho(\mu^2)}{1 + b^2(\mu-N)^2} \right\}^{\frac{1}{2}} \times$$

$$\times \left[\left\{ \int_{I_1(N)} |F(\mu^2)|^2 d\rho(\mu^2) \right\}^{\frac{1}{2}} + \left\{ \frac{4}{4 + b^2N^2} \int_{0+}^{\infty} |F(\mu^2)|^2 d\rho(\mu^2) \right\}^{\frac{1}{2}} \right] .$$

Since $F(\mu) \in L_{2,\rho}(-\infty,\infty)$, both terms in the square brackets tend to zero as $N \to \infty$. Hence, to complete the proof, it suffices to verify the inequality

$$\overline{\lim_{N\to+\infty}} \int_{0+}^{\infty} \frac{d\rho_2(\mu)}{1 + b^2(\mu-N)^2} < \infty .$$

But

$$\int_{0+}^{\infty} \frac{d\rho(\mu^2)}{1 + b_2(\mu-N)^2} = \int_{0+}^{\infty} \frac{d\{\rho(\mu^2) - \rho(0+)\}}{1 + b^2(\mu-N)^2} \leq$$

$$\leq \int_{-\infty}^{\infty} \frac{d\sigma_2(\mu)}{1 + b^2(\mu-N)^2} = \int_{-\infty}^{\infty} \frac{d\sigma_2(\mu+N)}{1 + b^2\mu^2} ,$$

where the function $\sigma_2(\mu)$ is defined by formula (2.4.5) and is related to $\sigma_1(\mu) = \frac{\mu}{\pi}$ through identity (2.4.6). Hence, upon using Lemma 2.4.1, we get

$$\overline{\lim_{N\to\infty}} \int_{0+}^{\infty} \frac{d\rho(\mu^2)}{1 + b^2(\mu-N)^2} \leq \overline{\lim_{N\to\infty}} \int_{-\infty}^{\infty} \frac{d\sigma_2(\mu+N)}{1 + b^2\mu^2} \leq 5\sigma_1^*(\tfrac{1}{b}) = \frac{5}{b} < \infty ,$$

as needed.

□

THEOREM 2.4.3. *Let* $f(x)$ *be an arbitrary function in* $L_2[0,\infty)$. *Let*

$$\omega(\sqrt{\mu},f) = \int_0^\infty f(x)\omega(\sqrt{\mu},x)dx ,$$

and

$$C(\sqrt{\mu},f) = \int_0^\infty f(x)\cos\sqrt{\mu}\, xdx$$

(the integrals converge in the norms of the spaces $L_{2,\rho}(-\infty,\infty)$ *and* $L_{2,\sqrt{\mu}}(0,\infty)$, *respectively).*

Then for every $b > 0$, *the integral*

$$\int_{-\infty}^{0+} \omega(\sqrt{\mu},f)\omega(\sqrt{\mu},x)d\rho(\mu)$$

converges absolutely and uniformly in $x \in [0,b]$, *and*

$$\lim_{N\to\infty} \sup_{0\leq x\leq b} \left| \int_{-\infty}^{N} \omega(\sqrt{\mu},f)\omega(\sqrt{\mu},x)d\rho(\mu) - \frac{2}{\pi}\int_0^N C(\sqrt{\mu},f)\cos\sqrt{\mu}\, xd\sqrt{\mu}\right| = 0 .$$

PROOF. The function

$$u_M(x,t) = \int_{-M}^{M} \omega(\sqrt{\mu},f)\omega(\sqrt{\mu},x)\omega(\sqrt{\mu},t)d\rho(\mu) \tag{2.4.21}$$

clearly satisfies the equation

$$u_{xx} - q(x)u = u_{tt} - q(t)u$$

and the initial data

$$u(x,0) = f_M(x) \quad , \quad u_t(x,0) = hf_M(x) ,$$

where

$$f_M(x) = \int_{-M}^{M} \omega(\sqrt{\mu},f)\omega(\sqrt{\mu},x)d\rho(\mu) .$$

Hence, upon using the Riemann formula (1.1.7) and observing that

$u_M(x,t) = u_M(t,x)$, we can represent it in the form

$$u_M(x,t) = \frac{f_M(x+t) + f_M(|x-t|)}{2} + \int_{|x-t|}^{x+t} W(x,t,u) f_M(u)\,du \ , \tag{2.4.22}$$

where the kernel $W(x,t,u)$ is bounded in any bounded domain of variation of its arguments. Next, upon applying the transformation operator which takes the function $\omega(\sqrt{\mu},t)$ into $\cos\sqrt{\mu}\,t$ to both sides of equality (2.4.21), we get

$$\int_{-M}^{M} \omega(\sqrt{\mu},f)\omega(\sqrt{\mu},x)\cos\sqrt{\mu}\,t\,d\rho(\mu) = u_M(x,t) + \int_0^t L(t,\xi)u_M(x,\xi)\,d\xi \ ,$$

which yields, in turn, the identity

$$\int_{-M}^{M} C(\sqrt{\mu},g)\omega(\sqrt{\mu},f)\omega(\sqrt{\mu},x)\,d\rho(\mu) =$$

$$= \int_0^\infty g(t)\left\{u_M(x,t) + \int_0^t L(t,\xi)u_M(x,\xi)\,d\xi\right\} dt \tag{2.4.23}$$

for all even infinitely differentiable functions $g(t)$ with compact support.

Now we let $M \to \infty$. By Theorem 2.2.3, the sequence $f_M(x)$ converges in the norm of $L_2[0,\infty)$ to the function $f(x)$. Hence, equalities (2.4.22), (2.4.23) yield

$$\int_{-\infty}^{\infty} C(\sqrt{\mu},g)\omega(\sqrt{\mu},f)\omega(\sqrt{\mu},x)\,d\rho(\mu) =$$

$$= \int_0^\infty g(t)\left\{\frac{f(x+t) + f(|x-t|)}{2} + A_1(x,t)\right\} dt \ , \tag{2.4.24}$$

where

$$A_1(x,t) = \int_{|x-t|}^{x+t} W(x,t,\xi)f(\xi)\,d\xi +$$

$$+ \int_0^t L(t,\xi)\left\{\frac{f(x+\xi) + f(|x-\xi|)}{2} + \int_{|x-\xi|}^{x+\xi} W(x,\xi,\eta)f(\eta)\,d\eta\right\} d\xi \ .$$

Since the functions $W(t,x,\xi)$ and $L(t,\xi)$ are bounded in every bounded domain of variation of their arguments, an application of the Cauchy-Bunyakovskii inequality gives

$$\sup_{0\leq x\leq b} |A_1(x,t)| \leq C_1 t^{\frac{1}{2}} \left\{ \int_0^\infty |f(\xi)|^2 d\xi \right\}^{\frac{1}{2}} \tag{2.4.25}$$

for all $t \in [0,b]$.

The integral (2.4.2) converges for all values of x , as we showed above, and from the existence of the transformation operators we derive the estimate

$$\sup_{0\leq x\leq b} |\omega(\sqrt{\mu},x)| \leq C \exp b\sqrt{\mu} ,$$

which holds uniformly in $\mu \in (-\infty,0]$. Consequently, the integral

$$\int_{-\infty}^{0+} \omega(\sqrt{\mu},f)\omega(\sqrt{\mu},x)d\rho(\mu)$$

converges absolutely and uniformly in $x \in [0,b]$, and the function

$$A_2(x,t) = \int_{-\infty}^{0+} \omega(\sqrt{\mu},f)\omega(\sqrt{\mu},x)[\cos\sqrt{\mu}\, t - 1]d\rho(\mu) \tag{2.4.26}$$

satisfies the inequality

$$\sup_{0\leq x\leq b} |A_2(x,t)| \leq C_2 t \tag{2.4.27}$$

for all $t \in [0,b]$. Since by the convolution theorem for Fourier integrals

$$\int_0^\infty g(t)\,\frac{f(x+t) + f(|x-t|)}{2}\,dt = \frac{2}{\pi}\int_0^\infty C(\sqrt{\mu},g)C(\sqrt{\mu},f)\cos\sqrt{\mu}\,x d\sqrt{\mu} ,$$

we can reexpress equality (2.4.24) as

$$\int_{0+}^\infty C(\sqrt{\mu},g)\omega(\sqrt{\mu},f)\omega(\sqrt{\mu},x)d\rho(\mu) - \frac{2}{\pi}\int_0^\infty C(\sqrt{\mu},g)C(\sqrt{\mu},f)\cos\sqrt{\mu}x\, d\sqrt{\mu} +$$

$$+ C(0,g)\int_{-\infty}^{0+} \omega(\sqrt{\mu},f)\omega(\sqrt{\mu},x)d\rho(\mu) = \int_0^\infty g(t)A_3(x,t)dt , \tag{2.4.28}$$

where $A_3(x,t) = A_1(x,t) - A_2(x,t)$ and $A_2(x,t)$ is given by formula (2.4.26). Moreover, in view of the estimates (2.4.25) and (2.4.27),

$$\sup_{0\leq x\leq b} |A_3(x,t)| \leq C_3 t^{\frac{1}{2}} \quad (0 \leq t \leq b) . \tag{2.4.29}$$

Now, upon setting

$$g(t) = g_N(t) = \frac{2}{\pi}\frac{\sin Nt}{t}\varphi(\frac{t}{a}) \quad (0 < a < b)$$

in identity (2.4.28), and invoking formula (2.4.15), we get

$$\int_{-\infty}^{N^2} \omega(\sqrt{\mu},f)\omega(\sqrt{\mu},x)d\rho(\mu) - \frac{2}{\pi}\int_0^{N^2} C(\sqrt{\mu},f)\cos\sqrt{\mu}\, xd\sqrt{\mu} =$$

$$= (1 - C(0,g_N)) \int_{-\infty}^{0+} \omega(\sqrt{\mu},f)\omega(\sqrt{\mu},x)d\rho(\mu) + \frac{1}{\pi}\int_0^a \frac{\sin Nt}{t}\varphi(\frac{t}{a})A_3(x,t)dt -$$

$$- \int_{0+}^{\infty} \omega(\sqrt{\mu},f)\omega(\sqrt{\mu},x)[S_\varphi(a(\sqrt{\mu} + N)) - S_\varphi(a(\sqrt{\mu} - N))]d\rho(\mu) +$$

$$+ \frac{2}{\pi}\int_0^{\infty} C(\sqrt{\mu},f)\cos\sqrt{\mu}\, x[S_\varphi(a(\sqrt{\mu} + N)) - S_\varphi(a(\sqrt{\mu} - N))]d\sqrt{\mu} .$$

Let us examine each term of the right-hand side of this identity separately. Since the function $\varphi(x)$ is infinitely differentiable, $\varphi(0) = 1$, and

$$1 - C(0,g_N) = \frac{2}{\pi}\int_0^{\infty} \frac{\sin Nt}{t} dt - \frac{2}{\pi}\int_0^{\infty} \frac{\sin Nt}{t}\varphi(\frac{t}{a})dt =$$

$$= \frac{2}{\pi}\int_0^{\infty} \frac{1 - \varphi(a^{-1}t)}{t} \sin Ntdt ,$$

the Riemann-Lebesgue theorem yields $\lim\limits_{N\to\infty} (1 - C(0,g_N)) = 0$. Consequently, the first term tends to zero as $N \to \infty$, uniformly in $x \in [0,b]$. The second and third terms also tend to zero uniformly in $x \in [0,b]$, as follows from Lemma 2.4.2 upon observing that

$$\sup_{\substack{0\le x\le b\\ 0\le\mu<\infty}} |\omega(\sqrt{\mu},x)| < \infty \quad , \quad |\cos\sqrt{\mu}\, x| < 1 ,$$

and that the function $S_\varphi(\lambda)$ satisfies the estimate (2.4.16). The fourth term can be estimated with the help of inequality (2.4.29):

$$\sup_{0\le x\le b} \left| \frac{1}{\pi}\int_0^a \frac{\sin Nt}{t}\varphi(a^{-1}t)A_3(x,t)dt\right| \le \frac{C_3}{\pi}\int_0^a t^{-\frac{1}{2}}dt = \frac{2C_3\sqrt{a}}{\pi} .$$

Hence,

$$\overline{\lim_{N\to\infty}} \sup_{0\leq x\leq b} \left| \int_{-\infty}^{N^2} \omega(\sqrt{\mu},f)\omega(\sqrt{\mu},x)d\rho(\mu) - \frac{2}{\pi}\int_0^{N^2} C(\sqrt{\mu},f)\cos\sqrt{\mu}x\, d\sqrt{\mu}\right| \leq \frac{2C_3\sqrt{a}}{\pi} ,$$

and, since a is arbitrarily small,

$$\overline{\lim_{N\to\infty}} \sup_{0\leq x\leq b} \left| \int_{-\infty}^{N^2} \omega(\sqrt{\mu},f)\omega(\sqrt{\mu},x)d\rho(\mu) - \frac{2}{\pi}\int_0^{N^2} C(\sqrt{\mu},f)\cos\sqrt{\mu}x\, d\sqrt{\mu}\right| = 0 ,$$

as desired.

PROBLEMS

1. Show that the nondecreasing function $\rho(\mu)$ $(-\infty < \mu < \infty)$ is a spectral function of a symmetric boundary value problem of the form (2.2.1), (2.2.2) if and only if the function

$$\Phi(x) = \int_{-\infty}^{\infty} \frac{1 - \cos\sqrt{\mu}\, x}{\mu} d\rho(\mu) \quad (0 < x < \infty)$$

is thrice continuously differentiable and $\Phi'(0+) = 1$.

<u>Hint</u>. According to Theorem 2.3.1, it suffices to verify that

$$(f(\lambda)\overline{f(\overline{\lambda})},R) = \int_{-\infty}^{\infty} |f(\sqrt{\mu})|^2 d\rho(\mu) > 0$$

if $f(\lambda) \in CK^2(\sigma)$ and $f(\lambda) \not\equiv 0$. Assuming the contrary, you have

$$0 = \int_{-\infty}^{\infty} |f(\sqrt{\mu})|^2 d\rho(\mu) \geq \int_{0+}^{\infty} |f(\sqrt{\mu})|^2 d\rho(\mu) = \int_{0+}^{\infty} |f(\lambda)|^2 d\rho(\mu^2) ,$$

which is possible if and only if the growth points of the function $\rho(\lambda^2)$ form a discrete set, consisting of the zeros of the function $f(\lambda)$. Let $n(a)$ denote the number of zeros of $f(\lambda)$ in the interval $(0,a)$. Then, since $f(\lambda)$ is an entire function of exponential type σ , $\overline{\lim_{a\to\infty}} \frac{n(a)}{a} < \infty$. On the other hand, if the conditions of the problem are satisfied, then, by Theorem 2.4.1, the function $\rho(\lambda^2)$ admits the asymptotics

$$\rho(\lambda^2) = \frac{2}{\pi}\lambda + c + o(1) \quad (\lambda \to +\infty) ,$$

from which it follows that for every $\delta > 0$, each interval $(b,b+\delta)$ contains growth points of $\rho(\lambda^2)$, provided that b is large enough. Hence, $\lim_{a\to\infty} \frac{m(a)}{a} = \infty$, where $m(a)$ is the number of growth points of the function $\rho(\lambda^2)$ in $(0,a)$, which is a contradiction.

2. Show that the spectral function $\rho(\mu)$ of the symmetric boundary value problem

$$-y'' + q(x)y = \lambda^2 y \quad (0 \leq x < a < \infty) \quad , \quad y(0)h - y'(0) = 0$$

($\operatorname{Im} q(x) = 0$, $\operatorname{Im} h = 0$; function $q(x)$ is continuous for $0 \leq x < a$) satisfies the conditions

$$\lim_{\mu\to-\infty} e^{\sqrt{|\mu|}x}\{\rho(\mu) - \rho(-\infty)\} > 0 \quad (0 \leq x < 2a)$$

and

$$\overline{\lim_{\mu\to+\infty}} \left|\rho(\mu) - \rho(-\infty) + \frac{2}{\pi}\sqrt{\mu} + h\right| \leq \frac{12}{a} .$$

Hint. This follows from Theorem 2.4.1 and its corollary (formula (2.4.20)) if you notice that in the present case equality (2.4.1) holds for all $x \in [0,2a)$.

3. Show that the spectral matrices (operator-valued measures) $\rho(\mu)$ of symmetric operator Sturm-Liouville boundary value problems satisfy the conditions

$$\lim_{\mu\to-\infty} e^{\sqrt{|\mu|}x}|\rho(\mu) - \rho(-\infty)| = 0 \quad (0 \leq x < \infty)$$

and

$$\lim_{\mu\to+\infty} \left|\rho(\mu) - \rho(-\mu) - \frac{2}{\pi}\sqrt{\mu}\, I + h\right| = 0 .$$

Here and in the following, $|A|$ designates the norm of the operator $A \in \mathcal{O}H$.

4. Show that the spectral matrices $\rho(\mu)$ of symmetric Dirac boundary value problems satisfy the condition

$$\lim_{|\mu|\to\infty} \left|\rho(\mu) - \frac{\lambda}{\pi} P - \frac{P\Omega(0)P}{\pi} \ln|\lambda| + \frac{\lambda}{2|\lambda|} PB\Omega(0)P\right| = 0$$

if for $x \to 0$,

$$|\Omega(x) - \Omega(0)| \leqslant Cx^{\alpha} \quad (\alpha > 0) . \tag{2.4.30}$$

Hint. It follows from the formula $R = \frac{1}{\pi}(P + \tilde{\omega}_0(L,P))$ that

$$(\omega_0(\lambda,f;P),R) = \frac{1}{\pi}\int_{-\infty}^{\infty} \omega_0(\lambda,f;P)Pd\lambda + \int_0^{\infty} f(x)L_p(x,0)Pdx .$$

After elementary transformations, this equality takes the form

$$\int_{-\infty}^{\infty} \tilde{f}(\lambda)d\rho(\lambda) = 2f(0)P - \int_{-\infty}^{\infty} f(x)\left\{P - i\frac{x}{|x|}PB\right\} L_p(|x|,0)Pdx ,$$

where $f(x)$ is an arbitrary infinitely differentiable function with compact support and $\tilde{f}(\lambda)$ designates its Fourier transform. Next, from the equality

$$L_p(x,0) + K_p(x,0) + \int_0^x K_p(x,t)L_p(t,0)dt = 0 ,$$

formulas (1.2.7), (1.2.9), (1.2.39'), the estimate (1.2.38') and condition (2.4.30), you find that $|L_p(x,0) + B\Omega(0)P| \leqslant Cx^{\alpha}$ $(x \to 0)$. Hence

$$\int_{-\infty}^{\infty} \tilde{f}(\lambda)d\rho(\lambda) = \frac{1}{\pi}\int_{-\infty}^{\infty} \tilde{f}(\lambda)Pd\lambda + \int_{-\infty}^{\infty} f(x)P\left\{M(x) - B\Omega(0) - i\frac{x}{|x|}\Omega(0)\right\} Pdx ,$$

where $|M(x)| \leqslant Cx^{\alpha}$ for $x \to 0$. The desired result now follows from Theorem 2.4.1 and Dini's test for the convergence of a Fourier series and of its conjugate.

5. Suppose the odd functions $\sigma_2(\lambda)$ and $\sigma_1(\lambda) = \lambda$ satisfy conditions I, II for every $b > 0$, and the function $G(x) = G(-x)$ is twice differentiable on the half line $0 < x < \infty$:

$$G(x) = G(0+) + xG'(0+) + \int_0^x (x-t)G''(t)dt \quad (0 < x < \infty) ,$$

where, in addition,

$$M_2(b) = \sup_{0<x<b} |G''(x)| < \infty \quad (0 < b < \infty) .$$

Show that under these assumptions Theorem 2.4.1 can be sharpened to

$$\sup_{\lambda \geqslant N > 0} |\sigma_2(\lambda) - \lambda - \tfrac{1}{2}G(0+)| \leqslant \frac{10^3}{b(N)}\left\{1 + \frac{G(0+) + G'(0+)b(N)}{N}\right\} ,$$

where $b(N)$ is the root of the equation $b^2M_2(b) = N$.

Hint. Set $T = 0$ in formula (2.4.18) and estimate its right-hand side by straightforward use of equality (2.4.12), in which you have to integrate the second right-hand term by parts once, using $e^{iTx} = \frac{1}{iT}\frac{d}{dx}e^{iTx}$. Then, in order to estimate the difference

$$\frac{1}{\pi}\int_{-\infty}^{\infty}\varphi(b^{-1}x)\,\frac{\sin Nx}{x}\,G(x)dx - G(0+) =$$

$$= -\frac{2}{\pi N}\int_{0}^{\infty}\varphi(b^{-1}x)\,\frac{G(x)-G(0+)}{x}\,d(\cos Nx) +$$

$$+ G(0+)\left\{\frac{1}{\pi}\int_{-\infty}^{\infty}\varphi(b^{-1}x)\,\frac{\sin Nx}{x}\,dx - 1\right\},$$

you integrate the first right-hand term by parts and use formula (2.4.15) to estimate the second. Finally, choose b so that the condition $b^2M_2(b) = N$ is satisfied.

6. Use the previous problem to estimate the quantity $\rho(\mu) - \frac{2}{\pi}\sqrt{\mu} - \rho(-\infty) + h$ for $\mu \to +\infty$, where $\rho(\mu)$ is a spectral function of a Sturm-Liouville boundary value problem in which the function $q(x)$ is continuously differentiable on the half line $(0,\infty)$.

CHAPTER 3

THE BOUNDARY VALUE PROBLEM OF SCATTERING THEORY

1. AUXILIARY PROPOSITIONS

This chapter is devoted to a detailed study of the boundary value problem generated on the half line $0 \leq x < \infty$ by the differential equation

$$-y'' + q(x)y = \lambda^2 y \tag{3.1.1}$$

and the boundary condition

$$y(0) = 0 \tag{3.1.2}$$

for the important special case where the function $q(x)$ is real and satisfies the condition

$$\int_0^\infty x|q(x)|dx < \infty , \tag{3.1.3}$$

which is assumed to hold throughout the chapter. From condition (3.1.3) it is clear that (3.1.1) reduces to the simpler equation $-y'' = \lambda^2 y$ when $x \to \infty$. This permits us a complete investigation of the properties of the solution to equation (3.1.1), and this is our goal in the present section.

We shall use the following notation:

$$\sigma(x) = \int_x^\infty |q(t)|dt \quad , \quad \sigma_1(x) = \int_x^\infty \sigma(t)dt . \tag{3.1.4}$$

One can easily verify that condition (1.3.3) is equivalent to the summability of the function $\sigma(x)$ over the entire half line $[0,\infty)$, i.e., to the inequality $\sigma_1(0) < \infty$.

LEMMA 3.1.1. *For any* λ *from the closed upper half plane, equation* (3.1.1) *has a solution* $e(\lambda,x)$ *that can be represented in the form*

$$e(\lambda,x) = e^{i\lambda x} + \int_x^\infty K(x,t)e^{i\lambda t}dt \ , \tag{3.1.5}$$

where the kernel $K(x,t)$ *satisfies the inequality*

$$|K(x,t)| \leqslant \frac{1}{2}\,\sigma\left(\frac{x+t}{2}\right) \exp\{\sigma_1(x) - \sigma_1\left(\frac{x+t}{2}\right)\} \ . \tag{3.1.6}$$

In addition,

$$K(x,x) = \frac{1}{2}\int_x^\infty q(t)dt \ . \tag{3.1.6'}$$

PROOF. Consider the integral equation

$$e(\lambda,x) = e^{i\lambda x} + \int_x^\infty \frac{\sin \lambda(t-x)}{\lambda}\, q(t)e(\lambda,t)dt \ , \tag{3.1.7}$$

which is equivalent to the differential equation (3.1.1) with the boundary condition $\lim\limits_{x\to\infty} e^{-i\lambda x}e(\lambda,x) = 1$, and look for a solution of the form (3.1.5). Substituting the expression (3.1.5) for $e(\lambda,x)$ in (3.1.7), we get

$$\int_x^\infty K(x,t)e^{i\lambda t}dt = \int_x^\infty \frac{\sin \lambda(s-x)}{\lambda}\, e^{i\lambda s}q(s)ds \ +$$

$$+ \int_x^\infty q(s)\left\{\int_s^\infty \frac{\sin \lambda(s-x)}{\lambda}\, e^{i\lambda u}K(s,u)du\right\} ds \ . \tag{3.1.7'}$$

Since

$$\frac{\sin \lambda(s-x)}{\lambda}\, e^{i\lambda u} = \frac{1}{2}\int_{-s+u+x}^{s+u-x} e^{i\lambda\xi}d\xi \ , \tag{3.1.8}$$

we have

$$\int_x^\infty \frac{\sin \lambda(s-x)}{\lambda}\, e^{i\lambda s}q(s)ds = \frac{1}{2}\int_x^\infty q(s)\left\{\int_x^{2s-x} e^{i\lambda t}dt\right\} ds \ ,$$

whence, upon changing the order of integration, we find that

$$\int_x^\infty \frac{\sin\lambda(s-x)}{\lambda} e^{i\lambda s} q(s)ds = \frac{1}{2}\int_x^\infty e^{i\lambda t}\{\int_{\frac{x+t}{2}}^\infty q(s)ds\}dt . \tag{3.1.9}$$

Using (3.1.8) once more, we obtain

$$\int_x^\infty q(s)\left\{\int_s^\infty \frac{\sin\lambda(x-s)}{\lambda} e^{i\lambda u}K(s,u)du\right\}ds =$$

$$= \frac{1}{2}\int_x^\infty q(s)\left\{\int_s^\infty K(s,u)\int_{-s+u+x}^{s+u-x} e^{i\lambda t}dtdu\right\}ds .$$

Now extending the function $K(s,u)$ by zero for $u < s$ yields for all $s \geq x$,

$$\int_s^\infty K(s,u)\int_{-s+u+x}^{s+u-x} e^{i\lambda t}dtdu = \int_{-\infty}^\infty K(s,u)\left\{\int_{-s+u+x}^{s+u-x} e^{i\lambda t}dt\right\}du =$$

$$= \int_{-\infty}^\infty e^{i\lambda t}\left\{\int_{t-(s-x)}^{t+(s-x)} K(s,u)du\right\}dt = \int_0^\infty e^{i\lambda t}\left\{\int_{t-(s-x)}^{t+(s-x)} K(s,u)du\right\}dt ,$$

because for $t < x$,

$$\int_{t-(s-x)}^{t+(s-x)} K(s,u)du = 0 .$$

Therefore,

$$\frac{1}{2}\int_x^\infty q(s)\left\{\int_s^\infty K(s,u)\int_{-s+u+x}^{s+u-x} e^{i\lambda t}dtdu\right\}ds =$$

$$= \frac{1}{2}\int_x^\infty e^{i\lambda t}\left\{\int_x^\infty q(s)\int_{t-(s-x)}^{t+(s-x)} K(s,u)duds\right\}dt . \tag{3.1.10}$$

From (3.1.9) and (3.1.10) it follows that equality (3.1.7) is satisfied, provided that the function $K(x,t)$ satisfies the equality

$$K(x,t) = \frac{1}{2}\int_{\frac{x+t}{2}}^\infty q(s)ds + \frac{1}{2}\int_x^\infty q(s)\int_{t-(s-x)}^{t+(s-x)} K(s,u)duds \tag{3.1.11}$$

and the condition $K(x,t) = 0$ for $t < x$. The double integral on the right-hand side of (3.1.11) is taken over the domain depicted in Figure 3, and consists of two parts: 1 and 2. However, $K(s,u) = 0$ in subdomain 2, since $s > u$. Therefore, the double integral in equation (3.1.11) need only be taken over subdomain 1.

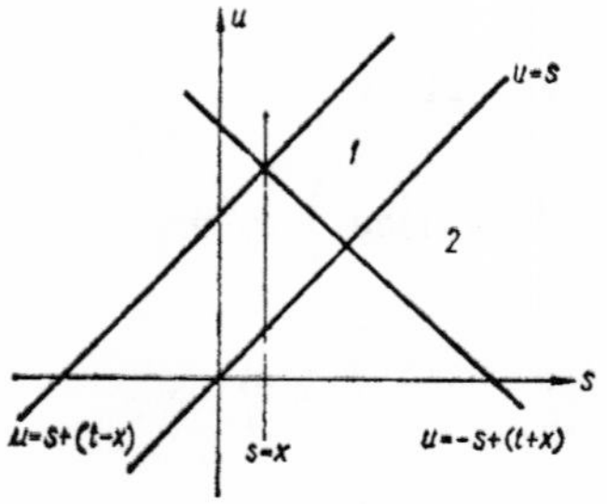

Figure 3

Changing the variables in this integral to $u + s = 2\alpha$, $u - s = 2\beta$, we are led to the following equation:

$$K(x,t) = \frac{1}{2}\int_{\frac{x+t}{2}}^{\infty} q(s)ds + \int_{\frac{x+t}{2}}^{\infty} d\alpha \int_{0}^{\frac{t-x}{2}} q(\alpha-\beta)K(\alpha-\beta,\alpha+\beta)d\beta \ , \tag{3.1.12}$$

where the identity $K(x,t) = 0$ for $x > t$ is automatically accounted for. Setting $H(\alpha,\beta) = K(\alpha-\beta,\alpha+\beta)$, $x + t = 2u$, and $t - x = 2v$, one can recast (3.1.12) in the form

$$H(u,v) = \frac{1}{2}\int_{u}^{\infty} q(s)ds + \int_{u}^{\infty} d\alpha \int_{0}^{v} q(\alpha-\beta)H(\alpha,\beta)d\beta \ . \tag{3.1.13}$$

Therefore, in order to complete the proof the lemma, it suffices to verify that equation (3.1.13) has a solution which satisfies the inequality

$$|H(u,v)| \leqslant \frac{1}{2}\sigma(u)\exp\{\sigma_1(u-v) - \sigma_1(u)\} \ . \tag{3.1.14}$$

Set

$$H_0(u,v) = \frac{1}{2}\int_{u}^{\infty} q(s)ds \ ,$$

$$H_n(u,v) = \int_{u}^{\infty} d\alpha \int_{0}^{v} q(\alpha-\beta)H_{n-1}(\alpha,\beta)d\beta \ ,$$

and let us show that

$$|H_n(u,v)| \leq \frac{1}{2}\sigma(u)\frac{\{\sigma_1(u-v) - \sigma_1(u)\}^n}{n!} . \tag{3.1.14'}$$

Obviously, this will imply that the series $H(u,v) = \sum_{n=0}^{\infty} H_n(u,v)$ converges, and its sum $H(u,v)$ satisfies inequality (3.1.14) and is a solution to equation (3.1.13).

We verify the bounds (3.1.14') by induction. Such a bound clearly holds for $n = 0$, and if it holds for $H_n(u,v)$, and therefore, using the fact that the functions $\sigma(x)$ and $\sigma_1(x)$ are monotonic, we get

$$|H_{n+1}(u,v)| \leq \frac{1}{2}\int_u^\infty d\alpha \int_0^v |q(\alpha-\beta)| \frac{\sigma(\alpha)\{\sigma_1(\alpha-\beta) - \sigma_1(\alpha)\}^n}{n!} d\beta \leq$$

$$\leq \frac{1}{2}\frac{\sigma(u)}{n!}\int_u^\infty \{\sigma_1(\alpha-v) - \sigma_1(\alpha)\}^n\{\sigma(\alpha-v) - \sigma(\alpha)\}d\alpha =$$

$$= \frac{\sigma(u)}{2}\frac{\{\sigma_1(u-v) - \sigma_1(u)\}^{n+1}}{(n+1)!} .$$

Formula (3.1.6') is a straightforward consequence of (3.1.11), when one takes $t = x$. The lemma is proved. □

The operator $\mathbb{I} + \mathbb{K}$, which is defined via the formula

$$\{(\mathbb{I} + \mathbb{K})f\}(x) = f(x) + \int_x^\infty K(x,t)f(t)dt ,$$

is called a transformation operator preserving the asymptotics of the solutions at infinity. Since these are the only operators of this kind which are dealt with in this chapter, they will be simply referred to as transformation operators.

Remark. Estimate (1.1.6) implies that the transformation operator $\mathbb{I} + \mathbb{K}$ defines a one-to-one mapping of each space $L_i(a,\infty)$ $(i = 1,2,\infty)$ onto itself, and has an inverse $(\mathbb{I} + \mathbb{K})^{-1} = \mathbb{I} + \mathbb{L}$ of the same form:

$$\{(\mathbb{I} + \mathbb{L})f\}(x) = f(x) + \int_x^\infty L(x,t)f(t)dt .$$

Using the equation

$$L(x,y) + K(x,y) + \int_x^y L(x,t)K(t,y)dt = 0 \ ,$$

one can easily estimate the kernel $L(x,t)$. It turns out that

$$|L(x,y)| \leq \frac{1}{2}\,\sigma(\tfrac{x+y}{2})\ \exp\{\sigma_1(\tfrac{x+y}{2}) - \sigma_1(y)\} \ ,$$

but we shall not need this estimate in the sequel.

LEMMA 3.1.2. *The function* $K(x,t)$ *has first-order partial derivatives with respect to both variables, and*

$$\left|\frac{K(x_1,x_2)}{x_j} + \frac{1}{4}\,q\left(\frac{x_1+x_2}{2}\right)\right| \leq$$

$$\leq \frac{1}{2}\,\sigma(x_1)\sigma\left(\frac{x_1+x_2}{2}\right)\exp\left\{\sigma_1(x_1) - \sigma_1\left(\frac{x_1+x_2}{2}\right)\right\} . \tag{3.1.15}$$

PROOF. Since $H(u,v) = K(u-v,u+v)$, it suffices to show the existence of partial derivatives for $H(u,v)$. But this follows at once from (3.1.13) and the estimate

$$|H(u,v)| \leq \frac{1}{2}\,\sigma(u)\ \exp\{\sigma_1(u-v) - \sigma_1(u)\} \ . \tag{3.1.16}$$

Moreover,

$$\left.\begin{aligned} &\frac{\partial H(u,v)}{\partial u} = -\frac{1}{2}\,q(u) - \int_0^v q(u-\beta)H(u,\beta)d\beta \ , \\ &\text{and} \\ &\frac{\partial H(u,v)}{\partial v} = \int_u^\infty q(\alpha-v)H(\alpha,v)d\alpha \ . \end{aligned}\right\} \tag{3.1.16'}$$

Using estimate (3.1.14), these equalities yield

$$\left|\frac{\partial H(u,v)}{\partial u} + \frac{1}{2}\,q(u)\right| \leq \frac{1}{2}\,\sigma(u)\sigma(u-v)\ \exp\{\sigma_1(u-v) - \sigma_1(u)\} \ ,$$

$$\left|\frac{\partial H(u,v)}{\partial v}\right| \leq \frac{1}{2}\,\sigma(u)\sigma(u-v)\ \exp\{\sigma_1(u-v) - \sigma_1(u)\} \ ,$$

whence, upon setting $2u = x + t$, $2v = t - x$,

$$\left|\frac{\partial K(x,t)}{\partial x} + \frac{1}{4}\,q(\tfrac{x+t}{2})\right| = \frac{1}{2}\left|\frac{\partial H(u,v)}{\partial u} - \frac{\partial H(u,v)}{\partial v} + \frac{1}{2}\,q(u)\right| \leq$$

$$\leq \frac{1}{2}\sigma(x)\sigma(\tfrac{x+t}{2}) \exp\{\sigma_1(x) - \sigma_1(\tfrac{x+t}{2})\} ,$$

and

$$\left|\frac{\partial K(x,t)}{\partial t} + \frac{1}{4} q(\tfrac{x+t}{2})\right| = \frac{1}{2}\left|\frac{\partial H(u,v)}{\partial u} + \frac{\partial H(u,v)}{\partial v} + \frac{1}{2} q(u)\right| \leq$$

$$\leq \frac{1}{2}\sigma(x)\sigma(\tfrac{x+t}{2}) \exp\{\sigma_1(x) - \sigma_1(\tfrac{x+t}{2})\} .$$

□

Remark. From equalities (3.1.16') it follows that the function $H(u,v)$ satisfies the equation

$$\frac{\partial^2 H(u,v)}{\partial u \partial v} = -q(u-v)H(u,v)$$

and the condition

$$\frac{\partial H(u,0)}{\partial u} = -\frac{1}{2} q(u) .$$

In addition, if $q(u)$ is differentiable, then $H(u,v)$ has all the second-order derivatives. Consequently, the kernel $K(x,t)$ is twice differentiable and satisfies both the equation

$$\frac{\partial^2 K(x,t)}{\partial x^2} - \frac{\partial^2 K(x,t)}{\partial t^2} = q(x)K(x,t) \tag{3.1.17}$$

and the conditions

$$\frac{d}{dx} K(x,x) = -\frac{1}{2} q(x) , \tag{3.1.18}$$

and

$$\lim_{x+t\to\infty} \frac{\partial K(x,t)}{\partial t} = \lim_{x+t\to\infty} \frac{\partial K(x,t)}{\partial t} = 0 , \tag{3.1.18'}$$

which define it uniquely.

Thus, in order for $K(x,t)$ to be the kernel of the transformation, it is necessary and sufficient that it satisfies (3.1.17), (3.1.18) and (3.1.18').

LEMMA 3.1.3. *The solution* $e(\lambda,x)$ *is an analytic function of* λ *in the upper half plane* $\text{Im}\,\lambda > 0$ *and is continuous on the real line. The following estimates hold through the half plane* $\operatorname{Im}\lambda \geqslant 0$:

$$|e(\lambda,x)| \leqslant \exp\{-\operatorname{Im}\lambda x + \sigma_1(x)\} , \tag{3.1.19}$$

$$|e(\lambda,x) - e^{i\lambda x}| \leqslant \{\sigma_1(x) - \sigma_1(x + \frac{1}{|\lambda|})\} \exp\{-\operatorname{Im}\lambda x + \sigma_1(x)\} , \tag{3.1.20}$$

and

$$|e'(\lambda,x) - i\lambda e^{i\lambda x}| \leqslant \sigma(x) \exp\{- \operatorname{Im}\lambda x + \sigma_1(x)\} . \tag{3.1.21}$$

For real $\lambda \neq 0$, *the functions* $e(\lambda,x)$ *and* $e(-\lambda,x)$ *form a fundamental system of solutions of equation* (3.1.1) *and their Wronskian is equal to* $2i\lambda$:

$$W\{e(\lambda,x),e(-\lambda,x)\} = e'(\lambda,x)e(-\lambda,x) - e(\lambda,x)e'(-\lambda,x) = 2i\lambda . \tag{3.1.22}$$

PROOF. The fact that $e(\lambda,x)$ is analytic in the half plane $\operatorname{Im}\lambda > 0$ is a straightforward consequence of formula (3.1.5) and the bound (3.1.6). Further,

$$|e(\lambda,x)| = \left|e^{i\lambda x} + \int_x^\infty K(x,t)e^{i\lambda t}dt\right| \leqslant e^{-\operatorname{Im}\lambda x}\left[1 + \int_x^\infty |K(x,t)|dt\right] ,$$

whence, by (3.1.6),

$$|e(\lambda,x)| \leqslant e^{-\operatorname{Im}\lambda x}\left[1 + \frac{1}{2}\int_x^\infty \sigma(\tfrac{x+t}{2}) \exp\{\sigma_1(x) - \sigma_1(\tfrac{x+t}{2})\}dt\right] =$$

$$= \exp\{-\operatorname{Im}\lambda x + \sigma_1(x)\} .$$

This proves (3.1.19).

Next we note that the function $e(\lambda,x)$ satisfies equation (3.1.7). Using this equation and (3.1.19), we find that

$$|e(\lambda,x) - e^{i\lambda x}| \leqslant \int_x^\infty \left|\frac{\sin\lambda(t-x)}{\lambda} q(t)\right| e^{-\operatorname{Im}\lambda(t-x)}dt \exp\{-\operatorname{Im}\lambda x + \sigma_1(x)\}$$

and

$$|e'(\lambda,x) - i\lambda e^{i\lambda x}| \leqslant \int_x^\infty |\cos\lambda(t-x)q(t)|e^{-\operatorname{Im}\lambda(t-x)}dt \cdot \exp\{-\operatorname{Im}\lambda x + \sigma_1(x)\} .$$

When λ belongs to the closed upper half plane,

$$\left|\frac{\sin\lambda y}{\lambda}\right| e^{-\operatorname{Im}\lambda y} \leqslant \frac{1}{|\lambda|} , \quad \left|\frac{\sin\lambda y}{\lambda}\right| e^{-\operatorname{Im}\lambda y} \leqslant y ,$$

and

$$|\cos \lambda y| e^{-\operatorname{Im} \lambda y} \leq 1$$

for all $y \geq 0$. Consequently,

$$\int_x^\infty \left|\frac{\sin \lambda(t-x)}{\lambda}\right| e^{-\operatorname{Im} \lambda(t-x)} |q(t)| dt \leq$$

$$\leq \int_x^{x+\frac{1}{|\lambda|}} (t-x)|q(t)|dt + \frac{1}{|\lambda|} \int_{x+\frac{1}{|\lambda|}}^\infty |q(t)|dt =$$

$$= -(t-x)\sigma(t) \Big|_x^{x+\frac{1}{|\lambda|}} + \int_x^{x+\frac{1}{|\lambda|}} \sigma(t)dt + \frac{1}{|\lambda|}\sigma(x + \frac{1}{|\lambda|}) =$$

$$= \sigma_1(x) - \sigma_1(x + \frac{1}{|\lambda|}) ,$$

and

$$\int_x^\infty |\cos \lambda(t-x)| e^{-\operatorname{Im} \lambda(t-x)} |q(t)| dt \leq \sigma(x) .$$

These inequalities imply the estimates (3.1.20) and (3.1.21). For real $\lambda \neq 0$ the functions $e(\lambda,x)$ and $e(-\lambda,x)$ are well-defined and satisfy one and the same equation (3.1.1). Consequently, their Wronskian $W\{e(\lambda,x),e(-\lambda,x)\}$ does not depend upon x, and in fact $W\{e(\lambda,x),e(-\lambda,x)\} = 2i\lambda$, because $\lim_{x\to\infty} W\{e(\lambda,x),e(-\lambda,x)\} = 0$, as follows from (3.1.20) and (3.1.21). □

LEMMA 3.1.4. *For any value of* λ, *equation* (3.1.1) *has a solution* $\omega(\lambda,x;\infty)$ *which satisfies the conditions*

$$\omega(\lambda,x;\infty) = x(1 + o(1)) \quad \text{and} \quad \omega'_x(\lambda,x;\infty) = 1 + o(1) \tag{3.1.23}$$

for $x \to \infty$. *This solution is an analytic function of* λ *and satisfies the inequality*

$$|(\lambda\omega(\lambda,x;\infty) - \sin \lambda x)e^{i\lambda x}| \leq [\sigma_1(0) - \sigma_1(\tfrac{1}{|\lambda|})] \exp\left\{ \int_x^0 t|q(t)|dt \right\} \tag{3.1.24}$$

for $\operatorname{Im} \lambda \geq 0$.

PROOF. Let the function $\omega(\lambda,x;\infty) = O(x)$ for $x \to 0$ satisfy the integral equation

$$\omega(\lambda,x;\infty) = \frac{\sin \lambda x}{\lambda} + \int_0^x \frac{\sin \lambda(x-t)}{\lambda} q(t)\omega(\lambda,t;\infty)dt . \tag{3.1.25}$$

Then $\omega(\lambda,x;\infty)$ is obviously a solution of equation (3.1.1). We seek the solution of (3.1.25) for $\text{Im}\,\lambda \geq 0$ in the form $\omega(\lambda,x;\infty) = xe^{-i\lambda x}z(\lambda,x)$. Then the resulting equation for $z(\lambda,x)$ is

$$z(\lambda,x) = \frac{\sin \lambda x}{x} e^{i\lambda x} + \int_0^x \frac{\sin \lambda(x-t)}{\lambda} e^{i\lambda(x-t)}tq(t)z(\lambda,t)dt , \tag{3.1.26}$$

which can be solved by successive approximation upon setting

$$z(\lambda,x) = \sum_{k=0}^{\infty} z_k(\lambda,x) ,$$

where

$$z_0(\lambda,x) = \frac{\sin \lambda x}{\lambda x} e^{i\lambda x} ,$$

and

$$z_k(\lambda,x) = \int_0^x \frac{\sin \lambda(x-t)}{\lambda} e^{i\lambda(x-t)}tq(t)z_{k-1}(\lambda,t)dt .$$

Since

$$\left|\frac{\sin \lambda x}{\lambda x} e^{i\lambda x}\right| = \left|\frac{1}{x}\int_0^x e^{2i\lambda\xi}d\xi\right| \leq 1$$

and

$$\left|\frac{\sin \lambda(x-t)}{\lambda x} e^{i\lambda(t-x)}\right| \leq 1 - \frac{t}{x} \leq 1 ,$$

for $\text{Im}\,\lambda \geq 0$ and $0 \leq t < x$, the series (3.1.27) is majorized by the series $\sum_{k=0}^{\infty} \zeta_k(x)$, where $\zeta_0(x) = 1$ and

$$\zeta_k(x) = \int_0^x t|q(t)|\zeta_{k-1}(t)dt .$$

The latter is clearly uniformly convergent on each finite interval of the half line $[0,\infty)$. Indeed, a simple induction shows that

$$0 \leqslant \zeta_k(x) \leqslant \frac{1}{k!}\left[\int_0^x t|q(t)|dt\right]^k .$$

It follows that for any $a > 0$, the series (3.1.27) converges uniformly in the domain $0 \leqslant x \leqslant a$, $\operatorname{Im}\lambda \geqslant 0$, and its sum $z(\lambda,x)$ satisfies equation (3.1.26) and the inequality

$$|z(\lambda,x)| \leqslant \exp\left\{\int_0^x t|q(t)|dt\right\} .$$

Moreover, $z(\lambda,x)$ is an analytic function of λ for $\operatorname{Im}\lambda > 0$, and is continuous in the closed upper half plane $\operatorname{Im}\lambda \geqslant 0$. But this means that the function $\omega(\lambda,x;\infty) = xz(\lambda,x)e^{-i\lambda x}$ satisfies both the equations (3.1.25) and (3.1.1) and the inequality

$$|\omega(\lambda,x;\infty)e^{i\lambda x}| \leqslant x\exp\left\{\int_0^x t|q(t)|dt\right\} ; \tag{3.1.28'}$$

also, $\omega(\lambda,x;\infty)$ is an analytic function of λ for $\operatorname{Im}\lambda > 0$ and continuous in the closed half plane $\operatorname{Im}\lambda \geqslant 0$.

Proceeding similarly, one proves that equation (3.1.25) is solvable for $\operatorname{Im}\lambda \leqslant 0$ and its solution $\omega(\lambda,x;\infty)$ is analytic in λ in the half plane $\operatorname{Im}\lambda \quad 0$ and continuous for $\operatorname{Im}\lambda \leqslant 0$. We see that $\omega(\lambda,x;\infty)$ satisfies equation (3.1.1), vanishes at $x = 0$, and is an entire function of λ.

Now, equation (3.1.25) and the one resulting from it upon differentiating with respect to x imply, together with estimate (3.1.28'), that

$$\left|\omega(\lambda,x;\infty) - \frac{\sin\lambda x}{\lambda}\right| \leqslant x\int_0^x t|q(t)|dt\,\exp\left\{|\operatorname{Im}\lambda x| + \int_0^x t|q(t)|dt\right\} ,$$

and

$$|\omega'(\lambda,x;\infty) - \cos\lambda x| \leqslant \int_0^x t|q(t)|dt\,\exp\left\{|\operatorname{Im}\lambda x| + \int_0^x t|q(t)|dt\right\} .$$

That is to say, $\omega(\lambda,x;\infty)$ satisfies conditions (3.1.23). Using equation (3.1.25) and estimate (3.1.28') once more, we obtain

$$|\lambda\omega(\lambda,x;\infty)e^{i\lambda x} - \sin\lambda x e^{i\lambda x}| \leqslant$$

$$\leqslant \int_0^x |\sin\lambda(x-t)e^{i\lambda(x-t)}tq(t)z(\lambda,t)|dt \leqslant$$

$$\leqslant \int_0^t t|q(t)| \exp\left\{\int_0^t \xi|q(\xi)|d\xi\right\} dt = \exp\left\{\int_0^x t|q(t)|dt\right\} - 1$$

for $\operatorname{Im}\lambda \geqslant 0$. In particular, this yields

$$|\lambda\omega(\lambda,x;\infty)e^{i\lambda x} - \sin\lambda x e^{i\lambda x}| \leqslant \int_0^x t|q(t)|dt \exp\left\{\int_0^x t|q(t)|dt\right\} \tag{3.1.29}$$

and

$$|\lambda\omega(\lambda,x;\infty)| \leqslant \exp\left\{\int_0^x t|q(t)|dt\right\} . \tag{3.1.30}$$

Now let $\operatorname{Im}\lambda \geqslant 0$ and $|\lambda|^{-1} < x$. Then, by (3.1.25), (3.1.28) and (3.1.30),

$$|\lambda\omega(\lambda,x;\infty)e^{i\lambda x} - \sin\lambda x e^{i\lambda x}| \leqslant$$

$$\leqslant \int_0^x |\sin\lambda(x-t)e^{i\lambda(x-t)}q(t)e^{i\lambda t}\omega(\lambda,t;\infty)|dt \leqslant$$

$$\leqslant \left\{\int_0^{|\lambda|^{-1}} t|q(t)|dt + |\lambda|^{-1}\int_{|\lambda|^{-1}}^x |q(t)|dt\right\} \exp\left\{\int_0^x t|q(t)|dt\right\} =$$

$$= \left\{-|\lambda|^{-1}\sigma(|\lambda|^{-1}) + \int_0^{|\lambda|^{-1}} \sigma(t)dt + |\lambda|^{-1}[\sigma(|\lambda|^{-1}) - \sigma(x)]\right\} \times$$

$$\times \exp\left\{\int_0^x t|q(t)|dt\right\} \leqslant [\sigma_1(0) - \sigma_1(|\lambda|^{-1})]\exp\left\{\int_0^x t|q(t)|dt\right\} ,$$

and so inequality (3.1.24) is proved for $|\lambda|^{-1} < x$. When $|\lambda|^{-1} \geqslant x$, this inequality follows from (3.1.29), because

$$\int_0^x t|q(t)|dt = -x\sigma(x) + \int_0^x \sigma(t)dt \leqslant \sigma_1(0) - \sigma_1(x) \leqslant \sigma_1(0) - \sigma_1(|\lambda|^{-1}) .$$ □

LEMMA 3.1.5. *The identity*

$$-\frac{2i\omega(\lambda,x;\infty)}{e(\lambda,0)} = e(-\lambda,x) - S(\lambda)e(\lambda,x) \tag{3.1.31}$$

holds for all $\lambda \neq 0$, *where*

$$S(\lambda) = \frac{e(-\lambda,0)}{e(\lambda,0)} = \overline{S(-\lambda)} = [S(-\lambda)]^{-1} . \tag{3.1.32}$$

PROOF. Since the two functions, $e(-\lambda,x)$ and $e(\lambda,x)$, form a fundamental system of solutions to equation (3.1.1) for all $\lambda \neq 0$, we can write

$\omega(\lambda,x;\infty) = A^{-}e(-\lambda,x) + A^{+}e(\lambda,x)$.

Formula (3.1.22) yields

$W\{\omega(\lambda,x;\infty),e(\mp\lambda,x)\} = \pm 2i\lambda A^{\pm}$.

Letting x approach 0 and using the estimates (3.1.21) and (3.1.23), we get $e(\mp\lambda,0) = \pm 2i\lambda A$, whence

$\omega(\lambda,x;\infty) = (2i\lambda)^{-1}[-e(\lambda,0)e(-\lambda,x) + e(-\lambda,0)e(\lambda,x)]$.

Since $q(x)$ is real, it follows that $e(-\lambda,0) = \overline{e(\lambda,0)}$, and hence that $e(\lambda,0) \neq 0$ for all real $\lambda \neq 0$. Therefore,

$$-\frac{2i\lambda\omega(\lambda,x;\infty)}{e(\lambda,0)} = e(-\lambda,x) - S(\lambda)e(\lambda,x) ,$$

with

$$S(\lambda) = \frac{e(-\lambda,0)}{e(\lambda,0)} = \overline{\left\{\frac{e(\lambda,0)}{e(-\lambda,0)}\right\}} = \left[\frac{e(\lambda,0)}{e(-\lambda,0)}\right]^{-1} ,$$

as claimed.

□

Next let us examine the right-hand side of (3.1.31) in more detail: it is clearly a meromorphic function in the upper half plane $\text{Im}\,\lambda > 0$, with poles at the zeros of the function $e(\lambda,0)$.

LEMMA 3.1.6. *The function* $e(\lambda,0)$ *may have only a finite number of zeros in the half plane* $\text{Im}\,\lambda > 0$. *They are all simple and lie on the imaginary axis. In addition, the function* $\lambda[e(\lambda,0)]^{-1}$ *is bounded in a neighborhood of the point* $\lambda = 0$.

PROOF. Since $e(\lambda,0) \neq 0$ for all real $\lambda \neq 0$, the point $\lambda = 0$ is the only possible real zero of the function $e(\lambda,0)$. Inequality (3.1.20) implies that $e(\lambda,0) \to 1$ as $|\lambda| \to \infty$, which shows that the zeros of $e(\lambda,0)$ form an at most countable set having 0 as the only possible limit point.

Let μ ($\text{Im}\,\mu > 0$ or $\mu = 0$) be one of the zeros of $e(\lambda,0)$. Then, by virtue of estimates (3.1.21) and (3.1.21), $W\{\omega(\mu,x;\infty),e(\mu,x)\} = e(\mu,0) = 0$, whence $e(\mu,x) = c\omega(\mu,x;\infty)$. Therefore, the limit $\lim_{x\to\infty} e'(\mu,x) = e'(\mu,0) = c$ exists, and

$$e(\mu,x) = e'(\mu,0)\omega(\mu,x;\infty) . \tag{3.1.33}$$

This formula yields

$$\lim_{x\to 0} W\{e(\mu_1,x),\overline{e(\mu_2,x)}\} = 0 \tag{3.1.34}$$

for two arbitrary zeros of $e(\lambda,0)$, μ_1 and μ_2 . Since $q(x)$ is real, the function $\overline{e(\mu_2,x)}$ satisfies the same differential equation as $e(\mu_1,x)$, but with μ_2^2 replacing $\overline{\mu_1^2}$. Thus

$$W\{e(\mu_1,x),\overline{e(\mu_2,x)}\}\Big|_a^b + (\mu_1^2 - \overline{\mu_2^2}) \int_a^b e(\mu_1,x)\overline{e(\mu_2,x)}dx = 0 ,$$

which, upon using (3.1.34) and the estimates (3.1.21) and (3.1.19), implies that

$$(\mu_1^2 - \overline{\mu_2^2}) \int_0^\infty e(\mu_1,x)\overline{e(\mu_2,x)}dx = 0 \tag{3.1.35}$$

whenever μ_1 and μ_2 are zeros of the function $e(\mu,0)$. In particular, the choice $\mu_2 = \mu_1$ implies that $\mu_1^2 - \overline{\mu_1^2} = 0$, or $\mu_1 = i\lambda_1$, where $\mu_1 \geq 0$. Therefore, the zeros of the function $e(\lambda,0)$ can lie only on the imaginary axis. Let us prove that there are only finitely many. This is obvious if $e(0,0) \neq 0$, because under this assumption, the set of zeros cannot have limit points. To verify that the number of zeros of $e(\lambda,0)$ is finite in the general case too, we show that the distance between neighboring zeros is bounded away from zero.

We let δ denote the infimum of the distances between two neighboring zeros of $e(\lambda,0)$, and show next that $\delta > 0$. Otherwise, we could exhibit a sequence of zeros, $\{i\hat{\lambda}_k\}$ and $\{i\lambda_k\}$, of the function $e(\lambda,0)$, such that $\lim_{k\to\infty} (\hat{\lambda}_k - \lambda_k) = 0$, $\hat{\lambda}_k \geq \lambda_k > 0$, and $\max_k \hat{\lambda}_k < M$. Then it follows from estimate (3.1.20) that, for A large enough, the inequality $e(i\lambda,x) > \frac{1}{2}\exp(-\lambda x)$ holds uniformly with respect to $x \in [A,\infty)$ and $\lambda \in [0,\infty)$, whence

$$\int_A^\infty e(i\hat{\lambda}_k,x)\overline{e(i\lambda_k,x)} > \frac{e^{-A(\hat{\lambda}_k+\lambda_k)}}{4(\hat{\lambda}_k + \lambda_k)} > \frac{e^{-2AM}}{8M} . \tag{3.1.36}$$

On the other hand, inequality (3.1.35) yields

$$0 = \int_0^\infty e(i\hat{\lambda}_k,x)\overline{e(i\lambda_k,x)}dx = \int_0^A e(i\hat{\lambda}_k,x)[\overline{e(i\lambda_k,x)} - \overline{e(i\hat{\lambda}_k,x)}]dx +$$

$$+ \int_0^A e(i\hat{\lambda}_k,x)\overline{e(i\hat{\lambda}_k,x)}dx + \int_A^\infty e(i\hat{\lambda}_k,x)\overline{e(i\lambda_k,x)}dx .$$

Letting $k \to \infty$, we get

$$0 \geq \varlimsup_{k\to\infty} \int_A^\infty e(i\hat{\lambda}_k,x)\overline{e(i\hat{\lambda}_k,x)}dx , \tag{3.1.36'}$$

since

$$\lim_{k\to\infty} [e(i\hat{\lambda}_k,x) - e(i\lambda_k,x)] = 0$$

uniformly with respect to $x \in [0,A]$. Comparing (3.1.36) and (3.1.36'), we reach a contradiction. We conclude that $\delta > 0$ and so the function $e(\lambda,0)$ has only a finite number of zeros.

Let us agree to denote differentiation with respect to λ and x with a dot and a prime, respectively:

$$\dot{e}(\lambda,x) = \frac{\partial}{\partial\lambda} e(\lambda,x) \quad , \quad e'(\lambda,x) = \frac{\partial}{\partial x} e(\lambda,x) .$$

Now differentiating the equation

$$-e''(\lambda,x) + q(x)e(\lambda,x) = \lambda^2 e(\lambda,x)$$

with respect to λ , one obtains the following equation for $\dot{e}(\lambda,x)$:

$$-\dot{e}''(\lambda,x) + q(x)\dot{e}(\lambda,x) = \lambda^2\dot{e}(\lambda,x) + 2\lambda e(\lambda,x) .$$

Therefore,

$$-W\{e(\lambda,x),\dot{e}(\lambda,x)\}\Big|_a^b + 2\lambda \int_a^b [e(\lambda,x)]^2 dx = 0 .$$

Let $\lambda = i\mu$ $(\mu > 0)$ be a zero of the function $e(\lambda,0)$. Then the function $e(i\lambda,x)$ is real-valued. On the other hand, by virtue of the formula (3.1.33) and the bounds (3.1.19) and (3.1.21), one has

$$\lim_{x\to 0} W\{e(i\mu,x),\dot{e}(i\mu,x)\} = e'(i\mu,0)\dot{e}(i\mu,0)$$

and

$$\lim_{x\to\infty} W\{e(i\mu,x),\dot{e}(i\mu,x)\} = 0 .$$

Consequently,

$$e'(i\mu,0)\dot{e}(i\mu,0) + 2i\mu \int_0^\infty |e(i\mu,x)|^2 dx = 0 . \tag{3.1.37}$$

Since $\int_0^\infty |e(i\mu,x)|^2 dx > 0$, we see that $\dot{e}(i\mu,0) \neq 0$, i.e., the zeros of $e(\lambda,0)$ are all simple.

It remains only to check that the function $\lambda[e(\lambda,0)]^{-1}$ is bounded in the half disc $D_\rho = \{\lambda : |\lambda| \leqslant \rho , \text{Im } \lambda \geqslant 0\}$, provided that ρ is small enough.

If $e(0,0) \neq 0$, then there is nothing to prove. Now let $e(0,0) = 0$. When $\rho < \frac{1}{2}\delta$ (δ the same as above), the function $e(\lambda,0)$ has no other zeros in D_ρ . It follows that the function $\Phi(\lambda,x) = \frac{\omega(\lambda,x;\infty)}{e(\lambda,0)}$ is regular in the interior of D_ρ and bounded on the semicircle $\{\lambda : |\lambda| = \rho , \text{Im } \lambda \geqslant 0\}$. In addition, $e(\lambda,x)$ is uniformly bounded on the segment $-\rho \leqslant \lambda \leqslant \rho$, as shown by formula (3.1.31) and the fact that $|S(\lambda)| = 1$. However, the maximum principle does not apply at once to the function $\Phi(\lambda,x)$ in D_ρ , because we do not know whether $\Phi(\lambda,x)$ is actually continuous as $\lambda \to 0$. One has therefore to consider the family of equations $-y'' + q_\beta(x)y = \lambda^2 y$, where

$$q_\beta(x) = \begin{cases} q(x) & , \text{ if } x < \beta , \\ 0 & , \text{ if } x \geqslant \beta , \end{cases}$$

and the corresponding solutions $e_\beta(\lambda,x)$ and $\omega_\beta(\lambda,x;\infty)$. Obviously, for $x < \beta$,

$$\omega_\beta(\lambda,x;\infty) = \omega(\lambda,x;\infty)$$

and

$$\Phi_\beta(\lambda,x) = \frac{\lambda\omega_\beta(\lambda,x;\infty)}{e_\beta(\lambda,0)} = \frac{\lambda\omega(\lambda,x;\infty)}{e_\beta(\lambda,0)} .$$

Furthermore, according to Lemma 3.1.1,

$$e_\beta(\lambda,x) = e^{i\lambda x} + \int_x^\infty K_\beta(x,t)e^{it\lambda}dt ,$$

where $K_\beta(x,t) = 0$ for $x + t > 2\beta$. Moreover, $\lim_{\beta\to\infty} K_\beta(x,t) = K(x,t)$ and

$$|K_\beta(x,t)| \leqslant \frac{1}{2}\,\sigma(\tfrac{x+t}{2}) \exp\{\sigma_1(x) - \sigma_1(\tfrac{x+t}{2})\}$$

uniformly in β. Consequently, $e_\beta(\lambda,x)$ are entire functions of λ and $\lim_{\beta\to\infty} e_\beta(\lambda,x) = e(\lambda,x)$ uniformly in the whole closed upper half plane $\operatorname{Im}\lambda \geqslant 0$.

Let δ_β denote the infimum of the distance between neighboring zeros of the function $e_\beta(\lambda,0)$. Repeating the proof carried out above to deduce the positivity of δ_β and noticing that the necessary estimates are uniform in β, we verify that $\inf_\beta \delta_\beta = \delta_0 > 0$. We see that in the half disc D_{ρ_0} (where $\rho_0 = \frac{1}{2}\delta_0$), each of the functions $e_\beta(\lambda,0)$ may have at most one zero (denoted by $i\lambda_\beta$; if there are no zeros, we put $\lambda_\beta = 0$).

Now consider the functions $\Phi_\beta(\lambda,x)\dfrac{\lambda - i\lambda_\beta}{\lambda + i\lambda_\beta}$ which, as shown by the above discussion, are meromorphic in the entire complex plane and regular in the interior of D_{ρ_0}. Since $|\lambda - i\lambda_\beta||\lambda + i\lambda_\beta|^{-1} \leqslant 1$ in the closed upper half plane, and the function $e_\beta(\lambda,0)$ cannot vanish on the arc $\{\lambda : |\lambda| = \rho_0 ,\ \operatorname{Im}\lambda \geqslant 0\}$, the considered functions are uniformly bounded on this arc for all sufficiently large values of β. Moreover, formula (3.1.31) shows that they are also uniformly bounded on the segment $-\rho_0 \leqslant \lambda \leqslant \rho_0$. We conclude that, for all sufficiently large values of β, these functions are regular in D_{ρ_0} (including the boundary), and

$$\sup_{\lambda\in\partial D_{\rho_0}} \left| \Phi_\beta(\lambda,x)\frac{\lambda - i\lambda_\beta}{\lambda + i\lambda_\beta} \right| = C(x) < \infty .$$

Therefore, by the maximum modulus principle,

$$\sup_{\lambda\in D_{\rho_0}} \left| \Phi_\beta(\lambda,x)\frac{\lambda - i\lambda_\beta}{\lambda + i\lambda_\beta} \right| = C(x) < \infty .$$

Letting $\beta \to \infty$ and observing that $\lim_{\beta\to\infty} \lambda_\beta = 0$ and $\lim_{\beta\to\infty} \Phi_\beta(\lambda,x) = \Phi(\lambda,x)$, we get

$$\sup_{\lambda \in D_{\rho_0}} |\Phi(\lambda,x)| \leqslant C(x) .$$

Finally, from inequality (3.1.28) it follows that, for sufficiently small $x = x_0$,

$$\inf_{\lambda \in D_{\rho_0}} |\omega(\lambda,x;\infty)| > \frac{1}{2} x_0 ,$$

and hence

$$\sup_{\lambda \in D_{\rho_0}} \left| \frac{\lambda}{e(\lambda,0)} \right| \leqslant \frac{2C(x_0)}{x_0} < \infty ,$$

as asserted. □

LEMMA 3.1.7. *The function* $1 - S(\lambda)$ *is the Fourier transform of a function* $F_S(x)$ *of the form*

$$F_S(x) = F_S^{(1)}(x) + F_S^{(2)}(x) ,$$

where $F_S^{(1)}(x) \in L_1(-\infty,\infty)$, *whereas* $F_S^{(2)}(x) \in L_2(-\infty,\infty)$ *and*

$$\sup_{-\infty<x<\infty} |F_S^{(2)}(x)| < \infty .$$

PROOF. Recall that the Fourier transform of the convolution

$$(f*g)(x) = \int_{-\infty}^{\infty} f(x-t)g(t)dt$$

of two functions from $L_1(-\infty,\infty)$ equals the product $\tilde{f}(\lambda)\tilde{g}(\lambda)$ of their Fourier transforms, and the norm of the convolution does not exceed the product of norms: $||f*g||_{L_1} \leqslant ||f||_{L_1}||g||_{L_1}$. Consequently, if $||f||_{L_1} < 1$, then the series

$$-f(x) + f*f(x) - f*f*f(x) + \dots$$

converges in the metric of $L_1(-\infty,\infty)$, its sum belongs to this space, and its Fourier transform is equal to

$$-\tilde{f}(\lambda) + \{\tilde{f}(\lambda)\}^2 - \{\tilde{f}(\lambda)\}^3 + \dots = \{1 + \tilde{f}(\lambda)\}^{-1} - 1 .$$

Next we note that

$$\tilde{h}(\lambda) = \begin{cases} 1 & , \text{ if } |\lambda| < 1 , \\ 2 - |\lambda| & , \text{ if } 1 \leq |\lambda| \leq 2 , \\ 0 & , \text{ if } 2 < |\lambda| , \end{cases}$$

is the Fourier transform of a function $h(x) \in L_1(-\infty,\infty)$. Also, $\tilde{h}(\lambda N^{-1})$ is the Fourier transform of the function $h_N(x) = Nh(xN)$, and

$$\lim_{N\to\infty} ||f - h_N * f||_{L_1} = 0 \tag{3.1.38}$$

for all $f(x) \in L_1(-\infty,\infty)$. Since the Fourier transform of $f(x) - h_N * f(x)$ is simply $\{1 - \tilde{h}(\lambda N^{-1})\}\tilde{f}(\lambda)$, we conclude, from (3.1.38) and the previous argument, that for N large enough, the function $[1 + \{1 - \tilde{h}(\lambda N^{-1})\}\tilde{f}(\lambda)]^{-1} - 1$ is the Fourier transform of a function from $L_1(-\infty,\infty)$.

Denoting $K(0,t) = k(t)$ for simplicity, we see that

$$e(\lambda,0) = 1 + \int_0^\infty K(0,t)e^{i\lambda t}dt = 1 + \tilde{k}(-\lambda)$$

and

$$1 - S(\lambda) = 1 - \frac{1 + \tilde{k}(\lambda)}{1 + k(-\lambda)} = \frac{\tilde{k}(-\lambda) - \tilde{k}(\lambda)}{1 + \tilde{k}(-\lambda)} .$$

Now let us rewrite the last equality as

$$1 - S(\lambda) = [\tilde{k}(-\lambda) - \tilde{k}(\lambda)][\{1 + (1 - \tilde{h}(\lambda N^{-1}))\tilde{k}(-\lambda)\}^{-1} - 1] + [\tilde{k}(-\lambda) - \tilde{k}(\lambda)] -$$

$$- [\tilde{k}(-\lambda) - \tilde{k}(\lambda)]\left\{ \frac{1}{1 + \{1 - \tilde{h}(\lambda N^{-1})\}k(-\lambda)} - \frac{1}{1 + \tilde{k}(-\lambda)} \right\} , \tag{3.1.39}$$

and observe that, for N large enough, $\{1 + \tilde{h}(\lambda N^{-1}))\tilde{k}(-\lambda)\}^{-1} - 1$ is the Fourier transform of a summable function. It follows that the sum of the first two terms in the right-hand side of (3.1.39) is also the Fourier transform of a summable function $F_S^{(1)}(x)$. Finally, since $\tilde{h}(\lambda N^{-1}) = 0$ for $|\lambda| > 2N$, the third term in the same formula vanishes for $|\lambda| > 2N$ and is bounded. As such, it is the Fourier transform of a bounded function $F_S^{(2)}(x) \in L_2(-\infty,\infty)$, and the lemma is proved. □

PROBLEMS

1. Show that the operator Dirac equation (1.2.37), in which $\int_x^\infty |\Omega(t)|dt = \sigma_1(x) < \infty$, admits a solution $e(\lambda,x)$ which, for all real values of λ, can be expressed as

$$e(\lambda,x) = e^{-B\lambda x} + \int_x^\infty K(x,t)e^{-B\lambda t}dt ,$$

where $K(x,t) \in \mathcal{OH}$, $K(x,t) = K_+(x,t) + K_-(x,t)$, $BK_+(x,t) + K_+(x,t)B \equiv 0$, $BK_-(x,t) - K_-(x,t)B \equiv 0$, and

$$\int_x^\infty |K_+(x,t)|dt \leqslant \sinh \sigma_1(x) \quad , \quad \int_x^\infty |K_-(x,t)|dt \leqslant \cosh \sigma_1(x) - 1 . \tag{3.1.40}$$

Moreover, $2K_+(x,x) = -B\Omega(x)$, and if $\left[\int_x^\infty |\Omega(t)|^2 dt \right]^{\frac{1}{2}} = \sigma_2(x) > \infty$, then

$$|K_-(x,t)| \leqslant \frac{1}{2} \sigma_2(x)\sigma_2(\tfrac{x+t}{2}) \cosh \sigma_1(x) ,$$
$$|K_+(x,t) + \frac{1}{2} B\Omega(\tfrac{x+t}{2})| \leqslant \frac{1}{2} \sigma_2(x)\sigma_2(\tfrac{x+t}{2}) \sinh \sigma_1(x) . \tag{3.1.40'}$$

<u>Hint</u>. Set $e(\lambda,x) = e^{-B\lambda x}u(\lambda,x)$ to obtain the integral equation

$$u(\lambda,x) = I - \int_x^\infty B\Omega(t)e^{-2B\lambda t}u(\lambda,t)dt$$

for $u(\lambda,x)$, and search for a solution of the form

$$u(\lambda,x) = I + \int_x^\infty e^{B\lambda x}K(x,t)e^{-B\lambda t}dt .$$

Then

$$\int_x^\infty K(x,t)e^{-B\lambda t}dt = - \int_x^\infty B\Omega(t)e^{-B\lambda(2t-x)}dt \; -$$

$$- \int_x^\infty B\Omega(t)e^{-B\lambda(t-x)} \int_t^\infty K(t,\xi)e^{-B\lambda\xi}d\xi dt ,$$

which yields the following system of integral equations:

$$K_+(x,t) = -\frac{1}{2}B\Omega\left(\frac{x+t}{2}\right) - \int_x^{\frac{x+t}{2}} B\Omega(\xi)K_-(\xi,t+x-\xi)d\xi ,$$

$$K_-(x,t) = -\int_x^{\infty} B\Omega(\xi)K_+(\xi,t-x+\xi)d\xi ,$$

for the operator functions

$$K_\pm(x,t) = \frac{1}{2}[K(x,t) \pm BK(x,t)B] \quad (K(x,t) = K_+(x,t) + K_-(x,t)) ;$$

here use the fact that the operators $K_+(x,t)$ $(K_-(x,t))$ commute (respectively, anticommute) with B. The solvability of this system and the bounds (3.1.40) and (3.1.40') are obtained through the method of successive approximations.

2. Show that the kernel $K(x,t)$ (of the transformation operator) constructed in Lemma 3.1.1 satisfies the inequality

$$|K(x,t)| \leq \frac{1}{2}w\left(\frac{x+t}{2}\right)\exp\left\{2\int_x^{\infty} w(\xi)d\xi\right\} , \tag{3.1.41}$$

where

$$w(x) = \sup_{x\leq\eta<\infty} \left|\int_\eta^{\infty} q(t)dt\right| .$$

<u>Hint</u>. Pass to equation (1.3.1) and find its solution in the form of a sum $H_1(u,v) + H_2(u,v)$ of functions that satisfy the system of equations

$$H_1(u,v) = \frac{1}{2}\int_u^{\infty} q(t)dt + \int_u^{\infty} d\alpha \int_0^{v} q(\alpha-\beta)H_2(\alpha,\beta)d\beta ,$$

$$H_2(u,v) = \int_u^{\infty} d\alpha \int_0^{v} q(\alpha-\beta)H_1(\alpha,\beta)d\beta$$

(see Chapter 1, Section 2, Problem 2).

3. Show that Lemma 3.1.1 and the estimate (3.1.4) are valid also for operator Sturm-Liouville equations which satisfy a condition analogous to (3.1.3).

4. Consider the operator Dirac equation

$$By' + \{mT + \Omega(x)\}y = \lambda y , \tag{3.1.42}$$

in which the constant operators B and T satisfy the conditions

$$-B^2 = T^2 = I \quad , \quad BT + TB = 0 \quad , \quad B\Omega(x) + \Omega(x)B = 0 \ ,$$

and the norm $|\Omega(x)|$ is integrable on the half line $[0,\infty)$. The general solution of this equation for $\lambda^2 > m^2$ is the operator-valued function $\{E_1(\lambda,x) + E_2(\lambda,x)\}C$, where C is an arbitrary constant operator, and the particular solutions $E_j(\lambda,x)$ admit the asymptotics

$$E_1(\lambda,x) = e^{ikx}(\frac{\lambda + m}{k} I + iB)(\frac{I + T}{2}) + o(1)$$

and

$$E_2(\lambda,x) = e^{-ikx}(\frac{\lambda + m}{k} I - iB)(\frac{I + T}{2})B + o(1) \ ,$$

respectively, as $x \to \infty$ (here $k = \sqrt{\lambda^2 - m^2}$).

If the operator-valued function (x) is continuously differentiable, then every solution of equation (3.1.42) is also a solution of the operator Sturm-Liouville equation

$$-y'' + q(x)y = k^2 y \quad (k^2 = \lambda^2 - m^2) \ ,$$

where

$$q(x) = B\Omega'(x) + m[T\Omega(x) + \Omega(x)T] + \Omega^2(x) \ .$$

Suppose that the norm $|q(x)|$ satisfies condition (3.1.3). Then, according to the analog of Lemma 3.1.1, the transformation operator exists, and

$$E_j(\lambda,x) = \left\{e^{ik_j x} + \int_x^\infty K(x,t)e^{ik_j t}dt\right\} A_j(\lambda) \ , \tag{3.1.43}$$

where $k_1 = \sqrt{\lambda^2 - m^2}$, $k_2 = -k_1$,

$$A_1(\lambda) = (\frac{\lambda + m}{k} I + iB)(\frac{I + T}{2}) \ ,$$

and

$$A_2(\lambda) = (\frac{\lambda + m}{k} I - iB)(\frac{I + T}{2})B \ .$$

Show that formula (3.1.43) holds also under the assumption that the functions $\sup_{x\le\xi<\infty} |\Omega(x)|$ and

$$w(x) = \sup_{x \le \xi < \infty} \left| B\Omega(\xi) + \int_{\xi}^{\infty} [\{T\Omega(t) + \Omega(t)T\}m + \Omega^2(t)]dt \right|$$

are integrable on $[0,\infty)$, and that in this case,

$$|K(x,t)| \le \frac{1}{2} w(\frac{x+t}{2}) \exp \left\{2 \int_x^{\infty} w(\xi)d\xi\right\} .$$

Hint. Approximate $\Omega(x)$ by the continuously differentiable function

$$\Omega_{\delta}(x) = \frac{1}{\delta} \int_x^{x+\delta} \Omega(t)dt ,$$

and then let $\delta \to 0$.

5. Let M be the infimum of the set of nonnegative numbers λ such that $\sigma_1(0) - \sigma_1(\lambda^{-1}) \le \frac{1}{2}$. Prove the following sharper forms of Lemma 3.1.6:

a) the function $e(\lambda,0)$ has no zeros on the half line $[iM,i\infty)$;

b) if $M > 0$, then any segment of length

$$\Delta = \frac{\exp \{-Mx_0 - \sigma_1(0)\}}{(x_0 + 3t_0)\sqrt{2Mx_0}}$$

of the imaginary half line $[0,i\infty)$ contains at most one zero of $e(\lambda,0)$. Here x_0 and t_0 are the solutions of the equations

$$\sigma_1(x_0) = \frac{1}{3} \quad \text{and} \quad \sigma_1(t_0) = \frac{\exp \{-Mx_0 - \sigma_1(0)\}}{\sqrt{2Mx_0}} \frac{t_0}{x_0 + 3t_0} . \tag{3.1.44}$$

Hint. Let $m(\lambda,x) = \sup_{x \le t < \infty} |e(\lambda,t)e^{-i\lambda t}|$. It follows from equation (3.1.7) that

$$m(\lambda,x) \le 1 + m(\lambda,x) \left\{ \int_x^{x+|\lambda|^{-1}} (t-x)|q(t)|dt + |\lambda|^{-1} \int_{x+|\lambda|^{-1}}^{\infty} |q(t)|dt \right\} =$$

$$= 1 + m(\lambda,x)\{\sigma_1(x) - \sigma_1(x + |\lambda|^{-1})\} .$$

Therefore, if $\sigma_1(x) - \sigma_1(x + |\lambda|^{-1}) < 1$, then

$$m(\lambda,x) \leq [1 - \{\sigma_1(x) - \sigma_1(x + |\lambda|^{-1})\}]^{-1} ,$$

whence

$$e(i\lambda,x) \geq e^{-\lambda x}\left\{1 - \frac{\sigma_1(x) - \sigma_1(x + |\lambda|^{-1})}{1 - \{\sigma_1(x) - \sigma_1(x + |\lambda|^{-1})\}}\right\} .$$

This clearly proves assertion a and establishes the estimate $e(i\lambda,x) \geq \frac{1}{2} e^{-\lambda x}$ $(x \geq x_0)$.

Now let $i\lambda$ and $i\mu = i(\lambda+\delta)$ be two neighboring zeros of the function $e(\lambda,0)$. Then

$$0 = \int_0^\infty e(i\lambda,x)e(i\mu,x)dx = \frac{1}{4}\int_0^{x_0} [e(i\lambda,x) + e(i\mu,x)]^2 dx -$$

$$- \frac{1}{4}\int_0^{x_0} [e(i\lambda,x) - e(i\mu,x)]^2 dx + \int_{x_0}^\infty e(i\lambda,x)e(i\mu,x)dx \geq$$

$$\geq \frac{1}{4}\left\{- x_0 \max_{0\leq x\leq x_0} [e(i\lambda,x) - e(i\mu,x)]^2 + \frac{\exp[-x_0(\lambda+\mu)]}{\lambda + \mu}\right\} ,$$

and hence

$$\max_{0\leq x\leq x_0} |e(i\lambda,x) - e(i\mu,x)| \geq \sqrt{\frac{\exp[-x_0(\lambda+\mu)]}{x_0(\lambda+\mu)}} > \frac{e^{-Mx_0}}{\sqrt{2Mx_0}} . \tag{3.1.45}$$

On the other hand,

$$|e(i\lambda,x) - e(i\mu,x)| = |e^{-\lambda x} - e^{-\mu x} + \int_x^\infty K(x,t)\{e^{-\lambda t} - e^{-\mu t}\}dt| \leq$$

$$\leq 1 - e^{-\delta x} + e^{\sigma_1(x)} \int_x^\infty \sigma(\tfrac{x+t}{2})e^{-\sigma_1(\frac{x+t}{2})} [1 - e^{-\delta t}]d\,\frac{t}{2} =$$

$$= 1 - e^{-\delta x} + e^{\sigma_1(x)}\left\{-[e^{-\sigma_1(x)} - 1][1 - e^{-\delta x}] - \delta\int_x^\infty e^{-\delta t}[e^{-\sigma_1(\frac{x+t}{2})} - 1]dt\right\} \leq$$

$$\leq e^{\sigma_1(x)}\left\{[1 - e^{-\delta x}] + 2\delta t_0 + [1 - e^{-\sigma_1(x+t_0)}]\delta \int_{x+2t_0}^\infty e^{-\delta t}dt\right\} \leq$$

$$\leq e^{\sigma_1(0)}\{\delta x + 2\delta t_0 + \sigma_1(t_0)\} ,$$

whence, in view of (3.1.45),

$$e^{\sigma_1(0)}\{(x_0 + 2t_0)\delta + \sigma_1(t_0)\} > \frac{e^{-Mx_0}}{\sqrt{2Mx_0}} .$$

Since t_0 satisfies equation (3.1.44), the last inequality yields

$$\delta > \frac{e^{-Mx_0}}{\sqrt{2Mx_0}} \frac{1}{x_0 + 3t_0} = \Delta .$$

Thus, the distance between neighboring zeros is larger than Δ , which is equivalent to assertion b.

6. Suppose that a symmetric ($q(x) = q^*(x)$) operator Sturm-Liouville equation satisfies condition (3.1.3) and the numbers M , x_0 , t_0 are defined as in the preceding problem. Show that

a) the operators $e(\lambda,0)$ ($\operatorname{Im} \lambda \geqslant 0$) are invertible for every $\lambda \notin [0,iM]$;

b) the orthogonal projections P_λ onto the kernels of the operators $e(i\lambda,0)$ ($\lambda \geqslant 0$) have the property that $|P_\lambda P_\mu| \leqslant \theta < 1$ if $0 < \lambda-\mu \leqslant \Delta$. In particular, if the space H has dimension $n < \infty$, then on any interval of length Δ on the half line $[0,i\infty)$, the determinant of the operator (matrix) $e(\lambda,0)$ has at most n zeros.

7. Lemma 3.1.1 is also valid for complex-valued functions $q(x)$ which satisfy the condition (3.1.3). Let $\mathbb{I} + \mathbb{L} = (\mathbb{I} + \mathbb{K})^{-1}$ and let $L(x,y)$ denote the kernel of the operator $\mathbb{L}$. Show that the function $\Phi(x,y)$, defined for all $x \geqslant 0$ and $y \geqslant 0$ by the formula

$$\Phi(x,y) = \begin{cases} L(x,y) + \int_y^\infty L(x,t)L(y,t)dt , & \text{if } 0 \leqslant x \leqslant y , \\ L(y,x) + \int_x^\infty L(x,t)L(y,t)dt , & \text{if } x \geqslant y \geqslant 0 , \end{cases}$$

depends only on $x + y$, i.e., $\Phi(x,y) = F(x+y)$, where

$$F(u) = L(0,u) + \int_u^\infty L(u,t)L(0,t)dt . \qquad (3.1.46)$$

Generalize this result to operator Sturm-Liouville and Dirac equations.

8. Prove the following analog of the Lemma 1.4.1: if in equation (3.1.1) the potential $q(x)$ has $n \geqslant 0$ derivatives which are subject to the bounds

$$\int_0^\infty (1+x)|q^{(k)}(x)|dx < \infty \quad (k = 0,1,\ldots,n) \ ,$$

then the solution $e(\lambda,x)$ of this equation admits the representation

$$e(\lambda,x) = e^{i\lambda x}\left[1 + \frac{u_1(x)}{2i\lambda} + \ldots + \frac{u_n(x)}{(2i\lambda)^n} + \frac{u_{n+1}(\lambda,x)}{(2i\lambda)^{n+1}}\right] ,$$

where

$$u_1(x) = -\int_x^\infty q(t)dt \ ,$$

$$u_k(x) = -\int_x^\infty [-u''_{k-1}(\xi) + q(\xi)u_{k-1}(\xi)]d\xi \ ,$$

$$u_{n+1}(\lambda,x) = u_{n+1}(x) - \frac{1}{2i\lambda}\int_x^\infty q(t)u_{n+1}(t)dt +$$

$$+ \int_0^\infty \{u'_{n+1}(x+\xi) + \frac{1}{2i\lambda}K^{(0)}_{n+1}(x,\xi)\}e^{2i\lambda\xi}d\xi \ ,$$

$$u'_{n+1}(\lambda,x) = -2i\lambda\int_0^\infty \{u'_{n+1}(x+\xi) + \frac{1}{2i\lambda}K^{(1)}_{n+1}(x,\xi)\}e^{2i\lambda\xi}d\xi \ ,$$

and, in addition,

$$\int_0^\infty \{|u'_{n+1}(x+\xi)| + |K^{(0)}_{n+1}(x,\xi)| + |K^{(1)}_{n+1}(x,\xi)|\}d\xi < \infty \ .$$

9. Prove the following analog of Lemma 1.4.2: if the conditions of the preceding problem are satisfied, then

$$e(\lambda,x) = \exp\left\{i\lambda x - \int_x^\infty \sigma(\lambda,t)dt\right\} ,$$

where

$$\sigma(\lambda,t) = \sum_{k=1}^n \frac{\sigma_k(t)}{(2i\lambda)^t} + \frac{\sigma_n(\lambda,t)}{(2i\lambda)^n} ,$$

the functions $\sigma_k(x)$ are determined from the recursion formulas (1.4.20),

$$\sigma_n(\lambda,x) = \int_0^\infty \sigma_{n+1}(x+\xi)e^{2i\lambda\xi}d\xi + \frac{1}{2i\lambda}\int_0^\infty \tilde{K}_{n+1}(x,\xi)e^{2i\lambda\xi} + o(\lambda^{-2}) ,$$

and

$$\int_0^\infty \{|\tilde{K}_{n+1}(x,\xi)| + |\sigma_{n+1}(x+\xi)|\}d\xi < \infty .$$

10. In the upper half plane, the function $e(\lambda,0)$ has a finite number of zeros, the squares of which, $\mu_1,\mu_2,\ldots,\mu_p$, are the discrete eigenvalues of the boundary value problem (3.1.1), (3.1.2). Show that under the assumptions of the preceding problems, the following analog of the trace formula holds for $n \geq 2m$:

$$\lim_{R\to\infty}\left\{\frac{1}{2\pi i}\int_0^R \lambda^{2m}d\ln S(\lambda) - \sum_{j=0}^{m-1}\frac{c_{2j+1}(2j+1)R^{2m-2j-1}}{2\pi(-4)^j(2m-2j-1)}\right\} = \frac{mc_{2m}}{(-4)^m} - \sum_{k=1}^{p}\mu_k^m ,$$

where $c_k = \int_0^\infty \sigma_k(t)dt$.

<u>Hint</u>. It follows from the analogs of Lemmas 1.4.1 and 1.4.2 (see Problems 8 and 9) that the following asymptotic expansions are valid for $|\lambda| \to \infty$ in the upper half plane:

$$\frac{d}{d\lambda}\ln e(\lambda,0) = \frac{d}{d\lambda}\ln\left[1 + \frac{u_1(0)}{2i\lambda} + \ldots + \frac{u_n(0)}{(2i\lambda)^n}\right] + o(\lambda^{-n-1}) ,$$

$$\frac{d}{d\lambda}\ln\left[1 + \frac{u_1(0)}{2i\lambda} + \ldots + \frac{u_n(0)}{(2i\lambda)^n}\right] =$$

$$= \int_0^\infty \sum_{k=1}^{n}\frac{k\sigma_k(t)}{(2i)^k\lambda^{k+1}}dt + o(\lambda^{-n-2}) = \sum_{k=1}^{n}\frac{kc_k}{(2i)^k}\lambda^{-k-1} + o(\lambda^{-n-2}) .$$

Since

$$S(\lambda) = e(-\lambda,0)[e(\lambda,0)]^{-1} ,$$

you have

$$\int_0^R \lambda^{2m}d\ln S(\lambda) = -\int_{-R}^{R}\lambda^{2m}d\ln e(\lambda,0) = -2\pi i\sum_{k=1}^{p}\mu_k^m + \int_{C(R)}\lambda^{2m}d\ln e(\lambda,0) ,$$

where $C(R)$ is the semicircle $C(R) = \{\lambda : |\lambda| = R , \operatorname{Im} \lambda \geq 0\}$. Therefore,

$$\frac{1}{2\pi i} \int_0^R \lambda^{2m} d \ln S(\lambda) = - \sum_{k=1}^{p} \mu_k^m + \frac{1}{2\pi i} \int_{C(R)} \lambda^{2m} d \ln e(\lambda,0) =$$

$$= - \sum_{k=1}^{p} \mu_k^m + \frac{1}{2\pi i} \int_{C(R)} \left\{ \sum_{k=1}^{n} \frac{kc_k}{(2i)^k} \lambda^{2m-k-1} + o(\lambda^{2m-n-1}) \right\} d\lambda =$$

$$= - \sum_{k=1}^{p} \mu_k^m - \frac{2}{2\pi i} \sum_{j=0}^{m-1} \frac{(2j+1)c_{2j+1} R^{2m-2j-1}}{(2i)^{2j+1}(2m-2j-1)} +$$

$$+ \frac{2mc_{2m}}{2\pi i(2i)^{2m}} \int_{C(R)} \lambda^{-1} d\lambda + o(R^{2m-n}) =$$

$$= - \sum_{k=1}^{p} \mu_k^m + \frac{1}{2\pi} \sum_{j=0}^{m-1} \frac{(2j+1)c_{2j+1}}{(-4)^j(2m-2j-1)} R^{2m-2j-1} + \frac{mc_{2m}}{(-4)^m} + o(R^{2m-n}) .$$

2. THE PARSEVAL EQUALITY AND THE FUNDAMENTAL EQUATION

Let $i\lambda_k$ $(k = 1,2,\ldots,n)$ be the zeros of the function $e(\lambda,0)$, numbered in the order of increase of their moduli $(0 < \lambda_1 < \lambda_2 < \ldots < \lambda_n)$, and let m_k^{-1} be the norm of the function $e(i\lambda_k,x)$ in $L_2[0,\infty)$. We remark that according to formula (3.1.37),

$$m_k^{-2} = \int_0^\infty |e(i\lambda_k,x)|^2 dx = - \frac{e'(i\lambda_k,0)\dot{e}(i\lambda_k,0)}{2i\lambda_k} . \qquad (3.2.1)$$

It follows from the results of the preceding section that the functions

$$u(\lambda,x) = e(-\lambda,x) - S(\lambda)e(\lambda,x) \quad (\lambda \in (0,\infty)) , \qquad (3.2.2)$$

and

$$u(i\lambda_k,x) = m_k e(i\lambda_k,x) \quad (k = 1,2,\ldots,n) \qquad (3.2.3)$$

are bounded solutions of the boundary value problem (3.1.1), (3.1.2). Let us show that they form a complete set of normalized eigenfunctions of this problem in the sense that, i.e., for every pair of functions $f(x)g(x) \in L_2(0,\infty)$, the Parseval equality holds:

$$(f,g) = \sum_{k=1}^{n} u(i\lambda_k,f)\overline{u(i\lambda_k,g)} + \frac{1}{2\pi}\int_0^\infty u(\lambda,f)\overline{u(\lambda,g)}d\lambda , \qquad (3.2.4)$$

where (f,g) designates the inner product in $L_2[0,\infty)$, and

$$u(\lambda,f) = \int_0^\infty f(x)u(\lambda,x)dx .$$

It follows from the estimate (3.1.6) that the transformation operator $\mathbb{I} + \mathbb{K}$ and its adjoint $\mathbb{I} + \mathbb{K}^*$ are bounded in $L_2[0,\infty)$. Hence, if $f(x) \in L_2[0,\infty)$, the function $f^*(x) = [(\mathbb{I} + \mathbb{K}^*)f](x)$ also belongs to $L_2[0,\infty)$, and by Lemma 3.1.1,

$$\int_0^\infty f(x)e(-\lambda,x)dx = \int_0^\infty f(x)\left[e^{-i\lambda x} + \int_x^\infty K(x,t)e^{-i\lambda t}dt\right]dx =$$

$$= \int_0^\infty \left[f(x) + \int_0^x f(\xi)K(\xi,x)d\xi\right]e^{-i\lambda x}dx = \int_x^\infty f^*(x)e^{-i\lambda x}dx ,$$

i.e.,

$$\int_0^\infty f(x)e(-\lambda,x)dx = \tilde{f}^*(\lambda) ,$$

where $\tilde{f}^*(\lambda)$ designates the ordinary Fourier transform of the function $f^*(x)$. Consequently,

$$u(\lambda,f) = \tilde{f}^*(\lambda) - S(\lambda)\tilde{f}^*(-\lambda)$$

and

$$u(i\lambda_k,f) = m_k \int_0^\infty f^*(x)e^{-\lambda_k x}dx .$$

Now let us substitute these expressions into the right-hand side of formula (3.2.4), which for brevity we shall denote by J . Since $S(\lambda) = \overline{S(-\lambda)} = [S(-\lambda)]^{-1}$, it is readily seen that

$$J = \int_0^\infty\int_0^\infty \left\{\sum_{k=1}^{n} m_k^2 e^{-\lambda_k(x+y)}\right\} f^*(x)\overline{g^*(y)}dxdy +$$

$$+ \frac{1}{2\pi}\int_{-\infty}^{\infty} \tilde{f}^*(\lambda)\overline{\tilde{g}^*(\lambda)}d\lambda - \frac{1}{2\pi}\int_{-\infty}^{\infty} S(\lambda)\tilde{f}^*(-\lambda)\overline{\tilde{g}^*(\lambda)}d\lambda . \qquad (3.2.5)$$

By the Parseval equality for the ordinary Fourier transforms,

$$\frac{1}{2\pi}\int_{-\infty}^{\infty}\tilde{f}^*(\lambda)\overline{\tilde{g}^*(\lambda)}d\lambda = \int_{-\infty}^{\infty} f^*(y)\overline{g^*(y)}dy = (f^*,g^*) \ ,$$

whereas

$$\frac{1}{2\pi}\int_{-\infty}^{\infty}\tilde{f}^*(-\lambda)\overline{\tilde{g}^*(\lambda)}d\lambda = \int_{-\infty}^{\infty} f^*(-y)\overline{g^*(y)}dy = 0 \ ,$$

because $f^*(y) = g^*(y) = 0$ for $y < 0$. Hence, formula (3.2.5) can be re-expressed in the form

$$J = (f^*,g^*) + \int_0^{\infty}\int_0^{\infty}\left\{\sum_{k=1}^{n} m_k^2 e^{-\lambda_k(x+y)}\right\} f^*(x)\overline{g^*(y)}dxdy +$$

$$+ \frac{1}{2\pi}\int_{-\infty}^{\infty}(1 - S(\lambda))\tilde{f}^*(-\lambda)\overline{\tilde{g}^*(\lambda)}d\lambda \ ,$$

or

$$J = (f^*,g^*) + \int_0^{\infty}\int_0^{\infty} F(x + y)f^*(x)\overline{g^*(y)}dxdy \ , \tag{3.2.6}$$

in which the function

$$F(x) = \sum_{k=1}^{n} m_k^2 e^{-\lambda_k x} + \frac{1}{2\pi}\int_{-\infty}^{\infty}(1 - S(\lambda))e^{i\lambda x}d\lambda \ , \tag{3.2.7}$$

whose existence is guaranteed by Lemma 3.1.7, is real, since $S(\lambda) = \overline{S(-\lambda)}$. It is readily verified that the operator $\mathbb{F}$, defined by the formula

$$(\mathbb{F}f)(x) = \int_0^{\infty} F(x + y)f(y)dy \ ,$$

is bounded and self-adjoint in $L_2[0,\infty)$. Using $\mathbb{F}$, we can rewrite formula (3.2.6) as

$$J = (\{\mathbb{I} + \mathbb{F}\}f^*,g^*) \ ,$$

whence, upon recalling that $f^* = (\mathbb{I} + \mathbb{K}^*)f$, it follows that

$$J = (\{\mathbb{I} + \mathbb{F}\}\{\mathbb{I} + \mathbb{K}^*\}f,\{\mathbb{I} + \mathbb{K}^*\}g) = (\{\mathbb{I} + \mathbb{K}\}\{\mathbb{I} + \mathbb{F}\}\{\mathbb{I} + \mathbb{K}^*\}f,g) \ .$$

Thus, for the Parseval equality (3.2.4) to hold, it is necessary and sufficient that the operators $\mathbb{K}$, $\mathbb{F}$, and $\mathbb{K}^*$, be related through the identity

$$(\mathbb{I} + \mathbb{K})(\mathbb{I} + \mathbb{F})(\mathbb{I} + \mathbb{K}^*) = \mathbb{I} , \tag{3.2.8}$$

or equivalently,

$$\mathbb{K} + \mathbb{K}^* + \mathbb{K}\mathbb{K}^* + \mathbb{F} + \mathbb{F}\mathbb{K}^* + \mathbb{K}\mathbb{F} + \mathbb{K}\mathbb{F}\mathbb{K}^* = 0 . \tag{3.2.8'}$$

The left-hand side of this equality is the integral operator with kernel

$$\Phi(x,y) = K(x,y) + K(y,x) + F(x+y) +$$

$$+ \int_x^\infty K(x,t)F(t+y)dt + \int_y^\infty F(x+\xi)K(y,\xi)d\xi +$$

$$+ \int_0^\infty K(x,\xi)K(y,\xi)d\xi + \int_y^\infty \int_x^\infty K(x,t)F(t+\xi)K(y,\xi)dtd\xi , \tag{3.2.9}$$

Since $\Phi(x,y) = \Phi(y,x)$, identities (3.2.8), (3.2.8') are equivalent to the equality $\Phi(x,y) = 0$ for $y > x$.

We define a function $\varphi_x(y)$ on the half line (x,∞) by the formula

$$\varphi_x(y) = K(x,y) + F(x+y) + \int_x^\infty K(x,t)F(y+t)dt .$$

Since $K(y,x) = 0$ for $y > x$, it follows from formula (3.2.9) that, for $y > x$,

$$\Phi(x,y) = \varphi_x(y) + \int_y^\infty K(y,\xi)\varphi_x(\xi)d\xi .$$

By the remark to Lemma 3.1.1, the operator $\mathbb{I} + \mathbb{K}$ is invertible. Therefore, the identity $\Phi(x,y) = 0$ $(y > x)$, and hence the identities (3.2.8), (3.2.8') are equivalent to the equality

$$\varphi_x(y) = F(x+y) + K(x,y) + \int_x^\infty K(x,t)F(y+t)dt = 0 \quad (0 \leqslant x < y < \infty).$$

We thus see that, for the identity (3.2.8), which is equivalent to the Parseval equality (3.2.4), to hold, it is necessary and sufficient that, for every $x \in [0,\infty)$, the kernel $K(x,y)$ of the transformation operator satisfy (as a function of the variable $y \in [x,\infty)$) the integral equation

$$F(x+y) + K(x,y) + \int_x^\infty K(x,t)F(y+t)dt = 0\ , \tag{3.2.10}$$

in which the function $F(x)$ is defined by formula (3.2.7). We shall refer to (3.2.10) as the fundamental equation.

To derive the fundamental equation, we use the equality

$$-\frac{2i\lambda\omega(\lambda,x;\infty)}{e(\lambda,0)} = e(-\lambda,x) - S(\lambda)e(\lambda,x)\ ,$$

which was obtained in Lemma 4.1.5. Here,

$$e(\lambda,x) = e^{i\lambda x} + \int_x^\infty K(x,t)e^{i\lambda t}dt\ ,$$

so that

$$-\frac{2i\lambda\omega(\lambda,x;\infty)}{e(\lambda,0)} = e^{-i\lambda x} - e^{i\lambda x} + \{1 - S(\lambda)\}\left\{e^{i\lambda x} + \int_x^\infty K(x,t)e^{i\lambda t}dt\right\} +$$

$$+ \int_x^\infty K(x,t)e^{-i\lambda t}dt - \int_x^\infty K(x,t)e^{i\lambda t}dt\ ,$$

or, equivalently,

$$-2i\lambda\omega(\lambda,x;\infty)\{\frac{1}{e(\lambda,0)} - 1\} + 2i\{\sin\lambda x - \lambda\omega(\lambda,x;\infty)\} =$$

$$= \{1 - S(\lambda)\}\left\{e^{i\lambda x} + \int_x^\infty K(x,t)e^{i\lambda t}dt\right\} +$$

$$+ \int_x^\infty K(x,\xi)e^{-i\lambda\xi}d\xi - \int_{-\infty}^{-x} K(x,-\xi)e^{-i\lambda\xi}d\xi\ . \tag{3.2.11}$$

As was shown in Lemma 1.3.7, $1 - S(\lambda)$ is the Fourier transform of the function

$$F_S(y) = \frac{1}{2\pi}\int_{-\infty}^{\infty} (1 - S(\lambda))e^{i\lambda y}dy\ . \tag{3.2.12}$$

The well-known convolution theorem permits us to conclude that the right (and hence the left) hand side of identity (3.2.11) is the Fourier transform of the function

$$F_S(x+y) + \int_{-\infty}^{\infty} F_S(y-t)K(x,-t)dt + K(x,y) - K(x,-y)\ . \tag{3.2.13}$$

Conseuently, the integral of the product of the left-hand side of identity (3.2.11) and $\frac{1}{2} e^{i\lambda y}$, taken over the real line $-\infty < \lambda < \infty$, must equal (3.2.13). Let us show that for $y > x$, this integral converges and can be calculated by contour integration. In fact, by Lemma 3.1.6, the first term in the integral,

$$-\frac{2i\lambda\omega(\lambda,x;\infty)}{2\omega}\left\{\frac{1}{e(\lambda,0)}-1\right\},$$

is regular everywhere in the upper half plane, apart from a finite number of points $i\lambda_k$ $(k = 1,2,\ldots,n)$, which are simple zeros of the function $e(\lambda,0)$; moreover, it is continuous for real $\lambda \neq 0$ and bounded in a neighborhood of $\lambda = 0$ $(\operatorname{Im}\lambda \geqslant 0)$. Furthermore, since the function $\lambda\omega(\lambda,x;\infty)e^{i\lambda x}$ is bounded in the half plane $\operatorname{Im}\lambda \geqslant 0$ by Lemma 3.1.4, and since the function $\frac{1}{e(\lambda,0)} - 1$ tends uniformly to zero in this half plane as $|\lambda| \to \infty$ by Lemma 3.1.3, an application of Jordan's lemma yields

$$-\frac{1}{2\pi}\int_{-\infty}^{\infty} 2i\lambda\omega(\lambda,x;\infty)e^{i\lambda y}\left\{\frac{1}{e(\lambda,0)}-1\right\}d\lambda = \sum_{k=1}^{n}\frac{2i\lambda_k\omega(i\lambda_k;x;\infty)e^{-\lambda_k y}}{\dot{e}(i\lambda_k,0)} \tag{3.2.14}$$

for $y > x$. Next, upon using formula (3.2.1) and (3.1.33), we can reexpress the right-hand side of this equality in the form

$$\sum_{k=1}^{n}\frac{2i\lambda_k\omega(i\lambda_k,x;\infty)e^{-\lambda_k y}}{\dot{e}(i\lambda_k,0)} = -\sum_{k=1}^{n} m_k^2 e(i\lambda_k,x)e^{-\lambda_k y} =$$

$$= -\sum_{k=1}^{n} m_k^2\left\{e^{-\lambda_k(x+y)} + \int_x^{\infty} K(x,t)e^{-\lambda_k(t+y)}dt\right\}. \tag{3.2.15}$$

The second term in the integral in question, $2i\{\sin\lambda x - \lambda\omega(\lambda,x;\infty)\}$, is an entire function of λ and, in view of estimate (3.1.24) and Jordan's lemma,

$$\int_{-\infty}^{\infty} 2i\{\sin\lambda x - \lambda\omega(\lambda,x;\infty)\}e^{-i\lambda y}dy = 0$$

for $y > x$.

Thus, for $y > x$, the integral of the product of the left-hand side of equality (3.2.11) and $\frac{1}{2\pi} e^{i\lambda y}$ exists and equals (3.2.15). Hence, for $y > x$,

$$- \sum_{k=1}^{n} m_k^2 \left\{ e^{-\lambda_k(x+y)} + \int_x^\infty K(x,t) e^{-\lambda_k(t+y)} dt \right\} =$$

$$= F_S(x+y) + \int_{-\infty}^{\infty} F_S(y+t) K(x,-t) dt + K(x,y) - K(x,-y) \,,$$

and, since $K(x,y) = 0$ for $x > y$,

$$- \sum_{k=1}^{n} m_k^2 \left\{ e^{-\lambda_k(x+y)} + \int_x^\infty K(x,t) e^{-\lambda_k(t+y)} dt \right\} =$$

$$= F_S(x+y) + \int_x^\infty K(x,t) F_S(y+t) dt + K(x,y) \,.$$

This finally yields

$$F(x+y) + K(x,y) + \int_x^\infty K(x,t) F(t+y) dt = 0 \quad (x < y < \infty) \,,$$

where

$$F(x) = \sum_{k=1}^{n} m_k^2 e^{-\lambda_k x} + F_S(x) = \sum_{k=1}^{n} m_k^2 e^{-\lambda_k x} + \frac{1}{2\pi} \int_{-\infty}^{\infty} (1 - S(\lambda)) e^{i\lambda x} d \,.$$

We have thus proved the following result:

THEOREM 3.2.1. *The kernel* $K(x,y)$ *of the transformation operator satisfies the fundamental equation*

$$F(x+y) + K(x,y) + \int_x^\infty K(x,t) F(t+y) dt = 0 \quad (x < y < \infty) \,,$$

where, for every $x \geq 0$,

$$F(x) = \sum_{k=1}^{n} m_k^2 e^{-\lambda_k x} + \frac{1}{2\pi} \int_{-\infty}^{\infty} (1 - S(\lambda)) e^{i\lambda x} d \,.$$

The basic equation yields the identity

$$(\mathbb{I} + \mathbb{K})(\mathbb{I} + \mathbb{F})(\mathbb{I} + \mathbb{K}^*) = \mathbb{I}$$

which in turn implies the validity of the Parseval identity (3.2.4). □

Now consider the operator $\mathbf{F}_a$ in $L_2[a,\infty)$, which is defined by the rule

$$(\mathbb{F}_a f)(y) = \int_a^\infty F(y+t)f(t)dt .$$

The theorem we just proved has the following

COROLLARY. *The operator* $\mathbb{I} + \mathbb{F}_a$ *is invertible for every* $a \geq 0$, *and*

$$(\mathbb{I} + \mathbb{F}_a)^{-1} = (\mathbb{I} + \mathbb{K}_a^*)(\mathbb{I} + \mathbb{K}_a) , \tag{3.2.16}$$

where the operators $\mathbb{K}_a$ *and* $\mathbb{K}_a^*$ *are defined by the formulas*

$$\left.\begin{aligned} &(\mathbb{K}_a f)(y) = \int_y^\infty K(y,t)f(t)dt , \\ &\text{and} \\ &(\mathbb{K}_a^* f)(y) = \int_a^y K(t,y)f(t)dt . \end{aligned}\right\} \tag{3.2.16'}$$

PROOF. Let $f(y) \in L_2[a,\infty)$ and

$$\hat{f}(y) = \begin{cases} f(y) , & a \leq y < \infty , \\ 0 , & 0 \leq y < a . \end{cases}$$

By identity (3.2.8),

$$(\mathbb{I} + \mathbb{K})(\mathbb{I} + \mathbb{F})(\mathbb{I} + \mathbb{K}^*)\hat{f} = \hat{f}$$

and, since

$$[(\mathbb{I} + \mathbb{K}^*)\hat{f}](y) = \begin{cases} [(\mathbb{I} + \mathbb{K}_a^*)\hat{f}](y) , & a \leq y < \infty , \\ 0 , & 0 \leq y < a , \end{cases}$$

we have, for $y \in [a,\infty)$,

$$(\mathbb{I} + \mathbb{F})(\mathbb{I} + \mathbb{K}^*)\hat{f} = (\mathbb{I} + \mathbb{F}_a)(\mathbb{I} + \mathbb{K}_a^*)f$$

whence

$$f(y) = \hat{f}(y) = \{(\mathbb{I} + \mathbb{K})(\mathbb{I} + \mathbb{F})(\mathbb{I} + \mathbb{K}^*)\hat{f}\}(y) = \{(\mathbb{I} + \mathbb{K}_a^*)(\mathbb{I} + \mathbb{F}_a)(\mathbb{I} + \mathbb{K}_a^*)f\}(y)$$

Thus,

$$(\mathbb{I} + \mathbb{K}_a)(\mathbb{I} + \mathbb{F}_a)(\mathbb{I} + \mathbb{K}_a^*) = \mathbb{I} ,$$

which is equivalent to identity (3.2.16). We remark also that both the operator appearing in the right-hand side of this identity and its inverse are bounded in every space $L_i[a,\infty)$, $i = 1,2,\infty$, and hence identity (3.2.16) holds in any of these spaces $(a \geq 0)$.

□

We now use the fundamental equation to extract more information on the function $F(x)$. First of all, we note that from the continuity of the kernel $K(x,y)$, it follows that $F(x)$ is continuous on $[0,\infty)$, and hence that the fundamental equation is valid also for $y = x$. Upon letting $y = x$ in the fundamental equation, and performing the substitution $t + x = 2\xi$, we get

$$F(2x) + K(x,x) + 2 \int_x^\infty K(x,2\xi-x)F(2\xi)d\xi = 0 .$$

From this equality and estimate (3.1.6), it follows that

$$|F(2x) + K(x,x)| \leq e^{\sigma_1(x)} \int_x^\infty \sigma(\xi)e^{-\sigma_1(\xi)} |F(2\xi)|d\xi . \tag{3.2.17}$$

This further implies that the function

$$z(x) = \int_x^\infty \sigma(\xi)e^{-\sigma_1(\xi)} |F(2\xi)|d\xi \tag{3.2.18}$$

satisfies the inequality

$$-z'(x) \leq \sigma(x)e^{-\sigma_1(x)} |K(x,x)| + \sigma(x)z(x) ,$$

which in turn can be reexpressed as

$$-\{z(x)e^{-\sigma_1(x)}\}' \leq \sigma(x)|K(x,x)|e^{-2\sigma_1(x)} \leq \frac{1}{2} \sigma^2(x)e^{-2\sigma_1(x)} .$$

Integrating this inequality, we get

$$z(x)e^{-\sigma_1(x)} \leq \frac{1}{2} \int_x^\infty \sigma^2(t)e^{-2\sigma_1(t)}dt \leq \frac{\sigma(x)}{4} \{1 - e^{-2\sigma_1(x)}\} ,$$

i.e.,

$$z(x) \leq \frac{1}{2} \sigma(x) \sinh \sigma_1(x) .$$

In view of (3.2.17) and (3.2.18), this yields

$$|F(2x) + K(x,x)| \leq \frac{1}{2} \sigma(x)e^{\sigma_1(x)} \sinh \sigma_1(x) , \tag{3.2.19}$$

and a fortiori

$$|F(2x)| \leqslant \frac{1}{2}\sigma(x)\left\{\frac{e^{2\sigma_1(x)}-1}{2}+1\right\} = \frac{1}{2}\sigma(x)e^{\sigma_1(x)}\cosh\sigma_1(x) . \qquad (3.2.20)$$

Next, it follows from the differentiability of the function $K(x,y)$ and the estimate (3.1.15) for $|K_x(x,y)|$ that the derivative $F'(x)$ exists for $x > 0$, and hence that the fundamental equation can be differentiated with respect to x. We thus get

$$0 = F'(x+y) + K_x(x,y) - K(x,x)F(x+y) + \int_x^\infty K_x(x,t)F(t+y)dt ,$$

whence, upon letting $y = x$ and observing that, by (3.1.12),

$$K_x(x,y)\big|_{y=x} = -\frac{q(x)}{4} - \frac{1}{2}\int_x^\infty q(t)K(t,t)dt = -\frac{1}{4}q(x) - \frac{1}{8}\left\{\int_x^\infty q(t)dt\right\}^2 ,$$

we obtain

$$F'(2x) - \frac{q(x)}{4} - \frac{1}{8}\left\{\int_x^\infty q(t)dt\right\}^2 - K(x,x)F(2x) + \int_x^\infty K_x'(x,t)F(t+x)dt = 0 .$$

To take advantage in the best way of the estimates derived earlier, we reexpress this identity as

$$F'(2x) - \frac{q(x)}{4} - \frac{1}{8}\left\{\int_x^\infty q(t)dt\right\}^2 + K^2(x,x) + \frac{1}{4}\int_x^\infty q\left(\frac{x+t}{2}\right)K\left(\frac{x+t}{2}, \frac{x+t}{2}\right)dt =$$

$$= K(x,x)\{F(2x) + K(x,x)\} - \int_x^\infty \left\{K_x'(x,t) + \frac{1}{4}q\left(\frac{x+t}{2}\right)\right\}F(t+x)dt +$$

$$+ \frac{1}{4}\int_x^\infty q\left(\frac{x+t}{2}\right)\left\{F(t+x) + K\left(\frac{x+t}{2}, \frac{x+t}{2}\right)\right\}dt .$$

Now, from this equality, formula (3.1.6), and the estimates (3.2.19), (3.1.15), it follows that

$$\left|F'(2x) - \frac{q(x)}{4} + \frac{1}{4}\left\{\int_x^\infty q(t)dt\right\}^2\right| \leqslant \frac{\sigma^2(x)}{4}e^{\sigma_1(x)}\sinh\sigma_1(x) +$$

$$+ \frac{\sigma^2(x)}{4}e^{\sigma_1(x)}\int_x^\infty \sigma\left(\frac{x+t}{2}\right)\cosh\sigma_1\left(\frac{x+t}{2}\right)dt +$$

$$+ \frac{\sigma(x)}{8}e^{\sigma_1(x)}\sinh\sigma_1(x)\int_x^\infty \left|q\left(\frac{x+t}{2}\right)\right|dt \leqslant \sigma^2(x)e^{\sigma_1(x)}\sinh\sigma_1(x) ,$$

i.e.,

$$\left| F'(2x) - \frac{q(x)}{4} + \frac{1}{4}\left\{ \int_x^\infty q(t)dt \right\}^2 \right| \leq \sigma^2(x)e^{\sigma_1(x)} \sinh \sigma_1(x) . \tag{3.2.21}$$

Since the functions $x|q(x)|$ and $\sigma(x)$ are summable on $[0,\infty)$, and $\sup_{0<x<\infty} x\sigma(x) < \infty$, inequality (3.2.21) shows that the function $x|F'(x)|$ is also summable on $[0,\infty)$, and therefore so is $F_S(x)$. Inequalities (3.2.19), (3.2.21), which indicate the close connection between the function $4F'(x)$ and $q(x)$, are interesting in their own right. However, in the ensuing analysis we shall need only the following cruder result: the function $F_S(x)$ is differentiable on $(0,\infty)$ and its derivative satisfies the same condition $\int_0^\infty x|F_S'(x)|dx < \infty$, as $q(x)$.

Let us prove also that the function $S(\lambda)$ is continuous at all real points λ , and the increment of its logarithm is related to the number n of negative eigenvalues of problem (3.1.1), (3.1.2) through the equality

$$n = \frac{\ln S(0+) - \ln S(+\infty)}{2\pi i} - \frac{1 - S(0)}{4} . \tag{3.3.22}$$

The continuity of the function $S(\lambda) = \frac{e(-\lambda,0)}{e(\lambda,0)}$ at all real points $\lambda \neq 0$ is a straightforward consequence of Lemma (3.1.3). It is also clear that if $e(0,0) \neq 0$, then $S(\lambda)$ is also continuous at zero, and $S(0) = 1$. Thus, it remains to examine the case where

$$e(0,0) = 1 + \int_0^\infty K(0,t)dt = 0 . \tag{3.2.23}$$

Upon letting $x = 0$ in the basic equation and subsequently integrating it with respect to y from z to ∞ , we get

$$\int_z^\infty F(y)dy + \int_z^\infty K(0,y)dy + \int_0^\infty K(0,t) \int_{t+z}^\infty F(\xi)d\xi dt = 0 ,$$

from which we further obtain, via an integration by parts,

$$\left\{1 + \int_0^\infty K(0,y)dy\right\} \int_z^\infty F(y)dy + \int_z^\infty K(0,y)dy - \int_0^\infty F(t+z) \int_t^\infty K(0,\xi)d\xi dt = 0 .$$

Therefore, if equality (3.2.23) holds, then the function

$$K_1(z) = \int_z^\infty K(0,y)dy \tag{3.2.24}$$

is a bounded solution of the equation

$$K_1(z) - \int_0^\infty K_1(t)F(t+z)dt = 0 \quad (0 \leq z < \infty) .$$

However, every bounded solution of this equation is automatically summable on the half line $[0,\infty)$. In fact, upon writing

$$K_1(z) - \int_N^\infty K_1(t)F(t+z)dt = f_N(z) \quad (N \leq z < \infty) , \tag{3.2.25}$$

where

$$f_N(z) = \int_0^N K_1(t)F(t+z)dt ,$$

we can choose N so large that the equation (3.2.25) may be solved by the method of successive approximations. The approximations will then converge to the solution simultaneously in the norms of the spaces $L_1[N,\infty)$ and $L_2[N,\infty)$, so that $K_1(z) \in L_1[N,\infty)$. Hence, in the present case,

$$e(\lambda,0) = 1 + \int_0^\infty K(0,t)e^{i\lambda t}dt =$$

$$= \left\{1 + \int_0^\infty K(0,t)dt\right\} + i\lambda \int_0^\infty K_1(t)e^{i\lambda t}dt = i\lambda\tilde{K}_1(-\lambda) \tag{3.2.26}$$

and

$$S(\lambda) = -\frac{\tilde{K}_1(\lambda)}{\tilde{K}_1(-\lambda)} , \tag{3.2.27}$$

where $\tilde{K}_1(\lambda)$ is the Fourier transform of the function $K_1(t)$, which vanishes for $t < 0$ and is summable on $[0,\infty)$. By (3.2.26) and identity (3.1.31), $2\omega(\lambda,x;\infty) = \tilde{K}_1(-\lambda)\{e(-\lambda,x) - S(\lambda)e(\lambda,x)\}$, from which it follows that $\tilde{K}_1(0) \neq 0$, and hence that $S(\lambda)$ is continuous at $\lambda = 0$; moreover, $S(0) = -1$.

To prove formula (3.2.22), we apply the argument principle to the function $e(\lambda,0)$. This function is regular in the upper half plane, continuous in the closed upper half plane $\operatorname{Im} \lambda \geq 0$, and tends uniformly to one as

$|\lambda| \to \infty$ therein. Moreover, $e(\lambda,0) = \overline{e(-\lambda,0)}$ for all real λ. Hence, the increment of its argument $\eta(\lambda)$ as λ runs over the real axis from $-\infty$ to $+\infty$, bypassing the point $\lambda = 0$ along a semicircle of sufficiently small radius ε in the upper half plane, is equal to the number of zeros of $e(\lambda,0)$ in the upper half plane multiplied by 2π:

$$2\pi n = \{\eta(\varepsilon-) - \eta(-\infty)\} + \{\eta(\varepsilon+) - \eta(\varepsilon-)\} + \{\eta(+\infty) - \eta(\varepsilon+)\} =$$

$$= 2\{\eta(+\infty) - \eta(\varepsilon+)\} + \{\eta(\varepsilon+) - \eta(\varepsilon-)\} .$$

If $e(0,0) \neq 0$, then $\lim_{\varepsilon\to 0} \{\eta(\varepsilon+) - \eta(\varepsilon-)\} = 0$. If, however, $e(0,0) = 0$, then by (3.2.26), $e(\lambda,0) = i\lambda\tilde{K}_1(-\lambda)$, where $\tilde{K}_1(0) \neq 0$, and hence $\lim_{\varepsilon\to 0} \{\eta(\varepsilon+) - \eta(\varepsilon-)\} = -\pi$. Therefore,

$$\frac{2\{\eta(+\infty) - \eta(0+)\}}{2\pi} = \begin{cases} n , & \text{if } e(0,0) \neq 0 , \\ n + \frac{1}{2} , & \text{if } e(0,0) = 0 . \end{cases}$$

On the other hand,

$$S(0) = \begin{cases} 1 , & \text{if } e(0,0) \neq 0 , \\ -1 , & \text{if } e(0,0) = 0 , \end{cases}$$

and since by formula (3.1.23),

$$\ln S(\lambda) = -2i \arg e(\lambda,0) = -2i\eta(\lambda) ,$$

we finally get

$$\frac{\ln S(0+) - \ln S(+\infty)}{2\pi i} = n + \frac{1 - S(0)}{4} ,$$

as claimed.

PROBLEMS

1. Prove the following strengthened form of Lemma 3.1.7: the function $F_S(x)$ whose Fourier transform is $1 - S(\lambda)$, is summable on the full real line.

Hint. Since

$$1 - S(\lambda) = \frac{\tilde{k}(-\lambda) - \tilde{k}(\lambda)}{1 + \tilde{k}(-\lambda)} , \quad \tilde{k}(\lambda) = \int_0^\infty K(0,t)e^{-i\lambda t}dt ,$$

and $1 + \tilde{k}(-\lambda) = e(\lambda,0) \neq 0$ for $\lambda \neq 0$, the desired result follows from the Wiener-Levy theorem when $1 + \tilde{k}(0) \neq 0$. If $1 + \tilde{k}(0) = 0$, you may use formula (3.2.27) and the following identity, which is a consequence of (3.2.27):

$$1 - S(\lambda) = (1 - S(\lambda))\tilde{h}(\lambda N^{-1}) + (1 - S(\lambda))(1 - \tilde{h}(\lambda N^{-1})) =$$

$$= \tilde{h}(\lambda N^{-1}) + \frac{\tilde{K}_1(\lambda)}{\tilde{K}_1(-\lambda)}\tilde{h}(\lambda N^{-1}) + \frac{\tilde{k}(-\lambda) - \tilde{k}(\lambda)}{1 + \tilde{k}(-\lambda)}(1 - \tilde{h}(\lambda N^{-1})) =$$

$$= \tilde{h}(\lambda N^{-1}) + \frac{\tilde{K}_1(\lambda)\overline{\tilde{K}_1(-\lambda)}\tilde{h}(\lambda N^{-1})}{1 - \{1 - \tilde{K}_1(-\lambda)\overline{\tilde{K}_1(-\lambda)}\}\tilde{h}(\frac{\lambda N^{-1}}{2})} + \frac{\{\tilde{k}(-\lambda) - \tilde{k}(\lambda)\}\{1 - \tilde{h}(\lambda N^{-1})\}}{1 + \{1 - \tilde{h}(2\lambda N^{-1})\}\tilde{k}(-\lambda)} ,$$

where

$$\tilde{h}(\lambda) = \begin{cases} 1 , & |\lambda| \leqslant 1 , \\ 2 - |\lambda| , & 1 \leqslant |\lambda| \leqslant 2 , \\ 0 , & |\lambda| \geqslant 2 . \end{cases}$$

Since

$$1 - \{1 - \tilde{K}_1(-\lambda)\overline{\tilde{K}_1(-\lambda)}\}\tilde{h}(\frac{\lambda N^{-1}}{2}) > 0 \quad , \quad -\infty < \lambda < \infty ,$$

and, for sufficiently large N ,

$$\sup_{-\infty<\lambda<\infty} |\tilde{k}(-\lambda)\{1 - \tilde{h}(2\lambda N^{-1})\}| < 1 ,$$

the desired result follows again from the Wiener-Levy theorem.

2. Suppose that the real-valued function $q(x)$ satisfies condition (3.1.3). Let $h(x)$ be an arbitrary real solution of the equation $y'' - q(x)y = 0$. Extend the results of the preceding sections to the Sturm-Liouville boundary value problem

$$-y'' + q(x)y = \lambda^2 y \quad (0 < x < \infty)$$

with the boundary condition

$$\lim_{x\to 0} W\{y(x),h(x)\} = \lim_{x\to 0} \{y'(x)h(x) - y(x)h'(x)\} = 0 ,$$

which reduces to the usual form

$$y'(0) - y(0)h = 0 \quad (h = h'(0)\{h(0)\}^{-1}) ,$$

when the function $q(x)$ is integrable in the neighborhood of zero.

3. Extend the results of the preceding sections to the Sturm-Liouville boundary value problem

$$-y'' + q(x)y = \lambda^2 y \quad , \quad y(0) = 0 \quad (0 < x < \infty) \, ,$$

in which the function $q(x)$ satisfies the condition

$$\int_0^1 x\left|q(x) - \frac{\ell_0(\ell_0+1)}{x^2}\right|dx + \int_1^\infty x\left|q(x) - \frac{\ell_\infty(\ell_\infty+1)}{x^2}\right|dx < \infty \, ,$$

where ℓ_0 and ℓ_∞ are positive integers. Show, in particular, that in the present case, formula (3.2.22) is replaced by

$$n = \frac{\ln S(0+) - \ln S(+\infty)}{2\pi i} + \frac{\ell_\infty - \ell_0}{2} \, .$$

Hint. In the last two problems it is convenient to use transformations operators of the form (2.3.25), (3.1.26) (see Chapter 2, Section 3, Problems 4-6) in order to reduce everything to the investigation of the boundary value problem (3.1.1), (3.1.2) with a function $q(x)$ satisfying condition (3.1.3).

4. Consider the Sturm-Liouville equation

$$-y'' + q(x)y = \lambda^2 y \quad (0 < x < \infty)$$

in which the complex-valued function $q(x)$ satisfies condition (3.1.3). Let $\mathbb{I} + \mathbb{K}$ denote the transformation operator which takes $e^{i\lambda x}$ into $e(\lambda,x)$, $\mathbb{I} + \mathbb{L}$ the inverse of $\mathbb{I} + \mathbb{K}$, and let $\mathbb{K}^+$ and $\mathbb{L}^+$ denote the transposes of the operators $\mathbb{K}$ and $\mathbb{L}$, respectively:

$$(\mathbb{K}^+ f)(x) = \int_0^x K(t,x)f(t)dt \, ,$$

$$(\mathbb{L}^+ f)(x) = \int_0^x L(t,x)f(t)dt \, .$$

Define the function $F(x)$ by

$$F(x) = L(0,x) + \int_x^\infty L(x,t)L(0,t)dt \quad (0 < x < \infty)$$

and denote its Fourier transform by

$$\tilde{F}(\lambda) = \int_0^\infty F(x)e^{-i\lambda x}dx .$$

Prove the identity

$$(\mathbb{I} + \mathbb{K})(\mathbb{I} + \mathbb{F})(\mathbb{I} + \mathbb{K}^+) = \mathbb{I} \quad \text{(where } (\mathbb{F}f)(x) = \int_0^\infty F(x+t)f(t)dt)$$

and show that

$$x_+(\lambda) = e(-\lambda,0) + \tilde{F}(\lambda)e(\lambda,0) - 1$$

is the Fourier transform of a function $x(t) \in L_1(-\infty,\infty)$ which vanishes for $t > 0$:

$$x_+(\lambda) = \int_{-\infty}^0 x(t)e^{-i\lambda t}dt .$$

Hint. Use the results of Problem 7, Section 1.

5. Let $Q(\lambda)$ be an arbitrary bounded measurable function on the real line such that $Q(\lambda)Q(-\lambda) = 1$. Let

$$u(\lambda,x) = e(-\lambda,x) - Q(\lambda)e(\lambda,x) ,$$

$$u(\lambda,f) = \int_0^\infty f(x)u(\lambda,x)dx$$

and

$$e(\lambda,f) = \int_0^\infty f(x)e(\lambda,x)dx .$$

Show that the following equalities hold for every pair of functions $f(x),g(x) \in L_2[0,\infty)$:

$$\int_0^\infty f(x)g(x)dx = \frac{1}{2\pi}\int_{-\infty}^\infty \{e(-\lambda,f) + \tilde{F}(\lambda)e(\lambda,f)\}e(\lambda,g)d\lambda ,$$

$$\int_0^\infty f(x)g(x)dx = \frac{1}{2\pi}\int_0^\infty u(\lambda,f)u(-\lambda,g)d\lambda - \frac{1}{2\pi}\int_{-\infty}^\infty \{Q(\lambda) + \tilde{F}(\lambda)\}e(\lambda,f)e(\lambda,g)d\lambda .$$

Note that if the function $e(\lambda,0)$ has no zeros on the real line, then it follows from these equalities, upon setting $Q(\lambda) = e(-\lambda,0)\{e(\lambda,0)\}^{-1}$, that

$$\int_0^\infty f(x)g(x)dx = \frac{1}{2\pi}\int_0^\infty u(\lambda,f)u(-\lambda,g)d\lambda - i\sum_{\operatorname{Im}\lambda>0} \operatorname{Res}\left\{\frac{x_+(\lambda)+1}{e(\lambda,0)}\,e(\lambda,f)e(\lambda,g)\right\} .$$

6. Generalize the results and the problems of the last two sections to operator Sturm-Liouville and Dirac equations. Examine separately the case

in which the space H is finite dimensional.

3. THE INVERSE PROBLEM OF QUANTUM SCATTERING THEORY

In "classical" quantum mechanics, the stationary state of a system consisting of two particles of masses m_1 and m_2 and energy E is described by the Ψ-function, which satisfies the Schrödinger equation

$$-\frac{\hbar^2}{2M}\Delta\Psi + V(x)\Psi = E \quad , \tag{3.3.1}$$

where $\hbar$ is Planck's constant, $M = \frac{m_1 m_2}{m_1 + m_2}$, $V(x)$ is the interaction potential, and $x = |\vec{x}|$ is the distance between the two particles. Since the potential $V(x)$ depends only on the distance $|\vec{x}|$, the variables in equation (3.3.1) separate upon setting

$$\Psi(\vec{x}) = x^{-1}u_l(E,x)Y_l^m(\theta,\varphi) \; ,$$

where $Y_l^m(\theta,\varphi)$ are spherical harmonics. The function $u_l(E,x)$ satisfies the equation

$$-\frac{\hbar^2}{2M}\{u_\ell'' - \ell(\ell+1)x^{-2}u_\ell\} + V(x)u_\ell = Eu_\ell \; ,$$

and the boundary condition $u_\ell(E,0) = 0$. Introducing the notations

$$q(x) = \frac{2M}{\hbar^2}V(x) \quad , \quad \lambda^2 = \frac{2ME}{\hbar^2} \quad , \quad u_\ell(\lambda,x) = u_\ell(E,x) \; , \tag{3.3.2}$$

for the sake of brevity, we are led to the boundary value problem

$$-u_\ell'' + q(x)u_\ell + \ell(\ell+1)x^{-2}u_\ell = \lambda^2 u_\ell \quad (0 < x < \infty) \; , \tag{3.3.3}$$

$$u_\ell(\lambda,0) = 0 \; .$$

The solutions of this boundary value problem which are bounded at infinity will be referred to as radial wave functions.

Throughout this section, we will assume that the potential $V(x)$ satisfies condition (3.1.3). Under this assumption, it follows from the results of the preceding sections that the boundary value problem (3.3.3), (3.3.4) with $\ell = 0$ has bounded solutions $u_0(\lambda,x)$ for $\lambda^2 > 0$ and $\lambda = i\lambda_k$ $(k = 1,2,\dots,n)$; moreover, as $x \to \infty$,

$$u_0(\lambda,x) = e^{-i\lambda x} - S(\lambda)e^{i\lambda x} + o(1) \quad (0 < \lambda^2 < \infty)$$

and

$$u_0(i\lambda_k,x) = m_k e^{-\lambda_k x}(1 + o(1)) \quad (k = 1,2,\dots,n) ,$$

respectively. Thus, the collection

$$\{S(\lambda)\ (-\infty < \lambda < \infty)\ ;\ \lambda_k, m_k\ (k = 1,2,\dots,n)$$

provides a complete description of the behavior at infinity of all radial wave functions $u_0(\lambda,x)$. A similar statement is valid for the other values of ℓ .

For a complete description of the system in question, it suffices to know the behavior at infinity of all radial wave functions, since the latter permits a description of all observable effects. It is therefore naturally to ask whether the behavior of the Ψ-functions at infinity determines the potential $V(x)$. In other words, can one recover the potential $V(x)$ from experimental data?

The problem of recovering the potential from the experimental data is known as the inverse problem of quantum scattering theory, because most (though not all) of the fundamental data is obtained in experiments on the scattering of particles (it would be more correct to refer to it as to the problem of recovering the potential from the behavior of all Ψ-functions at infinity). In accordance with this terminology, the collection of quantities $\{S(\lambda)\ (-\infty < \lambda < \infty)\ ;\ \lambda_k, m_k\ (k = 1,2,\dots,n)\}$ that specify the behavior of the radial wave functions $u_0(\lambda,x)$ at infinity, will be referred to as the scattering data of the boundary value problem (3.1.1), (3.1.2) (i.e., of problem (3.3.3), (3.3.4) with $\ell = 0$).

In terms of the scattering data, the potential $V(x)$ is recovered uniquely. This is a straightforward consequence of Theorem 3.2.1. In fact, given the scattering data, one can construct the function $F(x)$ via formula (3.2.7) and then write the fundamental equation (3.2.10). By the corollary to Theorem 3.2.1, the fundamental equation has a unique solution for every $x \geq 0$. Moreover, the solution is the kernel $K(x,y)$ of the transformation operator, and hence the potential $q(x)$ is $q(x) = -\frac{1}{2}\frac{d}{dx}K(x,x)$.

It remains to find those properties which a collection $\{S(\lambda)\ (-\infty < \lambda < \infty)\ ;\ \lambda_k, m_k\ (k = 1,2,\dots,n)\}$ must enjoy in order that it be the scattering data of a boundary value problem (3.1.1), (3.1.2) with potential subject to (3.1.3). It follows from the results of the preceding section that the scattering data always satisfy the following conditions.

I. *The function* $S(\lambda)$ *is continuous on the whole line,* $S(\lambda) = \overline{S(-\lambda)} = [S(-\lambda)]^{-1}$ *; also,* $1 - S(\lambda)$ *tends to zero as* $|\lambda| \to \infty$ *and is the Fourier transform of the function*

$$F_S(x) = \frac{1}{2\pi} \int_{-\infty}^{\infty} (1 - S(\lambda))e^{i\lambda x}d\lambda ,$$

which in turn can be written as a sum of two functions, one of which belongs to $L_1(-\infty,\infty)$ *, while the second is bounded and belongs to* $L_2(-\infty,\infty)$ *. On the positive half line,* $F_S(x)$ *has a derivative* $F_S'(x)$ *which satisfies the condition*

$$\int_0^{\infty} x|F_S'(x)|dx < \infty .$$

II. *The increment of the argument of* $S(\lambda)$ *and the number* n *of negative eigenvalues of the boundary value problem* (3.1.1), (3.1.2) *are related by the formula*

$$n = \frac{\ln S(0+) - \ln S(+\infty)}{2\pi i} - \frac{1 - S(0)}{4} .$$

We shall prove that conditions I and II are not only necessary, but also sufficient for the given collection $\{S(\lambda);\lambda_k,m_k\ (k = 1,\ldots,n)\}$ to be the scattering data of a boundary value problem (3.1.1), (3.1.2) with potential subject to condition (3.1.3). To prove this fundamental result we need a number of auxiliary lemmas.

Let the function $S(\lambda)$ satisfy condition I, and put

$$F_S(x) = \frac{1}{2\pi} \int_{-\infty}^{\infty} (1 - S(\lambda))e^{i\lambda x}d\lambda \tag{3.3.5}$$

and

$$F(x) = F_S(x) + \sum_{k=1}^{n} m_k^2 e^{-\lambda_k x} . \tag{3.3.6}$$

Next we write the fundamental equation

$$F(x+y) + K(x,y) + \int_x^{\infty} K(x,t)F(t+y)dt = 0 , \tag{3.3.7}$$

and reexpress it in the more convenient form

$$F(2x+y) + K(x,x+y) + \int_0^\infty K(x,x+t)F(t+y+2x)dt = 0 . \tag{3.3.7'}$$

We shall seek its solution $K(x,x+y)$ for every $x \geqslant 0$ in the same space $L_1[0,\infty)$.

We consider the operators $\mathbb{F}^+_{S,a}$ and $\mathbb{F}^-_S$ acting in the spaces $L_i[0,\infty)$ $(i = 1,2)$ and, respectively, $L_2(-\infty,0]$, by the rules

$$(\mathbb{F}^+_{S,a}f)(y) = \int_0^\infty F_S(t+y+2a)f(t)dt \tag{3.3.8}$$

and

$$(\mathbb{F}^-_S f)(y) = \int_{-\infty}^0 F_S(y+t)f(t)dt , \tag{3.3.9}$$

as well as the operator $\mathbb{F}^+_a$,

$$(\mathbb{F}^+_a f)(y) = \int_0^\infty F(t+y+2a)f(t)dt , \tag{3.3.10}$$

which appears in the fundamental equation.

LEMMA 3.3.1. *The operators* $\mathbb{F}^+_{S,a}$ *and* $\mathbb{F}^+_a$ *are compact in each of the spaces* $L_i[0,\infty)$, $i = 1,2$, *for every choice of* $a \geqslant 0$. *The operator* $\mathbb{F}^-_S$ *is compact in* $L_2(-\infty,0]$.

PROOF. From condition I it follows that the function $F_S(y)$ is summable on the positive half line. Hence, if $f(y) \in L_1[0,\infty)$ and $||f|| = 1$, then the following estimates hold for the function $g = {}^{+}_{S,a}f$:

$$||g|| = \int_0^\infty dy\left|\int_0^\infty f(t)F_S(t+y+2a)dt\right| \leqslant \int_0^\infty |f(t)| \int_0^\infty |F_S(t+y+2a)|dydt \leqslant$$

$$\leqslant \int_0^\infty |f(t)| \int_{2a}^\infty |F_S(u)|dudt \leqslant \int_0^\infty |F_S(u)|du ,$$

$$\int_0^\infty |g(y+h) - g(y)|dy = \int_0^\infty dy\left|\int_0^\infty f(t)\{F_S(t+y+h+2a) - F_S(t+y+2a)\}dt\right| \leqslant$$

$$\leqslant \int_0^\infty |f(t)| \int_{2a}^\infty |F_S(u+h) - F_S(u)|dudt \leqslant \int_0^\infty |F_S(u+h) - F_S(u)|du ,$$

$$\int_N^\infty |g(y)|dy = \int_N^\infty dy\left|\int_0^\infty f(t)F_S(t+y+2a)dt\right| \leqslant$$

$$\leq \int_0^\infty |f(t)| \int_{2a+N}^\infty |F_S(u)|dudt \leq \int_N^\infty |F_S(u)|du .$$

Therefore, since

$$\lim_{h\to 0} \int_0^\infty |F_S(u+h) - F_S(u)|du = 0$$

and

$$\lim_{N\to\infty} \int_N^\infty |F_S(u)|du ,$$

it follows that the image of the unit ball of $L_1[0,\infty)$ under the operator $\mathbf{F}^+_{S,a}$ is compact, and hence that the operator is compact.

To prove the compactness of the operator $\mathbf{F}^+_{S,a}$ in $L_2[0,\infty)$, we first of all note that $S(\lambda)$ is a continuous function. Thanks to this, one can find a sequence of smooth functions $\tilde{\Phi}_k(\lambda)$ such that

$$\max_{-\infty<\lambda<\infty} |\tilde{\Phi}_k(\lambda) - (1 - S(\lambda))| \leq \frac{1}{k}$$

and

$$\int_{-\infty}^\infty \{|\tilde{\Phi}_k(\lambda)|^2 + |\tilde{\Phi}_k'(\lambda)|^2\}d\lambda < \infty .$$

Let

$$\Phi_k(y) = \frac{1}{2\pi} \int_{-\infty}^\infty \tilde{\Phi}_k(\lambda)e^{i\lambda y}d\lambda .$$

Since

$$\int_0^\infty \int_0^\infty |\Phi_k(t+y+2a)|^2dydt = \int_0^\infty dy \int_{2a+y}^\infty |\Phi_k(u)|^2du =$$

$$= \int_0^\infty y|\Phi_k(y+2a)|^2dy \leq \int_0^\infty |\Phi_k(y+2a)|^2dy + \int_0^\infty (y+2a)^2|\Phi_k(y+2a)|^2dy \leq$$

$$\leq \frac{1}{2} \int_{-\infty}^\infty \{|\tilde{\Phi}_k(\lambda)|^2 + |\tilde{\Phi}_k'(\lambda)|^2\}d\lambda < \infty ,$$

the operators $\mathbf{\Phi}_k$:

$$(\mathbf{\Phi}_k f)(y) = \int_0^\infty f(t)\Phi_k(t+y+2a)dt ,$$

are Hilbert-Schmidt, and hence compact. Compact operators form a closed ideal in the algebra of all bounded operators. Therefore, to prove the compactness of $\mathbb{F}^+_{S,a}$, it suffices to check that the sequence of operators Φ_k converges to $\mathbb{F}^+_{S,a}$ as $k \to \infty$. Since

$$\{\Phi_k - \mathbb{F}^+_{S,a}\}f = \int_0^\infty f(t)\{\Phi_k(t+y+2a) - F_S(t+y+2a)\}dt =$$

$$= \frac{1}{2\pi}\int_{-\infty}^{\infty}\{\tilde{\Phi}_k(\lambda) - (1 - S(\lambda))\}\tilde{f}(-\lambda)e^{i\lambda(y+2a)}d\lambda ,$$

we have

$$||\{\Phi_k - \mathbb{F}^+_{S,a}\}f||^2 \leq \frac{1}{2\pi}\int_{-\infty}^{\infty}|\tilde{\Phi}_k(\lambda) - (1 - S(\lambda))|^2|\tilde{f}(-\lambda)|^2 d\lambda \leq \frac{1}{k^2}||f|| ,$$

and hence $\lim_{k\to\infty} ||\Phi_k - {}^{+}_{S,a}|| = 0$, which proves the compactness of the operator $\mathbb{F}^+_{S,a}$.

The compactness of the operator $\mathbb{F}^-_S$ in $L_2(-\infty,0]$ can be proved analogously.

To complete the proof the lemma, we remark that $\mathbb{F}^+_a$ differs from $\mathbb{F}^+_{S,a}$ by a finite-rank operator, and hence it is compact too. □

It follows from Lemma 3.3.1 that in order to prove the solvability of the fundamental equation, it suffices to verify that the homogeneous equation $f + \mathbb{F}^+_a f = 0$ has no nonnull solutions in the corresponding space. Notice that the solutions of this equation in the space $L_1[0,\infty)$ are bounded, and hence belong also to $L_2[0,\infty)$. In fact, the kernel $F(t+y+2a)$ of $\mathbb{F}^+_a$ can be approximated by a bounded function $\Phi(t+y+2a)$, so that

$$\int_0^\infty |F(t) - \Phi(t)|dt < 1 .$$

Upon rewriting the equation $f + \mathbb{F}^+_a f = 0$ in the form $f - \{\Phi - \mathbb{F}^+_a\}f = -\Phi f$, we obtain an equation with a bounded function on the right-hand side and an operator $\Phi - \mathbb{F}^+_a$, whose norm is smaller than one. Hence

$$f = -\Phi f - \sum_{n=1}^{\infty}(\Phi - \mathbb{F}^+_a)^n\Phi f ,$$

and the series converges in $L_1[0,\infty)$ as well as in $L[0,\infty)$. Consequently, $f(y) \in L_1[0,\infty) \cap L\ [0,\infty) \subset L_2[0,\infty)$.

Thus, it suffices to investigate the equation $f + \mathbb{F}_a^+ f = 0$ in the space $L_2[0,\infty)$.

LEMMA 3.3.2. *The operators* $\mathbb{I} + \mathbb{F}_a^+$ *and* $\mathbb{I} + \mathbb{F}_{S,a}^+$ *, acting in* $L_2[0,\infty)$ *, are nonnegative for every* $a \geq 0$:

$$(\{\mathbb{I} + \mathbb{F}_a^+\}f,f) \geq 0 , \tag{3.3.11}$$

and equality is attained if and only if

$$\left.\begin{aligned} &\tilde{f}(\lambda) - S(\lambda)e^{2i\lambda a}\tilde{f}(-\lambda) = 0 \quad (-\infty < \lambda < \infty) , \\ &\tilde{f}(-i\lambda_k) = 0 \quad (k = 1,2,\dots,n) . \end{aligned}\right\} \tag{3.3.12}$$

The operator $\mathbb{I} + \mathbb{F}_S^-$ *acting in* $L_2(-\infty,0]$ *is nonnegative:*

$$(\mathbb{I} - \mathbb{F}_S^- g,g) \geq 0 ,$$

and equality is attained if and only if

$$\tilde{g}(\lambda) + S(\lambda)\tilde{g}(-\lambda) = 0 \quad (-\infty < \lambda < \infty) . \tag{3.3.13}$$

PROOF. It follows from formulas (3.3.5), (3.3.6), (3.3.10) and the Parseval equality that

$$(\{\mathbb{I} + \mathbb{F}_a^+ \}f,f) = \frac{1}{2\pi} \int_{-\infty}^{\infty} |f(\lambda)|^2 d\lambda + \sum_{k=1}^{n} e^{-2a\lambda_k} m_k^2 |f(-i\lambda_k)|^2 +$$

$$+ \frac{1}{2\pi} \int_{-\infty}^{\infty} (1 - S(\lambda))e^{2i\lambda a}\tilde{f}(-\lambda)\overline{\tilde{f}(\lambda)}d\lambda .$$

Since $\tilde{f}(\lambda)$ is the Fourier transform of a function $f(y)$ which vanishes for $y < 0$, $\tilde{f}(-\lambda)e^{2i\lambda a}$ is the Fourier transform of a function $f(-y-2a)$, which vanishes for $y > -2a$. Hence

$$\frac{1}{2\pi} \int_{-\infty}^{\infty} e^{2i\lambda a}\tilde{f}(-\lambda)\overline{\tilde{f}(\lambda)}d\lambda = \int_0^{\infty} f(-y-2a)\overline{f(y)}dy ,$$

and

$$(\{\mathbb{I} + \mathbb{F}_a^+\}f,f) = \sum_{k=1}^{n} e^{-2a\lambda_k} m_k^2 |\tilde{f}(-i\lambda_k)|^2 +$$

$$+ \frac{1}{2\pi} \int_{-\infty}^{\infty} \{\tilde{f}(\lambda) - S(\lambda)e^{2ia\lambda}\tilde{f}(-\lambda)\}\overline{\tilde{f}(\lambda)}d\lambda . \tag{3.3.14}$$

Next, since $|S(\lambda)e^{2i\lambda a}| = 1$,

$$\left|\int_{-\infty}^{\infty} S(\lambda)e^{2i\lambda a}\tilde{f}(-\lambda)\overline{\tilde{f}(\lambda)}d\lambda\right|^2 \leq \int_{-\infty}^{\infty} |\tilde{f}(-\lambda)|^2 d\lambda \int_{-\infty}^{\infty} |\tilde{f}(\lambda)|^2 d\lambda$$

by the Cauchy-Bunyakovskii inequality or, equivalently,

$$\left|\int_{-\infty}^{\infty} S(\lambda)e^{2i\lambda a}\tilde{f}(-\lambda)\overline{\tilde{f}(\lambda)}d\lambda\right| \leq \int_{-\infty}^{\infty} |\tilde{f}(\lambda)|^2 d\lambda \; .$$

Therefore, the second term on the right-hand side of formula (3.3.14) is nonnegative. Since the first term is obviously nonnegative, inequality (3.3.11) holds, with equality, if and only if

$$\tilde{f}(-i\lambda_k) = 0 \quad (k = 1,2,\ldots,n)$$

and

$$\int_{-\infty}^{\infty} \{\tilde{f}(\lambda) - S(\lambda)e^{2i\lambda a}\tilde{f}(-\lambda)\}\overline{\tilde{f}(\lambda)}d\lambda = 0 \; .$$

This shows that the function $z(\lambda) = \tilde{f}(\lambda) - S(\lambda)e^{2i\lambda a}\tilde{f}(-\lambda)$ is orthogonal to $\tilde{f}(\lambda)$ in $L_2(-\infty,\infty)$. But then

$$||\tilde{f}(\lambda)||^2 = ||S(\lambda)e^{2i\lambda a}\tilde{f}(-\lambda)||^2 = ||\tilde{f}(\lambda) - z(\lambda)||^2 = ||\tilde{f}(\lambda)||^2 + ||z(\lambda)||^2 \; ,$$

which is possible if and only if $z(\lambda) = 0$. Thus, inequality (3.3.11) holds, with equality for those functions f whose Fourier transform $\tilde{f}(\lambda)$ satisfies conditions (3.3.12).

The analogous assertions for the operators $\mathbb{I} + \mathbb{F}^+_{S,a}$ and $\mathbb{I} - \mathbb{F}^-_S$ are proved in the same manner. □

LEMMA 3.3.3. *The operator* $\mathbb{I} + \mathbb{F}^+_a$, *acting in* $L_2[0,\infty)$, *is invertible for every* $a > 0$, *and*

$$\sup_{\varepsilon \leq a < \infty} ||\{\mathbb{I} + \mathbb{F}^+_a\}^{-1}|| = C(\varepsilon) \tag{3.3.15}$$

for all $\varepsilon > 0$. *If the operator* $\mathbb{I} + \mathbb{F}^+_a$ *is invertible for* $a = 0$, *then inequality* (3.3.15) *remains valid for* $\varepsilon = 0$ *too, i.e.,* $C(0) < \infty$.

PROOF. By the foregoing analysis, in order to establish the existence of the operator $\{\mathbb{I} + \mathbb{F}^+_a\}^{-1}$, it suffices to verify that the equation $f + \mathbb{F}^+_a f = 0$ has only the null solution in $L_2[0,\infty)$. But, by Lemma 3.3.2, the Fourier transform $\tilde{f}(\lambda)$ of any solution $f(y)$ of this equation satisfies the identity

$\tilde{f}(\lambda) - S(\lambda)e^{2i\lambda a}\tilde{f}(-\lambda) = 0$. Hence, upon setting $\tilde{\varphi}_h(\lambda) = \tilde{f}(\lambda)e^{-i\lambda a}\cos\lambda h$, $0 < h < a$, we get

$$\tilde{\varphi}_h(\lambda) - S(\lambda)\tilde{\varphi}_h(-\lambda) = 0 . \qquad (3.3.16)$$

Since $\tilde{\varphi}_h(\lambda)$ is the Fourier transform of the function

$$\varphi_h(t) = \frac{1}{2}\{f(t-a+h) + f(t-a-h)\} ,$$

which vanishes for $t < a-h$, identity (3.3.16) yields

$$\varphi_h + \mathbb{F}^+_{S,0}\varphi = 0 \qquad (3.3.17)$$

for all $h \in (0,a)$. Therefore, if the equation $f + \mathbb{F}^+_a f = 0$ has a nonzero solution, then equation (3.3.17) has infinitely many linearly independent solutions $\varphi_h(y)$, which in turn contradicts the compactness of the operator $\mathbb{F}^+_{S,0}$. Hence, $f(y) = 0$ and the operator $\mathbb{I} + \mathbb{F}^+_a$ is invertible for every $a > 0$.

It is readily verified that the norm of the operator $\mathbb{I} + \mathbb{F}^+_a$ depends continuously on a and tends to 1 as $a \to \infty$. It follows that $||\{\mathbb{I} + \mathbb{F}^+_a\}^{-1}||$ is a continuous function of $a \in (0,\infty)$ which tends to 1 as $a \to \infty$. Consequently, for every $\varepsilon > 0$,

$$\sup_{\varepsilon\leq a<\infty} ||\{\mathbb{I} + \mathbb{F}^+_a\}^{-1}|| = C(\varepsilon) < \infty .$$

Moreover, $C(0) < \infty$ if the operator $\mathbb{I} + \mathbb{F}^+_a$ is also invertible for $a = 0$. □

THEOREM 3.3.1. *If condition* I *is satisfied, then the fundamental equation* (3.3.7) *has a unique solution* $K(x,y) \in L_1[x,\infty)$ *for every* $x > 0$, *the function*

$$e(\lambda,x) = e^{i\lambda x} + \int_x^\infty K(x,y)e^{i\lambda y}dy \quad (\text{Im}\,\lambda \geq 0)$$

satisfies the equation

$$-y'' + q(x)y = \lambda^2 y \quad , \quad 0 < x < \infty ,$$

where $q(x) = -2\frac{d}{dx}K(x,x)$, *and for every* $\varepsilon > 0$,

$$\int_\varepsilon^\infty x|q(x)|dx < \infty . \qquad (3.3.18)$$

If, in addition, the operator $\mathbb{I} + \mathbb{F}_a^+$ *is also invertible for* $a = 0$ *, then this inequality remains valid for* $\varepsilon = 0$ *, i.e., the function* $q(x)$ *satisfies condition* (3.1.3).

PROOF. The solvability of the fundamental equation for all $x > 0$ was proved in Lemma 3.3.3. Since the function $F(x)$ is differentiable on the half line $(0,\infty)$, and its derivative belongs to $L_1[\varepsilon,\infty)$ by assumption, it is readily established by passage to the limit under the integral sign, that the function $K(x,y)$ has first-order partial derivatives and that equality (3.3.7) can be differentiated term-wise.

To estimate the function $K(x,y)$ and its derivatives, it is convenient to introduce the notations

$$\tau(x) = \int_x^\infty |F'(t)|dt \quad , \quad \tau_1(x) = \int_x^\infty \tau(t)dt \; . \tag{3.3.19}$$

We remark that

$$|F(x)| \leqslant \tau(x) \tag{3.3.19'}$$

and $\tau_1(0) < \infty$ (since $\int_0^\infty x|F'(x)|dx < \infty$ by assumption). From equation (3.3.7) we find that

$$K(x,x+y) = - \{\mathbb{I} + \mathbb{F}_x^+\}^{-1}F(y+2x)$$

whence, in view of inequality (3.3.15) of Lemma 3.3.3,

$$\int_0^\infty |K(x,x+y)|dy \leqslant ||\{\mathbb{I} + \mathbb{F}_x^+\}^{-1}|| \int_0^\infty |F(y+2x)|dy \leqslant C(x)\tau_1(2x) \; .$$

Using this estimate, we get

$$|K(x,x+y)| \leqslant |F(y+2x)| + \int_0^\infty |K(x,x+t)||F(t+y+2x)|dt \leqslant \tau(y+2x)\{1 + C(x)\tau_1(2x)\} \; .$$

Next, since differentiation of equation (3.3.7') with respect to x yields

$$2F'(y+2x) + K_x(x,x+y) + \int_0^\infty K_x(x,x+t)F(t+y+2x)dt + 2\int_0^\infty K(x,x+t)F'(t+y+2x)dt = 0 \; ,$$

we see that

$$\int_0^\infty |K_x(x,x+y)|dy \leqslant$$

$$\leqslant 2||\{\mathbb{I} + \mathbb{F}_x^+\}^{-1}|| \int_0^\infty \left|F'(y+2x) + \int_0^\infty K(x,x+t)F'(t+y+2x)dt\right|dy \leqslant$$

$$\leqslant 2C(x)\tau(2x)\{1 + C(x)\tau_1(2x)\} .$$

Hence,

$$|K_x(x,x+y) + 2F'(y+2x)| \leqslant$$

$$\leqslant \int_0^\infty |K_x(x,x+t)F(t+y+2x)|dt + 2\int_0^\infty |K(x,x+t)F'(t+y+2x)|dt \leqslant$$

$$\leqslant 2\tau(y+2x)C(x)\tau(2x)\{1 + C(x)\tau_1(2x)\} + 2\tau(2x)\{1 + C(x)\tau_1(2x)\}\tau(y+2x) ,$$

i.e., the function $q(x) = -2\frac{d}{dx}K(x,x)$ satisfies the inequality

$$|q(x)| \leqslant 4|F'(2x)| + 8C(x)\{1 + C(x)\tau_1(2x)\}\tau^2(2x) . \tag{3.3.20}$$

Since

$$x\tau(x) = x\int_x^\infty |F'(t)|dt \leqslant \int_x^\infty t|F'(t)|dt \leqslant \int_0^\infty t|F'(t)|dt < \infty ,$$

we have

$$\int_\varepsilon^\infty x|q(x)|dx \leqslant 4\int_0^\infty x|F'(2x)|dx + 4C(\varepsilon)\{1 + C(\varepsilon)\tau_1(0)\}\int_0^\infty t|F'(t)|dt \int_\varepsilon^\infty \tau(2x)dx \leqslant$$

$$\leqslant \{1 + 2C(\varepsilon)[1 + C(\varepsilon)\tau_1(0)]\tau_1(0)\}\int_0^\infty t|F'(t)|dt ,$$

which proves inequality (3.3.18). If the operator $\mathbb{I} + \mathbb{F}_a^+$ is also invertible for $a = 0$, then $C(0) < \infty$ and (3.3.18) remains valid for $\varepsilon = 0$.

We now turn to the proof of the main assertion of the theorem, assuming first that the function $F(x)$ is twice continuously differentiable and $F''(x) \in L_1[\varepsilon,\infty)$ for every $\varepsilon > 0$. Then the solution $K(x,y)$ of the fundamental equation is also twice continuously differentiable and its second-order derivatives are summable on the half line $[\varepsilon,\infty)$ for every $\varepsilon > 0$. Upon differentiating the fundamental equation twice with respect to y, we get

$$F''(x+y) + K_{yy}(x,y) + \int_x^\infty K(x,t)F''(t+y)dt = 0 ,$$

whence, by an integration by parts,

$$F''(x+y) + K_{yy}(x,y) - K(x,x)F'(x+y) +$$

$$+ K_t(x,t)F(t+y)\Big|_{t=x} + \int_0^\infty K_{tt}(x,t)F(t+y)dt = 0 .$$

On the other hand, differentiation of the fundamental equation twice with respect to x yields

$$F''(x+y) + K_{xx}(x,y) - \frac{\partial}{\partial x}\{K(x,x)F(x+y)\} -$$

$$- K_x(x,t)F(t+y)\Big|_{t=x} + \int_x^\infty K_{xx}(x,t)F(t+y)dt = 0 .$$

Substracting this equality from the preceding one, we get

$$K_{xx}(x,y) - K_{yy}(x,y) + q(x)F(x+y) + \int_x^\infty \{K_{xx}(x,t) - K_{tt}(x,t)\}F(t+y)dt = 0 ,$$

where $q(x) = -2 \frac{d}{dx} K(x,x)$. But by the fundamental equation,

$$q(x)F(x+y) = -q(x)K(x,y) - \int_x^\infty q(x)K(x,t)F(t+y)dt .$$

Therefore,

$$K_{xx}(x,y) - K_{yy}(x,y) - q(x)K(x,y) +$$

$$+ \int_x^\infty \{K_{xx}(x,t) - K_{tt}(x,t) - q(x)K(x,t)\}F(t+y)dt = 0 ,$$

i.e., the function

$$\varphi(y) = K_{xx}(x,y) - K_{yy}(x,y) - q(x)K(x,y)$$

is a summable solution of the homogeneous equation

$$\varphi(y) + \int_x^\infty \varphi(t)F(t+y)dt = 0 \quad (x \leqslant y < \infty) .$$

By Lemma 3.3.3, this equation has only the zero solution for every $x > 0$. Hence, $\varphi(y) = 0$, i.e., the function $K(x,y)$ is a solution of the equation

$$K_{xx}(x,y) - K_{yy}(x,y) - q(x)K(x,y) = 0 ,$$

which satisfies the conditions

$$q(x) = -2 \frac{d}{dx} K(x,x)$$

and

$$\lim_{x+y\to\infty} K_x(x,y) = \lim_{x+y\to\infty} K_y(x,y) = 0 .$$

From this it follows, in view of the remark to Lemma 3.1.2, that $K(x,y)$ is the kernel of the transformation operator, i.e., the function

$$e(\lambda,x) = e^{i\lambda x} + \int_x^\infty K(x,t)e^{i\lambda t}dt \tag{3.3.21}$$

satisfies the equation

$$-y'' + q(x)y = \lambda^2 y \quad (0 < x < \infty) . \tag{3.3.22}$$

Now suppose that only condition I is satisfied, so that the function $F(x)$ does not necessarily have a second derivative. Let $F_n(x)$ be a sequence of twice continuously differentiable functions such that, for every $\varepsilon > 0$,

$$\lim_{n\to\infty} \int_\varepsilon^\infty |F_n(x) - F(x)|dx = 0$$

and

$$\lim_{n\to\infty} \int_\varepsilon^\infty x|F_n'(x) - F_n'(x)|dx = 0 .$$

Then, for sufficiently large n , each of the equations

$$F_n(x+y) + K_n(x,y) + \int_x^\infty K_n(x,t)F_n(t+y)dt = 0$$

has a unique solution, and

$$\lim_{n\to\infty} \sup_{\varepsilon\leqslant x\leqslant\infty} \int_x^\infty |K_n(x,y) - K(x,y)|dy = 0$$

and

$$\lim_{n\to\infty} \int_\varepsilon^\infty |K_n'(x,x) - K'(x,x)|dx = 0$$

for every $\varepsilon > 0$. Moreover, as was already shown, the functions

$$e_n(\lambda,x) = e^{i\lambda x} + \int_x^\infty K_n(x,t)e^{i\lambda t}dt$$

satisfy the equations

$$-y'' + q_n(x)y = \lambda^2 y \quad (q_n(x) = -2 \frac{d}{dx} K_n(x,x)) .$$

Letting $n \to \infty$ in these formulas, we conclude that the functions (3.3.21) must satisfy equation (3.3.22), as asserted. □

THEOREM 3.3.2. *Suppose that condition* I *is satisfied, and*

a) the equation $f + \mathbb{F}_{S,0}^+ f = 0$ *has* n *linearly independent solutions in* $L_2[0,\infty)$,

b) the equation $g - \mathbb{F}_S^- g = 0$ *has only the zero solution in* $L_2(-\infty,0]$.

Then the collection $\{S(\lambda)\ (-\infty < \lambda < \infty)\ ;\ \lambda_k, m_k\ (k = 1,2,\ldots,n)\}$ *is the scattering data of a boundary value problem* (3.1.1), (3.1.2), *whose potential* $q(t)$ *satisfies condition* (3.1.3)

PROOF. By Theorem 3.3.1, condition I guarantees the solvability of the fundamental equation for $x > 0$, and the functions (3.3.21) satisfy equation (3.3.22). If the fundamental equation is solvable for $x = 0$ too (i.e., the equation $f + \mathbb{F}_0^+ f = 0$ has only the zero solution in $L_2[0,\infty)$, then the potential $q(x)$ in equation (3.3.22) satisfies condition (3.1.3). Hence, to prove the theorem, it suffices to show that conditions a) and b) imply that: 1) the equation $f + \mathbb{F}_0^+ f = 0$ has no nonzero solutions; 2) $e(i\lambda_k, 0) = 0$ $(k = 1,2,\ldots,n)$; 3) $e(-\lambda,0) - S(\lambda)e(\lambda,0) = 0$ $(-\infty < \lambda < \infty)$. We prove these implications successively.

1. If the function $f(y) \in L_2[0,\infty)$ satisfies the equation $f + \mathbb{F}_0^+ f = 0$, then, by Lemma 3.3.2, its Fourier transform $\tilde{f}(\lambda)$ is such that

$$\tilde{f}(-i\lambda_k) = 0 \quad (k = 1,2,\ldots,n)$$

and

$$\tilde{f}(\lambda) - S(\lambda)\tilde{f}(-\lambda) = 0 \quad (-\infty < \lambda < \infty)\ .$$

Hence

$$\tilde{z}_1(\lambda) = \tilde{f}(\lambda)\ ,$$
$$\tilde{z}_{k+1}(\lambda) = (\lambda^2 + \lambda_k^2)^{-1}\tilde{f}(\lambda) \quad , \quad k = 1,2,\ldots,n\ ,$$

are the Fourier transforms of the functions

$$z_p(y) = \frac{1}{2\pi}\int_{-\infty}^{\infty} \tilde{z}_p(\lambda) e^{i\lambda y} d\lambda \quad (p = 1,2,\ldots,n+1)\ ,$$

which obviously vanish for $y < 0$ and belong to $L_2[0,\infty)$. Moreover, the functions $\tilde{z}_p(\lambda)$ also satisfy the identity

$$\tilde{z}_p(\lambda) - S(\lambda)\tilde{z}_p(-\lambda) = 0\ .$$

Therefore, $z_p + \mathbb{F}_{S,0}^+ z_p = 0$, and hence the equation $z + \mathbf{F}_{S,0}^+ z = 0$ has $n+1$ solutions $z_1(y), z_2(y), \ldots, z_{n+1}(y)$, which are linearly independent if $\tilde{f}(\lambda) \neq 0$. But, in view of condition a), this equation has only n linearly independent solutions. Therefore, $\tilde{f}(\lambda) \equiv 0$, the equation $f + {}_0^+ f = 0$ has only the null solution, and the fundamental equation is solvable for $x = 0$ also.

2. We now consider the fundamental equation for $x = 0$, which, upon substituting expression (3.3.6) for $F(y)$, becomes

$$\sum_{k=1}^{n} m_k^2 e^{-\lambda_k y} + F_S(y) + K(0,y) +$$

$$+ \sum_{k=1}^{n} m_k^2 e^{-\lambda_k y} \int_0^\infty K(0,t) e^{-\lambda_k t} dt + \int_0^\infty K(0,t) F_S(t+y) dt = 0 ,$$

or, equivalently,

$$\sum_{k=1}^{n} m_k^2 e(-i\lambda_k,0) e^{-\lambda_k y} + F_S(y) + K(0,y) + \int_0^\infty K(0,t) F_S(t+y) dt = 0 . \qquad (3.3.23)$$

By condition a, the equation $f + \mathbb{F}_{S,0}^{+} f = 0$ has n linearly independent solutions $f_1(y), f_2(y), \ldots, f_n(y)$. Upon multiplying both sides of equality (3.3.23) by $f_j(y)$ and integrating, we get

$$\sum_{k=1}^{n} m_k^2 e(-i\lambda_k,0) \tilde{f}_j(-i\lambda_k) + \int_0^\infty F_S(y) f_j(y) dy = 0 .$$

As we showed in the proof Lemma 3.3.2, the Fourier transforms $f_j(\lambda)$ of these solutions must satisfy the equality $\tilde{f}_j(\lambda) - S(\lambda)\tilde{f}_j(-\lambda) = 0$. Hence,

$$\int_0^\infty F_S(y) f_j(y) dy = \frac{1}{2\pi} \int_{-\infty}^{\infty} (1 - S(\lambda)) \tilde{f}_j(-\lambda) d\lambda = \frac{1}{2\pi} \int_{-\infty}^{\infty} \{\tilde{f}_j(-\lambda) - \tilde{f}_j(\lambda)\} d\lambda = 0$$

and

$$\sum_{k=1}^{n} m_k^2 e(-i\lambda_k,0) \tilde{f}_j(-i\lambda_k) = 0 \quad (j = 1,2,\ldots,n) .$$

Therefore, to prove the equalities $e(-i\lambda_k,0) = 0$, it suffices to verify that the determinant of the matrix $||f_j(-i\lambda_k)||_{j,k=1}^{n}$ is different from zero. But if it were not different from zero, then one could find numbers c_j such that $\sum |c_j| \neq 0$ and

$$\sum_{j=1}^{n} c_j \tilde{f}_j(-\lambda_k) = 0 \quad (k = 1,2,\ldots,n) .$$

Consequently, the function $\tilde{z}(\lambda) = \sum_{j=1}^{n} c_j \tilde{f}_j(\lambda)$ would satisfy the equalities

(3.3.12), i.e., $z(y) = \sum_{j=1}^{n} c_j f_j(y)$ would be a nonzero solution of the equation $z + \mathbb{F}_0^+ z = 0$ which, as we showed earlier, is impossible. Thus,

$$\operatorname{Det}\|\tilde{f}_j(-i\lambda_k)\| \neq 0$$

and

$$e(-i\lambda_k,0) = 0 \quad (k = 1,2,\dots,n) .$$

3. Since

$$e(-\lambda,0) - S(\lambda)e(\lambda,0) = 1 + \tilde{k}(\lambda) - S(\lambda)[1 + \tilde{k}(-\lambda)] =$$
$$= 1 - S(\lambda) + \tilde{k}(\lambda) - \tilde{k}(-\lambda) + [1 - S(\lambda)]\tilde{k}(-\lambda) ,$$

where

$$\tilde{k}(\lambda) = \int_0^\infty K(0,t)e^{-i\lambda t}dt ,$$

the equality that we have to prove is equivalent to

$$1 - S(\lambda) + \tilde{k}(-\lambda) - \tilde{k}(-\lambda) + [1 - S(\lambda)]\tilde{k}(-\lambda) = 0 \quad (-\infty < \lambda < \infty) .$$

Since the left-hand side is the Fourier transform of the function

$$\Phi(y) = F_S(y) + K(0,y) - K(0,-y) + \int_0^\infty K(0,t)F_S(t+y)dt , \tag{3.3.24}$$

we must check that $\Phi(y) \equiv 0$. We first take $y > 0$. From equation (3.3.23) and the equalities $e(-i\lambda_k,0) = 0$ $(k = 1,2,\dots,n)$ established above, it follows that for $y > 0$,

$$F_S(y) + K(0,y) + \int_0^\infty K(0,t)F_S(t+y)dt = 0 ,$$

and therefore, since $K(0,-y) = 0$, we conclude that $\Phi(y) = 0$ for all $0 < y < \infty$. Next, upon multiplying both sides of equality (3.3.24) by $F_S(x+y)$ and integrating with respect to y , we get

$$\int_{-\infty}^{0} \Phi(y)F_S(x+y)dy = \int_{-\infty}^{\infty} F_S(y)F_S(y+x)dy + \int_0^\infty K(0,y)F_S(y+x)dy -$$
$$- \int_{-\infty}^{0} K(0,-y)F_S(y+x)dy + \int_0^\infty K(0,t) \int_{-\infty}^{\infty} F_S(t+y)F_S(y+x)dydt .$$

Since the Fourier transform of the function $F_S(y)$ equals $1 - S(\lambda)$, the theorem on convolution shows that the Fourier transform of the function

$\int_{-\infty}^{\infty} F_S(y)F_S(y+x)dy$ is

$\{1 - S(-\lambda)\}\{1 - S(\lambda)\} = 1 - S(-\lambda) - S(\lambda) + 1$.

Therefore,

$$\int_{-\infty}^{\infty} F_S(y)F_S(y+x)dy = F_S(-x) + F_S(x)$$

and

$$\int_{-\infty}^{\infty} F_S(t+y)F_S(y+x)dy = F_S(t-x) + F_S(x-t) .$$

It follows that

$$\int_{-\infty}^{0} \Phi(y)F_S(y+x)dy = F_S(-x) + F_S(x) + \int_0^{\infty} K(0,y)F_S(y+x)dy -$$

$$- \int_0^{\infty} K(0,\xi)F_S(-\xi+x)d\xi + \int_0^{\infty} K(0,t)\{F_S(t-x) + F_S(x-t)\}dt =$$

$$= F_S(-x) + K(0,-x) - K(0,x) + \int_0^{\infty} K(0,y)F_S(y-x)dy +$$

$$+ F_S(x) + K(0,x) - K(0,-x) + \int_0^{\infty} K(0,y)F_S(y+x)dy =$$

$$= \Phi(-x) + \Phi(x) ,$$

whence, upon recalling that $\Phi(-x) = 0$ for $x < 0$, we get

$$\int_{-\infty}^{0} \Phi(y)F_S(y+x)dy = \Phi(x) \quad (-\infty < x < 0) .$$

Hence, the function $\Phi(x)$ is a solution of the homogeneous equation $\Phi - \mathbb{F}_S^-\Phi = 0$. It follows from formula (3.2.24) and condition I that $\Phi(x)$ can be expressed as the sum of two functions, one of which is summable, while the second is bounded and belongs to $L_2(-\infty,0]$. From this we conclude, just as above, that $\Phi(x)$ is a solution of the equation $\Phi - \mathbb{F}_S^-\Phi = 0$ in $L_2(-\infty,0]$. Thus, if condition b is satisfied, then $\Phi(x) \equiv 0$, which completes the proof. □

LEMMA 3.3.4. *If the function* $S(\lambda)$ *satisfies condition* I, *then for every* $a > 0$, *the function*

$$S_1(\lambda) = S(\lambda) \frac{\lambda + ia}{\lambda - ia}$$

also satisfies this condition.

PROOF. The continuity of $S_1(\lambda)$ and the equalities $\lim_{|\lambda|\to\infty} S_1(\lambda) = 1$ and $S_1(\lambda) = \overline{S_1(-\lambda)} = [S_1(-\lambda)]^{-1}$ are obvious consequences of the fact that they are enjoyed by $S(\lambda)$ also. Next, since

$$1 - S_1(\lambda) = 1 - S(\lambda) + \frac{2ai}{\lambda - ia}\{1 - S(\lambda)\} - \frac{2ai}{\lambda - ia}$$

and

$$-\frac{1}{\lambda - ia} = \int_0^\infty e^{-ay}e^{-i\lambda y}dy ,$$

the convolution theorem implies that the function $1 - S_1(\lambda)$ is the Fourier transform of the function

$$F_{S_1} = \begin{cases} F_S(y) - 2aG(y) + 2ae^{-ay} , & y > 0 , \\ F_S(y) - 2aG(y) , & y < 0 , \end{cases}$$

where

$$G(y) = \int_0^\infty F_S(y-t)e^{-at}dt = e^{-ay}\int_{-\infty}^y F_S(\xi)e^{a\xi}d\xi = \frac{i}{2\pi}\int_{-\infty}^\infty \frac{1 - S(\lambda)}{\lambda - ia}e^{i\lambda y}d\lambda .$$

A straightforward consequence of this formula and of the properties of the function $F_S(y)$ is that $G(y) \in L_2(-\infty,\infty) \cap L(-\infty,\infty)$. Hence, it remains to verify that

$$\int_0^\infty y|G'(y)|dy < \infty . \tag{3.3.25}$$

Since

$$G'(y) = -ae^{-ay}\int_{-\infty}^y F_S(\xi)e^{a\xi}d\xi + F_S(y) =$$

$$= -ae^{-ay}A + e^{a(1-y)}F_S(1) + e^{-ay}\int_1^y F_S'(\xi)d\xi$$

for $y > 0$, where

$$A = \int_{-\infty}^1 F_S(\xi)e^{a\xi}d\xi ,$$

the inequality will be established if we prove that

$$\int_0^\infty ye^{-ay}\left|\int_1^y F_S'(\xi)e^{a\xi}d\xi\right|dy < \infty .$$

But

$$\int_1^\infty ye^{-ay}\left|\int_1^y F_S'(\xi)e^{a\xi}d\xi\right|dy \leq \int_1^\infty \left\{\int_1^y |F_S'(\xi)|e^{a\xi}d\xi\right\} d\left\{-\int_y^\infty \xi e^{-a}\ d\xi\right\} \leq$$

$$\leq \int_1^\infty \{a^{-1}ye^{-ay} + a^{-2}e^{-ay}\}|F_S'(y)|e^{-ay}dy < \infty ,$$

because

$$\int_1^\infty y|F_S'(y)|dy < \infty .$$

Similarly,

$$\int_0^1 ye^{-ay}\left|\int_1^\infty F_S'(\xi)e^{a\xi}d\xi\right|dy \leq e\int_0^1 dy\int_y^1 |F_S'(\xi)|d\xi < \infty .$$

The lemma is proved. □

THEOREM 3.3.3. *For the collection* $\{S(\lambda)\ (-\infty < \lambda < \infty)\ ;\ \lambda_k, m_k\ (k = 1,2,\ldots,n)\}$ *to be the scattering data of a boundary value problem* (3.1.1), (3.1.2) *with a potential* $q(t)$ *subject to condition* (3.1.3), *it is necessary and sufficient that conditions* I *and* II *be satisfied.*

PROOF. The necessity of these conditions was proved in Section 2. By Theorem 3.3.2, their sufficiency will be established if we show that conditions I and II imply conditions a and b of Theorem 3.3.2, i.e., that: $\tilde{a})\ \kappa^+ = \kappa$, and $\tilde{b})\ \kappa^- = 0$, where κ^+ (κ^-) denotes the number of linearly independent solutions of the equation

$$f + \mathbb{F}_{S,0}^+ f = 0 \quad (\text{respectively, } g - \mathbb{F}_S^- g = 0) ;$$

and

$$\kappa = \frac{1}{2\pi i}\{\ln S(0+) - \ln S(+\infty)\} - \frac{1 - S(0)}{4} .$$

To prove this, we introduce the function

$$S_1(\lambda) = S(\lambda)\left(\frac{\lambda + i}{\lambda - i}\right)^{2\kappa^-} , \tag{3.3.26}$$

which, by Lemma 3.3.4, satisfies condition I. It also satisfies condition b), i.e., the equation $g - \mathbb{F}_{S_1}^- g = 0$ has no nonzero solutions in $L_2(-\infty,0]$. In fact, by Lemma 3.3.2, the Fourier transform $\tilde{g}(\lambda)$ of any solution of this equation satisfies the identity

$\tilde{g}(\lambda) + S_1(\lambda)\tilde{g}(-\lambda) = 0 \quad (-\infty < \lambda < \infty)$,

from which it further follows, in view of formula (3.3.26), that the function

$$\tilde{z}_k(\lambda) = \frac{\tilde{g}(\lambda)}{(\lambda^2 + 1)^k} \left(\frac{\lambda - i}{\lambda + i}\right)^{\kappa^-} \quad (k = 0,1,\dots,\kappa^-)$$

satisfies the identity

$$\tilde{z}_k(\lambda) + S(\lambda)\tilde{z}_k(-\lambda) = 0 \ . \tag{3.3.27}$$

Since the functions

$$(\lambda^2 + 1)^{-k}\left(\frac{\lambda - i}{\lambda + i}\right)^{\kappa^-} \quad (k = 0,1,\dots,\kappa^-)$$

are holomorphic and bounded in the upper half plane, the $\tilde{z}_k(\lambda)$ are, like $\tilde{g}(\lambda)$, Fourier transforms of functions $z_k(y) \in L_2(-\infty,0]$ that vanish for $y > 0$. But then (3.3.27) implies that these functions must satisfy the equation $z_k - \mathbf{F}_S^- z_k = 0$. Therefore, the latter has not κ^- , but at least κ^-+1 linearly independent solutions $z_0, z_1, \dots, z_{\kappa^-}$ if $\tilde{g}(\lambda) \not\equiv 0$, which is a contradiction. Hence, $\tilde{g}(\lambda) \equiv 0$ and $S_1(\lambda)$ satisfies condition b).

Let $\overset{+}{1}$ denote the number of linearly independent solutions of the equation $f + \mathbf{F}_{S_1}^+ f = 0$. The collection $\{S_1(\lambda), \lambda_k, m_k \ (k = 1,2,\dots,\kappa_1^+)\}$ satisfies conditions I, a) and b). Hence, by Theorem 3.3.2, it is the scattering data of a boundary value problem (3.3.1), (3.1.2), with a potential which is subject to condition (3.1.3). Therefore,

$$S_1(\lambda) = \frac{e_1(-\lambda,0)}{e_1(\lambda,0)} \ , \tag{3.3.28}$$

where the function $e_1(\lambda,0)$ is holomorphic in the upper half plane, tends to 1 as $|\lambda| \to \infty$, and has κ_1^+ zeros in $i\lambda_1, i\lambda_2, \dots, i\lambda_{\kappa_1^+}$. Moreover,

$$\kappa_1^+ = \frac{1}{2\pi i}\{\ln S_1(0+) - \ln S_1(+\infty)\} - \frac{1 - S_1(0)}{4} \ .$$

Using this equality and formula (3.3.26), we get

$$\kappa_1^+ = \kappa + \kappa^- \ . \tag{3.3.29}$$

If $\kappa^- > 0$, then the equation $g - \mathbf{F}_S^- g = 0$ has linearly independent solutions $g_1, \dots, g_{\kappa^-}$. Thus, one can obviously construct a nonzero solution $g = \sum c_k g_k$ of this equation, whose Fourier transform $\tilde{g}(\lambda)$ has a zero of order κ^--1 at the point i. Since $\tilde{g}(\lambda)$ must satisfy the identity $\tilde{g}(\lambda) + S(\lambda)\tilde{g}(-\lambda) = 0$, the formulas (3.3.26) and (3.3.28) yield

$$\tilde{g}(\lambda) + \frac{e_1(-\lambda,0)}{e_1(\lambda,0)} \left(\frac{\lambda - i}{\lambda + i}\right)^{2\kappa^-} \tilde{g}(-\lambda) = 0$$

or, equivalently,

$$e_1(\lambda,0)(\lambda+i)^2\tilde{g}(\lambda)\left(\frac{\lambda + i}{\lambda - i}\right)^{\kappa^- -1} = -e_1(-\lambda,0)(-\lambda+i)^2 g(-\lambda)\left(\frac{-\lambda + i}{-\lambda - i}\right)^{\kappa^- -1} .$$

The function on the left (right) hand side of this identity is holomorphic in the upper (respectively, lower) half plane. Hence

$$\varphi(\lambda) = (\lambda+i)^2\tilde{g}(\lambda)\left(\frac{\lambda + i}{\lambda - i}\right)^{\kappa^- -1} e_1(\lambda,0)$$

has a continuation to the whole complex plane as an entire function. Moreover, $|\varphi(\lambda)| = o(|\lambda|^2)$ as $|\lambda| \to \infty$. But then $\varphi(\lambda) = c\lambda$, and if $e_1(\lambda,0)$ has at least one zero in the upper half plane (i.e., $\kappa_1^+ > 0$), then $\varphi(\lambda) \equiv 0$, and hence $\tilde{g}(\lambda) \equiv 0$, contrary to assumption. Therefore, $\kappa_1^+ = 0$, and formula (3.3.29) yields $0 = \kappa + \kappa^-$ and $\kappa^- = 0$, since $\kappa \geqslant 0$ by condition II. Thus, the function $S(\lambda)$ satisfies conditions I and b). Adjoining to it an arbitrary collection of numbers $\{\lambda_1,\ldots,\lambda_{\kappa^+},m_1,\ldots,m_{\kappa^+}\}$, we obtain, by Theorem 3.3.2, the scattering data of a boundary value problem (3.1.1), (3.1.2), with a potential that satisfies condition (3.1.3). Hence,

$$\kappa^+ = \frac{1}{2\pi i}\{\ln S(0+) - \ln S(+\infty)\} - \frac{1 - S(0)}{4} = \kappa ,$$

i.e., condition a) is satisfied, too. This completes the proof of the theorem. □

We remark that it follows from Theorems 3.3.2 and 3.3.3 that the two sets of conditions: I, II, and I, a), b) are equivalent.

We conclude this section by examining a particular case of Theorem 3.3.3 which plays an important role in the investigation of the inverse Sturm-Liouville problems on a finite interval. Specifically, we are interested here in boundary value problems (3.3.1), (3.3.2) for which: 1) the potential $q(x)$ vanishes for $x > T$ and is square summable on the interval $(0,T)$; 2) there is no discrete spectrum. The scattering data for such a problem reduces to a single scattering function $S(\lambda)$ which, of course, possesses properties I and II with $n = 0$. The function S is also subject to the following condition:

III. *The function* $F_S(x)$ *vanishes for* $x > 2T$ *and* $F_S' \in L_2[0,2T]$.

In fact, since in the present case $F(x) = F_S(x)$ and the function $\sigma(x)$ is bounded and vanishes for $x > T$, it follows from the estimates (3.2.20), (3.2.21) that $F_S(2x) = 0$ for $x > T$ and $F_S'(2x) \in {}_2[0,T]$.

Conversely, if the function $S(\lambda)$ possesses properties I, II with $n = 0$, and III, then it is the scattering function of a boundary value problem which satisfies conditions 1 and 2. In fact, by Theorem 3.3.3, $S(\lambda)$ is the scattering function of a boundary value problem without discrete spectrum; moreover, the estimates (3.3.20) imply that for this problem, the potential $q(x)$ vanishes for $x > T$ and is square summable on $(0,T)$.

Thus, Theorem 3.3.3 admits the following

COROLLARY. *For* $S(\lambda)$ $(-\infty < \lambda < \infty)$ *to be the scattering function of a boundary value problem which satisfies conditions* 1 *and* 2, *it is necessary and sufficient that it possess properties* I, II *with* $n = 0$, *and* III.

□

PROBLEMS

1. If the real-valued function $q(x)$ satisfies condition (3.1.3) and $h(x)$ is a solution of the equation $y'' - q(x)y = 0$, such that $h(0) \neq 0$, then the normalized eigenfunctions $u(\lambda,x)$ of the boundary value problem

$$-y'' + q(x)y = \lambda^2 y , \quad \lim_{x\to 0} W\{y(x),h(x)\} = 0 , \tag{3.3.30}$$

which generate the Parseval identity (3.2.4), have the asymptotics

$$\begin{aligned} u(\lambda,x) &= e^{-i\lambda x} + S_h(\lambda)e^{i\lambda x} + o(1) \quad (-\infty < \lambda < \infty) , \\ u(i\lambda_k,x) &= m_k e^{-\lambda_k x}(1 + o(1)) \quad (k = 1,2,\dots,n) \end{aligned} \tag{3.3.31}$$

as $x \to \infty$ (see Section 2, Problem 2). Show that, for a given collection $\{S(\lambda)\ (-\infty < \lambda < \infty);\ \lambda_k,m_k\ (k = 1,2,\dots,n)\}$, there is a boundary value problem (3.3.30) whose normalized eigenfunctions have the asymptotics (3.3.31), if and only if it satisfies condition I and

II'. $n = \dfrac{\ln S(0+) - \ln S(+\ )}{2\pi i} - \dfrac{1 - S(0)}{4}$.

Hint. If conditions I and II' are satisfied, then the collection $\{S_1(\lambda);\lambda_k,m_k\}$, where

$$S_1(\lambda) = S(\lambda)\frac{\lambda + iA}{\lambda - iA} \quad (A > \max_{1\leq k\leq n} \lambda_k)$$

satisfies conditions I, II, and hence, by Theorem 3.3.3, it is the scattering data of a boundary value problem (3.1.1), (3.1.2) with a potential $q_1(x)$ subject to condition (3.1.3). The operator $u_1(\lambda,x) \to$

$$u(\lambda,x) = -\frac{W\, e_1(iA,x),u_1(\lambda,x)\}}{e_1(iA,x)(A^2 + \lambda^2)}$$

takes the normalized eigenfunctions $u_1(\lambda,x)$ of this boundary value problem into the eigenfunctions $u(\lambda,x)$ of a boundary value problem of the form (3.3.30); moreover, following an appropriate normalization, the asymptotics (3.3.31) hold. The necessity of conditions I and II is proved in an analogous manner.

2. Consider the boundary value problem (3.3.3), (3.3.4) with a potential $q(x)$ subject to (3.1.3) and a positive integer ℓ (which hereafter will be fixed). The normalized eigenfunctions (radial wave functions) $u_\ell(\lambda,x)$ of this problem have the asymptotics

$$u_\ell(\lambda,x) = e^{-i\lambda x} + (-1)^\ell S(\lambda)e^{i\lambda x} + o(1) \quad (-\infty < \lambda < \infty) ,$$

$$u_\ell(i\lambda_k,x) = m_k e^{-\lambda_k x}(1 + o(1)) \quad (k = 1,2,\dots,n-1) ,$$

$$u_\ell(i\lambda_n,x) = m_n x^{-\ell}(1 + o(1)) \ , \text{ if } \lambda_n = 0 ,$$

as $x \to \infty$. The collection $\{S(\lambda);\lambda_1 > \lambda_2 > \dots > \lambda_n \geq 0;m_1,m_2,\dots,m_n\}$ is called the scattering data of the given problem. Show that $\{S(\lambda);\lambda_1 > \lambda_2 > > \dots > \lambda_n \geq 0;m_1,\dots,m_n\}$ $(m_k > 0)$ is the scattering data of a boundary value problem of the form (3.3.3), (3.3.4) if and only if $S(0) = 1$ and conditions I, II of Theorem 3.3.3 are satisfied.

Hint. You may use the transformation operators considered in problems 4 and 5 of Section 2, Chapter 2, and reduce the question to Theorem 3.3.3, as you did in the preceding problem.

3. Find the characteristic properties of the scattering data of the problem

$$-y'' + q(x)y = \lambda^2 y \quad , \quad y(0) = 0 \quad (0 \leq x < \infty) \; , \qquad (3.3.32)$$

in which the potential $q(x)$ satisfies the condition

$$\int_0^1 x|q(x) - \ell_0(\ell_0+1)x^{-2}|dx + \int_1^\infty x|q(x) - \ell_\infty(\ell_\infty+1)x^{-2}|dx < \infty \; ,$$

where ℓ_0 and ℓ_∞ are integers, and $\ell_0 + \ell_\infty > \infty$.

4. Consider the matrix Sturm-Liouville problem (3.3.32), in which the $n \times n$ self-adjoint matrix (the potential) $q(x)$ satisfies the condition

$$\int_0^\infty x|q(x)|dx < \infty \; .$$

The bounded matrix-valued solutions $u(\lambda,x)$ of this problem have the asymptotic form

$$u(\lambda,x) = Ie^{-i\lambda x} - S(\lambda)e^{i\lambda x} + o(1) \quad (-\infty < \lambda < \infty) \; ,$$

$$u(i\lambda_k,x) = e^{-\lambda_k x}(m_k + o(1)) \quad (k = 1,2,\ldots,n) \; ,$$

as $x \to \infty$, where $S(\lambda) = [S(-\lambda)]^{-1} = S(-\lambda)^*$ is a unitary matrix and the m_k are nonnegative self-adjoint matrices. The collection $\{S(\lambda);\lambda_1,\ldots,\lambda_n; m_1,\ldots,m_n\}$ is called the scattering data of the given problem. Show that the collection $\{S(\lambda);\lambda_1,\ldots,\lambda_n;m_1,\ldots,m_n\}$, where $S(\lambda) = [S(-\lambda)]^{-1} = S(-\lambda)^*$, $\lambda_k > 0$, and m_k are nonnegative self-adjoint matrices, is the scattering data of a boundary value problem (3.3.2) with self-adjoint matrix potential $q(x)$ subject to (3.3.33) if and only if it enjoys the following properties:

I. The entries of the matrix-valued function

$$F_S(x) = \frac{1}{2\pi} \int_{-\infty}^{\infty} (I - S(\lambda))e^{i\lambda x}d\lambda$$

are summable over the real line and differentiable on the positive half line, and

$$\int_0^\infty x|F_S'(x)|dx < \infty \; .$$

II. The equation

$$f(y) - \int_{-\infty}^{0} f(t)F_S(t+y)dt = 0 \quad (-\infty < y \leq 0)$$

has no nonzero solutions in the space $L_2(-\infty,0]$ (by definition, a matrix- or vector-valued function $f(y)$ belongs to $L_2(-\infty,0]$ if all its entries belong to this space).

III. The equation

$$f(y) + \int_{0}^{\infty} f(t)F(t+y)dt = 0 \quad (0 \leq y < \infty) ,$$

where

$$F(x) = F_S(x) + \sum_{k=1}^{n} m_k^2 e^{-\lambda_k x} ,$$

has no nonzero solutions in the space $L_2[0,\infty)$.

IV. The number of linearly independent vector-valued solutions $g(y)$ of the equation

$$g(y) + \int_{0}^{\infty} g(t)F_S(t+y)dt = 0 \quad (0 \leq y < \infty)$$

in $L_2[0,\infty)$, is equal to the sum of the ranks of the matrices m_k .

Show that conditions I-IV are equivalent to conditions I, IV, and

$$r = \frac{\ln \operatorname{Det} S(0+) - \ln \operatorname{Det} S(+\infty)}{2\pi i} - \frac{s}{2} ,$$

where r denotes the sum of the ranks of the matrices m_k , and s denotes the rank of the matrix $I - S(0)$.

4. INVERSE STURM-LIOUVILLE PROBLEMS ON A BOUNDED INTERVAL

In this section we consider boundary value problems which are defined by the Sturm-Liouville equation

$$-y'' + q(x)y = \lambda^2 \tag{3.4.1}$$

on the interval $0 \leq x \leq \pi$, with real-valued potential $q(x) \in L_2[0,\pi]$, and either separated boundary conditions of the form

$$y(0) = 0 \quad , \quad y(\pi) = 0 , \tag{3.4.2}$$

or

$$y(0) = 0 \quad , \quad y'(\pi) = 0 \ , \tag{3.4.3}$$

or the periodic boundary conditions

$$y(0) - y(\pi) = y'(0) - y'(\pi) = 0 \ , \tag{3.4.4}$$

or the anti-periodic boundary conditions

$$y(0) + y(\pi) = y'(0) + y'(\pi) = 0 \ . \tag{3.4.4'}$$

All these problems are self-adjoint, and their eigenvalues are real. Following Section 1, Chapter 1, we denote the eigenvalues of these boundary value problems by $\lambda_1 \leq \lambda_2 \leq \lambda_3 \leq \dots$; $\nu_1 \leq \nu_2 \leq \dots$; $\mu_0 \leq \mu_2^- \leq \mu_2^+ \leq \dots$; and $\mu_1^- \leq \mu_1^+ \leq \mu_3^- \leq \mu_3^+ \leq \dots$, respectively. We deduce from Theorems 1.5.1 and 1.5.2 that that the following asymptotic formulas hold for $k \to \infty$:

$$\lambda_k = k^2 + a_1 + \alpha_k \ ,$$

$$\nu_k = (k - \tfrac{1}{2}) + a_1 + \beta_k \ , \tag{3.4.5}$$

$$\mu_k^{\pm} = k^2 + a_1 + \varepsilon_k^{\pm} \ ,$$

where $\sum_{k=1}^{\infty} \{|\alpha_k|^2 + |\beta_k|^2 + |\varepsilon_k^{\pm}|^2\} < \infty$. Furthermore, it follows from the classical oscillation theorems for the solutions of Sturm-Liouville equations with real-valued potential, that these eigenvalues interlace as indicated below:

$$-\infty < \nu_1 < \lambda_1 < \nu_2 < \lambda_2 < \nu_3 < \dots \ , \tag{3.4.6}$$

$$-\infty < \mu_0 < \mu_1^- \leq \lambda_1 \leq \mu_1^+ < \mu_2^- \leq \lambda_2 \leq \mu_2^+ \leq \dots \tag{3.4.7}$$

(see Problems 1 and 2 to this section).

The inverse problems considered in this section are formulated as follows:

A. To find necessary and sufficient conditions for two sequences of real numbers to be the spectra of the boundary value problems generated by an equation (3.4.1) with real-valued potential $q(x) \in W_2^n[0,\infty]$ and boundary conditions (3.4.2), (3.4.3) respectively, and also to provide a method for obtaining this equation.

B. To find necessary and sufficient conditions for two sequences of real numbers to be the spectra of the periodic and anti-periodic boundary value problems generated by the same equation (3.4.1) with real-valued potential $q(x) \in \tilde{W}_2^n[0,\pi]$, and also to provide a method for constructing all such equations. We solve these inverse problems using the results of the preceding section and the following lemmas.

LEMMA 3.4.1. *If in equation* (3.1.1) *the potential* $q(x)$ *is summable on the interval* $(0,\pi)$ *and vanishes for* $x > \pi$, *then*

$$e(\lambda,0) = e^{i\lambda\pi}[s'(\lambda,\pi) - i\lambda s(\lambda,\pi)] .$$

PROOF. Since the functions $s(\lambda,x)$ and $c(\lambda,x)$ form a fundamental system of solutions of equation (3.1.1), the particular solution $e(\lambda,x)$ of this equation is a linear combination thereof: $e(\lambda,x) = C_1 s(\lambda,x) + C_2 c(\lambda,x)$. Since the potential vanishes for $x > \pi$, $e(\lambda,x) = e^{i\lambda x}$ for $x \geqslant \pi$, and the following relations hold at the point $x = \pi$:

$$e^{i\lambda\pi} = C_1 s(\lambda,\pi) + C_2 c(\lambda,\pi)$$

and

$$i\lambda e^{i\lambda x} = C_1 s'(\lambda,\pi) + C_2 c'(\lambda,\pi) .$$

Solving for the coefficients C_1 and C_2 , we get

$$C_1 = e^{i\lambda\pi}[i\lambda c(\lambda,\pi) - c'(\lambda,\pi)]$$

and

$$C_2 = e^{i\lambda\pi}[s'(\lambda,\pi) - i\lambda s(\lambda,\pi)] .$$

Therefore, in the present case,

$$e(\lambda,x) = e^{i\lambda\pi}\{[i\lambda c(\lambda,\pi) - c'(\lambda,\pi)]s(\lambda,x) + [s'(\lambda,\pi) - i\lambda s(\lambda,\pi)]c(\lambda,x)\} ,$$

whence

$$e(\lambda,0) = e^{i\lambda\pi}[s'(\lambda,\pi) - i\lambda s(\lambda,\pi)] ,$$

as claimed. □

LEMMA 3.4.2. *In order that the functions* $u(z),v(z)$ *admit the representations*

$$u(z) = \sin \pi z + A\pi \frac{4z}{4z^2 - 1} \cos \pi z + \frac{f(z)}{z} \tag{3.4.8}$$

and

and

$$v(z) = \cos \pi z - B\pi \frac{\sin \pi z}{z} + \frac{g(z)}{z}, \tag{3.4.8'}$$

where

$$f(z) = \int_0^\pi \tilde{f}(t) \cos ztdt \quad , \quad \tilde{f}(t) \in L_2[0,\pi] \quad , \quad \int_0^\pi \tilde{f}(t)dt = 0$$

and

$$g(z) = \int_0^\pi \tilde{g}(z) \sin ztdt \quad , \quad \tilde{g}(t) \in L_2[0,\pi] \ ,$$

it is necessary and sufficient that

$$u(z) = \pi z \prod_{k=1}^{\infty} k^{-2}(u_k^2 - z^2) \ , \ \textit{where} \quad u_k = k - \frac{A}{k} + \frac{\alpha_k}{k} \ , \tag{3.4.9}$$

and

$$v(z) = \prod_{k=1}^{\infty} (k - \tfrac{1}{2})^{-2}(v_k^2 - z^2) \ , \ \textit{where} \quad v_k = k - \frac{1}{2} - \frac{B}{k} + \frac{\beta_k}{k} \ , \tag{3.4.9'}$$

where α_k *and* β_k *are arbitrary sequences of numbers subject only to the conditions*

$$\sum_{k=1}^{\infty} |\alpha k|^2 < \infty \quad \textit{and} \quad \sum_{k=1}^{\infty} |\beta_k|^2 < \infty \ .$$

PROOF. It follows from the method that was used repeatedly in Section 5 of Chapter 1, that the zeros of a function $u(z)$, which admits the representation (3.4.8), form a sequence $\ldots -u_k, -u_{k-1}, \ldots, -u_1, u_0 = 0 \ , \ u_1, \ldots, u_{k-1}, u_k, \ldots,$ where

$$u_k = k - \frac{A}{k} + \frac{\alpha_k}{k}$$

and

$$\alpha_k = \frac{(-1)^{k+1}}{\pi}\left\{f(k) - \frac{Af'(k)}{k} + \frac{(-1)^{k+1}f(k)f'(k)}{\pi k} + o(k^{-2})\right\} .$$

Hence, $\sum_{k=1}^{\infty} |\alpha_k|^2 < \infty$. Next, since $u(z)$ is an odd entire function of exponential type, it follows from the same asymptotic formula for its zeros that

$$u(z) = Cz \prod_{k=1}^{\infty} k^{-2}(u_k^2 - z) \ .$$

To find the constant C, we can use the equality $\lim_{y\to\infty} u(iy)(\sin \pi yi)^{-1} = 1$, which yields $C = \pi$. We have thus proved the necessity of the conditions given in the lemma for the function $u(z)$.

Now suppose that the function $u(z)$ admits a representation (3.4.9). We consider the auxiliary function

$$u_1(z) = \sin \varphi z + A\pi \frac{4z}{4z^2 - 1} \cos \pi z + \frac{f_1(z)}{z} ,$$

where

$$f_1(z) = \varphi(z) + A \frac{\varphi'(z)}{z} , \quad \varphi(z) = \int_0^\pi \varphi(t) \cos ztdt ,$$

$$\tilde{\varphi}(t) = c_1 \cos t + 2 \sum_{k=2}^\infty (-1)^{k+1} \alpha^k \cos kt ,$$

and the constant c_1 is chosen so that

$$\int_0^\pi t^2 \tilde{\varphi}(t)dt = -\varphi''(0) = 0 .$$

Since $\sum_{k=2}^\infty |\alpha_k|^2 < \infty$, it follows that $\tilde{\varphi}(t) \in L_2[0,\pi]$ and

$$f_1(z) = \int_0^\pi \tilde{\varphi}(t) \cos ztdt - A \int_0^\pi \tilde{\varphi}(t) t \frac{\sin zt}{z} dt =$$

$$= \int_0^\pi \left[\tilde{\varphi}(t) - A \int_t^\pi \tilde{\varphi}(\xi)d\xi\right] \cos ztdt = \int_0^\pi \tilde{f}_1(t) \cos ztdt ,$$

where

$$\tilde{f}_1(t) = \tilde{\varphi}(t) - A \int_t^\pi \tilde{\varphi}(\xi)\xi d\xi \in L_2[0,\pi] ,$$

and

$$\int_0^\pi \tilde{f}_1(t)dt = \int_0^\pi \tilde{\varphi}(t)dt - A \int_0^\pi dt \int_t^\pi \tilde{\varphi}(\xi)\xi d\xi = \int_0^\pi \tilde{\varphi}(t)[1 - At^2]dt = 0 .$$

Therefore, the function $u_1(z)$ admits a representation of the form (3.4.8), and by the foregoing discussion,

$$u_1(z) = \pi z \sum_{k=1}^\infty k^{-2}([u_k^{(1)}]^2 - z^2) , \quad u_k^{(1)} = k - \frac{A}{k} + \frac{\alpha_k^{(1)}}{k} , \tag{3.4.10}$$

where

$$\alpha_k^{(1)} = \frac{(-1)^{k+1}}{\pi} \left\{f_1(k) - \frac{A}{k} f_1'(k) + \frac{(-1)^{k+1} f_1(k) f_1'(k)}{\pi k} + O(k^{-2})\right\} . \tag{3.4.10'}$$

Now, upon expressing $f_1(k)$ and $f_1'(k)$ in terms of $\varphi(z)$ in (3.4.10'), and taking into account the fact that $\varphi(k) = \pi(-1)^{k+1}\alpha_k$, we obtain

$$\alpha_k^{(1)} = \alpha_k + \frac{f_1(k)f_1'(k)}{\pi^2 k} + o(k^{-2}) .$$

We see that the following equality holds for the difference between the zeros u_k and $u_k^{(1)}$ of the functions $u(z)$ and $u_1(z)$:

$$u_k - u_k^{(1)} = -\frac{f_1(k)f_1'(k)}{\pi^2 k^2} + o(k^{-3})$$

or, equivalently,

$$u_k - u_k^{(1)} = k^{-2}[\delta_k + k^{-1}d_k] , \tag{3.4.11}$$

where $\sup\limits_k |d_k| = d < \infty$ and

$$\sum_{k=1}^{n} |\delta_k| = \frac{1}{\pi^2}\sum_{k=1}^{\infty} |f_1(k)||f_1'(k)| \leqslant \frac{1}{\pi^2}\sum_{k=1}^{n}\{|f_1(k)|^2 + |f_1'(k)|^2\} < \infty . \tag{3.4.11'}$$

In order to show that the function $u(z) = u_1(z) + z^{-1}\{z[u(z) - u_1(z)]\}$ admits a representation of the form (3.4.8), it suffices to verify that $z[u(z) - u_1(z)] \in CK^2(\pi)$ (see Chapter 2, Section 1). It follows from the asymptotic formulas (3.4.10), (3.4.10') that

$$|[u_k^{(1)}]^2 - z^2| \geqslant \frac{|z|}{2}$$

on every circle $|z| = n - \frac{1}{2}$ starting with some $n = n_0$, for all $k = 1,2,\dots$. Hence, if $|z| = n - \frac{1}{2}$ and $n \geqslant n_0$, then

$$|z[u(z) - u_1(z)]| = |zu_1(z)|\left|\prod_{k=1}^{\infty}\left(\frac{u_k^2 - z^2}{[u_k^{(1)}]^2 - z^2}\right) - 1\right| =$$

$$= |zu_1(z)|\left|\prod_{k=1}^{\infty}(1 + \varepsilon_k(z)) - 1\right| ,$$

where

$$|\varepsilon_k(z)| = \left|\frac{u_k^2 - [u_k^{(1)}]^2}{[u_k^{(1)}]^2 - z^2}\right| \leqslant \frac{Ck|u_k - u_k^{(1)}|}{|z|} \leqslant \frac{C\{k^{-1}|\delta_k| + k^{-2}|d_k|\}}{|z|} .$$

This further implies that

$$|z[u(z) - u_1(z)]| \leqslant |z||u_1(z)|C\sum_{k=1}^{\infty}|\varepsilon_k(z)| \leqslant$$

$$\leqslant C|u_1(z)| \sum_{k=1}^{\infty} \{k^{-1}|\delta_k| + k^{-2}|d_k|\} \leqslant C|u_1(z)|$$

(here and in the rest of the proof, C denotes various constants that do not depend on k and z). It follows from this estimate and the inequality $\sup_{\operatorname{Im} z \geqslant 0} |e^{iz\pi}u_1(z)| = M < \infty$, which obviously holds for functions admitting a representation of the form (3.4.8), that on the semicircles $z = (n - \frac{1}{2})e^{i\varphi}$, $0 \leqslant \varphi \leqslant \pi$, $n \geqslant n_0$,

$$|e^{iz\pi}z[u(z) - u_1(z)]| \leqslant C < \infty , \tag{3.4.12}$$

where the constant C does not depend on n.

Now let us show that on the real line $-\infty < x < \infty$,

$$|x[u(x) - u_1(x)]| \leqslant C(1 + |x|)^{-1} \quad (C < \infty) . \tag{3.4.13}$$

To this end, it clearly suffices to show that

$$\varlimsup_{x \to +\infty} |x^2[u(x) - u_1(x)]| < \infty . \tag{3.4.13'}$$

Let $x \in [n - \frac{1}{2}, n + \frac{1}{2}]$. Then

$$u(x) - u_1(x) = u(x_1)\left\{ \prod_{k=1}^{\infty} \frac{u_k^2 - x^2}{[u_k^{(1)}]^2 - x^2} - 1 \right\} =$$

$$= \frac{u_1(x)}{u_n^{(1)} - x} \, \frac{u_n^2 - [u_n^{(1)}]^2}{u_n^{(1)} + x} (1 + \delta_n(x)) + u_1(x)\delta_n(x) ,$$

where

$$\delta_n(x) = \prod_{k \neq n} \frac{u_k^2 - x^2}{[u_k^{(1)}]^2 - x^2} - 1 = \prod_{k \neq n} (1 + \varepsilon_k(n)) - 1$$

and

$$\varepsilon_k(x) = \frac{u_k^2 - [u_k^{(1)}]^2}{[u_k^{(1)}]^2 - x^2} .$$

Next, it follows from the asymptotic formulas (3.4.10), (3.4.11), that

$$|[u_k^{(1)}]^2 - u_k|^2 \leqslant C\{k^{-1}|\delta_k| + k^{-2}d\} ,$$

and starting with some $n = n_0$,

$$|[u_k^{(1)}]^2 - x^2| \geqslant Cx|k - n| \quad , \quad |u_n^{(1)} + x| \geqslant x ,$$

for all $x \in [n - \frac{1}{2}, n + \frac{1}{2}]$ and $k \neq n$. Consequently,

$$\left|\frac{u_n^2 - [u_n^{(1)}]^2}{u_n^{(1)} + x}\right| \leq \frac{C}{x^2},$$

$$|\varepsilon_k(x)| \leq \frac{C}{xk|k-n|}\{|\delta_n| + k^{-1}d\} = \frac{C}{xn}\left|\frac{1}{k} - \frac{1}{k-n}\right|\{|\delta_k| + k^{-1}d\} \leq$$

$$\leq \frac{C}{x^2}\{|\delta_k| + d[k^{-2} + (k-n)^2]\} \quad (k \neq n)$$

and

$$\delta_n(x) \leq C \sum_{k\neq n} |\varepsilon_k(x)| \leq \frac{C}{x^2}\sum_{k\neq n}\{|\delta_k| + d[k^{-2} + (k-n)^{-2}]\} \leq \frac{C}{x^2}.$$

Therefore, if $x \in [n - \frac{1}{2}, n + \frac{1}{2}]$ and $n \geq n_0$, then

$$|u(x) - u_1(x)| \leq \frac{C}{x^2}\left\{\left|\frac{u_1(x)}{u_n^{(1)} - x}\right| + |u_1(x)|\right\},$$

where the constant C does not depend on n . Since every function which admits a representation of the form (3.4.8) is bounded together with its derivative in every strip $|\text{Im } z| \leq b < \infty$, we have

$$\sup_{-\infty<x<\infty}\left\{\left|\frac{u_1(x)}{u_n^{(1)}(x) - x}\right| + |u_1(x)|\right\} =$$

$$= \sup_{-\infty<x<\infty}\left\{\left|\frac{1}{u_n^{(1)} - x}\int_{u_n^{(1)}}^{x} u_1'(\xi)d\xi\right| + |u_1(x)|\right\} \leq C < \infty,$$

and

$$|u(x) - u_1(x)| \leq Cx^{-2}$$

$(x \in [n - \frac{1}{2}, n + \frac{1}{2}]$, $n \geq n_0)$, which is equivalent to inequality (3.4.13').

The estimate (3.4.13) shows that the function $z[u(z) - u_1(z)]$ is square summable and bounded on the real line. Hence, using (3.4.12) and the maximum modulus principle for analytic functions, we conclude that

$$\sup_{\text{Im } z\geq 0} |e^{iz\pi}z[u(z) - u_1(z)]| \leq C < \infty$$

and

$$|z[u(z) - u_1(z)]| \leq C \exp|\text{Im } z\pi|$$

As we showed in Chapter 2, Section 1, this implies that

$z[u(z) - u_1(z)] \in CK^2(\pi)$, i.e.,

$$z[u(z) - u_1(z)] = \int_0^\pi f_2(t) \cos ztdt \quad , \quad f_2(t) \in L_2[0,\pi] .$$

Hence, $u(z)$ can be represented in the form (3.4.8).

The proof of the lemma for the function $v(z)$ is carried out in much the same way. □

THEOREM 3.4.3. *In order that two sequences of numbers* $\{\lambda_k\}$ *and* $\{\nu_k\}$ $(k = 1,2,\dots)$ *be the spectra of the boundary values problems generated by one and the same Sturm-Liouville equation* $-y'' + q(x)y = \lambda^2 y$ $(0 \leq x \leq \pi)$ *with real-valued potential* $q(x) \in L_2[0,\pi]$ *and boundary conditions* $y(0) = y(\pi) = 0$ *and* $y(0) = y'(\pi) = 0$, *respectively, it is necessary and sufficient that they interlace:* $-\infty < \nu_1 < \lambda_1 < \nu_2 < \lambda_2 < \nu_3 < \dots$, *and that they satisfy the asymptotic formulas*

$$\begin{aligned} \lambda_k &= k^2 - 2A + \alpha_k , \\ \nu_k &= (k - \tfrac{1}{2})^2 - 2A + \beta_k , \end{aligned} \tag{3.4.14}$$

where A *is an arbitrary real number and* $\sum_{k=1}^{\infty} \alpha_k^2 < \infty$, $\sum_{k=1}^{\infty} \beta_k^2 < \infty$.

PROOF. The necessity of these conditions was established in the beginning of this section. To prove their sufficiency, we may obviously assume, without loss of generality, that all λ_k and ν_k are positive. It is clear also that the asymptotic formulas (3.4.14) are equivalent to the formulas

$$\sqrt{\lambda_k} = k - \frac{A}{k} + \frac{\tilde{\alpha}_k}{k} \quad , \quad \sqrt{\nu_k} = k - \frac{1}{2} - \frac{A}{k} + \frac{\tilde{\beta}_k}{k} , \tag{3.4.14'}$$

where $\sum_{k=1}^{\infty} \tilde{\alpha}_k^2 < \infty$ and $\sum_{k=1}^{\infty} \tilde{\beta}_k^2 < \infty$. From the given sequences $\{\lambda_k\}$ and $\{\nu_k\}$ we construct the functions

$$zs(z) = \pi z \prod_{k=1}^{\infty} k^{-2}(\lambda_k - z^2) \quad , \quad s_1(z) = \prod_{k=1}^{\infty} (k - \tfrac{1}{2})^{-2}(\nu_k - z^2) , \tag{3.4.15}$$

and put

$$e(z) = e^{iz\pi}[s_1(z) - izs(z)] . \tag{3.4.16}$$

It follows from the asymptotic formulas (3.4.14') and Lemma 3.4.2 that

$$zs(z) = \sin \pi z + A\pi \frac{4z}{4z^2 - 1} \cos \pi z + \frac{f(z)}{z}$$

and

$$s_1(z) = \cos \pi z - A\pi \frac{\sin \pi z}{z} + \frac{g(z)}{z} ,$$

where

$$f(z) = \int_0^{\pi} \tilde{f}(t) \cos ztdt \quad , \quad f(0) = 0 \quad , \quad \tilde{f}(t) \in L_2[0,\pi]$$

and

$$g(z) = \int_0^{\pi} \tilde{g}(t) \sin ztdt \quad , \quad \tilde{g}(t) \in L_2[0,\pi] .$$

Hence,

$$e(z) = 1 - \frac{iA\pi}{z} - \frac{e^{iz\pi}}{z} \left\{ \frac{iA\pi}{4z^2 - 1} \cos \pi z + g(z) - if(z) \right\} , \tag{3.4.17}$$

and, by Lemma 1.3.1,

$$e(z) = 1 + (z^{-1}) \tag{3.4.18}$$

as $|z| \to \infty$ in the upper half plane.

We calculate the increment of the argument of the function $e(z)$ when z runs through the real line. It follows from the fact that the roots of the functions $s_1(x)$ and $xs(x)$ interlace, and formulas (3.4.15), that the increment of the argument of the function $s_1(x) - ixs(x)$ on the segments $[-\sqrt{\lambda_k}, -\sqrt{\lambda_{k-1}}]$ and $[\sqrt{\lambda_{k-1}}, \sqrt{\lambda_k}]$ is $-\pi$. Hence, when x runs through $[-\sqrt{\lambda_k}, \sqrt{\lambda_k}]$, the argument of this function increases by $-2k\pi$, whereas the increment of the argument of $e^{ix\pi}$ on the same segment is obviously $2\sqrt{\lambda_k}\,\pi$. Thus, the increment of the argument of $e(x)$ on the segment $[-\sqrt{\lambda_k}, \sqrt{\lambda_k}]$ equals $2\pi(\sqrt{\lambda_k} - k)$, and tends to zero as $k \to \infty$. Since $\lim_{x\to\pm\infty} e(x) = 1$, we conclude that the increment of the argument of the function $e(x)$ when x changes from $-\infty$ to $+\infty$ is equal to zero:

$$\arg e(+\infty) - \arg e(-\infty) = 0 . \tag{3.4.19}$$

By the argument principle, this equality and (3.4.18) imply that $e(z)$ has no roots in the closed upper half plane.

Now we show that the function

$$S(\lambda) = \frac{e(-\lambda)}{e(\lambda)} \quad (-\infty < \lambda < \infty) \tag{3.4.20}$$

satisfies I, II with $n = 0$, and III, and, accordingly, that it is the scattering function of a boundary value problem (3.1.1), (3.1.2) with properties 1 and 2 (see the Corollary to Theorem 3.3.3). In fact, $S(\lambda)$ is continuous, $S(\lambda) = \overline{S(-\lambda)} = [S(-\lambda)]^{-1}$, $S(0) = 1$, and

$$1 - S(\lambda) = -\frac{2iA\pi}{\lambda} + \frac{2i}{\lambda}\,\mathrm{Im}\, e^{i\lambda\pi}[g(\lambda) + if(\lambda)] + \;(\lambda^{-2}) \tag{3.4.21}$$

as $\lambda \to \infty$. Since the functions $g(\lambda)$ and $f(\lambda)$ are bounded and belong to $L_2(-\infty,\infty)$, the function $F_S(x) = \frac{1}{2\pi}\int_{-\infty}^{\infty}(1 - S(\lambda))e^{i\lambda x}d\lambda$ is bounded and belongs to $L_2(-\infty,\infty)$. Furthermore, it follows from formula (3.4.21) that

$$1 - S(\lambda) = \frac{-2iA\pi}{\lambda + i} + \frac{\psi(\lambda)}{1 + |\lambda|} ,$$

where $\psi(\lambda) \in L_2(-\infty,\infty)$. Therefore, for x positive ,

$$F_S(x) = \frac{1}{2\pi}\int_{-\infty}^{\infty}\left\{\frac{-2iA\pi}{\lambda + i} + \frac{\psi(\lambda)}{1 + |\lambda|}\right\} e^{i\lambda x}d\lambda = \frac{1}{2\pi}\int_{-\infty}^{\infty}\frac{\psi(\lambda)}{1 + |\lambda|}\, e^{i\lambda x}d\lambda ,$$

which shows that $F_S(x)$ is absolutely continuous on the positive half line and $F_S' \in L_2[0,\infty)$. Finally, since $e(-z)$ and $e(z)$ are entire functions, and since $e(z)$ has no roots in the closed upper half plane, the function $S(z)$ is holomorphic in this half plane. Moreover, by (3.4.17), (3.4.20), and Lemma 1.3.1,

$$|1 - S(z)| \leqslant \frac{C}{|z|}\, e^{2\pi\,\mathrm{Im}\,z} \quad (\mathrm{Im}\, z \geqslant 0) ,$$

whence, by Jordan's lemma, $F_S(x) = 0$ for $x > 2\pi$. Therefore, $S(\lambda)$ satisfies conditions I and III. Condition II with $n = 0$ is also satisfied, because by (3.4.20) and (3.4.19),

$$\arg S(0) - \arg S(+\infty) = \arg e(0) - \arg e(-\infty) - \arg e(0) + \arg e(+\infty) =$$
$$= \arg e(+\infty) - \arg e(-\infty) = 0 .$$

Consequently, there exists a real-valued potential $q(x) \in L_2[0,\pi]$, $q(x) = 0$ for $x > \pi$, such that $S(\lambda)$ is the scattering function of the boundary value problem

$$-y'' + q(x)y = \lambda^2 y \quad , \quad y(0) = 0 \quad (0 \leqslant x < \infty) .$$

Combining the definition of the scattering function and formula (3.4.20), we get

$$\frac{e(-\lambda,0)}{e(\lambda,0)} = \frac{e(-\lambda)}{e(\lambda)} ,$$

or, equivalently,

$$\frac{e(-\lambda,0)}{e(-\lambda)} = \frac{e(\lambda,0)}{e(\lambda)} .$$

The function on the right (left) hand side of the last equality is holomorphic in the upper (respectively, lower) half plane and tends uniformly in it to one as $|\lambda| \to \infty$. Therefore, $e(\lambda) \equiv e(\lambda,0)$, which, in view of (3.4.16) and Lemma 3.4.1, implies that

$$s_1(z) - izs(z) = s'(z,\pi) - izs(z,\pi)$$

and hence that

$$s(z) = s(z,\pi) \quad , \quad s_1(z) = s'(z,\pi) .$$

These equalities show that the sequences $\{\lambda_k\}$ and $\{\nu_k\}$ consist of the squares of the roots of the functions $s(z,\pi)$ and $s'(z,\pi)$, i.e., they are the spectra of the boundary value problems generated by the equation $-y'' + q(x)y = \mu y$ $(0 \leqslant x \leqslant \pi)$ and the boundary conditions $y(0) = y(\pi) = 0$ and $y(0) = y'(\pi) = 0$, as claimed. The proof given above also provides a method for recovering the potential $q(x)$ from the spectra $\{\lambda_k\}$ and $\{\nu_k\}$. □

COROLLARY. *In order that the two sequences of real numbers* $\{\lambda_k\}$ *and* $\{\nu_k\}$ *be the spectra of a boundary value problem generated by one and the same equation* (3.4.1) *with real-valued potential* $q(x) \in W_2^n[0,\pi]$ *and boundary conditions* (3.4.2) *and* (3.4.4) *respectively, it is necessary and sufficient that they interlace and satisfy the asymptotic formulas*

$$\sqrt{\lambda_k} = k + \sum_{1\leqslant 2j+1\leqslant n+2} a_{2j+1}(2k)^{-2j-1} + k^{-n-1}\alpha_k$$

and

$$\sqrt{\nu_k} = k - \frac{1}{2} + \sum_{1\leqslant 2j+1\leqslant n+2} b_{2j+1}(2k-1)^{-2j-1} + k^{-n-1}\beta_k ,$$

respectively, where $a_1 = b_1$ *and*

$$\sum_{k=1}^{\infty} |\alpha_k|^2 < \infty \quad , \quad \sum_{k=1}^{\infty} |\beta_k|^2 < \infty .$$

PROOF. The necessity of these conditions follows from Theorem 1.5.1 and the fact that the eigenvalues of the boundary value problems in question interlace. Conversely, if these conditions are satisfied, then, by Theorem 3.4.1, the sequences $\{\lambda_k\}$ and $\{\nu_k\}$ are the spectra of boundary value problems (3.4.1), (3.4.2) and (3.4.1), (3.4.3), with the same real-valued potential $q(x) \in L_2[0,\pi]$; moreover, by the Corollary to Theorem 1.5.1, $q(x) \in W_2^n[0,\pi]$. □

Now we turn to the resolution of problem B. First of all we remark that if the sequence $-\infty < \mu_0 < \mu_1^- \leqslant \mu_1^+ < \mu_2^- \leqslant \mu_2^+ < \mu_3^- \leqslant \mu_3^+ < \dots$ consists of the eigenvalues of the periodic $(\mu_{2k}^{\pm})$ and anti-periodic $(\mu_{2k+1}^{\pm})$ boundary value problems generated by an equation (3.4.1) with real-valued potential $q(x) \in \tilde{W}_2^n[0,\pi]$, then the sequence $-\infty < 0 < (\mu_1^0 - \mu_0) \leqslant (\mu_1^+ - \mu_0) < (\mu_2^- - \mu_0) \leqslant (\mu_2^+ - \mu_0) < \dots$ consists of the eigenvalues of the periodic $(\mu_{2k}^{\pm} - \mu_0)$ and anti-periodic $(\mu_{2k+1}^{\pm} - \mu_0)$ boundary value problems generated by the equation $-y'' + [q(x) - \mu_0]y = \mu y$ with real-valued potential $q(x) - \mu_0$, which also belongs to $\tilde{W}_2^n[0,\pi]$. Therefore, in the following, we will assume, without loss of generality, that $\mu_0 = 0$.

Calculating by formula (1.3.4) the characteristic functions $\chi_p(\lambda)$ and $\chi_a(\lambda)$ of the periodic and anti-periodic boundary value problems, we get

$$\chi_p(\lambda) = 2[1 - u_+(\lambda)] \quad , \quad \chi_a(\lambda) = 2[1 + u_+(\lambda)] \ , \tag{3.4.22}$$

where

$$u_+(\lambda) = \frac{1}{2} [c(\lambda,\pi) + s'(\lambda,\pi)] \tag{3.4.23}$$

is the Lyapunov function, known also as the Hill discriminant. The Hill discriminant is readily recovered from either of the sequences $\mu_0, \mu_2^-, \mu_2^+, \dots$ or $\mu_1^-, \mu_1^+, \mu_3^-, \mu_3^+, \dots$, in the form of an infinite product. However, it does not determine uniquely the potential, since, for example, the equations

$$-y'' + q(x)y = \lambda^2 y \quad \text{and} \quad -y'' + \tilde{q}(x+c)y = \lambda^2 y$$

$(0 \leqslant x \leqslant \pi)$ have the same Hill discriminant for every $c \in (-\infty,\infty)$, if $\tilde{q}(x)$ is a π-periodic extension of the function $q(x)$ to the entire real line. On the other hand, the function $s(\lambda,\pi)$, $u_+(\lambda)$, and

$$u_-(\lambda) = \frac{1}{2} [c(\lambda,\pi) - s'(\lambda,\pi)] \tag{3.4.24}$$

determine the potential uniquely, since $s'(\lambda,\pi) = u_+(\lambda) - u_-(\lambda)$ and, by Theorem 3.4.1, knowledge of the functions $s(\lambda,\pi)$ and $s'(\lambda,\pi)$ is sufficient to recover the potential.

By formulas (3.4.23), (3.4.24), and (1.3.11),

$$u_+(\lambda) = \cos \lambda\pi + \frac{\sin \lambda\pi}{2} \int_0^{\pi} q(t)dt -$$

$$- \frac{1}{2} \int_0^{\pi} [K_t(\pi,t;0) - K_x(\pi,t;\infty)] \frac{\sin \lambda t}{\lambda} dt \tag{3.4.23'}$$

and

$$u_-(\lambda) = -\frac{1}{2}\int_0^\pi [K_t(\pi,t;0) + K_x(\pi,t;\infty)] \frac{\sin \lambda t}{\lambda} dt , \qquad (3.4.24')$$

from which we deduce, using Lemma 1.3.1, that

$$u_-(z) = o(z^{-1} \exp |\text{Im } z\pi|) , \qquad (3.4.25)$$

$$u_+(z) = \cos \pi z + O(z^{-1} \exp |\text{Im } z\pi|) , \qquad (3.4.26)$$

and

$$u_+'(z) = -\pi \sin \pi z + O(z^{-1} \exp |\text{Im } z\pi|) , \qquad (3.4.26')$$

as $|z| \to \infty$. Next, it follows from the estimate (3.4.25) and the inequality $|s(z,\pi)| \geq C|z|^{-1} \exp |\text{Im } z\pi|$, which holds on the contours $K_n = C(n + \frac{1}{2})$ (see Lemma 1.3.2) that

$$0 = \lim_{n\to\infty} \frac{1}{2\pi i} \int_{K_n} \frac{u_-(\xi)}{s(\xi,\pi)(\xi - z)} d\xi =$$

$$= \lim_{n\to\infty} \left\{ \frac{u_-(z)}{s(z,\pi)} - \sum_{k=1}^{n} \frac{2\sqrt{\lambda_k}\, u_-(\sqrt{\lambda_k})}{\dot{s}(\sqrt{\lambda_k},\pi)(z^2 - \lambda_k)} \right\} ,$$

i.e.,

$$u_-(z) = s(z,\pi) \sum_{k=1}^{\infty} \frac{2\sqrt{\lambda_k}\, u_-(\sqrt{\lambda_k})}{\dot{s}(\sqrt{\lambda_k},\pi)(z^2 - \lambda_k)} .$$

Here $\{\lambda_k\}$ is the spectrum of the boundary value problem (3.4.1), (3.4.2), i.e., the sequence of squares of the roots of its characteristic function $s(\lambda,\pi)$, and the dot denotes differentiation with respect to λ . Since the Wronskian of the solutions $c(\lambda,x)$ and $s(\lambda,x)$ is identically equal to one: $c(\lambda,\pi)s'(\lambda,\pi) - c'(\lambda,\pi)s(\lambda,\pi) = 1$, and since $c(\lambda,\pi)s'(\lambda,\pi) = u_+^2(\lambda) - u_-^2(\lambda)$, we have

$$-u_-^2(\lambda) - c'(\lambda,\pi)s(\lambda,\pi) = 1 - u_+^2(\lambda) . \qquad (3.4.27)$$

Letting $\lambda = \sqrt{\lambda_k}$, we obtain

$$-u_-^2(\sqrt{\lambda_k}) = 1 - u_+^2(\sqrt{\lambda_k}) ,$$

whence

$$u_-(\sqrt{\lambda_k}) = [\text{sign } u_-(\sqrt{\lambda_k})]\sqrt{u_+^2(\sqrt{\lambda_k}) - 1} .$$

Therefore, the function $u_-(z)$ is uniquely determined by the spectrum $\{\lambda_k\}$ of the boundary value problem (3.4.1), (3.4.2), the Hill discriminant, and the sequence $\{\text{sign } u_-(\sqrt{\lambda_k})\}$ via the formula

$$u_-(z) = s(z,\pi) \sum_{k=1}^{\infty} \frac{2\sqrt{\lambda_k}\,[\text{sign } u_-(\sqrt{\lambda_k})]\sqrt{u^2(\sqrt{\lambda_k}) - 1}}{(z^2 - \lambda_k)\dot{s}(\sqrt{\lambda_k},\pi)}, \tag{3.4.28}$$

and the function $s(z,\pi)$ is in turn uniquely determined by the spectrum $\{\lambda_k\}$ since

$$s(z,\pi) = \pi \prod_{k=1}^{\infty} k^{-2}(\lambda_k - z^2) . \tag{3.4.28'}$$

Therefore, in order to recover the potential uniquely, it suffices to know the spectrum of the periodic $(\mu_0,\mu_2^-,\mu_2^+,\ldots)$ or anti-periodic $(\mu_1^-,\mu_1^+,\mu_3^-,\mu_3^+,\ldots)$ boundary value problem, the spectrum $\lambda_1,\lambda_2,\lambda_3,\ldots$ of the boundary value problem (3.4.1), (3.4.2), and the sequence sign $u_-(\sqrt{\lambda_1})$, sign $u_-(\sqrt{\lambda_2})$, sign $u_-(\sqrt{\lambda_3}),\ldots$.

In order to describe the characteristic properties of the spectra of the periodic and anti-periodic boundary value problems, we must investigate first the properties of the Hill discriminant $u_+(z)$. As we remarked above, we can assume without loss of generality that $\mu_0 = 0$.

It is easily seen from formulas (3.4.23') and (3.4.22) that $u_+(z)$ is an even entire function of exponential type π which takes the value +1 only at the points

$$\alpha_0 = \sqrt{\mu_0} = 0 \ , \quad \alpha_{2k}^{\pm} = \sqrt{\mu_{2k}^{\pm}} \ , \quad \alpha_{-2k}^{\pm} = -\alpha_{2k}^{\mp} \quad (k = 1,2,\ldots) ,$$

and the value -1 only at the points

$$\alpha_{2k-1}^{\pm} = \sqrt{\mu_{2k-1}^{\pm}} \ , \quad \alpha_{-(2k-1)}^{\pm} = -\alpha_{2k-1}^{\mp} \quad (k = 1,2,\ldots) .$$

Moreover, by (3.4.7),

$$\ldots -\alpha_2^+ \leqslant -\alpha_2^- < -\alpha_1^+ \leqslant -\alpha_1^- < \alpha_0^- = 0 = \alpha_0^+ < \alpha_1^- \leqslant \alpha_1^+ < \alpha_2^- \leqslant \alpha_2^+ < \ldots .$$

Since the function $u_+(z)$ takes equal values $(-1)^k$ at the endpoints of the segments $[\alpha_k^-,\alpha_k^+]$, its derivative has a root $\gamma_k = -\gamma_{-k}$ in each such interval. Using the estimate (3.4.26') and Rouche's theorem, it is readily established that for n large, $u'(z)$ has exactly $2n+1$ roots in each strip $|\text{Re} z| \leqslant$ $\leqslant n + \frac{1}{2}$, and from the asymptotic formulas (3.4.5) it follows that

$$-(n + \tfrac{1}{2}) < \gamma_{-n} < \gamma_{-n+1} < \ldots < \gamma_0 = 0 < \gamma_1 < \ldots < \gamma_n < n + \tfrac{1}{2} .$$

Hence, the roots γ_k are all simple, the derivative $u_+'(z)$ has no other roots, and the graph of the function $u_+(\lambda)$ $(-\infty < \lambda < \infty)$ has the shape pictured in Fig. 4.

Figure 4

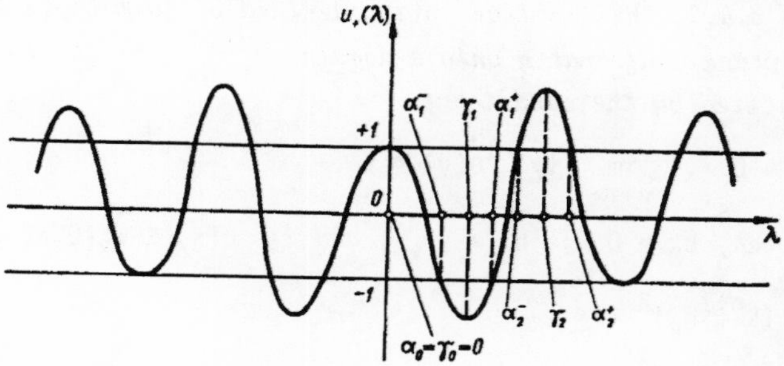

Figure 5

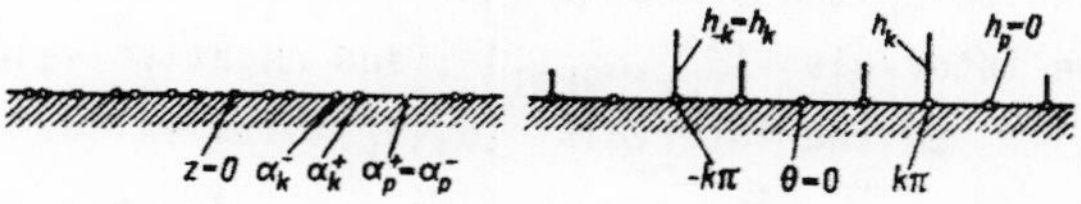

We choose a single-valued branch of the square root $\sqrt{1 - u_+^2(z)}$ in the upper half plane so that it will be positive on the interval $(0,\alpha_1^-)$, and we put

$$\theta(z) = \int_0^z \frac{-u_+'(z)}{\sqrt{1 - u_+^2(\xi)}}\, d\xi \quad (\operatorname{Im} z > 0) . \tag{3.4.29}$$

Obviously,

$$u_+(z) = \cos \theta(z) , \tag{3.4.30}$$

and this equality holds for all values of z if $\theta(z)$ is extended by continuity to the boundary of the upper half plane, and then continued analytically into the lower half plane by the rule $\theta(\bar{z}) = \overline{\theta(z)}$. We remark that the function $\theta(z)$ continued in this manner is single-valued and holomorphic in the entire complex plane, with slits along the segments $[\alpha_k^-,\alpha_k]$ $(k = 1, 2,\ldots)$, and is defined there by the same equality (3.4.29). We next show that $\theta(z)$ maps the upper half plane conformally (Fig. 5,a) into the following domain $\Theta_+\{h_k\}$ (Fig. 5,b):

$$\Theta_+\{h_k\} = \{\theta : \operatorname{Im} \theta > 0\} \setminus \bigcup_{k=-\infty}^{\infty} \{\theta : \operatorname{Re} \theta = k\pi , 0 \leqslant \operatorname{Im} \theta \leqslant h_k\} . \tag{3.4.31}$$

From this it follows that its analytic continuation $(\theta(\bar{z}) = \overline{\theta(z)})$ maps the z-plane with horizontal slits $[\alpha_k^-,\alpha_k^+]$ conformally into the θ-plane , with the vertical slits $[k\pi - ih_k , k\pi + ih_k]$.

LEMMA 3.4.3. *The function* $\theta(z)$ *defined by formula* (3.4.29) *maps the upper half plane conformally onto a domain* $\Theta_+\{h_k\}$ *of the form* (3.4.31), *which is normalized by the conditions*

$$\theta(0) = 0 \quad , \quad \lim_{y\to+\infty} (iy)^{-1}\theta(y) = \pi \ .$$

Moreover, $h_0 = 0$, $h_k = -h_k$, *and if* $q(x) \in \tilde{W}_2^n[0,\pi]$, *then*

$$\sum_{k=-\infty}^{k} (k^{n+1}h_k)^2 < \infty \ .$$

PROOF. Our choice of the branch of the function $\sqrt{1 - u_+^2(z)}$ assumes positive values on the intervals $(\alpha_{2k}^+,\alpha_{2k+1}^-)$, and negative values on the intervals $(\alpha_{2k+1}^+,\alpha_{2k+2}^-)$. On the intervals (α_k^-,α_k^+) its values are purely imaginary; specifically, $\operatorname{Im}\sqrt{1 - u_+^2(z)} > 0$ on $(\alpha_{2k+1}^-,\alpha_{2k+1}^+)$, and $\operatorname{Im}\sqrt{1 - u_+^2(z)} < 0$ on $(\alpha_{2k}^-,\alpha_{2k}^+)$. Since the $\gamma_k \in [\alpha_k^-,\alpha_k^+]$ are simple roots of the derivative $u_+'(z)$, the function $-u_+'(z)$ is positive (negative) on the intervals $(\gamma_{2k},\gamma_{2k+1})$ (respectively, $(\gamma_{2k+1},\gamma_{2k+2})$). Consequently, the limit values of the integrand in formula (3.4.29) on the boundary $(z = \lambda + i0 , -\infty < \lambda < \infty)$ of the upper half plane satisfy the inequalities

$$\left.\begin{aligned} &\frac{-u_+'(\lambda)}{\sqrt{1 - u_+^2(\lambda)}} > 0 \ , \quad \alpha_k^+ < \lambda < \alpha_{k+1}^- \ , \\ &\frac{-u_+'(\lambda)}{i\sqrt{1 - u_+^2(\lambda)}} > 0 \ , \quad \alpha_k^- < \lambda < \gamma_k \ , \\ &\frac{-u_+'(\lambda)}{i\sqrt{1 - u_+^2(\lambda)}} < 0 \ , \quad \gamma_k < \lambda < \alpha_k^+ \ . \end{aligned}\right\} \tag{3.4.32}$$

Hence, when λ varies from 0 to α_1^- , $\theta(\lambda)$ increases monotonically from 0 to $\theta(\alpha_1^-)$, and since under this variation of λ , $u_+(\lambda)$ decreases monotonically from +1 to -1 , $\theta(\alpha_1^-) = \pi$. As λ increases further from α_1^- to γ_1 , the real part of $\theta(\lambda)$ remains equal to π , whereas the imaginary part first increases from 0 to $h_1 = \frac{1}{i}\,\theta(\gamma_1)$, and then decreases from h_1 to zero, since $\cos\theta(\alpha_1^+) = u_+(\alpha_1^+) = -1$, and $u_+(\gamma_1) = \cos\theta(\gamma_1) = \cos(\pi + ih_1) = -\cosh h_1$. Thus, when λ runs through the segments $[0,\alpha_1^-]$ and $[\alpha_1^-,\alpha_1^+]$ in the positive direction, the point $\theta(\lambda)$ runs through the part of the boundary of the domain $\Theta_+\{h_k\}$ which consists of the segment $[0,\pi]$ and the slit $\mathrm{Re}\,\theta = \pi$, $0 \leqslant \mathrm{Im}\,\theta \leqslant h_1$ in the positive direction.

Furthermore, using (3.4.32), it is established by induction that, as λ runs through the segments $[\alpha_k^+, \alpha_{k+1}^-]$ and $[\alpha_{k+1}^-, \alpha_{k+1}^+]$ $(k = \pm 1, \pm 2, \ldots)$ in the positive direction, the point $\theta(\lambda)$ runs through the part of the boundary of $\Theta_+\{h_k\}$ which consists of the segment $[k\pi, (k+1)\pi]$ and the slit $\mathrm{Re}\,\theta = (k+1)\pi$, $0 \leqslant \mathrm{Im}\,\theta \leqslant h_{k+1}$ in the positive direction; moreover,

$$u_+(\gamma_{k+1}) = (-1)^{k+1} \cosh h_{k+1} . \tag{3.4.33}$$

It follows that $\theta(z)$ induces a one-to-one direction-preserving mapping of the boundary of the upper half plane onto the boundary of the domain $\Theta_+\{h_k\}$. Moreover, as the estimates (3.4.26) and (3.4.26') show, the integrand in (3.4.29) is $\pi(1 + O(z^{-1}))$ on the semicircles $C_n^+ = \{z : z = (n + \frac{1}{2})e^{i\varphi}, 0 \leqslant \varphi \leqslant \pi\}$ for n large. Hence, $\theta(z)$ induces a one-to-one direction-preserving map of C_n onto the curve $\Gamma_n^+ = \{\theta : \theta = \pi(n + \frac{1}{2})e^{i\varphi}(1 + \varepsilon_n(\varphi)), 0 \leqslant \varphi \leqslant \pi\}$, where $|\varepsilon_n(\varphi)| = O(n^{-1})$. This implies, by the principle of correspondence of boundaries, that $\theta(z)$ maps the semidiscs $\{z : |z| < n + \frac{1}{2}, \mathrm{Im}\, z > 0\}$ conformally into the subdomains of $\Theta_+\{h_k\}$ which are bounded in the upper half plane by the curves Γ_n^+. Since n is arbitrary, this implies that $\theta(z)$ maps the entire upper half plane conformally onto $\Theta_+\{h_k\}$; moreover, $\theta(0) = 0$ and $\lim_{|z|\to\infty} z^{-1}\theta(z) = \pi$. The equalities $h_0 = 0$, $h_k = h_{-k}$ are obvious consequences of formulas (3.4.33) and the evenness of the function $u_+(z)$.

To complete the proof of the lemma, we still need to estimate the lengths h_k of the slits $\mathrm{Re}\,\theta = k\pi$, $0 \leqslant \mathrm{Im}\,\theta \leqslant h_k$. Since $u_+(\alpha_k^\pm) = (-1)^k$ and $u_+'(\gamma_k) = 0$, Taylor's formula yields

$$(-1)^k = u_+(\alpha_k^\pm) = u_+(\gamma_k) + \frac{1}{2}(\alpha_k^\pm - \gamma_k)^2 u_+''(\gamma_k^\pm) .$$

Hence, by (3.4.33),

$$\cosh h_k = 1 + \frac{(-1)^k}{2}(\alpha_k^\pm - \gamma_k)^2 u_+''(\gamma_k^\pm) ,$$

where the points $\gamma_k^\pm$ lie between $\alpha_k^\pm$ and γ_k. Consequently,

$$1 + \frac{1}{2}h_k^2 \leqslant 1 + \frac{1}{2}(\alpha_k^\pm - \gamma_k)^2 M_2(k) ,$$

where $M_2(k) = \max_{\alpha_k^- \leq \lambda \leq \alpha_k^+} |u_+''(\lambda)|$,

and hence

$$h_k \leq \frac{1}{2}(\alpha_k^+ - \alpha_k^-)\sqrt{M_2(k)} \quad ,$$

since one of the numbers $|\alpha_k^- - \gamma_k|$, $|\alpha_k^+ - \gamma_k|$ does not exceed $\frac{1}{2}|\alpha_k^+ - \alpha_k^-|$. By the Bernstein inequality for the derivatives of bounded entire functions of exponential type,

$$\sup_{-\infty<\lambda<\infty} |u_+''(\lambda)| \leq \pi^2 M \ ,$$

where

$$M = \sup_{-\infty<\lambda<\infty} |u_+(\lambda)| < \infty \ .$$

Therefore,

$$h_k \leq \frac{\pi\sqrt{M}}{2}(\alpha_k^+ - \alpha_k^-) = \frac{\pi\sqrt{M}}{2}(\sqrt{\mu_k^+} - \sqrt{\mu_k^-}) \ , \tag{3.4.34}$$

and, by Theorem 1.5.2,

$$\sum_{k=-\infty}^{\infty} (k^{n+1}h_k)^2 < \infty$$

if $q(x) \in \tilde{W}_2^n[0,\pi]$. The lemma is proven. □

<u>Remark</u>. The expression (3.4.30) for the Hill discriminant $u_+(z)$ in terms of $\theta(z)$ was obtained for the case $\mu_0 = 0$. It is readily verified that in the general case ($\mu_0 \neq 0$), formula (3.4.30) must be replaced by

$$u_+(z) = \cos\theta(\sqrt{x^2 - \mu_0}) \ . \tag{3.4.30'}$$

The inverse function $z(\theta)$ to $\theta(z)$ maps the domain $\Theta_+\{h_k\}$ onto the upper half plane. Also, $z(k\pi \pm 0) = \alpha_k^{\pm}$ and $\lim_{\theta\to+\infty}(i\theta)^{-1}z(\theta) = \pi^{-1}$. Hence, Lemma 3.4.3 admits the following simple corollary.

COROLLARY. *In order that the sequence* $-\infty < \mu_0 < \mu_1^- \leq \mu_1^+ < \mu_2^- \leq \mu_2^+ < \ldots$ *be the union of the spectra of the periodic* $(\mu_0, \mu_2^-, \mu_2^+, \ldots)$ *and anti-periodic* $(\mu_1^-, \mu_1^+, \mu_3^-, \mu_3^+, \ldots)$ *boundary value problems, generated on the interval* $(0,\pi)$ *by one and the same equation* (3.4.1) *with real-valued potential*

$q(x) \in \tilde{W}_2^n[0,\pi]$, *it is necessary and sufficient that*

$\mu_k^{\pm} = \mu_0 + z^2(k\pi \pm 0) \quad (k = 1,2,\ldots)$

where $z(\theta)$ *is a function which maps a domain* $\Theta_+\{h_k\}$ *of the form* (3.4.31) *conformally onto the upper half plane, and satisfies the conditions*

$z(0) = 0 \quad , \quad \lim_{\theta\to+\infty} (i\theta)^{-1}z(i\theta) = \pi^{-1}$.

If this is the case, then $h_0 = 0$, $h_k = h_{-k}$, *and* $\sum_{k=-\infty}^{k} (k^{n+1}h_k)^2 < \infty$. □

In point of fact, the condition of the corollary is not only necessary, but is also sufficient. To prove the sufficiency, we need to establish some properties of functions $\theta(z)$ which map the upper half plane conformally onto a domain $\Theta_+\{h_k\}$ of the form (3.4.31).

LEMMA 3.4.4. *Suppose that* $\theta(z)$ *maps the upper half plane conformally onto a domain* $\Theta_+\{h_+\}$ *of the form* (3.4.1), *and satisfies the conditions* $\theta(0) = i h_0$, $\lim_{y\to+\infty} (iy)^{-1}\theta(iy) = \pi$, *and let* $\sup_{-\infty<k<\infty} h_k = H < \infty$. *Then*

1) $u(z) = \cos\theta(z)$ *is a real entire function of exponential type* π *and* $\sup_{-\infty<x<\infty} |u(x)| = \cosh H$;

2) *the preimages* $\alpha_k^{\pm}$ *of the points* $k\pi \pm 0$ *under* $\theta(z)$ *satisfy the inequalities*

$$1 \geq \alpha_k^- - \alpha_{k-1}^+ \geq 2(\pi\cosh H)^{-1} , \tag{3.4.35}$$

$$2\pi^{-1}h_k \geq \alpha_k^+ - \alpha_k^- \geq 2(\pi\sqrt{\cosh H})^{-1}h_k , \tag{3.4.36}$$

and

$$2(\pi\cosh H)^{-1}k \leq |\alpha_{\pm k}^{\pm}| \leq k(1 + 2\pi^{-1}H) + 2\pi^{-1}h_0 ; \tag{3.4.37}$$

3) *the function* $\mathrm{Im}\,\theta(z)$, *which obviously vanishes on the segments* $[\alpha_k^+,\alpha_{k+1}^-]$, *satisfies the inequalities*

$$0 < \mathrm{Im}\,\theta(z) \leq \pi\sqrt{\cosh H}\sqrt{(x - \alpha_k^-)(\alpha_k^+ - x)} \tag{3.4.38}$$

on the intervals (α_k^-,α_k^+) .

PROOF. The function $u(z) = \cos\theta(z)$ is obviously holomorphic in the upper half plane and continuous on the real line. Since $\operatorname{Im}\cos\theta = 0$ whenever $\operatorname{Im}\theta = 0$ or $\operatorname{Re}\theta = k\pi$, $u(z)$ takes only real values on the real line, and by the symmetry principle $(u(\overline{z}) = \overline{u(z)})$ it is a real entire function. The function $\theta(z)$ is holomorphic in the upper half plane, continuous on the real line, $\operatorname{Im}\theta(z) > 0$ for $\operatorname{Im} z > 0$, and $\lim_{y\to+\infty} (iy)^{-1}\theta(iy) = \pi$. Hence, Nevanlinna's formula holds for $\theta(z)$:

$$\theta(z) = \pi z + d + \frac{1}{\pi}\int_{-\infty}^{\infty} \frac{1+tz}{t-z}\,\frac{\operatorname{Im}\theta(t)}{1+t^2}\,dt \tag{3.4.39}$$

($\operatorname{Im} d = 0$), and yields the following representation for $\operatorname{Im}\theta(z)$:

$$\operatorname{Im}\theta(z) = \pi y + \frac{y}{\pi}\int_{-\infty}^{\infty} \frac{\operatorname{Im}\theta(t)}{(x-t)^2 + y^2}\,dz \quad (z = x + iy)\,.$$

Since

$$0 \leqslant \operatorname{Im}\theta(x) \leqslant \sup_{-\infty<x<\infty} \operatorname{Im}\theta(x) = \sup_{-\infty<k<\infty} h_k = H\,,$$

we have

$$0 \leqslant \frac{y}{\pi}\int_{-\infty}^{\infty} \frac{\operatorname{Im}\theta(t)}{(x-t)^2 + y^2}\,dt \leqslant \frac{H}{\pi}\int_{-\infty}^{\infty} \frac{y}{(x^2-t^2) + y^2}\,dt = H$$

and

$$\pi \operatorname{Im} z \leqslant \operatorname{Im}\theta(z) \leqslant z \operatorname{Im} z + H\,. \tag{3.4.40}$$

Consequently,

$$|u(z)| = |\cos\theta(z)| \leqslant \cosh|\operatorname{Im}\theta(z)| \leqslant \cosh(\pi|\operatorname{Im} z| + H)\,,$$

and hence $u(z)$ is of exponential type π; moreover,

$$\sup_{-\infty<x<\infty} |u(x)| = \sup_{-\infty<x<\infty} |\cos\theta(x)| = \sup_{-\infty<k<\infty} \cosh h_k = \cosh H < \infty\,.$$

Now let us establish the right-hand sides of the double inequalities (3.4.35) and (3.4.36). Since $\theta(\alpha_k^-) = k\pi$ and $\theta(\alpha_{k-1}^+) = (k-1)\pi$, we have $u(\alpha_k^-) - u(\alpha_{k-1}^+) = \cos\theta(\alpha_k^-) - \cos\theta(\alpha_{k-1}^+) = (-1)^k - (-1)^{k-1} = (-1)^k 2$, and by the mean value theorem,

$$u(\alpha_k^-) - u(\alpha_{k-1}^+) = (\alpha_k^- - \alpha_{k-1}^+)u'(\beta_k) \quad (\alpha_{k-1}^+ < \beta_k < \alpha_k^-)\,.$$

Moreover, by Bernstein's inequality,

$$|u'(\beta_k)| \leqslant \pi \sup_{-\infty<x<\infty} |u(x)| \leqslant \pi\cosh H\,.$$

Therefore,

$$2 = |u(\alpha_k^-) - u(\alpha_k^+)| = (\alpha_k^- - \alpha_{k-1}^+)|u'(\beta_k)| \leq (\alpha_k^- - \alpha_{k-1}^+)\pi \cosh H ,$$

which is equivalent to the right-hand side of inequality (3.4.35). The right-hand side of inequality (3.4.36) is proved in exavtly the same way that inequality (3.4.34) was verified in the preceding lemma.

To prove inequality (3.4.38), we introduce the function

$$f(x) = (-1)^k u(x) - [1 + (x-\alpha_k^-)(\alpha_k^+-x) \frac{\pi^2}{2} \cosh H] .$$

At the endpoints of the segment $[\alpha_k^-,\alpha_k^+]$, this function vanishes:

$$f(\alpha_k^\pm) = (-1)^k u(\alpha_k^\pm) - 1 = (-1)^k \cos\theta(\alpha_k^\pm) - 1 = (-1)^k \cos k\pi - 1 = 0 .$$

Moreover, its graph is downward convex, since

$$f''(x) = (-1)^k u''(x) + \pi^2 \cosh H \geq 0$$

by Bernstein's inequality. Consequently, the function $f(x)$ is negative on the segment $[\alpha_k^-,\alpha_k^+]$ and

$$(-1)^k u(x) \leq 1 + (x-\alpha_k^-)(\alpha_k^+-x) \frac{\pi^2}{2} \cosh H \leq \cosh \{\pi(\cosh H(x-\alpha_k^-)(\alpha_k^+-x))^{\frac{1}{2}}\}$$

therein. On the other hand, for $x \in [\alpha_k^-,\alpha_k^+]$,

$$\theta(x) = k\pi + i \operatorname{Im} \theta(x) \quad , \quad u(x) = \cos\theta(x) = (-1)^k \cosh \operatorname{Im} \theta(x) ,$$

and

$$\cosh \operatorname{Im} \theta(x) \leq \cosh \{\pi(\cosh H(x-\alpha_k^-)(\alpha_k^+-x))^{\frac{1}{2}}\} ,$$

which is equivalent to (3.4.38).

The left-hand sides of inequalities (3.4.35) and (3.4.36) are proved by using variational principles for conformal mappings. It follows from the inequalities (3.4.40) that the harmonic function $\operatorname{Im} \theta(z) - \pi \operatorname{Im} z$ is nonnegative in the upper half plane, and since it vanishes on the segment $[\alpha_{k-1}^+,\alpha_k^-]$,

$$\frac{\partial}{\partial y} \{\operatorname{Im} \theta(x+iy) - \pi y\}\Big|_{y=0} = \frac{\partial}{\partial y} \operatorname{Im} \theta(x+iy)\Big|_{y=0} - \pi \geq 0 ,$$

whence, upon using the Cauchy-Riemann equation, we find that

$$\frac{\partial}{\partial x} \operatorname{Re} \theta(x) = \frac{\partial}{\partial x} \theta(x) \geq \pi \quad (\alpha_{k-1}^+ \leq x \leq \alpha_k^-) .$$

Therefore,

$$\pi = \theta(\alpha_k^-) - \theta(\alpha_{k-1}^+) = \int_{\alpha_{k-1}^+}^{\alpha_k^-} \frac{\partial}{\partial x}\theta(x)dx \geqslant \pi(\alpha_k^- - \alpha_{k-1}^+) ,$$

which is equivalent to the left-hand side of inequality (3.4.35).

To prove the left-hand side of inequality (3.4.36), we consider the inverse function $z(\theta)$ to $\theta(z)$, and the auxiliary function

$$z_k(\theta) = \sqrt{(\theta-k\pi)^2 + h_k^2} , \tag{3.4.41}$$

which conformally maps the domain

$$\Theta_k = \{\theta : \operatorname{Im}\theta > 0\} \setminus \{\theta ; \operatorname{Re}\theta = k\pi , 0 \leqslant \operatorname{Im}\theta \leqslant h_k\}$$

onto the upper half plane. Obviously, $\operatorname{Im} z_k(\theta) - \operatorname{Im} z(\theta) \geqslant 0$ on the whole boundary of $\Theta_+\{h_k\}$, and from (3.4.40) and (3.4.41) we find that $\operatorname{Im} z(\theta) \leqslant \pi^{-1}\operatorname{Im}\theta$ and $\operatorname{Im} z_k(\theta) = \pi^{-1}\operatorname{Im}\theta + O(|\theta|^{-1})$ as $|\theta| \to \infty$ $(\theta \in \Theta_+\{h_k\})$, and hence that $\lim_{|\theta|\to\infty} \{\operatorname{Im} z_k(\theta) - \operatorname{Im} z(\theta)\} \geqslant 0$. By the minimum principle for harmonic functions, this implies that the inequality

$$\operatorname{Im} z_k(\theta) - \operatorname{Im} z(\theta) \geqslant 0 \tag{3.4.42}$$

holds throughout $\Theta_+\{h_k\}$. Let L_k denote the common part of the boundaries of the domains $\Theta_+\{h_k\}$ and Θ_k, which consists of the two lips of the slit $\operatorname{Re}\theta = k\pi$, $0 \leqslant \operatorname{Im}\theta \leqslant h_k$, and let n denote the inward normal to L_k. Since $\operatorname{Im} z_k(\theta) = \operatorname{Im} z(\theta) = 0$ on L_k, it follows from (3.4.42) that

$$\frac{\partial}{\partial n}\{\operatorname{Im} z_k(\theta) - \operatorname{Im} z(\theta)\} \geqslant 0 , \quad \frac{\partial}{\partial n}\operatorname{Im} z(\theta) \geqslant 0 \quad (\theta \in L_k) ,$$

and hence that

$$\left|\frac{\partial \operatorname{Im} z_k(\theta)}{\partial n}\right| \geqslant \left|\frac{\partial \operatorname{Im} z(\theta)}{\partial n}\right| \quad (\theta \in L_k) .$$

By the Cauchy-Riemann equations,

$$\left|\frac{\partial z_k(\theta)}{\partial\theta}\right| = \left|\frac{\partial \operatorname{Im} z_k(\theta)}{\partial n}\right| \quad (\theta \in L_k)$$

and

$$\left|\frac{\partial z(\theta)}{\partial\theta}\right| = \left|\frac{\partial \operatorname{Im} z(\theta)}{\partial\theta}\right| \quad (\theta \in L_k) .$$

Therefore,

$$\alpha_k^+ - \alpha_k^- = \int_{L_k} \left|\frac{\partial z(\theta)}{\partial\theta}\right| |d\theta| \leq \int_{L_k} \left|\frac{\partial z_k(\theta)}{\partial\theta}\right| |d\theta| = \frac{2}{\pi} h_k ,$$

which proves the left-hand side of inequality (3.4.36).

The inequalities (3.4.37) follow from the already established inequalities (3.4.35), (3.4.36) and the identities

$$\alpha_k^+ = \sum_{j=1}^{k} \{(\alpha_j^+ - \alpha_j^-) + (\alpha_j^- - \alpha_{j-1}^+)\} + \alpha_0^+ ,$$

and

$$\alpha_k^- = \sum_{j=-1}^{-k} \{(\alpha_j^- - \alpha_j^+) + (\alpha_j^+ - \alpha_{j+1}^-)\} + \alpha_0^- ,$$

since $\alpha_k^+ \geq 0$ and $\alpha_{-k}^- \leq 0$ $(k = 0,1,\dots)$. □

LEMMA 3.4.5. *If the assumptions of the preceding lemma are in force with* $h_0 = 0$, $h_k = h_{-k}$, *and* $\sum_{k=-\infty}^{\infty} (kh_k)^2 < \infty$, *then*

1) *the function* $u(z) = \cos\theta(z)$ *is even and admits the representation*

$$u(z) = \cos\pi z - z^{-1}d_1 \sin\pi z + z^{-1}g(z) , \tag{3.4.43}$$

where

$$g(z) = \int_0^\pi \tilde{g}(t) \sin ztdt \quad , \quad \tilde{g}(t) \in L_2[0,\pi] ;$$

2) *the preimages* $\alpha_k^\pm$ *of the points* $k\pi \pm 0$ *satisfy the equalities*

$$\alpha_k^\pm = -\alpha_{-k}^\mp \quad , \quad \alpha_k^\pm = k - \frac{d_1}{\pi k} + \frac{\varepsilon_k^\pm}{k} ,$$

where $\sum_{k=1}^{\infty} |\varepsilon_k|^2 < \infty$.

PROOF. In the present case, the domain $\Theta_+\{h_k\}$ is symmetric relative to the imaginary axis. Hence, upon continuing the function $\theta(z)$ into the lower half plane by the symmetry principle ($\theta(\overline{z}) = \overline{\theta(z)}$), it turns out that the function $-\theta(-z)$ also effects a conformal mapping of the upper half plane onto the domain $\Theta_+\{h_k\}$, which is fixed by the same conditions: $-\theta(-0) = 0$, $\lim_{y\to+\infty} (iy)^{-1}\{-\theta(-iy)\} = \lim_{y\to+\infty} \overline{(iy)^{-1}\theta(iy)} = \pi$. Consequently, $\theta(z) = -\theta(-z)$, and $\theta(z)$ is odd, which in turn implies that the function $u(z) = \cos\theta(z)$

is even and that $\alpha_k^{\pm} = -\alpha_{-k}^{\mp}$.

We next derive the asymptotic formula for $\theta(z)$ as $|z| \to \infty$. Since $\theta(z)$ maps the segments $[\alpha_k^-,\alpha_k^+]$ into the slits $\operatorname{Re}\theta = k\pi$, $0 \leq \operatorname{Im}\theta(z) \leq h_k$, the inequalities $0 \leq \operatorname{Im}\theta(t) \leq h_k$ and

$$\int_{\alpha_k^-}^{\alpha_k^+} |t|^s \operatorname{Im}\theta(t)dt \leq (m_k)^s \int_{\alpha_k^-}^{\alpha_k^+} h_k dt = (m_k)^s h_k(\alpha_k^+ - \alpha_k^-) ,$$

where

$$m_k = \max\{|\alpha_k^+|,|\alpha_k^-|\} = \begin{cases} \alpha_k^+ , & k > 0 , \\ |\alpha_k^-| , & k < 0 , \end{cases}$$

must hold on these segments. From this we further obtain

$$\int_{\alpha_k^-}^{\alpha_k^+} |t|^s \operatorname{Im}\theta(t)dt \leq \frac{2}{\pi}\left(1 + \frac{2}{\pi}H\right)^s |k|^s h_k^2$$

with the help of the estimates (3.4.36), (3.4.37). Therefore, if $\sum_{k=-\infty}^{\infty} (kh_k)^2 < \infty$, then

$$\int_{-\infty}^{\infty} |t|^s \operatorname{Im}\theta(t)dt = \sum_{k=-\infty}^{\infty} \int_{\alpha_k^-}^{\alpha_k^+} |t|^s \operatorname{Im}\theta(t)dt \leq$$

$$\leq \frac{2}{\pi}\left(1 + \frac{2}{\pi}H\right)^s \sum_{k=-\infty}^{\infty} |k|^s h_k^2 = C_s < \infty \qquad (3.4.44)$$

for $s = 0,1,$ and 2. This allows us to reexpress (3.4.39) in the form

$$\theta(z) = \pi z + d - \frac{1}{\pi}\int_{-\infty}^{\infty} \frac{t}{1 + t^2} \operatorname{Im}\theta(t)dt -$$

$$- \frac{1}{\pi z}\int_{-\infty}^{\infty} \operatorname{Im}\theta(t)dt + \frac{1}{\pi z}\int_{-\infty}^{\infty} \frac{1}{t-z} \operatorname{Im}\theta(t)dt .$$

It follows from the oddness of the function $\theta(z) = \overline{\theta(\overline{z})}$ that for real values of t the function $\operatorname{Im}\theta(t) = \operatorname{Im}\theta(t+i0)$ is even. Thus the integral on the right-hand side of this equality that does not depend on z vanishes, whereas each of the three terms which depend on z is an odd function of z. Hence,

$$\theta(z) = \pi z + z^{-1}\{d_1 + \psi(z)\}, \qquad (3.4.45)$$

where

$$d_1 = -\frac{1}{\pi}\int_{-\infty}^{\infty} \operatorname{Im}\theta(t)dt$$

and

$$\psi(z) = \frac{1}{\pi}\int_{-\infty}^{\infty} \frac{t}{t-z}\operatorname{Im}\theta(t)dt \quad , \quad \psi(z) = \psi(-z) \ .$$

To estimate

$$\max_{\alpha_{n-1}^+ \leqslant x \leqslant \alpha_n^-} |\psi(x)| = \max_{\alpha_{-n}^+ \leqslant x \leqslant \alpha_{-n+1}^-} |\psi(x)| \ ,$$

we use the equality

$$\psi(x) = \frac{1}{\pi}\int_{-\infty}^{\infty} \frac{t}{t-x}\operatorname{Im}\theta(t)dt = \frac{1}{\pi}\sum_{k=-\infty}^{\infty}\int_{\alpha_k^-}^{\alpha_k^+} \frac{t}{t-x}\operatorname{Im}\theta(t)dt =$$

$$= \frac{1}{\pi}\sum{}'' \int_{\alpha_k^-}^{\alpha_k^+} \frac{t}{t-x}\operatorname{Im}\theta(t)dt + \frac{1}{\pi}\left\{\int_{\alpha_{n-1}^-}^{\alpha_{n-1}^+} \frac{t}{t-x}\operatorname{Im}\theta(t)dt + \int_{\alpha_n^-}^{\alpha_n^+} \frac{t}{t-x}\operatorname{Im}\theta(t)dt\right\} ,$$

in which the double prime superscipt indicates that the sum is taken over all $k \neq n-1,n$. By (3.4.35), for $x \in [\alpha_{n-1}^+, \alpha_n^-]$ and $t \in [\alpha_k^-, \alpha_k^+]$, $k \neq n-1,n$,

$$|t-x| \geqslant 2\pi(\cosh H)^{-1} ,$$

and hence

$$\frac{1}{\pi}\left|\int_{\alpha_k^-}^{\alpha_k^+} \frac{t}{t-x}\operatorname{Im}\theta(t)dt\right| = \frac{1}{\pi x}\left|\int_{\alpha_k^-}^{\alpha_k^+}\left(\frac{t^2}{t-x} - t\right)\operatorname{Im}\theta(t)dt\right| \leqslant$$

$$\leqslant \frac{1}{\pi x}\int_{\alpha_k^-}^{\alpha_k^+}\left(\frac{t^2}{2}\pi\cosh H + |t|\right)\operatorname{Im}\theta(t)dt \ .$$

Upon invoking inequality (3.4.44), this in turn yields

$$\frac{1}{\pi}\left|\sum{}''\right| \leqslant \frac{1}{\pi x}\left(\frac{C_2\pi\cosh H}{2} + C_1\right) .$$

Next, by inequalities (3.4.38), (3.4.36), and (3.4.37), it follows readily that, for $x \in [\alpha_{n-1}^+, \alpha_n^-]$,

$$\frac{1}{\pi}\left|\int_{\alpha_{n-1}^-}^{\alpha_{n-1}^+} \frac{t}{t-x}\operatorname{Im}\theta(t)dt + \int_{\alpha_n^-}^{\alpha_n^+} \frac{t}{t-x}\operatorname{Im}\theta(t)dt\right| \leqslant$$

$$\leq \sqrt{\cosh H}\left\{\alpha_{n-1}^{+}\int_{\alpha_{n-1}^{-}}^{\alpha_{n-1}^{+}}\frac{[(t-\alpha_{n-1}^{-})(\alpha_{n}^{+}-t)]^{\frac{1}{2}}}{\alpha_{n-1}^{+}-t}\,dt\, +\right.$$

$$\left. + \alpha_{n}^{+}\int_{\alpha_{n}^{-}}^{\alpha_{n}^{+}}\frac{[(t-\alpha_{n}^{-})(\alpha_{n}^{+}-t)]^{\frac{1}{2}}}{t-\alpha_{n}^{-}}\,dt\right\} = \sqrt{\cosh H}\left\{\alpha_{n-1}^{+}\frac{\alpha_{n-1}^{-}-\alpha_{n-1}^{-}}{2}+\alpha_{n}^{+}\frac{\alpha_{n}^{+}-\alpha_{n}^{-}}{2}\right\} \leq$$

$$\leq \sqrt{\cosh H}\,(1+\tfrac{2}{\pi}H)[(n-1)h_{n-1}+nh_n] .$$

Hence, there is a constant $B_1 < \infty$, independent of n, such that

$$|\psi(x)| \leq B_1[x^{-1}+(n-1)h_{n-1}+nh_n]$$

for all $x \in [\alpha_{n-1}^{+},\alpha_{n}^{-}]$.

To estimate $|\psi(x)|$ for $x \in [\alpha_n^-,\alpha_n^+]$, we use the equality

$$\psi(x) = x[\theta(x)-\pi x - x^{-1}d_1] = x[k\pi + i\,\mathrm{Im}\,\theta(x) - \pi x - x^{-1}d_1] =$$
$$= x[\theta(\alpha_n^-) + i\,\mathrm{Im}\,\theta(x) - \pi x - x^{-1}d_1] =$$
$$= x[\pi(\alpha_n^- - x) + (\alpha_n^- x)^{-1}d_1(x-\alpha_n^-) + (\alpha_n^-)^{-1}\psi(\alpha_n^-) + i\,\mathrm{Im}\,\theta(x)] ,$$

from which it follows that

$$|\psi(x)| \leq 3\alpha_n^+h_n + (\alpha_n^-)^{-1}\{|d_1|h_n + \alpha_n^+|\psi(\alpha_n^-)|\} ,$$

since for the values of x considered

$$|\alpha_n^- - x| \leq \alpha_n^+ - \alpha_n^- \leq 2\pi^{-1}h_n$$

and

$$0 \leq \mathrm{Im}\,\theta(x) \leq h_n .$$

Comparing the estimates of $|\psi(x)|$ for $x \in [\alpha_{n-1}^+,\alpha_n^-]$ and $x \in [\alpha_n^-,\alpha_n^+]$ with inequalities (3.4.37), we deduce the existence of a constant $B < \infty$, which does not depend on n, such that

$$|\psi(x)| \leq B[x^{-1} + (n-1)h_n + nh_n]$$

for every $x \in [\alpha_{n-1}^+,\alpha_n^+]$, whence

$$\psi_n = \max_{\alpha_{n-1}^+<x<\alpha_n^+}|\psi(x)| \leq B[\frac{\pi\cosh H}{2(n-1)} + (n-1)h_{n-1} + nh_n] .$$

Therefore, if $\sum_{k=-\infty}^{\infty} (kh_k)^2 < \infty$, then

$$\sum_{k\neq 0} |\psi(\alpha_k^{\pm})|^2 < \infty \ , \quad \sum_{n=2}^{\infty} |\psi_n|^2 < \infty \ , \quad \lim_{x\to\infty} \psi(x) = 0 \ , \tag{3.4.46}$$

and

$$\int_{-\infty}^{-\alpha_1^+} |\psi(x)|^2 dx = \int_{\alpha_1^+}^{\infty} |\psi(x)|^2 dx = \sum_{k=2}^{\infty} \int_{\alpha_{k-1}^+}^{\alpha_k^+} |\psi(x)|^2 dx \leqslant$$

$$\leqslant \sum_{k=2}^{\infty} |\psi_k|^2 (\alpha_k^+ - \alpha_{k-1}^+) \leqslant (1 + \frac{2}{\pi} H) \sum_{k=2}^{\infty} |\psi_k|^2 < \infty \ .$$

The main assertions of Lemma 3.4.5 follow from these properties of the function $\psi(x)$. In fact, by the Paley-Wiener Theorem, in order to establish formula (3.4.43), it suffices to show that the function $g(x) = x[u(x) - \cos \pi x + x^{-1} d_1 \sin \pi x]$ belongs to $L_2(-\infty,\infty)$. But, in view of (3.4.45) and (3.4.46),

$$u(x) = \cos \theta(x) = \cos(\pi x + x^{-1} d_1 + x^{-1}\psi(x)) =$$
$$= \cos \pi x \cos x^{-1}(d_1 + \psi(x)) - \sin \pi x \sin x^{-1}(d_1 + \psi(x)) =$$
$$= \cos \pi x[1 + O(x^{-2})] - \sin \pi x[x^{-1}d_1 + x^{-1}\psi(x) + O(x^{-3})] =$$
$$= \cos \pi x - x^{-1}d_1 \sin \pi x - x^{-1}\psi(x) \sin \pi x + O(x^{-2})$$

for large values of $|x|$, and hence,

$$g(x) = -\psi(x) \sin \pi x + O(x^{-1})$$

as $|x| \to \infty$. Thus, since $g(x)$ is continuous and the right-hand side of the last equality is square integrable over the intervals $(-\infty,-\alpha_1^+)$ and (α_1^+,∞) , $g(x)$ clearly belongs to $L_2(-\infty,\infty)$. Finally, by (3.4.45), the equalities $\theta(\alpha_k^{\pm}) = k\pi$ imply that

$$k\pi = \alpha_k^{\pm} + (\alpha_k^{\pm})^{-1}\{d_1 + \psi(\alpha_k^{\pm})\} \ ,$$

whence

$$\alpha_k^{\pm} = k - \frac{d_1}{\pi\alpha_k^{\pm}} - \frac{\psi(\alpha_k^{\pm})}{\pi\alpha_k^{\pm}} = k - \frac{d_1}{\pi k} + \frac{\varepsilon_k^{\pm}}{k} \ ,$$

where

$$\varepsilon_k^{\pm} = \frac{d_1(\alpha_k^{\pm} - k)}{\pi\alpha_k^{\pm}} - \frac{k\psi(\alpha_k^{\pm})}{\pi\alpha_k^{\pm}} = - \frac{\psi(\alpha_k^{\pm})}{\pi} + O(k^{-2}) .$$

Using the first of equalities (3.4.46), we obtain $\sum_{k=-\infty}^{\infty} |\varepsilon_k^{\pm}|^2 < \infty$. □

THEOREM 3.4.2. *The sequence* $-\infty < \mu_0 < \mu_1^- \leqslant \mu_1^+ < \mu_2^- \leqslant \mu_2^+ < \dots$ *is the union of the spectrum* $\mu_0, \mu_2^-, \mu_2^+, \dots$ *of the periodic boundary value problem and the spectrum* $\mu_1^-, \mu_1^+, \mu_3^-, \mu_3^+, \dots$ *of the antiperiodic boundary value problem generated on the interval* $(0,\pi)$ *by one and the same equation* (3.4.1) *with real-valued potential* $q(x) \in \tilde{W}_2^n[0,\pi]$ *if and only if the entries in the sequence are given by the formula*

$$\mu_k^{\pm} = \mu_0 + z^2(k\pi \pm 0) ,$$

where $z(\theta)$ *is a conformal mapping of a domain* $\Theta_+\{h_k\}$ *of the form* (3.4.31) *onto the upper half plane with* $h_0 = 0$, $h_k = h_{-k}$,

$$\sum_{k=-\infty}^{\infty} (k^{n+1}h_k)^2 < \infty , \tag{3.4.47}$$

and $z(\theta)$ *is also subject to the conditions*

$$z(0) = 0 \quad , \quad \lim_{\theta\to+\infty} (i\theta)^{-1} z(i\theta) = \pi^{-1} .$$

PROOF. The necessity of these conditions was established in Lemma 3.4.3 and its Corollary. To prove their sufficiency we may obviously take $\mu_0 = 0$ without loss of generality. Let $\theta(z)$ denote the inverse of the function $z(\theta)$, and put $u_1(z) = \cos\theta(z)$. Then, by Lemma 3.4.5,

$$u_1(z) = \cos \pi z - z^{-1}\{d_1 \sin \pi z - g(z)\} , \tag{3.4.48}$$

where

$$g(z) = \int_0^{\pi} \tilde{g}(t) \sin zt dt \quad , \quad \tilde{g}(t) \in L_2[0,\pi] ,$$

and the sequences $\mu_0 = 0$, $\pm\sqrt{\mu_2^-}$, $\pm\sqrt{\mu_2^+}$,...; $\pm\sqrt{\mu_1^-}$, $\pm\sqrt{\mu_1^+}$, $\pm\sqrt{\mu_3^-}$, $\pm\sqrt{\mu_3^+}$, ,..., of roots of the equations $u_1(z) - 1 = 0$ and $u_1(z) + 1 = 0$, respectively, satisfy the asymptotic formulas

$$\sqrt{\mu_k^{\pm}} = k - (\pi k)^{-1} d_1 + k^{-1}\varepsilon_k^{\pm} \quad , \quad \sum_{k=1}^{\infty} (\varepsilon_k^{\pm})^2 < \infty . \tag{3.4.49}$$

After analytic continuation to the lower half plane, the function $z(\theta) = \overline{z(\overline{\theta})}$

maps the θ-plane with the vertical slits $\operatorname{Re}\theta = k\pi$, $-h_k \leq \operatorname{Im}\theta \leq h_k$, into the z-plane with horizontal slits $\sqrt{\mu_k^-} \leq z \leq \sqrt{\mu_k^+}$, $-\sqrt{\mu_k^+} \leq z \leq -\sqrt{\mu_k^-}$ on the real axis. Let $\theta_k = k\pi + ih_k'$ $(-h_k \leq h_k' \leq h_k)$ be an arbitrary point on one of the lips of the vertical slit $\operatorname{Re}\theta = k\pi$, $-h_k \leq \operatorname{Im}\theta \leq h_k$ $(k = 1,2,\dots)$, and set $\sqrt{\mu_k} = z(\theta_k)$. Then $\sqrt{\mu_k^-} \leq \sqrt{\mu_k} \leq \sqrt{\mu_k^+}$, and hence

$$\sqrt{\mu_k} = k - (\pi k)^{-1}d_1 + k^{-1}\delta_k \quad , \quad \sum_{k=1}^{\infty} \delta_k^2 < \infty . \tag{3.4.50}$$

From this it follows, by Lemma 3.4.2, that the function

$$zs(z) = \pi z \prod_{k=1}^{\infty} k^{-1}(\lambda_k - z^2) \tag{3.4.51}$$

admits the representation

$$zs(z) = \sin z + \frac{4d_1 z}{4z^2 - 1} \cos \pi z + \frac{f(z)}{z} , \tag{3.4.52}$$

where

$$f(z) = \int_0^{\pi} \tilde{f}(t) \cos zt dt \quad , \quad \tilde{f}(t) \in L_2[0,\pi] \quad , \quad f(0) = \int_0^{\pi} \tilde{f}(t)dt = 0 .$$

Let us show that $u_1(z)$ and $s(z)$ are the Hill discriminant and the characteristic function, respectively, of the boundary value problem (3.4.2) for an equation (3.4.1) with real-valued potential $q(x) \in \tilde{W}_2^n[0,\pi]$. As we remarked above, to specify such a problem, we must give, besides $u_1(z)$ and $s(z)$, a function $u_2(z)$ of the form

$$u_2(z) = \int_0^{\pi} k(t) \frac{\sin zt}{z} dt \quad , \quad k(t) \in L_2[0,\pi] , \tag{3.4.53}$$

which satisfies the conditions $|u_2(\sqrt{\lambda_k})| = \sqrt{u_2(\sqrt{\lambda_k})^2 - 1}$ and hence will serve as the function $u_-(z)$ of the sought-for equation. Since

$$u_1(\sqrt{\lambda_k}) = \cos(k\pi + ih_k') = (-1)^k \cosh h_k'$$

and

$$\sqrt{u_1(\sqrt{\lambda_k})^2 - 1} = |\sinh h_k'| , \tag{3.4.54}$$

we require that the equalities

$$u_2(\sqrt{\lambda_k}) = \sinh h_k' , \tag{3.4.55}$$

which determine not only $|u_2(\sqrt{\lambda_k})|$, but also $\operatorname{sign} u_2(\sqrt{\lambda_k}) = \operatorname{sign} h_k'$, be satisfied. Therefore, upon choosing an arbitrary sequence of points

$\theta_k = k\pi + ih'_k$ on one lip of each of the vertical slits $\mathrm{Re}\,\theta = k\pi$, $-h_k \leqslant \leqslant \mathrm{Im}\,\theta \leqslant h_k$, we can arbitrarily specify both the candidates for the eigenvalues $\lambda_k = z^2(\theta_k)$ $(\mu_k^- \leqslant \lambda_k \leqslant \mu_k^+)$ and the sequence $\mathrm{sign}\, u_2(\sqrt{\lambda_k})$. Hence, by the previous remark, in this way we can find all equations (3.4.1) with given spectra of the periodic and the anti-periodic boundary value problems, i.e., we can provide the complete solution to problem B.

In agreement with formula (3.4.28), we put

$$u_2(z) = s(z) \sum_{k=1}^{\infty} \frac{2\sqrt{\lambda_k}\,\sinh h'_k}{(z^2 - \lambda_k)\dot{s}(\sqrt{\lambda_k})} . \tag{3.4.56}$$

Let us verify that this function satisfies equalities (3.4.55) and admits a representation of the form (3.4.53). It follows from (3.4.50) and (3.4.47) that

$$\sum_{k=1}^{\infty} (\sqrt{\lambda_k}\,\sinh h'_k)^2 \leqslant C \sum_{k=1}^{\infty} (kh'_k)^2 \leqslant C \sum_{k=1}^{\infty} (kh'_k)^2 < \infty ,$$

whereas formula (3.4.52) yields

$$\lim_{k\to\infty} |\sqrt{\lambda_k}\,\dot{s}(\sqrt{\lambda_k})| = \pi ,$$

whence

$$\max_{1\leqslant k<\infty} |\sqrt{\lambda_k}\,\dot{s}(\sqrt{\lambda_k})|^{-1} = p < \infty , \tag{3.4.57}$$

since $\dot{s}(\sqrt{\lambda_k}) \neq 0$. Consequently, the series (3.4.56) converges uniformly on every compact set and its sum $u_2(z)$ is an entire function which obviously satisfies conditions (3.4.55). Moreover, the function

$$h(x) = \frac{2}{\pi} \sum_{k=1}^{\infty} k \sinh h'_k \cdot \sin kx$$

belongs to $L_2[0,\pi]$, and

$$\varphi(z) = \int_0^{\pi} h(t)\, \frac{\sin zt}{z}\, dt$$

admits the interpolation series

$$\varphi(z) = s(z) \sum_{k=1}^{\infty} \frac{2\sqrt{\lambda_k}\,\varphi(\sqrt{\lambda_k})}{(z^2 - \lambda_k)\dot{s}(\sqrt{\lambda_k})} .$$

This is proved in exactly the same way as formula (3.4.28). Therefore,

$$z[u_2(z) - \varphi(z)] = \sum_{k=1}^{\infty} \frac{\sinh h_k' - \varphi(\sqrt{\lambda_k})}{\dot{s}(\sqrt{\lambda_k})} \frac{2zs(z)\sqrt{\lambda_k}}{z^2 - \lambda_k} = \sum_{k=1}^{\infty} c_k \frac{2zs(z)\sqrt{\lambda_k}}{z^2 - \lambda_k},$$

where

$$c_k = \frac{\sinh h_k' - \varphi(\sqrt{\lambda_k})}{\dot{s}(\sqrt{\lambda_k})} = \frac{k\varphi(k) - \sqrt{\lambda_k}\,\varphi(\sqrt{\lambda_k}) + (\sqrt{\lambda_k} - k)\sinh h_k'}{\dot{s}(\sqrt{\lambda_k})\sqrt{\lambda_k}},$$

since

$$k\varphi(k) = \int_0^{\pi} h(t)\sin kt\,dt = k\sinh h_k'.$$

It follows from the estimate (3.4.57), the asymptotic formula (3.4.50), and Lemma 1.4.3, that $c_k = k^{-1}\eta_k$, $\sum_{k=1}^{\infty} |\eta_k|^2 < \infty$, which implies the absolute convergence of the series $\sum c_k$. Further, by the Paley-Wiener theorem,

$$\frac{2\sqrt{\lambda_k}\,zs(z)}{z^2 - \lambda_k} = \int_0^{\pi} s_k(t)\sin zt\,dt, \quad s_k(t) \in L_2[0,\pi],$$

and, by the Parseval equality,

$$||s_k||^2 = \int_0^{\pi} |s_k(t)|^2 dt = \frac{1}{\pi}\int_{-\infty}^{\infty} \left(\frac{2\sqrt{\lambda_k}\,xs(x)}{x^2 - \lambda_k}\right)^2 dx =$$

$$= \frac{1}{\pi}\int_{-\infty}^{\infty} \left(\frac{2\sqrt{\lambda_k}\,(x+i)s(x+i)}{(x+i)^2 - \lambda_k}\right)^2 dx \leqslant \frac{4\lambda_k m^2}{\pi}\int_{-\infty}^{\infty} \frac{dx}{|(x+i)^2 - \lambda_k|^2} =$$

$$= \frac{4\lambda_k m^2}{\pi}\int_{-\infty}^{\infty} \frac{dx}{[(x - \sqrt{\lambda_k})^2 + 1][x + \sqrt{\lambda_k})^2 + 1]} = \frac{2\lambda_k m^2}{\lambda_k + 1} < 2m^2,$$

where $m = \sup_{-\infty<x<\infty} |(x+i)s(x+i)| < \infty$. Hence, the series $\sum_{k=1}^{\infty} c_k s_k(t)$ converges in the norm of $L_2[0,\pi]$ to a function $h_1(t) \in L_2[0,\pi]$, and

$$z[u_2(z) - \varphi(z)] = \sum_{k=1}^{\infty} c_k \int_0^{\pi} s_k(t)\sin zt\,dt = \int_0^{\pi} h_1(t)\sin zt\,dt,$$

so that the function

$$u_2(z) = \varphi(z) + \int_0^{\pi} h_1(t)\frac{\sin zt}{z}\,dt = \int_0^{\pi} [h(t) + h_1(t)]\frac{\sin zt}{z}\,dt$$

indeed admits the representation (3.4.53).

Finally, we put

$$s_1(z) = u_1(z) - u_2(z) \ . \tag{3.4.58}$$

It follows from formulas (3.4.48), (3.4.53), and Lemma 3.4.2, that the roots $\pm\sqrt{\nu_1}, \pm\sqrt{\nu_2}, \ldots$ of this function satisfy the asymptotic equalities

$$\sqrt{\nu_k} = k - \frac{1}{2} - (\pi k)^{-1} d_1 + k^{-1}\beta_k \ , \quad \sum_{k=1}^{\infty} |\beta_k|^2 < \infty \ . \tag{3.4.59}$$

On the other hand, $\lim\limits_{z\to-\infty} s_1(\sqrt{z}) = +\infty$, and according to (3.4.54) and (3.4.55),

$$s_1(\sqrt{\lambda_k}) = u_1(\sqrt{\lambda_k}) - u_2(\sqrt{\lambda_k}) = (-1)^k \cosh h_k' - \sinh h_k' =$$

$$= (-1)^k \cosh h_k'[1 - (-1)^k \tanh h_k'] \ .$$

Thus, since $\tanh h_k' < 1$, it follows that $\operatorname{sign} s_1(\sqrt{\lambda_k}) = (-1)^k$. Hence, in each $(-\infty,\lambda_1),(\lambda_1,\lambda_2),(\lambda_2,\lambda_3),\ldots$ there is one and (by (3.4.59)) only one root of the function $s_1(\sqrt{z})$. This implies that the roots $\nu_1,\nu_2,\ldots$ of the function $s_1(\sqrt{z})$ and the roots $\lambda_1,\lambda_2,\ldots$ of the function $s(\sqrt{z})$ interlace. Using the asymptotic formulas (3.4.50) and (3.4.59), this permits us to conclude, by 3.4.1, that the sequences of roots in question are the spectra of the boundary value problems (3.4.2), (3.4.3) generated on the interval $(0,\pi)$ by one and the same equation (3.4.1) with real-valued potential $q(x) \in L_2[0,\pi]$, and that

$$s(\lambda) = s(\lambda,\pi) \quad , \quad s_1(\lambda) = s'(\lambda,\pi) \ , \tag{3.4.60}$$

where $s(\lambda,x)$ is the solution of this equation with initial data $s(\lambda,0) = 0$, $s'(\lambda,0) = 1$.

We next show that for the solution $c(\lambda,x)$ $(c(\lambda,0) = 1 \ , \ c'(\lambda,0) = 0)$ of this equation ,

$$c(\lambda,\pi) = u_1(\lambda) + u_2(\lambda) \ . \tag{3.4.61}$$

In fact, it follows from the identity $c(\lambda,\pi)s'(\lambda,\pi) - c'(\lambda,\pi)s(\lambda,\pi) = 1$ and (3.4.60) that

$$c(\pm\sqrt{\lambda_k},\pi) = [s'(\pm\sqrt{\lambda_k},\pi)]^{-1} = [s_1(\pm\sqrt{\lambda_k})]^{-1} \ ,$$

by (3.4.54), (3.4.55), (3.4.58),

$$u_1(\pm\sqrt{\lambda_k}) + u_2(\pm\sqrt{\lambda_k}) = [u_1(\pm\sqrt{\lambda_k})^2 - u_2(\pm\sqrt{\lambda_k})^2][u_1(\pm\sqrt{\lambda_k}) - u_2(\pm\sqrt{\lambda_k})]^{-1} =$$

$$= [\cosh^2 h_k' - \sinh^2 h_k'][s_1(\pm\sqrt{\lambda_k})]^{-1} = [s_1(\pm\sqrt{\lambda_k})]^{-1} \ .$$

Hence, the function

$$v(z) = c(z,\pi) - [u_1(z) + u_2(z)] = c(z,\pi) - s'(z,\pi) - [u_1(z) + u_2(z) - s_1(z)] =$$
$$= c(z,\pi) - s'(z,\pi) + 2u_2(z)$$

vanishes at the roots $\pm\sqrt{\lambda_k}$ of the function $s(z) = s(z,\pi)$ and, by (3.4.24) and (3.4.53), $|v(z)| = |z^{-1}| \exp\{\pi|\mathrm{Im}\, z|\} (1)$ as $|z| \to \infty$. Therefore, the function $v(z)[s(z)]^{-1}$ is also entire, and tends to zero on the sequence of contours $K_n = C(n + \frac{1}{2})$. By Liouville's theorem, $v(z)[s(z)]^{-1} \equiv 0$, which is equivalent to the claimed equality (3.4.61). Using formulas (3.4.60), (3.4.61) and (3.4.58), it is readily verified that the Hill discrimiant $u_+(z)$ of the equation obtained above is equal to

$$\frac{1}{2}[c(z,\pi) + s'(z,\pi)] = \frac{1}{2}[u_1(z) + u_2(z) + u_1(z) - u_2(z)] = u_1(z) ,$$

which in turn shows that the sequence $0 = \mu_0 < \mu_1^- \leq \mu_1^+ < \mu_2^- \leq \mu_2^+ < \ldots$ is indeed constructed from the spectra of the periodic and anti-periodic boundary value problems generated by this equation.

It remains to show that the potential belongs to $\tilde{W}_2^n[0,\pi]$. But, in view of inequalities (3.4.36) and the Corollary to Theorem 1.5.2, this is a straightforward consequence of condition (3.4.47).

Thus, we have completely solved problem B. As a matter of fact we have established a one-to-one correspondence between the set of real-valued potentials $q(x) \in \tilde{W}_2^n[0,\pi]$ and the set of all sequences of the form

$$\mu_0, h_1 ; \theta_1, h_2 ; \theta_2, \ldots h_k ; \theta_k, \ldots , \tag{3.4.62}$$

where μ_0 is an arbitrary real number, the h_k are arbitrary nonnegative numbers subject to the constraint $\sum_{k=1}^{\infty} (k^{n+1}h_k)^2 < \infty$, and θ_k is an arbitrary point on the two-sided boundary of the slit $\mathrm{Re}\,\theta_k = k\pi$, $-h_k \leq \mathrm{Im}\,\theta \leq h_k$. This correspondence is completely specified by the relations

$$\left.\begin{aligned} \cosh h_k &= \max_{\alpha_k^- \leq x \leq \alpha_k^+} |u_+(x)| , \\ \theta_k &= \lim_{\varepsilon \downarrow 0} \theta(\sqrt{\lambda_k - \mu_0} + i\varepsilon u_-(\sqrt{\lambda_k - \mu_0})) , \end{aligned}\right\} \tag{3.4.63}$$

where $\alpha_k^{\pm} = \sqrt{\mu_k^{\pm} - \mu_0}$, μ_0 and $\mu_k^{\pm}$ are the eigenvalues of the periodic and anti-periodic boundary value problems on the interval $(0,\pi)$ generated by the equation $-y'' + q(x)y = \mu y$, λ_k are the eigenvalues of the boundary value problem generated by the same equation and the boundary conditions (3.4.2), the function $\theta(z)$ is defined by formula (3.4.29),

$u_+(z) = \frac{1}{2}[c(z,\pi) + s'(z,\pi)]$, $u_-(z) = \frac{1}{2}[c(z,\pi) - s'(z,\pi)]$,

and $c(z,x)$, $s(z,x)$ is the fundamental system of solutions of the equation $-y'' + (q(x) - \mu_0)y = z^2 y$. The elements of the sequence (3.4.62) are obviously (nonlinear) functionals of the potentials $q(x)$:

$$\mu_0 = \mu_0(q) \quad , \quad h_k = h_k(q) \quad , \quad \theta_k = \theta_k(q) \ . \qquad (3.4.62')$$

They are continuous in the topology of $L_2[0,\pi]$. In fact, if the sequence $q_n(x)$ converges to $q(x)$ in the norm of $L_2[0,\pi]$, then the solutions $c_n(\lambda,x)$ and $s_n(\lambda,x)$ of the equation $-y'' + q_n(x)y = \lambda^2 y$, and their first-order derivatives converge to the corresponding solutions of the equation $-y'' + q(x)y = \lambda^2 y$ and their first-order derivatives, respectively, uniformly on every compact set $0 \leqslant x \leqslant \pi$, $|\lambda| \leqslant C$. Consequently,

$$\lim_{n\to\infty} \mu_0(q_n) = \mu_0(q) \quad , \quad \lim_{n\to\infty} \mu_k^{\pm}(q_n) = \mu_k^{\pm}(q) \quad , \quad \lim_{n\to\infty} \lambda_k(q_n) = \lambda_k(q) \ ,$$

and hence

$$\lim_{n\to\infty} u_+(z,q_n) = u_+(z,q) \quad , \quad \lim_{n\to\infty} u_-(z,q_n) = u_-(z,q)$$

uniformly on every compact set $|z| \leqslant C$. By (3.4.29), this implies that

$$\lim_{n\to\infty} \theta(z,q_n) = \theta(z,q)$$

uniformly on every compact set $|z| \leqslant C$. The continuity of the functionals (3.4.62') is now seen to be an obvious consequence of equalities (3.4.63).

The characterization, obtained in Theorem 3.4.2, of the spectra of the periodic and anti-periodic boundary value problems for equation (3.4.1) on $(0,\pi)$ also provides a complete description of the domain of stability of the Hill equation. The Hill equation is

$$-y'' + q(x)y = zy \quad (-\infty < x < \infty) \ , \qquad (3.4.64)$$

considered on the full real line, with real-valued periodic $(q(x+\pi) \equiv q(x))$ potential $q(x) \in \tilde{W}_2^n[0,\pi]$. The domain of stability of this equation is, by definition, the set of values of the parameter z for which all its solutions are bounded on the whole real line. It follows from the expression

$$y(x) = y(0)c(\sqrt{z}\ x) + y'(0)s(\sqrt{z},x)$$

for the general solution of equation (3.4.64), that

$$\begin{bmatrix} y(\pi) \\ y'(\pi) \end{bmatrix} = \begin{bmatrix} c(\sqrt{z},\pi) & s(\sqrt{z},\pi) \\ c'(\sqrt{z},\pi) & s'(\sqrt{z}, \ ) \end{bmatrix} \begin{bmatrix} y(0) \\ y'(0) \end{bmatrix} . \tag{3.4.65}$$

The matrix

$$U(z) = \begin{bmatrix} c(\sqrt{z},\pi) & s(\sqrt{z},\pi) \\ c'(\sqrt{z},\pi) & s'(\sqrt{z},\pi) \end{bmatrix}$$

is called the monodromy matrix of the Hill equation. The eigenvalues $\rho_1(z)$ and $\rho_2(z)$ are found from the equation

$$0 = \mathrm{Det}\,(U(z) - \rho I) =$$

$$= c(\sqrt{z},\pi)s'(\sqrt{z},\pi) - c'(\sqrt{z},\pi)s(\sqrt{z},\pi) - (c(\sqrt{z},\pi) + s'(\sqrt{z},\pi)) + \rho^2 =$$

$$= 1 - 2u_+(\sqrt{z})\rho + \rho^2 .$$

Therefore,

$$\rho_1(z) = u_+(\sqrt{z}) + \sqrt{u_+^2(\sqrt{z}) - 1} \ , \quad \rho_2(z) = u_+(\sqrt{z}) - \sqrt{u_-^2(\sqrt{z}) - 1} \ ,$$

whence, by (3.4.30'),

$$\rho_1(z) = \exp\{i\theta\sqrt{z - \mu_0}\} \ , \quad \rho_2(z) = \exp\{-i\theta\sqrt{z - \mu_0}\} .$$

If $u_+^2(\sqrt{z}) - 1 \neq 0$, i.e., if z is not an eigenvalue of the periodic or anti-periodic boundary value problems generated by equation (3.4.61) on the interval $(0,\pi)$, then $\rho_1(z) \neq \rho_2(z)$, and the corresponding eigenvectors $\begin{bmatrix} a_1 \\ b_1 \end{bmatrix}$ and $\begin{bmatrix} a_2 \\ b_2 \end{bmatrix}$ of the monodromy matrix are linearly independent. Consequently, the solutions $\psi_i(x,z)$, $i = 1,2$, of the Hill equation with initial conditions $\psi_i(x,0) = a_i$, $\psi'(x_i,0) = b_i$, are linearly independent. By (3.4.65), $\psi_i(z,\pi) = \rho_i(z)\psi_i(z,0)$ and $\psi_i'(z,\pi) = \rho_i(z)\psi_i'(z,0)$, whence, in view of the periodicity of the potential $(q(x+\pi) \equiv q(x))$,

$$\psi_i(z,x+\pi) \equiv \rho_i(z)\psi(z,x) ,$$

or, equivalently,

$$\varphi_i(z,x+\pi) \equiv \varphi_i(z,x) ,$$

where

$$\varphi_i(z,x) = \psi_i(z,x)\exp\{-\pi^{-1}x \ln \rho_i(z)\} .$$

Therefore, for the values of z considered, the Hill equation has a fundamental system of solutions of the form

$$\psi_1(z,x) = \varphi_1(z,x) \exp \{-\pi^{-1} x\theta(\sqrt{z - \mu_0})\}$$

and

$$\psi_2(z,x) = \varphi_2(z,x) \exp \{\pi^{-1} x\theta(\sqrt{z - \mu_0})\} ,$$

where $\varphi_i(z,x)$ are continuous periodic $(\varphi_i(z,x+\pi) \equiv \varphi_i(z,\pi))$ functions. These formulas show that the solutions $\psi_i(z,x)$ are bounded on the whole real line if and only if $\operatorname{Im} \theta(\sqrt{z - \mu_0}) = 0$. By the definition of the function $\theta(\sqrt{z - \mu_0})$, this implies that the intervals $(\mu_0,\mu_1^-),(\mu_1^+,\mu_2^-),\ldots,(\mu_k^+,\mu_{k+1}^-)$, ,... , are included in the domain of stability of the Hill equation, whereas all the other values $z \neq \mu_k^\pm$ belong to the domain of instability.

Now let us examine the points $\mu_k^\pm$. If $\mu_k^- = \mu_k^+$, then the function $u_+^2(\sqrt{z}) - 1$ has a double root at this point, and since $\mu_k^- \leqslant \lambda_k \leqslant \mu_k^+$ always, $s(\sqrt{\mu_k^\pm},\pi) = 0$ in this case. Hence, in view of (3.4.27), $c'(\sqrt{\mu_k^\pm}, \ ) = 0$, too. It follows that the functions $s(\sqrt{\mu_k^-},x)$ and $c(\sqrt{\mu_k^-},x)$ are periodic (for $k = 2m + 1$) solutions of the Hill equation, and hence are bounded on the real line. Therefore, the double roots $\mu_k^- = \mu_k^+$ of the equation $u_+^2(\sqrt{z}) - 1 = 0$ belong to the domain of stability, and for such roots the intervals of stability $(\mu_{k-1}^+,\mu_k^-),(\mu_k^+,\mu_{k+1}^-)$ merge into the interval $(\mu_{k-1}^+,\mu_{k+1}^-)$, all the points of which belong to the domain of stability. If $\mu_k^- < \mu_k^+$, then $u_+^2(\sqrt{z}) - 1$ has simple roots at these points, and it is readily verified that for $z = \mu_k^-$ and $z = \mu_k^+$, equation (3.4.46) has, besides the periodic (for $k = 2m$) or anti-periodic (for $k = 2m + 1$) solution $\psi(\mu_k^-,x)$ (respectively, $\psi(\mu_k^+,x)$), an unbounded solution $y(\mu_k^-,x)$ (respectively, $y(\mu_k^+,x)$) of the form

$$y(\mu_k^\pm,x) = \chi(\mu_k^\pm,x) + x\psi(\mu_k^\pm,x) ,$$

where $\chi(\mu_k^+,x)$ (respectively, $\chi(\mu_k^-,x)$) is periodic (respectively, anti-periodic). Therefore, the simple roots of the equation $u_+^2(\sqrt{z}) - 1 = 0$ belong to the domain of instability.

Summing up, the domain of stability of the Hill equation is the union of the intervals $(\mu_0,\mu_k^-),(\mu_{k_1}^+,\mu_{k_2}^-),(\mu_{k_2}^+,\mu_{k_3}^-),\ldots$, the endpoints of which coincide with the sequence $\mu_0 < \mu_{k_1}^- < \mu_{k_1}^+ < \mu_{k_2}^- < \mu_{k_2}^+ < \ldots$ of simple roots of the equation $u_+^2(\sqrt{z}) - 1 = 0$. These intervals are called zones of stability, and the segments $(-\infty,\mu_0),(\mu_{k_1}^-,\mu_{k_1}^+),\ldots$ that separate them are referred to, variously, as zones of instability, gaps, lacunae, or forbidden bonds. From this point of view, Theorem 3.4.2 obviously furnishes necessary and sufficient conditions for a given finite system of intervals $(a_0,a_1^-),(a_1^+,a_2^-),\ldots,(a_{N-1}^+,a_N^-),(a_N^+,\infty)$, to coincide with the domain of stability of a Hill equation (see Problem 4). Such equations and the corresponding potentials are termed finite-zone. Finite-zone potentials are obviously obtained when and only when the boundary of the domain $\Theta_+\{h_k\}$ contains only a finite number of vertical slits, i.e., when $h_k \neq 0$ only for finitely many values of k . It follows that such potentials are infinitely differentiable and the set $V[0,\pi]$, that they (or, more precisely, their restrictions to the segment $[0,\pi]$) form, is contained in the intersection of all the sets $\widetilde{W}_2^n[0,\pi]$:

$$V[0,\pi] \subset \bigcap_{n=0}^{\infty} \widetilde{W}_2^n[0,\pi] = W\,[0,\pi] .$$

THEOREM 3.4.3. *Every real-valued potential* $q(x) \in \widetilde{W}_2^n[0,\pi]$ *is the limit in* $\widetilde{W}_2^n[0,\pi]$ *of a sequence* $q_N(x)$ *of finite-zone potentials, i.e., the set* $V[0,\pi]$ *is dense in each of the spaces* $\widetilde{W}_2^n[0,\pi]$.

PROOF. Since $\widetilde{W}^\infty[0,\pi]$ is obviously dense in every space $\widetilde{W}_2^n[0,\pi]$, it suffices to show that for every real-valued potential $q(x) \in \widetilde{W}^\infty[0,\pi]$, there is a sequence $q_N(x)$ of finite-zone potentials which converges for $N \to \infty$ to $q(x)$ together with the corresponding derivatives of any order. By the foregoing discussion, the potential $q(x) \in \widetilde{W}^\infty[0,\pi]$ is uniquely determined by the sequence $\{\mu_0(q),h_k(q);\theta_k(q)\}$, which satisfies the inequalities

$$\sum_{k=1}^{\infty} (k^{n+1}h_k)^2 = A_n < \infty \tag{3.4.66}$$

for $n = 0,1,2,\ldots$. Starting with this sequence, we construct the truncated sequences

$$\{\mu_0(q),h_1(q)\ ;\ \theta_1(q),\ldots,h_N(q)\ ;\ \theta_N(q),0\ ;\ (N+1)\pi,0\ ;\ (N+2)\pi,\ldots\}$$

in which $h_k = 0$ for $k > N$. By Theorem 3.4.2, to each such sequence there corresponds a finite-zone potential $q_N(x)$ such that

$$\left.\begin{aligned} &\mu_0(q_N) = \mu_0(q)\ ,\quad h_k(q_N) = h_k(q)\ ,\quad \theta_k(q_N) = \theta_k(q)\quad (1 \leq k \leq N)\ ,\\ &h_k(q_N) = 0\quad (k = N+1,N+2,\ldots)\ . \end{aligned}\right\} \tag{3.4.67}$$

We will show below that the sequence $q_N(x)$ is compact in $\tilde{W}^{\infty}[0,\pi]$, i.e., it satisfies the inequalities

$$\sup_{-\infty<x<\infty} |q_N^{(n)}(x)| = \sup_{0\leq x\leq\pi} |q_N^{(n)}(x)| = C_n \quad (0 \leq n < \infty) , \tag{3.4.68}$$

in which the constants C_n are independent of N. Hence, one can extract from it a subsequence $q_{N'}(x)$ which converges uniformly, together with its derivatives of any order, to a function $\hat{q}(x) \in \tilde{W}^{\infty}[0,\pi]$. It follows from (3.4.67) and the continuity of the functionals $\mu_0(q)$, h_k, and $\theta_k(q)$, that

$$\mu_0(\hat{q}) = \lim_{N'\to\infty} \mu_0(q_{N'}) = \mu_0(q) \quad , \quad h_k(\hat{q}) = \lim_{N'\to\infty} h_k(q_{N'}) = h_k(q) ,$$

and

$$\theta_k(\hat{q}) = \lim_{N'\to\infty} \theta_k(q_{N'}) = \theta_k(q) .$$

Therefore, the potentials $\hat{q}(x)$ and $q(x)$ generate the same sequence $\{\mu_0(q),h_k(q);\theta_k(q)\}$. Hence, $\hat{q}(x) \equiv q(x)$, and the finite-zone potentials $q_{N'}(x)$ converge uniformly to $q(x)$ as $N' \to \infty$, together with the derivatives of any order.

Now let us prove the inequalities (3.4.68). Since $q_N(x) \in \tilde{W}^{\infty}[0,\pi]$ and

$$\int_0^{\pi} q_N^{(m)}(x)dx = q_N^{(m-1)}(\pi) - q_N^{(m-1)}(0) = 0$$

for every $m \geq 1$, there exist points $x_m \in [0,\pi]$ at which $q_N^{(m)}(x_m) = 0$. Consequently,

$$q_N^{(m)}(x) = \int_{x_m}^{x} q_N^{(m+1)}(t)dt$$

and

$$\max_{-\infty<x<\infty} |q_N^{(m)}(x)| \leq \pi \max_{-\infty<x<\infty} |q_N^{(m+1)}(x)| .$$

This shows that in order for the inequalities (3.4.68) to be valid, it suffices that the functions $\sigma_{2j-1}(N,x)$, constructed from the potentials $q_N(x)$ by means of the formulas (1.4.20), satisfy the inequalities

$$\max_{-\infty<x<\infty} |\sigma_{2j-1}(N,x)| \leq A_j \quad (j = 1,2,\ldots) , \tag{3.4.69}$$

in which the constants $A_j < \infty$ do not depend on N. It follows from the periodicity of the potential $q_N(x) \equiv q_N(x+\pi)$ that the eigenvalues $\mu_0(q_N) = \mu_0(q)$, $\mu_k^{\pm}(q_N)$ of the periodic and anti-periodic boundary value problems generated by the equation

$$-y'' + q_N(x+t)y = \mu y \tag{3.4.70}$$

on the interval $0 \leqslant x \leqslant \pi$, do not depend on the parameter $t \in (-\infty,\infty)$. Consequently, the elements μ_0, h_k of the sequences corresponding to the potentials $q_N(x+t)$ are also independent of t. The elements θ_k and the eigenvalues $\lambda_k(N,t)$ of the boundary value problem (3.4.2) for equation (3.4.70) depend on t, but since the eigenvalues interlace,

$$\mu_k^-(q_N) \leqslant \lambda_k(N,t) \leqslant \mu_k^+(q_N) \tag{3.4.71}$$

for all $t \in (-\infty,\infty)$.

Inequalities (3.4.69) are proved using the trace formulas (1.5.35) and (1.5.35') which, for equation (3.4.70), take the form

$$\Delta_m(N,t) = (\mu_0(q))^m + \sum_{k=1}^{\infty} \{(\mu_k^+(q_N))^m + (\mu_k^-(q_N))^m - 2(\lambda_k(N,t))^m\},$$

and

$$\Delta_m(N,t) = -2\left[\sum_{j=0}^{m-2} \sigma_{2j+1}(N,t)(-4)^{-j-1}\Delta_{m-j-1}(N,t) + 2m\sigma_{2m-1}(N,t)(-4)^m\right]. \tag{3.4.72}$$

The recursion formulas (3.4.72) show that the functions $\sigma_{2j-1}(N,t)$ are polynomials in $\Delta_1(N,t),\dots,\Delta_j(N,t)$, from which it follows that for inequalities (3.4.69) to hold, it suffices that

$$\sup_{-\infty<t<\infty} |\Delta_m(N,t)| < D_m \quad (m = 1,2,\dots)$$

with constants $D_m < \infty$ that do not depend on N. By (3.4.71),

$$|(\mu_k^+(q_N))^m + (\mu_k^-(q_N))^m - 2(\lambda_k(N,t))^m| \leqslant |(\mu_k^+(q_N))^m - (\mu_k^-(q_N))^m| \leqslant$$

$$\leqslant |\mu_k^+(q_N) - \mu_k^-(q_N)| \sum_{j=0}^{m-1} |\mu_k^+(q_N)|^{m-j-1}|\mu_k^-(q_N)|^j,$$

where

$$\mu_k^{\pm}(q_N) = \mu_0(q_N) + [\alpha_k^{\pm}(q_N)]^2 = \mu_0(q) + [\alpha_k^{\pm}(q_N)]^2$$

and

$$\alpha_k^{\pm}(q_N) = z_{q_N}(k\pi \pm 0).$$

Using the estimates (3.4.36), (3.4.37) proved in Lemma 3.4.4, we get

$$|\mu_q^+(q_N) - \mu_k^-(q_N)| = |\alpha_k^+(q_N) - \alpha_k^-(q_N)||\alpha_k^+(q_N) + \alpha_k^-(q_N)| \leq$$
$$\leq \frac{1}{\pi}(1 + \frac{2}{\pi} H(q_N))kh_k(q_N)$$

and

$$|\mu_k^\pm(q_N)| \leq |\mu_0(q)| + |\alpha_k^\pm(q_N)|^2 \leq |\mu_0(q)| + (1 + \frac{2}{\pi} H(q_N))^2k^2 \leq$$
$$\leq k^2\{|\mu_0(q)| + (1 + \frac{2}{\pi} H(q_N))^2\} .$$

Therefore, since $h_k(q_N) = h_k(q)$ $(1 \leq k \leq N)$, $h_k(q_N) = 0$ $(k = N+1,N+2,\ldots)$, and $H(q_N) = \max h_k(q_N) \leq \max h_k(q) = H(q)$, it follows that

$$|(\mu_k^+(q_N))^m + (\mu_k^-(q_N))^m - 2(\lambda_k(N,t))^m| \leq$$
$$\leq \frac{m}{2} (1 + \frac{2}{\pi} H(q))\{|\mu_0(q)| + (1 + \frac{2}{\pi} H(q))^2\}^{m-1}k^{2m-1}h_k(q) \leq$$
$$\leq \frac{m}{2} (1 + \frac{2}{\pi} H(q))^{2m-1}(1 + |\mu_0(q)|^{m-1})k^{2m-1}h_k(q) .$$

Hence,

$$|\Delta_m(N,t)| \leq |\mu_0(q)|^m + \frac{m}{2} (1 + \frac{2}{\pi} H(q))^{2m-1}(1 + |\mu_0(q)|) \sum_{k=1}^{\infty} k^{2m-1}h_k(q) = D_m ,$$

where the constants D_m are obviously independent of N and finite, because by (3.4.66),

$$\sum_{k=1}^{\infty} h_k(q)k^{2m-1} \leq \left\{ \sum_{k=1}^{\infty} (h_k(q)k^{2m})^2 \right\}^{\frac{1}{2}} \left\{ \sum_{k=1}^{\infty} k^{-2} \right\}^{\frac{1}{2}} = \pi \left(\frac{A_{2m-1}}{6} \right)^{\frac{1}{2}} .$$

This completes the proof of the theorem. □

PROBLEMS

1. Show that the eigenvalues $\mu_k(\beta_1)$ and $\mu_k(\beta_2)$ of the boundary value problems generated by the equation

$$-y'' + q(x)y = zy \quad (0 \leq x \leq \pi \quad , \quad \text{Im } q(x) = 0)$$

and the boundary conditions

$$y'(0) \sin \alpha - y(0) \cos \alpha = 0 \quad , \quad y'(\pi) \sin \beta_1 + y(\pi) \cos \beta_1 = 0 ,$$

and

$$y'(0) \sin \alpha - y(0) \cos \alpha = 0 \quad , \quad y'(\pi) \sin \beta_2 + y(\pi) \cos \beta_2 = 0 ,$$

where $0 \leq \beta_1 < \beta_2 < \pi$, $0 \leq \alpha < \pi$, interlace:

$$-\infty < \mu_1(\beta_2) < \mu_1(\beta_1) < \mu_2(\beta_2) < \mu_2(\beta_1) < \dots . \tag{3.4.73}$$

Hint. The characteristic functions of the indicated boundary value problems satisfy the equalities

$$\chi_i(z) = \varphi'(\pi,z) \sin \beta_i + \varphi(\pi,z) \cos \beta_i \quad (i = 1,2) ,$$

where

$$\varphi(x,z) = c(\sqrt{z},x) \sin \alpha + s(\sqrt{z},x) \cos \alpha .$$

Since φ and $\dot{\varphi} = \frac{\partial}{\partial z}\varphi(x,z)$ are solutions of the equations

$$-\varphi'' + q(x)\varphi = z\varphi \quad , \quad -\dot{\varphi}'' + q(x)\dot{\varphi} = z\dot{\varphi} + \varphi ,$$

it follows readily that

$$\int_0^\pi [\varphi(x,z)]^2 dx = \dot{\varphi}(\pi,z)\varphi'(\pi,z) - \dot{\varphi}'(\pi,z)\varphi(\pi,z) =$$

$$= \frac{\dot{\chi}_1(z)\chi_2(z) - \chi_1(z)\dot{\chi}_2(z)}{\sin(\beta_2 - \beta_1)} . \tag{3.4.74}$$

Moreover, the self-adjointness of the indicated boundary value problems implies that their eigenvalues, i.e., the roots of the characteristic functions, are real. Since for real values of z the left-hand side of equality (3.4.74) is strictly positive, the roots of the characteristic functions are all simple. Upon rewriting (3.4.74) in the form

$$\frac{\sin(\beta_2 - \beta_1)}{[\chi_2(z)]^2} \int_0^\pi [\varphi(x,z)]^2 dx = \frac{d}{dx}\left\{ \frac{\chi_1(x)}{\chi_2(x)} \right\} ,$$

and observing that, for real values of z , the left-hand side of the last equality is positive, you deduce that the function $\chi_1(z)/\chi_2(x)$ increases strictly from $\lim_{z\to-\infty} \chi_1(z)/\chi_2(z)$ to $+\infty$ in the interval $(-\infty,\mu_1(\beta_2))$, and from $-\infty$ to $+\infty$ in the intervals $(\mu_k(\beta_2),\mu_{k+1}(\beta_2))$. Since $\lim_{z\to-\infty} \frac{\varphi(\pi,z)}{\varphi'(\pi,z)} =$ $= 0$, you have $\lim_{z\to-\infty} \frac{\chi_1(z)}{\chi_2(z)} = \frac{\sin \beta_1}{\sin \beta_2} \geq 0$. Hence, the function $\chi_1(z)$ has no roots in the interval $(-\infty,\mu_1(\beta_2))$, and has a single root $\mu_k(\beta_1)$ in each interval $(\mu_2(\beta_2),\mu_{k+1}(\beta_2))$, which implies that the eigenvalues of the indicated boundary value problems indeed satisfy the inequalities (3.4.73).

2. Prove inequalities (3.4.7).

<u>Hint</u>. The eigenvalues λ_k of the boundary value problem (3.4.2) are the roots of the functions $s(\sqrt{z},\pi)$, while the eigenvalues $\mu_k^{\pm}$ of the periodic and anti-periodic boundary value problems are the roots of the equation

$u_+(\sqrt{z}) = \frac{1}{2}[c(\sqrt{z},\pi) + s'(\sqrt{z},\pi)] = (-1)^k$.

Letting $z = \lambda_k$ in the identity

$$c(\sqrt{z},\pi)s'(\sqrt{z},\pi) - c'(\sqrt{z},\pi)s(\sqrt{z},\pi) = 1 ,$$

you get

$$c(\sqrt{\lambda_k},\pi)s'(\sqrt{\lambda_k},\pi) = 1$$

and

$$u_+(\sqrt{\lambda_k}) = \frac{1}{2}\left[\frac{1}{s'(\sqrt{\lambda_k},\pi)} + s'(\sqrt{\lambda_k},\pi)\right] ,$$

which in turn implies that

$$u_+(\sqrt{\lambda_k}) \operatorname{sign} s'(\sqrt{\lambda_k},\pi) \geqslant 1 .$$

From equalities (3.4.74) with $\alpha = 0$, $\beta_1 = 0$, $\beta_2 = \frac{\pi}{2}$, $z = \lambda_k$, it follows that

$$0 < \int_0^{\pi} [s(\sqrt{\lambda_k},x)]^2 dx = \dot{s}(\sqrt{\lambda_k},\pi)s'(\sqrt{\lambda_k},\pi) .$$

Hence, $\operatorname{sign} s'(\sqrt{\lambda_k},\pi) = \operatorname{sign} \dot{s}(\sqrt{\lambda_k},\pi)$, and since the function $s(\sqrt{z},\pi)$ has only simple roots $\lambda_1 < \lambda_2 < \dots$, and $\lim_{z\to-\infty} s(\sqrt{z},\pi) = +\infty$, you see that $\dot{s}(\sqrt{\lambda_1},\pi) \quad 0$, $\dot{s}(\sqrt{\lambda_2},\pi) \quad 0$, $\dot{s}(\sqrt{\lambda_3},\pi) < 0$,... . Thus, $\operatorname{sign} s'(\sqrt{\lambda_k},\pi) = \operatorname{sign} \dot{s}(\sqrt{\lambda_k},\pi) = (-1)^k$ and $u_+(\sqrt{\lambda_k})(-1)^k \geqslant 1$, i.e., $u_+(\sqrt{\lambda_1}) \leqslant -1$, $u_+(\sqrt{\lambda_2}) \geqslant 1$, $u_+(\sqrt{\lambda_3}) \leqslant -1$,... . Since $\lim_{z\to-\infty} u_+(\sqrt{z}) = +\infty$, these inequalities show that in each of the intervals $(-\infty,\lambda_1]$, $[\lambda_1,\lambda_2]$, $[\lambda_2,\lambda_3]$,... , the function $u_+(\sqrt{z})$ takes at least one of the values +1 and -1 . Letting $\mu_k^+ \in [\lambda_k,\lambda_{k-1}]$ $(\mu_k^- \in [\lambda_{k-1},\lambda_k])$ denote the root of the equation $u_+(\sqrt{z}) = (-1)^k$ which is closest to λ_k , you obtain

$$\mu_0 < \mu_1^- \leqslant \lambda_1 \leqslant \mu_1^+ < \mu_2^- \leqslant \lambda_2 \leqslant \mu_2^+ < \dots .$$

Finally, it follows from the asymptotic formulas for the eigenvalues of the periodic and anti-periodic boundary value problems (Theorem 1.5.2) that the sequence $\{\mu_k^{\pm}\}$ includes all the eigenvalues of these problems.

3. Generalize Theorem 3.4.1 as follows: *The sequences* $\{\mu_k(1)\}$ *and* $\{\mu_k(2)\}$ *are the spectra of the boundary value problems generated by the same equation* $-y'' + q(x)y = \mu y$ $(0 \leq x \leq \pi)$ *with real-valued potential* $q(x) \in W_2^n[0,\pi]$ *and the boundary conditions*

$$y'(0) - hy(0) = 0 \quad , \quad y'(\pi) + H_1 y(\pi) = 0$$

and

$$y'(0) - hy(0) = 0 \quad , \quad y'(\pi) + H_2 y(\pi) = 0 \ ,$$

respectively ($\operatorname{Im} h = \operatorname{Im} H_1 = \operatorname{Im} H_2 = 0$), *if and only if they interlace and satisfy the asymptotic relations of* Problem 2, Section 5, Chapter 1.

Hint. The necessity of these conditions follows from Problem 1 in this section and Problem 2, Section 5, Chapter 1. The sufficiency part can be proved, as in the proof of Theorem 3.4.1, by reduction to the inverse scattering problem for the equation $-y'' + q(x)y = \lambda^2 y$, $y'(0) - hy(0) = 0$, $0 \leq x < \infty$ (see Problem 1, Section 3, Chapter 3). Here, instead of Lemma 3.4.1, you should use the formula

$$S_h(\lambda) = -\frac{\omega'(\lambda,\pi;h) + i\lambda\omega(\lambda,\pi;h)}{\omega'(\lambda,\pi;h) - i\lambda\omega(\lambda,\pi;h)} e^{-2i\lambda\pi} \ .$$

4. Find necessary and sufficient conditions for a sequence of intervals $(a_0,a_1^-),(a_1^+,a_2^-),\ldots,(a_{N-1}^+,a_N^-),(a_N^+,\infty)$ to be the domain of a Hill equation.

Hint. From the given sequence, build the polynomials

$$T(z) = \prod_{k=1}^{N} (z^2 - c_k^-)(z^2 - c_k^+) \quad , \quad P(z) = \prod_{k=1}^{N} (z^2 - p_k) \ ,$$

where $c_k^\pm = a_k^\pm - a_0$ and the numbers $p_k \in [c_k^-,c_k^+]$ are determined from the equation

$$\int_{\sqrt{c_k^-}}^{\sqrt{c_k^+}} \frac{P(x)}{\sqrt{T(x)}}\, dx = 0 \quad (k = 1,2,\ldots,N) \ .$$

Then the function

$$\theta(z) = \pi \int_0^z \frac{P(\xi)}{\sqrt{T(\xi)}}\, d\xi$$

will map the upper half plane conformally onto the domain

$$\{\theta : \operatorname{Im}\theta > 0\} \setminus \bigcup_{k=1}^{N} \{\theta : \operatorname{Re}\theta = \pm m_k \ , \ 0 \leqslant \operatorname{Im}\theta \leqslant h_k\} \ ,$$

where

$$m_k = \int_0^{\sqrt{c_k^-}} \frac{P(x)}{\sqrt{T(x)}} \ , \quad h_k = \frac{\pi}{i} \int_{\sqrt{c_k^-}}^{\sqrt{p_k}} \frac{P(x)}{\sqrt{T(x)}} dx \ .$$

5. THE INVERSE PROBLEM OF SCATTERING THEORY ON THE FULL LINE

We next consider the differential equation

$$-y'' + q(x)y = \lambda^2 y \quad (-\infty < x < \infty) \tag{3.5.1}$$

with a real-valued potential $q(x)$ which satisfies the condition

$$\int_{-\infty}^{\infty} (1 + |x|)|q(x)|dx < \infty \ . \tag{3.5.2}$$

By Lemmas 3.1.1 and 3.1.2, this equation has solutions $e^+(\lambda,x)$ and $e^-(-\lambda,x)$ which can be represented in the form

$$e^+(\lambda,x) = e^{i\lambda x} + \int_x^{\infty} K^+(x,t)e^{i\lambda t}dt$$

and

$$e^-(-\lambda,x) = e^{-i\lambda x} + \int_{-\infty}^{x} K^-(x,t)e^{-i\lambda t}dt \ , \tag{3.5.3}$$

for every choice of λ in the closed upper half plane. Here the kernels $K^{\pm}(x,t)$ and their derivatives are subject to the estimates

$$|K^{\pm}(x,t)| \leqslant \frac{1}{2}\sigma^{\pm}\left(\frac{x+t}{2}\right) \exp\left\{\sigma_1^{\pm}(x) - \sigma_1^{\pm}\left(\frac{x+t}{2}\right)\right\}$$

and

$$\left|\frac{\partial K\ (x_1,x_2)}{x_i} \pm \frac{1}{4} q\left(\frac{x_1 + x_2}{2}\right)\right| \leqslant \frac{1}{2}\sigma_1^{\pm}(x_1)\sigma^{\pm}\left(\frac{x_1 + x_2}{2}\right) \exp \sigma_1^{\pm}(x_1) \ ,$$

where

$$\sigma^+(x) = \int_x^{\infty} |q(t)|dt \ , \quad \sigma_1^+(x) = \int_x^{\infty} \sigma^+(t)dt \ , \quad \sigma^-(x) = \int_{-\infty}^{x} |q(t)|dt \ ,$$

and

$$\sigma_1^-(x) = \int_{-\infty}^{x} \sigma^-(t)dt \ .$$

Moreover,

$$K^+(x,x) = \frac{1}{2}\int_x^{\infty} q(t)dt \quad , \quad K^-(x,x) = \frac{1}{2}\int_{-\infty}^{x} q(t)dt .$$

Since for real $\lambda \neq 0$ both pairs of functions $e^+(\lambda,x)$, $e^+(-\lambda,x)$ and $e^-(\lambda,x)$, $e^-(-\lambda,x)$ form a fundamental system of solutions of equation (3.5.1) with Wronskian equal to $2i\lambda$, we can write

$$e^+(\lambda,x) = b(\lambda)e^-(-\lambda,x) + a(\lambda)e^-(\lambda,x) \tag{3.5.4}$$

and

$$e^-(-\lambda,x) = -b(-\lambda)e^+(\lambda,x) + a(\lambda)e^+(-\lambda,x) , \tag{3.5.4'}$$

where

$$a(\lambda) = \frac{W\{e^+(\lambda,x),e^-(-\lambda,x)\}}{2i\lambda} = \frac{e^+(\lambda,0)'e^-(-\lambda,0) - e^+(\lambda,0)e^-(-\lambda,0)'}{2i\lambda} \tag{3.5.5}$$

and

$$b(\lambda) = \frac{W\{e^-(\lambda,x),e^+(\lambda,x)\}}{2i\lambda} = \frac{e^-(\lambda,0)'e^+(\lambda,0) - e^-(\lambda,0)e^+(\lambda,0)'}{2i\lambda} , \tag{3.5.6}$$

because the Wronskians are independent of x . Furthermore, it follows from the equalities $\overline{e^{\pm}(\lambda,x)} = e^{\pm}(-\lambda,x)$, which are consequences of the real-valuedness of the potential $q(x)$, that $a(\lambda) = \overline{a(-\lambda)}$ and $b(\lambda) = \overline{b(-\lambda)}$. Hence,

$$2i\lambda = W\{e^+(\lambda,x),e^+(-\lambda,x)\} = -2i\lambda\{b(\lambda)b(-\lambda) - a(\lambda)a(-\lambda)\} =$$
$$= 2i\lambda\{|a(\lambda)|^2 - |b(\lambda)|^2\}$$

and

$$|a(\lambda)|^2 = 1 + |b(\lambda)|^2 = 1 + |b(-\lambda)|^2 . \tag{3.5.7}$$

Upon dividing both sides of equalities (3.5.4) and (3.5.4') by $a(\lambda)$, we obtain the following solutions of equation (3.5.1), which are defined for all real $\lambda \neq 0$:

$$u^-(\lambda,x) = t(\lambda)e^+(\lambda,x) = r^-(\lambda)e^-(-\lambda,x) + e^-(\lambda,x) , \tag{3.5.8}$$

and

$$u^+(\lambda,x) = t(\lambda)e^-(-\lambda,x) = r^+(\lambda)e^+(\lambda,x) + e^+(-\lambda,x) , \tag{3.5.8'}$$

where

$$r^-(\lambda) = \frac{b(\lambda)}{a(\lambda)} \quad , \quad r^+(\lambda) = -\frac{b(-\lambda)}{a(\lambda)} \quad , \quad t(\lambda) = \frac{1}{a(\lambda)} . \tag{3.5.9}$$

These solutions have the asymptotics

$$u^-(\lambda,x) \simeq r^-(\lambda)e^{-i\lambda x} + e^{i\lambda x} \quad (x \to -\infty) ,$$

$$u^-(\lambda,x) \simeq t(\lambda)e^{i\lambda x} \quad (x \to +\infty) ,$$

$$u^+(\lambda,x) \simeq t(\lambda)e^{-i\lambda x} \quad (x \to -\infty) ,$$

and

$$u^+(\lambda,x) \simeq r^+(\lambda)e^{i\lambda x} + e^{-i\lambda x} \quad (x \to +\infty) ;$$

$u^-(\lambda,x)$ $(u^+(\lambda,x))$ are called the eigenfunctions of the left (respectively, right) scattering problem, and the coefficients $r^-(\lambda)$, $r^+(\lambda)$, and $t(\lambda)$ are called the left and right reflection coefficients and the transmission coefficient, respectively.

For complex values of λ in the upper half-plane, the function $e^+(\lambda,x)$ $(e^-(-\lambda,x))$ is the unique solution of equation (3.5.1) which belongs to $L_2[0,\infty)$ (respectively, $L_2(-\infty,0]$). Therefore, (3.5.1) has a solution belonging to $L_2(-\infty,\infty)$ only for those values of λ $(\mathrm{Im}\,\lambda > 0)$ for which the functions $e^+(\lambda,x)$ and $e^-(-\lambda,x)$ are linearly dependent, i.e., for which $W\{e^+(\lambda,x),e^-(-\lambda,x)\} = 2i\lambda a(\lambda) = 0$. It follows from formula (3.5.5) that the function $a(\lambda)$ is analytic in the upper half plane, and its zeros form a discrete set. If $a(\lambda_k) = 0$ $(\mathrm{Im}\,\lambda_k > 0)$, then

$$W\{e^+(\lambda_k,x),e^-(-\lambda_k,x)\} = 0 ,$$

the solutions

$$u^-(\lambda_k,x) = e^-(-\lambda_k,x) \quad , \quad u^+(\lambda_k,x) = e^+(\lambda_k,x) , \tag{3.5.10}$$

are linearly independent:

$$u^-(\lambda_k,x) = c_k^- u^+(\lambda_k,x) \quad , \quad u^+(\lambda_k,x) = c_k^+ u^-(\lambda_k,x) \quad , \quad c_k^- c_k^+ = 1 , \tag{3.5.10'}$$

and belong to $L_2(-\infty,\infty)$. The functions $u^-(\lambda_k,x)$ and $u^+(\lambda_k,x)$ are called the left and right eigenfunctions of the discrete spectrum, which in turn consists of the eigenvalues $\mu_k = \lambda_k^2$. Proceeding in the standard manner, it is readily verified, using the formal self-adjointness of equation (3.5.1) (the reality of $q(x)$), that the eigenvalues μ_k are real. Consequently, $\lambda_k = i\kappa_k$, $\kappa_k > 0$, and the zeros of the function $a(\lambda)$ must all lie on the imaginary semi-axis $i\kappa$, $\kappa > 0$. Furthermore, by formula (3.1.37),

$$\int_{-\infty}^{\infty} |u^-(i\kappa_k,x)|^2 dx = \int_{-\infty}^{0} |e^-(-i\kappa_k,x)|^2 dx + (c_k^-)^2 \int_0^{\infty} |e^+(i\kappa_k,x)|^2 dx =$$

$$= \frac{1}{2i\kappa_k} W\{e^-(-\lambda,0),\dot{e}^-(-\lambda,0)\}\Big|_{\lambda=i\kappa_k} - \frac{(c_k^-)^2}{2i\kappa_k} W\{e^+(\lambda,0),\dot{e}^+(\lambda,0)\}\Big|_{\lambda=i\kappa_k} =$$

$$= \frac{c_k^-}{2i\kappa_k} (W\{e^+(\lambda,0),\dot{e}^-(-\lambda,0)\} + W\{\dot{e}^+(\lambda,0),e^-(-\lambda,0)\})\Big|_{\lambda=i\kappa_k} =$$

$$= \frac{c_k^-}{2i\kappa_k} \frac{\partial}{\partial\lambda} W\{e^+(\lambda,0),\dot{e}^-(-\lambda,0)\}\Big|_{\lambda=i\kappa_k} = \frac{c_k^-}{2i\kappa_k}\frac{\partial}{\partial\lambda}\{2i\lambda a(\lambda)\}\Big|_{\lambda=i\kappa_k} ,$$

whence, since $a(i\kappa_k) = 0$,

$$\int_{-\infty}^{\infty} |u^-(i\kappa_k,x)|^2 dx = ic_k^-\dot{a}(i\kappa_k) \quad , \quad \int_{-\infty}^{\infty} |u^+(i\kappa_k,x)|^2 dx = ic_k^+\dot{a}(i\kappa_k) . \tag{3.5.11}$$

Therefore, $\dot{a}(i\kappa_k) \neq 0$, and hence the zeros of the function $a(z)$ are simple. Hereafter, the reciprocals of the norms of the eigenfunctions $u^-(i\kappa_k,x) = e^-(-i\kappa_k,x)$ and $u^+(i\kappa_k,x) = e^+(i\kappa_k,x)$ of the discrete spectrum will be denoted by m_k^- and m_k^+ , respectively, so that

$$(m_k^-)^{-2} = \int_{-\infty}^{\infty} |e^-(i\kappa_k,x)|^2 dx \quad , \quad (m_k^+)^{-2} = \int_{-\infty}^{\infty} |e^+(i\kappa_k,x)|^2 dx . \tag{3.5.12}$$

The collections $\{r^-(\lambda),i\kappa_k,m_k^-\}$ and $\{r^+(\lambda),i\kappa_k,m_k^+\}$ are called the left, and right scattering data, respectively, of equation (3.5.1). The inverse scattering problem for this equation is to recover the potential $q(x)$ given either the left or the right scattering data, and to find necessary and sufficient conditions for an arbitrary collection $\{r(\lambda),i\kappa_k,m_k\}$ to be the left (or right) scattering data of an equation of the form (3.5.1) with a real-valued potential q which is subject to the constraint (3.5.2).

The transmission coefficient $t(\lambda) = [a(\lambda)]^{-1}$ does not figure in the scattering data. It nevertheless plays an essential role in the investigation of these data and in solving the inverse problem. For this reason we first study the properties of this coefficient (for the sake of convenience we formulate them in terms of the function $a(z)$).

LEMMA 3.5.1. *The coefficients* $a(\lambda)$ *and* $b(\lambda)$, *which are defined by formulas* (3.5.5) *and* (3.5.6) *respectively, admit the representations*

$$a(\lambda) = 1 - (2i\lambda)^{-1} \left\{ \int_{-\infty}^{\infty} q(x)dx + \int_{-\infty}^{0} A(t)e^{-i\lambda t}dt \right\} \tag{3.5.13}$$

and

$$b(\lambda) = (2i\lambda)^{-1} \int_{-\infty}^{\infty} B(t)e^{-i\lambda t}dt \ , \tag{3.5.13'}$$

where $A(t) \in L_1(-\infty,0]$ *and* $B(t) \in L_1(-\infty,\infty)$, *respectively.*

PROOF. By Lemmas 3.1.1 and 3.1.2,

$$e^+(\lambda,0)'e^-(-\lambda,0) = i\lambda + i\lambda \int_{-}^{0} K^-(0,t)e^{-i\lambda t}dt - K^+(0,0) -$$

$$- K^+(0,0) \int_{-\infty}^{0} K^-(0,t)e^{-i\lambda t}dt + \int_{0}^{\infty} K_x^+(0,t)e^{i\lambda t}dt \left[1 + \int_{-\infty}^{0} K^-(0,t)e^{-i\lambda t}dt\right] =$$

$$= i\lambda - K^-(0,0) - K^+(0,0) + \int_{-\infty}^{0} K_t^-(0,t)e^{-i\lambda t}dt - K^+(0,0) \int_{-\infty}^{0} K^-(0,t)e^{-i\lambda t}dt +$$

$$+ \int_{-\infty}^{0} K_x^+(0,-t)e^{-i\lambda t}dt \left[1 + \int_{-\infty}^{0} K^-(0,t)e^{-i\lambda t}dt\right] dt \ .$$

Moreover, the functions $K^{\pm}(0,\mp t)$, $K_x^{\pm}(0,\mp t)$, and $K_t^-(0,t)$ belong to $L_1(-\infty,\infty)$ and vanish for $t > 0$. Since the product of the Fourier transforms of two summable functions that vanish on the positive half-line is equal to the Fourier transform of their convolution, which is also summable and vanishes on the positive half-line, this implies that

$$e^+(\lambda,0)'e^-(-\lambda,0) = i\lambda - K^-(0,0) - K^+(0,0) + \int_{-\infty}^{0} A_1(t)e^{-i\lambda t}dt \ ,$$

where $A_1(t) \in L_1(-\infty,0]$. The equality

$$e^+(\lambda,0)e^-(-\lambda,0)' = -i\lambda + K^-(0,0) + K^+(0,0) + \int_{-\infty}^{0} A_2(t)e^{-i\lambda t}dt \ ,$$

in which $A_2(t) \in L_1(-\infty,0]$, is proved in a similar manner. Now, upon substituting these expressions into (3.5.5) and observing that

$$2\{K^-(0,0) + K^+(0,0)\} = \int_{-\infty}^{\infty} q(x)dx \ ,$$

we obtain representation (3.5.13), in which $A(t) = A_1(t) + A_2(t) \in L_1(-\infty,0]$.

The representation (3.5.13') is derived in much the same way. □

It follows from (3.5.13) that $a(\lambda) = 1 + o(\lambda^{-1})$ for $|\lambda| \to \infty$ ($\mathrm{Im}\,\lambda \geq 0$), while equality (3.5.7) implies that $|a(\lambda)|^{-1} \leq 1$ for all real $\lambda \neq 0$. These estimates permit us to prove the following analog of Lemma 3.1.6.

LEMMA 3.5.2. *The function* $a(\lambda)$ *may have only finitely many zeros in the half-plane* $\operatorname{Im}\lambda > 0$. *All these zeros are simple and lie on the imaginary half-line. The function* $[a(\lambda)]^{-1}$ *is bounded in a neighborhood of zero.*

These assertions are proved by the same method that was used to prove Lemma 3.1.6 (see Problem 1 at the end of this section).

□

We next deduce the fundamental integral equation, which allows us to recover the potential $q(x)$ from the given left or right scattering data. We first remark that, by (3.5.7) and (3.5.9), $|r^{\pm}(\lambda)| < 1$ for all real $\lambda \neq 0$, whereas by (3.5.13) and (3.5.13'), $r^{\pm}(\lambda) = \ (\lambda^{-1})$ for $\lambda \to \pm\infty$. It follows that $r^{\pm}(\lambda) \in L_2(-\infty,\infty)$, and hence that the functions

$$R^{\pm}(y) = \frac{1}{2\pi}\int_{-\infty}^{\infty} r^{\pm}(\lambda)e^{\pm i\lambda y}d\lambda \tag{3.5.14}$$

also belong to $L_2(-\infty,\infty)$.

To derive the fundamental equation, we use the identity (3.5.8'), which we reexpress as

$$\left(\frac{1}{a(\lambda)} - 1\right)e^{-}(-\lambda,x) = r^{+}(\lambda)e^{+}(\lambda,x) + e^{+}(-\lambda,x) - e^{-}(-\lambda,x) \ . \tag{3.5.15}$$

It follows from (3.5.3) and (3.5.14) that the right-hand side of this identity is, for each fixed $x \in (-\infty,\infty)$, the Fourier transform of the function

$$R^{+}(x+y) + \int_{x}^{\infty} R^{+}(y+t)K^{+}(x,t)dt + K^{+}(x,y) - K^{-}(x,y) \ . \tag{3.5.16}$$

Therefore, upon multiplying the left-hand side of identity (3.5.15) by $(2\pi)^{-1}e^{i\lambda y}$ and integrating it over $-\infty < \lambda < \infty$, we should get (3.5.16). Since the function $[a(\lambda)]^{-1} - 1$ is analytic in the upper half-plane appart from a finite number of simple poles $i\kappa_k$, it tends to zero as $|\lambda| \to \infty$ ($\operatorname{Im}\lambda \geqslant 0$), and is bounded in a neighborhood of zero, and since the function $e^{-}(-\lambda,x)e^{i\lambda y}$ is uniformly bounded in the half plane $\operatorname{Im}\lambda \geqslant 0$ for $y > x$, an application of Jordan's lemma gives, for $y > x$,

$$\frac{1}{2\pi}\int_{-\infty}^{\infty}\left(\frac{1}{a(\lambda)} - 1\right)e^{-}(-\lambda,x)e^{i\lambda y}dy =$$

$$= i\sum_{k=1}^{n}\frac{e^{-}(-i\kappa_k,x)e^{-\kappa_k y}}{\dot{a}(i\kappa_k)} = -\sum_{k=1}^{n}\frac{e^{+}(i\kappa_k,x)e^{-\kappa_k y}}{ic_k^{+}\dot{a}(i\kappa_k)} =$$

$$= -\sum_{k=1}^{n} (m_k^+)^2 \left\{ e^{-\kappa_k(x+y)} + \int_x^{\infty} K^+(x,t) e^{-\kappa_k(t+y)} dt \right\} . \tag{3.5.17}$$

Hence, for $y > x$, the functions (3.5.16) and (3.5.17) coincide, and therefore, since $K^-(x,y) = 0$ for $y > x$, we get

$$F^+(x+y) + K^+(x+y) + \int_x^{\infty} F^+(y+t) K^+(x,t) dt = 0 , \tag{3.5.18}$$

in which the function

$$F^+(x) = \sum_{k=1}^{n} (m_k^+)^2 e^{-\kappa_k x} + R^+(x) = \sum_{k=1}^{n} (m_k^+)^2 e^{-\kappa_k x} + \frac{1}{2\pi} \int_{-\infty}^{\infty} r^+(\lambda) e^{i\lambda x} d\lambda \tag{3.5.19}$$

is completely specified by the right scattering data.

Proceeding in a similar manner with identity (3.5.8), we obtain the equality

$$F^-(x+y) + K^-(x,y) + \int_{-\infty}^{x} F^-(y+t) K^-(x,t) dt = 0 \quad (y < x) , \tag{3.5.18'}$$

in which the function

$$F^-(x) = \sum_{k=1}^{n} (m_k^-)^2 e^{\kappa_k x} + \frac{1}{2\pi} \int_{-\infty}^{\infty} r^-(\lambda) e^{-i\lambda x} d\lambda \tag{3.5.19'}$$

is completely specified by the left scattering data.

Equalities (3.5.18) and (3.5.18') are identical in form with (3.2.10). They can therefore be used to sharpen the properties of the functions $F^{\pm}(x)$ in the same way that we dealt with $F(x)$ in Section 2. We thus deduce that the functions $F^{\pm}(x)$ are absolutely continuous and that for every $a > -\infty$,

$$\int_a^{\infty} |F^{\pm}(\pm x)| dx < \infty \quad , \quad \int_a^{\infty} (1 + |x|) |F^{\pm\prime}(\pm x)| dx < \infty .$$

These properties are clearly also enjoyed by the functions $R^{\pm}(x)$.

Thus, the scattering data of the problem in question satisfy the following set of conditions, which shall be referred to as "Condition I".

I. *For every real value of* $\lambda \neq 0$, *the reflection coefficients* $r^{\pm}(\lambda)$ *are continuous,* $r^{\pm}(-\lambda) = \overline{r^{\pm}(\lambda)}$, $|r^{\pm}(\lambda)| < 1$, *and* $r^{\pm}(\lambda) = O(\lambda^{-1})$ *as* $\lambda \to \pm\infty$. *Their Fourier transforms,*

$$R^{\pm}(x) = \frac{1}{2\pi} \int_{-\infty}^{\infty} r^{\pm}(\lambda) e^{\pm i\lambda x} d\lambda ,$$

are real, absolutely continuous, belong to $L_2(-\infty,\infty)$ *, and for every* $a > 0$ *, satisfy the inequalities*

$$\int_a^{\infty} |R^{\pm}(\pm x)| dx < \infty , \quad \int_a^{\infty} (1 + |x|) |R^{\pm\prime}(\pm x)| dx < \infty .$$

Now let us consider two arbitrary collections, $\{r^+(\lambda), i\kappa_k, m_k^+\}$ and $\{r^-(\lambda), i\kappa_k, m_k^-\}$ $(-\infty < \lambda < \infty$, $k = 1,2,\dots,n$, $\kappa_k > 0$, $m_k^{\pm} > 0)$, in which the functions $r^+(\lambda)$ and $r^-(\lambda)$ satisfy Condition I. From these collections we construct the functions $F^{\pm}(x)$ by recipes (3.5.19) and (3.5.19') and the families of integral equations (3.5.18) and (3.5.18') for the unknown functions $K^{\pm}(x,y)$, in which x plays the role of a parameter, and the equations are considered in $L_1[x,\infty)$ and $L_1(-\infty,x]$, respectively.

LEMMA 3.5.3. *If* Condition I *is satisfied, then for every* $x > -\infty$ *, equations* (3.5.18) *and* (3.5.18') *have unique solutions* $K^+(x,y) \in L_1[x,\infty)$ *and* $K^-(x,y) \in L_1(-\infty,x]$ *, respectively. Moreover, for every* λ *in the closed upper half-plane, the functions*

$$e^+(\lambda,x) = e^{i\lambda x} + \int_x^{\infty} K^+(x,t) e^{i\lambda t} dt$$

and

$$e^-(-\lambda,x) = e^{-i\lambda x} + \int_{-\infty}^{x} K^-(x,t) e^{-i\lambda t} dt \tag{3.5.20}$$

satisfy the equations

$$-e^{\pm}(\pm\lambda,x)'' + q^{\pm}(x) e^{\pm}(\pm\lambda,x) = \lambda^2 e^{\pm}(\pm\lambda,x) , \tag{3.5.21}$$

where

$$q^+(x) = -2 \frac{d}{dx} K^+(x,x) , \quad q^-(x) = -2 \frac{d}{dx} K^-(x,x) ,$$

and

$$\int_a^{\infty} (1 + |x|) |q^{\pm}(\pm x)| dx < \infty \tag{3.5.22}$$

for all $a > -\infty$.

PROOF. By Lemma 3.3.1, the operators figuring in the integral equations of interest are compact, and therefore, in order to prove the unique solvability of these equations, it suffices to verify that the corresponding homogeneous equations have only the null solution. Consider, for example, the equation

$$f(y) + \int_x^\infty F^+(y+t)f(t)dt = 0 \quad (f(y) \in L_1[x,\infty)) \ . \tag{3.5.23}$$

By Condition I, $F^+(y)$ is bounded on the half-line $x \leq y < \infty$, and hence so is any solution $f(y)$ of (3.5.23). Consequently, $f(y) \in L_2[x,\infty)$, and

$$0 = \int_x^\infty f(y)\overline{f(y)}dy + \int_x^\infty\int_x^\infty F^+(y+t)f(t)\overline{f(y)}dtdy =$$

$$= \frac{1}{2\pi}\int_{-\infty}^\infty |\tilde{f}(\lambda)|^2 d\lambda + \sum_{k=1}^n (m_k^+)^2|\tilde{f}(-i\kappa_k)|^2 + \frac{1}{2\pi}\int_{-\infty}^\infty r^+(\lambda)\tilde{f}(-\lambda)\overline{\tilde{f}(\lambda)}d\lambda \ ,$$

where

$$\tilde{f}(\lambda) = \int_x^\infty f(y)e^{-i\lambda y}dy \ .$$

Since $|r^+(-\lambda)| = |r^+(\lambda)|$ and $2|\tilde{f}(-\lambda)\overline{\tilde{f}(\lambda)}| \leq |\tilde{f}(-\lambda)|^2 + |\tilde{f}(\lambda)|^2$, we have

$$\int_{-\infty}^\infty |r^+(\lambda)\tilde{f}(-\lambda)\overline{\tilde{f}(\lambda)}|d\lambda \leq \int_{-\infty}^\infty |r^+(\lambda)||\tilde{f}(\lambda)|^2 d\lambda \ ,$$

and hence

$$\frac{1}{2\pi}\int_{-\infty}^\infty |\tilde{f}(\lambda)|^2 d\lambda = -\sum_{k=1}^n (m_k^+)^2|\tilde{f}(-i\kappa_k)|^2 - \frac{1}{2\pi}\int_{-\infty}^\infty r^+(\lambda)f(-\lambda)\overline{\tilde{f}(\lambda)}d\lambda \leq$$

$$\leq \frac{1}{2\pi}\int_{-\infty}^\infty |r^+(\lambda)\tilde{f}(-\lambda)\overline{\tilde{f}(\lambda)}|d\lambda \leq \frac{1}{2\pi}\int_{-\infty}^\infty |r^+(\lambda)||\tilde{f}(\lambda)|^2 d\lambda \ ,$$

i.e.,

$$\int_{-\infty}^\infty (1 - |r^+(\lambda)|)|\tilde{f}(\lambda)|^2 d\lambda \leq 0 \ .$$

Since $1 - |r^+(\lambda)| > 0$ for all $\lambda \neq 0$, this implies that $f(\lambda) \equiv 0$. Thus, the homogeneous equation (3.5.23) has only the null solution. The unique solvability of equations (3.5.18') is proved in a similar manner. The remaining assertions of the lemma are corollaries of Theorem 3.3.1, since the proof of the latter used only the unique solvability of the corresponding equations. □

To this point we have not been interested in the relationships between the left and right scattering data. It turns out that the left scattering data are uniquely determined by the right ones, and viceversa. In fact, it follows from formulas (3.5.9)-(3.5.12) that

$$r^-(\lambda) = -r^+(-\lambda)\,\frac{a(-\lambda)}{a(\lambda)}\ , \quad (m_k^-)^{-2} = -(m_k^+)^2[\dot a(i\kappa_k)]^2\ . \tag{3.5.24}$$

These equalities show that, in order to express one kind of scattering data in terms of the other, it suffices to recover the function $a(z)$ from the given data. But, by Lemma 3.5.2, the function

$$g(z) = \frac{1}{a(z)}\prod_{k=1}^{n}\frac{z-i\kappa_k}{z+i\kappa_k}$$

in the upper half plane is holomorphic, uniformly bounded, has no zeros, and behaves like $1 + O(z^{-1})$ as $|z| \to \infty$. Moreover, by (3.5.7) and (3.5.9),

$$|g(\lambda)| = |a(\lambda)|^{-1} = \sqrt{1 - |r^\pm(\lambda)|^2}$$

for real values of $\lambda \neq 0$. This permits us to reconstruct the function $\ln g(z)$ $(\mathrm{Im}\, z > 0)$ from its real part $\ln |g(\lambda)|$ $(-\infty < \lambda < \infty)$ via the Poisson-Schwarz formula:

$$\ln g(z) = \frac{1}{\pi i}\int_{-\infty}^{\infty}\frac{\ln|g(\lambda)|}{\lambda - z}\,d\lambda = \frac{1}{2\pi i}\int_{-\infty}^{\infty}\frac{\ln(1-|r^\pm(\lambda)|^2)}{\lambda - z}\,d\lambda\ .$$

This yields the following representation for the function $a(z)$, in terms of $|r^+(\lambda)| = |r^-(\lambda)|$:

$$a(z) = \exp\left\{-\frac{1}{2\pi i}\int_{-\infty}^{\infty}\frac{\ln(1-|r^+(\lambda)|^2)}{\lambda - z}\,d\lambda\right\}\prod_{k=1}^{n}\frac{z-i\kappa_k}{z+i\kappa_k}\ . \tag{3.5.25}$$

A rigorous justification of this formula is given in Problem 2. Here we merely notice that it follows from formula (3.5.13) that the function $za(z)$ is continuous in the closed upper half plane, and that the function $\lambda^2(1+\lambda^2)^{-1}|a(\lambda)|^2 = \lambda^2(1+\lambda^2)^{-1}(1-|r^+(\lambda)|^2)^{-1}$ is bounded on the real line, which in turn implies the inequality

$$1 > (1 - |r^+(\lambda)|^2) > C\lambda^2(1+\lambda^2)^{-1}\ .$$

The latter, in conjunction with the estimate $|r^+(\lambda)| = O(\lambda^{-1})$ $(\lambda \to \pm\infty)$, guarantees the convergence of the integral in (3.5.25).

The behaviour of the function $za(z)$ for $z \to \infty$ is closely related to that of the functions $r^\pm(\lambda)$ for $\lambda \to 0$. In fact, upon reexpressing equalities (3.5.8), (3.5.8') in the form

$$\lambda e^{\pm}(\pm\lambda,x) = \lambda a(\lambda)[r^{\pm}(\lambda) + 1]e^{\mp}(\mp\lambda,x) + e^{\mp}(\pm\lambda,x) - e^{\mp}(\mp\lambda,x) ,$$

and then letting $\lambda \to 0$, we obtain

$$0 \equiv e\ (0,x) \lim_{\lambda\to 0} \lambda a(\lambda)[r^{\pm}(\lambda) + 1] .$$

Consequently,

$$\lim_{\lambda\to 0} \lambda a(\lambda)[r^{\pm}(\lambda) + 1] = 0$$

which, in particular, shows that if $\lim_{z\to 0} za(z) \neq 0$, then the functions $r^{\pm}(\lambda)$ are continuous on the whole real line and $r^{\pm}(0) = -1$.

Thus, in addition to I, the scattering data satisfy the following set of conditions, which will be referred to as "Condition II".

II. *The reflection coefficients* $r^{+}(\lambda)$ *and* $r^{-}(\lambda)$ *and the norming constants* m_k^{+} *and* m_k^{-} *are connected by equalities* (3.5.24), *in which the function* $a(z)$ *is expressible in terms of the eigenvalues* $\mu_k = (i\kappa_k)^2$, *and the modulus of the reflection coefficients,* $|r^{+}(\lambda)| = |r^{-}(\lambda)|$, *by formula* (3.5.25). *The function* $za(z)$ *is continuous in the closed upper half-plane and* $\lim_{\lambda\to 0} \lambda a(\lambda)[r^{\pm}(\lambda) + 1] = 0.$

Conditions I and II are both necessary and sufficient for two collections of the type considered above to be the right and left scattering data of one and the same equation (3.5.1). This follows from the following result:

THEOREM 3.5.1. *In order for the collection* $\{r^{+}(\lambda), i\kappa_k, m_k^{+}\}$ $(-\infty < \lambda < \infty$, $k = 1,2,\ldots,k$, $\kappa_k > 0$, $m_k^{+} > 0)$ *to be the right scattering data of an equation of the form* (3.5.1) *with real-valued potential* $q(x)$ *which is subject to inequality* (3.5.2), *it is necessary and sufficient that the following conditions be satisfied:*

1) *for real* $\lambda \neq 0$ *the function* $r^{+}(\lambda) = \overline{r^{+}(-\lambda)}$ *is continuous,* $|r^{+}(\lambda)| \leq 1 - C\lambda^2(1+\lambda^2)^{-1}$, *and* $r^{+}(\lambda) = o(\lambda^{-1})$ *as* $\lambda \to \pm\infty$;

2) *the function*

$$R^{+}(x) = \frac{1}{2\pi} \int_{-\infty}^{\infty} r^{+}(\lambda)e^{i\lambda x}d\lambda$$

is absolutely continuous and, for every $a > -\infty$, *its derivative* $R^{+\prime}(x)$ *satisfies the bound*

$$\int_a^{\infty} (1 + |x|)|R^{+\prime}(x)|dx < \infty ;$$

3) the function $za(z)$, *where*

$$a(z) = \exp\left\{-\frac{1}{2\pi i}\int_{-\infty}^{\infty} \frac{\ln(1 - |r^+(\lambda)|^2}{\lambda - z} d\lambda\right\} \prod_{k=1}^{n} \frac{z - i\kappa_k}{z + i\kappa_k}$$

is continuous in the closed upper half-plane, and

$$\lim_{\lambda\to 0} \lambda a(\lambda)[r^+(\lambda) + 1] = 0 ;$$

4) the function

$$R^-(x) = -\frac{1}{2\pi}\int_{-\infty}^{\infty} r^+(-\lambda)\frac{a(-\lambda)}{a(\lambda)} e^{-i\lambda x} d\lambda$$

is absolutely continuous and, for every $a > -\infty$, *its derivative* $R^{-\prime}(x)$ *satisfies the bound*

$$\int_a^{\infty} (1 + |x|)|R^{-\prime}(-x)|dx < \infty .$$

PROOF. The necessity of these conditions has been established earlier. To prove their sufficiency, we construct from the given collection a new collection $\{r^-(\lambda) , i\kappa_k , m_k^-\}$, by setting

$$r^-(\lambda) = -r^+(-\lambda)\frac{a(-\lambda)}{a(\lambda)} , \quad (m_k^-)^{-2} = -(m_k^+)^2[\dot{a}(i\kappa_k)]^2 , \tag{3.5.26}$$

and then show that $\{r^+(\lambda) , i\kappa_k , m_k^+\}$ and $\{r^-(\lambda) , i\kappa_k , m_k^-\}$ are the right and left scattering data of one and the same equation (3.5.1) with real-valued potential q , which is subject to inequality (3.5.2). It follows from conditions 1, 2, and 4, that these two collections meet the requirements of Lemma 3.5.3. Hence, for every fixed $x \neq \pm\infty$, equations (3.5.18) and (3.5.18'), constructed from these collections, have unique solutions $K^{\pm}(x,y)$, and the functions $e^{\pm}(\pm\lambda,x)$, defined by formulas (3.5.20), are solutions of equations (3.5.21) with real-valued potentials $q^+(x)$ and $q^-(x)$, respectively, which satisfy the bounds (3.5.22). To prove the theorem, it obviously suffices to show that for real values of λ , the functions $e^+(\lambda,x)$ and $e^-(-\lambda,x)$ are connected by the relations

$$r^+(\lambda)e^+(\lambda,x) + e^+(-\lambda,x) = [a(\lambda)]^{-1}e^-(-\lambda,x) \tag{3.5.27}$$

and

$$r^-(\lambda)e^-(-\lambda,x) + e^-(\lambda,x) = [a(\lambda)]^{-1}e^+(\lambda,x) \ , \tag{3.5.27}$$

and that, in addition,

$$(m_k^-)^{-2} = \int_{-\infty}^{\infty} |e^-(-i\kappa_k,x)|^2 dx \quad , \quad (m_k^+)^{-2} = \int_{-\infty}^{\infty} |e^+(i\kappa_k,x)|^2 dx \ . \tag{3.5.28}$$

Since $R^{\pm}(y) \in L_2(-\infty,\infty)$, the functions

$$\Phi^+(x,y) = R^+(x+y) + \int_x^{\infty} R^+(y+t)K^+(x,t)dt$$

and

$$\Phi^-(x,y) = R^-(x+y) + \int_{-\infty}^{x} R^-(y+t)K^-(x,t)dt$$

belong to $L_2(-\infty,\infty)$ for every fixed x , and

$$\underset{N\to\infty}{\text{l.i.m.}} \int_{-N}^{N} \Phi^+(x,y)e^{-i\lambda y}dy = r^+(\lambda)\left[e^{i\lambda x} + \int_x^{\infty} K^+(x,t)e^{i\lambda t}dt\right] = r^+(\lambda)e^+(\lambda,x) \ ,$$

$$\underset{N\to\infty}{\text{l.i.m.}} \int_{-N}^{N} \Phi^-(x,y)e^{i\lambda y}dy = r^-(\lambda)\left[e^{-i\lambda x} + \int_{-\infty}^{x} K^-(x,t)e^{-i\lambda t}dt\right] = r^-(\lambda)e^-(-\lambda,x) \ .$$

On the other hand, by equations (3.5.18) and (3.5.18'),

$$\Phi^+(x,y) = -K^+(x,y) - \sum_{k=1}^{n} (m_k^+)^2 \left\{e^{-\kappa_k(x+y)} + \int_x^{\infty} K^+(x,t)e^{-\kappa_k(x+t)}dt\right\} =$$

$$= -K^+(x,y) - \sum_{k=1}^{n} (m_k^+)^2 e^{-\kappa_k y} e^+(i\kappa_k,x) \quad (x < y < \infty) \ ,$$

and

$$\Phi^-(x,y) = -K^-(x,y) - \sum_{k=1}^{n} (m_k^-)^2 \left\{e^{\kappa_k(x+y)} + \int_{-\infty}^{x} K^-(x,t)e^{\kappa_y(x+t)}dt\right\} =$$

$$= -K^-(x,y) - \sum_{k=1}^{n} (m_k^-)^2 e^{\kappa_k y} e^-(-i\kappa_k,x) \quad (-\infty < y < x) \ .$$

Therefore,

$$\underset{N\to\infty}{\text{l.i.m.}} \int_{-N}^{N} \Phi^+(x,y)e^{-i\lambda y}dy = \underset{N\to\infty}{\text{l.i.m.}} \int_{-N}^{x} \Phi^+(x,y)e^{-i\lambda y}dy +$$

$$+ e^{-i\lambda x} - e^+(-\lambda,x) - \sum_{k=1}^{n} \frac{(m_k^+)^2}{\kappa_k + i\lambda} e^{-i\lambda x}e^{-\kappa_k x}e^+(i\kappa_k,x)$$

and

$$\underset{N\to\infty}{\text{l.i.m.}} \int_{-N}^{N} \Phi^-(x,y)e^{i\lambda y}dy = \underset{N\to\infty}{\text{l.i.m.}} \int_{x}^{\infty} \Phi^-(x,y)e^{i\lambda y}dy +$$

$$+ e^{i\lambda x} - e^{-}(\lambda,x) - \sum_{k=1}^{n} \frac{(m_k^-)^2}{\kappa_k + i\lambda} e^{i\lambda x} e^{\kappa_k x} e^{-}(-i\kappa_k,x) .$$

Comparing these expressions for the Fourier transforms of the functions $\Phi^{\pm}(x,y)$, we obtain the equalities

$$r^+(\lambda)e^+(\lambda,x) + e^+(-\lambda,x) = [a(\lambda)]^{-1}h^-(-\lambda,x) \tag{3.5.29}$$

and

$$r^-(\lambda)e^-(-\lambda,x) + e^-(\lambda,x) = [a(\lambda)]^{-1}h^+(\lambda,x) , \tag{3.5.29'}$$

where $-\infty < \lambda < \infty$,

$$h^-(-\lambda,x) = e^{-i\lambda x}a(\lambda) \Big[1 + \underset{N\to\infty}{\text{l.i.m.}} \int_{-N}^{x} \Phi^+(x,y)e^{-i\lambda(y-x)}dy -$$

$$- \sum_{k=1}^{n} \frac{(m_k^+)^2}{\kappa_k + i\lambda} e^{-\kappa_k x} e^+(i\kappa_k,x)\Big] , \tag{3.5.30}$$

and

$$h^+(\lambda,x) = e^{i\lambda x}a(\lambda) \Big[1 + \underset{N\to\infty}{\text{l.i.m.}} \int_{x}^{N} \Phi^-(x,y)e^{i\lambda(y-x)}dy -$$

$$- \sum_{k=1}^{n} \frac{(m_k^-)^2}{\kappa_k + i\lambda} e^{\kappa_k x} e^-(-i\kappa_k,x)\Big] . \tag{3.5.30'}$$

We next list those properties of the functions $h^+(\lambda,x)$ and $h^-(\lambda,x)$ that will be used below. It follows from equalities (3.5.29), (3.5.29'), and condition 3 that, for real values of $\lambda \neq 0$, the functions $h^+(\lambda,x)$ and $h^-(-\lambda,x)$ are continuous,

$$h^+(\lambda,x) = \overline{h^+(-\lambda,x)} \quad , \quad h^-(-\lambda,x) = \overline{h^-(\lambda,x)} ,$$

$$W\{e^+(\lambda,x),h^-(-\lambda,x)\} = W\{h^+(\lambda,x),e^-(-\lambda,x)\} = 2i\lambda a(\lambda) ,$$

$$\sup_{\lambda\neq 0} |[a(\lambda)]^{-1}h^-(-\lambda,x)| < \infty \quad , \quad \sup_{\lambda\neq 0} |[a(\lambda)]^{-1}h^+(\lambda,x)| < \infty ,$$

and

$$\lim_{\lambda\to 0} \lambda h^-(-\lambda,x) = \lim_{\lambda\to 0} \lambda h^+(\lambda,x) = 0 .$$

Furthermore, formulas (3.5.30), (3.5.30'), and condition 3 guarantee that $h^+(\lambda,x)$ and $h^-(-\lambda,x)$ admit analytic continuations to the upper half plane, and

$$\lim_{|z|\to\infty} e^{izx}h^-(-z,x) = \lim_{|z|\to\infty} e^{-izx}h^+(z,x) = 1 , \tag{3.5.31}$$

$$\lim_{z\to 0} zh^-(-z,x) = \lim_{z\to 0} zh^+(z,x) = 0 , \tag{3.5.32}$$

$$\overline{\lim_{z\to 0}}\ |[a(z)]^{-1}h^-(-z,x)| < \infty , \quad \overline{\lim_{z\to 0}}\ |[a(z)]^{-1}h^+(z,x)| < \infty , \tag{3.5.33}$$

$$\left.\begin{aligned} h^-(-i\kappa_k,x) &= i\dot{a}(i\kappa_k)(m_k^+)^2e^+(i\kappa_k,x) , \\ h^+(i\kappa_k,x) &= i\dot{a}(i\kappa_k)(m_k^-)^2e^-(-i\kappa_k,x) , \end{aligned}\right\} \tag{3.5.34}$$

and

$$W\{e^+(z,x),h^-(-z,x)\} = W\{h^+(z,x),e^-(-z,x)\} = 2iza(z) . \tag{3.5.35}$$

Solving the system

$$\begin{cases} r^+(\lambda)e^+(\lambda,x) + e^+(-\lambda,x) = [a(\lambda)]^{-1}h^-(-\lambda,x) \\ r^+(-\lambda)e^+(-\lambda,x) + e^+(\lambda,x) = [a(-\lambda)]^{-1}h^-(\lambda,x) \end{cases}$$

for $e^+(\lambda,x)$ and $e^+(-\lambda,x)$, we get

$$e^+(\lambda,x)[1 - r^+(\lambda)r^+(-\lambda)] = \frac{h^-(\lambda,x)}{a(-\lambda)} - \frac{r^+(-\lambda)h^-(-\lambda,x)}{a(\lambda)} .$$

Since $1 - r^+(\lambda)r^+(-\lambda) = 1 - |r^+(\lambda)|^2 = |a(\lambda)|^{-2} = [a(\lambda)a(-\lambda)]^{-1}$, this yields the equality

$$\frac{e^+(\lambda,x)}{a(\lambda)} = -r^+(-\lambda)\frac{a(-\lambda)}{a(\lambda)}h^-(-\lambda,x) + h^-(\lambda,x)$$

which, in view of the definition of $r^-(\lambda)$, is equivalent to

$$r^-(\lambda)h^-(-\lambda,x) + h^-(\lambda,x) = \frac{e^+(\lambda,x)}{a(\lambda)} . \tag{3.5.36}$$

Upon eliminating the function $r^-(\lambda)$ from equalities (3.5.29') and (3.5.26), we obtain the identity

$$e^-(\lambda,x)h^-(-\lambda,x) - h^-(\lambda,x)e^-(-\lambda,x) = \frac{h^+(\lambda,x)h^-(-\lambda,x) - e^+(\lambda,x)e^-(-\lambda,x)}{a(\lambda)} \tag{3.5.37}$$

which holds for all real $\lambda \neq 0$. The right-hand side of this identity is the ratio of two functions which are holomorphic in the upper half plane; moreover, by (3.5.30), (3.5.30'), and (3.5.26), the numerator vanishes at the zeros $i\kappa_k$ of the denominator:

$$h^+(i\kappa_k,x)h^-(-i\kappa_k,x) - e^+(i\kappa_k,x)e^-(-i\kappa_k,x) =$$

$$= -e^+(i\kappa_k,x)e^-(-i\kappa_k,x)\{1 + [m_k^+m_k^-a(i\kappa_k)]^2\} = 0 .$$

Consequently, the function

$$g(z) = \frac{h^+(z,x)h^-(-z,x) - e^+(z,x)e^-(-z,x)}{a(z)}$$

is holomorphic in the upper half plane. It also tends to zero as $|z| \to \infty$ ($\operatorname{Im} z \geq 0$) since, by (3.5.31) and (3.5.20),

$$\lim_{|z|\to\infty} h^+(z,x)h^-(-z,x) = \lim_{|z|\to\infty} e^+(z,x)e^-(-z,x) = 1 .$$

Now, since the left-hand side of identity (3.5.37) is an odd function of λ, we have $g(\lambda) = -g(-\lambda)$ $(-\infty < \lambda < \infty\ ,\ \lambda \neq 0)$. Hence, upon extending the function $g(z)$ to the lower half plane by the rule $g(z) = -g(-z)$, we obtain a single-valued function $g(z)$, which is holomorphic for all $z \neq 0$, and tends to zero as $|z| \to \infty$. Next, it follows from the equalities (3.5.32), (3.5.33) and the boundedness of the function $[a(z)]^{-1}$ in a neighborhood of zero, which is an obvious consequence of formula (3.5.25), that

$$\lim_{z\to 0} g(z)z = \lim_{z\to 0} \frac{zh^+(z,x)h^-(-z,x)}{a(z)} - \lim_{z\to 0} \frac{ze^+(z,x)e^-(-z,x)}{a(z)} = 0 .$$

Therefore, the point $z = 0$ is a removable singularity of the function $g(z)$, and hence, by Liouville's theorem, $g(z) \equiv 0$ and

$$h^+(z,x)h^-(-z,x) - e^+(z,x)e^-(-z,x) \equiv 0 , \tag{3.5.38}$$

$$e^-(\lambda,x)h^-(-\lambda,x) - h^-(\lambda,x)e^-(-\lambda,x) \equiv 0 . \tag{3.5.39}$$

Since $e^-(-z,x)$ are nonzero solutions of the equation $-y'' + q^-(x)y = z^2y$, the set of values of z for which $e^-(-z,x) = 0$ is discrete. Hence, the set $\mathcal{O}\{0,i\kappa_1,\ldots,i\kappa_n\}$ of values of x for which at least one of the equalities $e^-(0,x) = 0$, $e^-(-i\kappa_1,x) = 0$, ..., $e^-(-i\kappa_n,x) = 0$ holds, is also discrete. Let us show that, for every $x \notin \mathcal{O}\{0,i\kappa_1,\ldots,i\kappa_n\}$, the function

$$p(z) = \frac{h^-(-z,x)}{e^-(-z,x)}$$

is holomorphic in the upper half plane. Since the zeros of the function

$e^-(-z,x)$ are simple, it suffices to verify that each such zero z_k is also a zero of the function $h^-(-z,x)$. But, if $e^-(-z_k,x) = 0$ and $x \notin \mathcal{O}\{0,i\kappa_1,\dots,i\kappa_n\}$, then $2iz_k a(z_k) \neq 0$, and by (3.5.35), $h^+(z_k,x) \neq 0$. On the other hand, it follows from identity (3.5.38) that $0 = h^+(z_k,x)h^-(-z_k,x)$, and hence $h^-(-z_k,x) = 0$. Thus, $p(z)$ is indeed holomorphic in the upper half plane. Moreover, it tends to 1 as $|z| \to \infty$, as follows from (3.5.20) and (3.5.31). Next, the identity (3.5.9) implies that $p(\lambda) = p(-\lambda)$ for all λ, $-\infty < \lambda < \infty$, $\lambda \neq 0$. Hence, the function obtained by extending $p(z)$ to the lower half plane by the rule $p(z) = p(-z)$ is single-valued and holomorphic for all $z \neq 0$. Since $e^-(0,x) \neq 0$ for the considered values of x and, by (3.5.32), $\lim_{z\to 0} zh^-(-z,x) = 0$, it follows that $\lim_{z\to 0} zp(z) = 0$. Therefore, the point $z = 0$ is a removable singularity of $p(z)$ and, by Liouville's theorem, $p(z) \equiv 1$. Hence, $e^-(-z,x) \equiv h^-(-z,x)$ for all $x \notin \mathcal{O}\{0,i\kappa_1,\dots,i\kappa_n\}$. This identity can be extended by continuity for all the values of x, since the set $\mathcal{O}\{0,i\kappa_1,\dots,i\kappa_n\}$ is discrete. Therefore, upon replacing $h^-(-z,x)$ by $e^-(-z,x)$ in (3.5.29) and (3.5.36), we obtain the equalities (3.5.27).

It remains to verify formulas (3.5.28). By (3.5.27), $q^+(x) = q^-(x)$, and hence, in computing $||e^-(-i\kappa_k,x)||^2$ and $||e^+(i\kappa_k,x)||^2$, we may use formulas (3.5.11), in which we must put, according to (3.5.34), $c_k^- = i\dot{a}(i\kappa_k)(m_k^+)^2$ and $c_k^+ = i\dot{a}(i\kappa_k)(m_k^-)^2$. This yields

$$\int_{-\infty}^{\infty} |e^-(-i\kappa_k,x)|^2 dx = -[\dot{a}(i\kappa_k)m_k^+]^2 \quad , \quad \int_{-\infty}^{\infty} |e^+(i\kappa_k,x)|^2 dx = -[\dot{a}(i\kappa_k)m_k^-]^2 ,$$

which, in view of (3.5.26), proves formulas (3.5.28).

PROBLEMS

1. Prove the assertions of Lemma 3.5.2.

Hint. As was shown on page 286, the zeros $i\kappa_k$ of the function $a(z)$ lie on a bounded segment of the imaginary half line: $0 \leqslant \kappa_k \leqslant M$. Let δ denote the infimum of the distances between neighboring zeros. The next step is to show that $\delta > 0$. Assuming the contrary, let $i\hat{\kappa}_k$ and $i\kappa_k$ be sequences of zeros such that $\hat{\kappa}_k > \kappa_k$ and $\lim_{k\to\infty} \hat{\kappa}_k = \lim_{k\to\infty} \kappa_k = \kappa_0$. Choose $A > 0$ such that, for every $\kappa \in [0,\infty)$,

$$e^+(i\kappa,x) > \frac{1}{2}\exp(-\kappa x) \quad \text{for all} \quad A \leqslant x < \infty ,$$

and

$$e^-(-i\kappa,x) > \frac{1}{2}\exp(-\kappa x) \quad \text{for all} \quad -\infty < x \leqslant -A .$$

By (3.1.36), this yields the estimates

$$\int_A^\infty e^+(i\hat{\kappa}_k,x)e^+(i\kappa_k,x)dx > \frac{e^{-2AM}}{8M} \quad , \quad \int_{-\infty}^{-A} e^-(-i\hat{\kappa}_k,x)e^-(-i\kappa_k,x)dx > \frac{e^{-2AM}}{8M} .$$

It follows from the orthogonality of the eigenvalues of the discrete spectrum and the equalities

$$e^+(i\hat{\kappa}_k,x) = c^+(\hat{\kappa}_k)e^-(-i\hat{\kappa}_k,x) \quad , \quad e^+(i\kappa_k,x) = c^+(\kappa_k)e^-(-i\kappa_k,x)$$

that

$$0 = \int_{-\infty}^{\infty} e^+(i\hat{\kappa}_k,x)e^+(i\kappa_k,x)dx = c^+(i\hat{\kappa}_k)c^+(i\kappa_k) \int_{-\infty}^{A} e^-(i\hat{\kappa}_k,x)e^-(-i\kappa_k,x)dx +$$

$$+ \int_{-A}^{A} e^+(i\hat{\kappa}_k,x)e^+(i\kappa_k,x)dx + \int_A^\infty e^+(i\hat{\kappa}_k,x)e^+(i\kappa_k,x)dx . \tag{3.5.40}$$

Since

$$\lim_{k\to\infty} e^\pm(\pm i\hat{\kappa}_k,x) = \lim_{k\to\infty} e^\pm(i\kappa_k,x) = e^\pm(i\kappa_0,x) ,$$

you have

$$\lim_{k\to\infty} c^+(i\hat{\kappa}_k)c^+(i\kappa_k) = [e^+(i\kappa_0,x)]^2[e^-(-i\kappa_0,x)]^2 > 0$$

and

$$\lim_{k\to\infty} \int_{-A}^{A} e^+(i\hat{\kappa}_k,x)e^+(i\kappa_k,x)dx = \int_{-A}^{A} [e^+(i\kappa_0,x)]^2dx .$$

Hence, upon letting $k \to \infty$ in both sides of (3.5.40), you obtain the absurd inequality $0 > e^{-2AM}(8M)^{-1}$, which proves that your assumption that $\delta = 0$, is false. Thus, the function $a(z)$ may have only finitely many zeros $i\kappa_1,\dots,i\kappa_n$.

The boundedness of the function $[a(z)]^{-1}$ in a neighborhood of zero follows from the inequality

$$\left|[a(z)]^{-1} \prod_{k=1}^{n} (z - i\kappa_k)(z + i\kappa_k)^{-1}\right| \leqslant 1 \quad (\operatorname{Im} z \geqslant 0) . \tag{3.5.41}$$

To prove (3.4.51), consider the functions $a_N(z)$ corresponding to the potentials

$$q_N(x) = \begin{cases} q(x) , & \text{if } |x| < N , \\ 0 , & \text{elsewhere} . \end{cases}$$

Since for such potentials, $e^+(z,x)$ and $e^-(z,x)$ are entire functions, $za_N(z)$ are also entire, and the inequalities

$$\left| [a_N(z)]^{-1} \prod_{k=1}^{n} (z - i\kappa_k)(z + i\kappa_k)^{-1} \right| \leqslant 1 \quad (\text{Im } z \geqslant 0) \tag{3.5.41'}$$

are strightforward consequences of the maximum modulus principle and the estimates

$$|a_N(\lambda)|^{-1} \leqslant 1 \quad (-\infty < \lambda < \infty)$$

and

$$a_N(z) = 1 + O(z^{-1}) \quad (|z| \to \infty \quad , \quad \text{Im } z \geqslant 0) .$$

Let δ_N be the smallest distance between neighboring zeros of the function $a_N(z)$. Then $\inf_N \delta_N = \delta > 0$, as follows from the uniformity in N of the estimates used above to prove that $\delta_N > 0$. You can now derive inequality (3.5.41) from (3.5.41') by letting $N \to \infty$.

2. Prove formula (3.5.25).

Hint. Let

$$g_N(z) = [a_N(z)]^{-1} \prod_{k=1}^{n_N} (z - i\kappa_k(N))(z + i\kappa_k(N))^{-1} ,$$

where $a_N(z)$ are the functions corresponding to the potentials q_N , which were introduced in the preceding problem, and the notations n_N and $\kappa_k(N)$ are self-evident. The functions $\ln g_N(z)$ are obviously single-valued and holomorphic in the upper half plane, and $\ln g_N(z) = O(z^{-1})$ for $|z| \to \infty$ and $\text{Im } z \geqslant 0$. Moreover,

$$|\ln g_N(z)| \leqslant A_N \ln |z| + B_N \quad (\text{Im } z \geqslant 0)$$

in the neighborhood of zero, since the function $za_N(z)$ is entire. Hence, $\ln g_N(z)$ admits the Cauchy integral representation

$$\ln g_N(z) = \frac{1}{2\pi i}\int_{-\infty}^{\infty}\frac{\ln g_N(\lambda)}{\lambda - z}\,d\lambda \quad (\operatorname{Im} z > 0)$$

from which you can obtain the Poisson-Schwarz formula

$$\ln g_N(z) = \frac{1}{\pi i}\int_{-\infty}^{\infty}\frac{\ln |g_N(\lambda)|}{\lambda - z}\,d\lambda \quad (\operatorname{Im} z > 0)$$

in the usual way. Since $|g_N(\lambda)| = |a_N(\lambda)|^{-1} = \sqrt{1 - |r_N^+(\lambda)|^2}$, this in turn implies that

$$a_N(z) = \exp\left\{\frac{-1}{2\pi i}\int_{-\infty}^{\infty}\frac{\ln(1 - |r_N^+(\lambda)|^2}{\lambda - z}\,d\lambda\right\}\prod_{k=1}^{n_N}\frac{z - i\kappa_k(N)}{z + i\kappa_k(N)}\,. \qquad (3.5.42)$$

Next, it follows from Lemma 3.5.1 and the identity $|a_N(\lambda)|^{-2} = 1 - |r_N^+(\lambda)|^2$ that

$$1 > 1 - |r_N^+(\lambda)|^2 > 1 - C_1\lambda^2(1 + \lambda^2)^{-1}\,, \quad |r_N^+(\lambda)|^2 < C_2\lambda^{-2}\,,$$

where the constants C_1 and C_2 are independent of N. These estimates permit you to let $N \to \infty$ under the integral sign in (3.5.42), and thus obtain the representation (3.5.25) for $a(z)$.

3. As we remarked above, if

$$\lim_{\lambda\to 0}\lambda a(\lambda) = (2i)^{-1}W\{e^+(0,x), e^-(0,x)\} \neq 0\,,$$

then the reflection coefficients $r^+(\lambda)$ and $r^-(\lambda)$ are continuous on the whole real line, and $r^+(0) = r^-(0) = -1$. It seems reasonable to assume that they are continuous in all cases. Show that this is true for potentials $q(x)$ which satisfy the inequality

$$\int_{-\infty}^{\infty}(1 + x^2)|q(x)|\,dx < \infty\,. \qquad (3.4.53)$$

<u>Hint</u>. It follows from (3.4.53) and the estimates for the kernels $K^+(x,y)$ and $K^-(x,y)$ of the transformation operators, that the functions $e^+(\lambda,x)$ and $e^-(-\lambda,x)$ are differentiable with respect to λ for all values of λ. Moreover, the functions $\dot{e}^+(0,x)$ and $\dot{e}^-(0,x)$ are solutions of equation (3.5.1) with the asymptotics

$$\begin{cases} \dot{e}^+(0,x) = ix + o(1)\,, \\ \dfrac{d}{dx}\,\dot{e}^+(0,x) = i + o(1)\,, \end{cases} \quad (x \to +\infty)\,,$$

and

$$\begin{cases} \dot{e}^-(0,x) = -ix + o(1) \ , \\ \frac{d}{dx}\dot{e}^-(0,x) = -i + o(1) \ , \end{cases} \quad (x \to -\infty)$$

respectively.

Now suppose that $W\{e^+(0,x),e^-(0,x)\} = 0$, and hence

$$e^+(0,x) = c^+e^-(0,x) \quad , \quad e^-(0,x) = c^-e^+(0,x) \quad , \quad c^+c^- = 1 \ .$$

Then

$$2i \frac{\partial}{\partial\lambda} [\lambda a(\lambda)]\Big|_{\lambda=0} = \frac{\partial}{\partial\lambda} W\{e^+(\lambda,x),e^-(-\lambda,x)\}\Big|_{\lambda=0} =$$

$$= W\{\dot{e}^+(0,x),e^-(0,x)\} + W\{e^+(0,x),\dot{e}^-(0,x)\} =$$

$$= c^-W\{\dot{e}^+(0,x),e^+(0,x)\} + c^+W\{e^-(0,x),\dot{e}^-(0,x)\} = i(c^- + c^+) \neq 0 \ ,$$

since condition (3.5.43) guarantees that

$$\frac{d}{dx} e^+(0,x) = o(x^{-1}) \quad (x \to +\infty)$$

and

$$\frac{d}{dx} e^-(0,x) = o(x^{-1}) \quad (x \to -\infty) \ .$$

Proceeding similarly, you may check that

$$\frac{\partial}{\partial\lambda} [2i\lambda b(\lambda)]\Big|_{\lambda=0} = \frac{\partial}{\partial\lambda} W\{e^-(\lambda,x),e^+(\lambda,x)\}\Big|_{\lambda=0} = i(c^+ - c^-) \ .$$

Therefore,

$$\lim_{\lambda\to0} r^+(\lambda) = \lim_{\lambda\to0} \frac{2i\lambda b(\lambda)}{2i\lambda a(\lambda)} = \frac{c^+ - c^-}{c^+ + c^-} = \tanh \gamma^+$$

and

$$\lim_{\lambda\to0} r^-(\lambda) = \lim_{\lambda\to0} \frac{-2i\lambda b(-\lambda)}{2i\lambda a(\lambda)} = \frac{c^- - c^+}{c^+ + c^-} = -\tanh \gamma^+ \ ,$$

where $\gamma^+ = \ln |c^+|$.

4. Prove the validity of the following expansions in terms of the eigenfunctions of the scattering problem: for every $f(x) \in L_2(-\infty,\infty)$,

$$f(x) = \underset{N\to\infty}{\text{l.i.m.}} \frac{1}{2\pi} \int_{-N}^{N} u^-(\lambda,f)u^+(\lambda,x)a(\lambda)d\lambda + \sum_{k=1}^{n} (m_k^-)^2 u^-(i\kappa_k,f)u^-(i\kappa_k,x) \ ,$$

where

$$u^{\pm}(\lambda,f) = \underset{N\to\infty}{\text{l.i.m.}} \int_{-N}^{N} u^{\pm}(\lambda,x)f(x)dx \;.$$

<u>Hint</u>. It suffices to prove the Parseval inequality

$$\int_{-\infty}^{\infty} f(x)g(x)dx = \frac{1}{2\pi}\int_{-\infty}^{\infty} u^-(\lambda,f)u^+(\lambda,g)a(\lambda)d\lambda + \sum_{k=1}^{n} (m_k^-)^2 u^-(i\kappa_k,f)u^-(i\kappa_k,g)$$

$$+ \sum_{k=1}^{n} (m_k^-)^2 u^-(i\kappa_k,f)u^-(i\kappa_k,g) \tag{3.5.44}$$

for arbitrary functions $f(x),g(x) \in L_2(-\infty,\infty)$ with compact support.

Suppose $f(x) = g(x) = 0$ for $|x| > N$. Then

$$\int_{-\infty}^{\infty} f(x)e^-(-\lambda,x)dx = \int_{-\infty}^{\infty}\left[f(x) + \int_{x}^{N} f(\xi)K^-(\xi,x)d\xi\right] e^{-i\lambda x}dx = \tilde{f}^*(\lambda) \;,$$

where

$$f^*(x) = \{(\mathbb{I} + \mathbb{K}^{-*})f\}(x), \quad \tilde{f}^*(\lambda) = \int_{-\infty}^{\infty} f^*(x)e^{-i\lambda x}dx \;,$$

and $\mathbb{I} + \mathbb{K}^{-*}$ designates the adjoint of the transformation operator $\mathbb{I} + \mathbb{K}^-$. From this, it follows, by (3.5.8) and (3.5.8'), that

$$\frac{1}{2\pi}\int_{-\infty}^{\infty} u^-(\lambda,f)u^+(\lambda,g)a(\lambda)d\lambda + \sum_{k=1}^{n} (m_k^-)^2 u^-(i\kappa_k,f)u^-(i\kappa_k,g) =$$

$$= \frac{1}{2\pi}\int_{-\infty}^{\infty} \{r^-(\lambda)\tilde{f}^*(\lambda)g^*(\lambda) + \tilde{f}^*(-\lambda)g^*(\lambda)\}d\lambda +$$

$$+ \sum_{k=1}^{n} (m_k^-)^2 \int_{-\infty}^{\infty}\int_{-\infty}^{\infty} e^{\kappa_k(x+y)} f^*(x)g^*(y)dx\,dy =$$

$$= \int_{-\infty}^{\infty}\int_{-\infty}^{\infty} F^-(x+y)f^*(x)g^*(y)dxdy + \int_{-\infty}^{\infty} f^*(t)g^*(t)dt =$$

$$= ((\mathbb{I} + \mathbb{F}^-)(\mathbb{I} + \mathbb{K}^{-*})f,(\mathbb{I} + \mathbb{K}^{-*})\bar{g}) = ((\mathbb{I} + \mathbb{K}^-)(\mathbb{I} + \mathbb{F}^-)(\mathbb{I} + \mathbb{K}^{-*})f,\bar{g}) \;,$$

where $\mathbb{F}^-$ designates the operator acting in $L_2(-\infty,N]$ according to the rule

$$(\mathbb{F}^- f)(x) = \int_{-\infty}^{N} F^-(x+t)f(t)dt \;.$$

Hence, for the Parseval equality (3.5.44) to hold, it is necessary and sufficient that

$$(\mathbb{I} + \mathbb{K}^-)(\mathbb{I} + \mathbb{F}^-)(\mathbb{I} + \mathbb{K}^{-*}) = \mathbb{I} \ .$$

The last identity is equivalent to the fundamental equation (3.5.18'), which may be established just as in Theorem 3.2.1.

CHAPTER 4

NONLINEAR EQUATIONS

1. TRANSFORMATION OPERATORS OF A SPECIAL FORM

In 1967 Gardner, Green, Kruskal, and Miura [6] discovered a number of profound connections between the Korteweg-deVries (KdV) equation

$$v_t - 6vv_x + v_{xxx} = 0 \ , \tag{4.1.1}$$

which describes the motion of waves in shallow water, and the spectral properties of the family

$$L = -\frac{d^2}{dx^2} + v(x,t) \quad (-\infty < x < \infty) \tag{4.1.2}$$

of Sturm-Liouville operators generated by a solution $v(x,t)$ of equation (4.1.1). These connections allowed them to find the solution to the Cauchy problem

$$v(x,0) = v_0 \ , \quad \lim_{x\to\pm\infty} v(x,t) = 0$$

for the KdV equation using the inverse problem of scattering theory (this is presently known as the "inverse scattering method" (ISM); translator's note). The fundamental idea of their method was developed further in the work of Lax [12], in which the notion of an operator L-A (or Lax) pair was introduced.

The operators L and A depending on the parameter t are said to form a Lax pair if their commutator $[A,L] = AL - LA$ and the derivative $\frac{dL}{dt}$ are both multiplication operators. For example, the operators (4.1.2) and

$$A = -4 \frac{d^3}{dx^3} + 3(v \frac{d}{dx} + \frac{d}{dx} v) \tag{4.1.3}$$

form a Lax pair. Indeed, $\frac{dL}{dt}$ is the operator of multiplication by $v_t(x,t) = \frac{\partial v}{\partial t}(x,t)$, while the commutator $[A,L]$ is equal to the operator of multiplication by the function $6vv_x - v_{xxx}$. The KdV equation is equivalent to the operator equation

$$\frac{dL}{dt} = [A,L] \tag{4.1.4}$$

for the Lax pair (4.1.2), (4.1.3).

Lax showed that partial differential equations which are equivalent to equations of the form (4.1.4) for some Lax pair possess an infinite set of first integrals: in fact, it follows from equation (4.1.4) that the spectrum of the operator $L = L(t)$ does not depend on t. Gardner discovered that such equations may be treated as Hamiltonian systems with infinitely many degrees of freedom. Zakharov and Faddeev [26] showed that the KdV equation, considered in the class of functions $v(x,t)$ of rapid decrease for $x \to \pm\infty$, is a completely integrable Hamiltonian system, and that the scattering data of the operator (4.1.2) serve as canonical action-angle type variables for this system.

The next important step in this direction was made by Novikov [23], who considered the KdV equation in the class of periodic and almost-periodic functions which are also finite-zone potentials. In this class, the KdV equation represents a Hamiltonian system with a finite number of degrees of freedom, for which a complete set of commuting first integrals was found in [23]. Using a method proposed by Akhiezer [1], Its and Matreev [10] found the explicit form of the finite-zone potentials which, in conjunction with the results of Dubrovin and Novikow [4], led to explicit formulas for the solutions of the KdV equation with finite-zone initial data. A survey of the results obtained in this direction is given in [3] (see also the supplementary list of references; translator's note).

Besides the KdV equation, there are many examples of equations which have physical meaning and can be integrated by analogous methods. Historically, the first example of such an equation that followed the KdV equation is the nonlinear Schrödinger equation

$$iu_t + u_{xx} \pm |u|^2 u = 0 ,$$

for which Zakharov and Shabat [27] found a Lax pair, which allowed them to solve this equation in the class of rapidly decreasing potentials by the ISM.

This chapter is devoted to the integration of the KdV equation in the classes of decreasing and periodic functions. We use the approach proposed in [21], which is based on the same circle of ideas that was used in [12, 6].

Consider a family of Sturm-Liouville operators (4.1.2) and a family

$$A = 2N \frac{\partial}{\partial x} + B \tag{4.1.5}$$

of first-order differential operators, where the functions $N = N(z,x,t)$ and $B = B(z,x,t)$, are, for the moment, arbitrary. It will be assumed that all the functions encountered in this section are differentiable as many times as necessary with respect to all variables. In this chapter it will be often convenient to denote derivatives with respect to x and t by a prime and a dot, respectively. The letter z designates a complex parameter. It is checked directly that the operators A take functions $y = y(z,x,t)$ satisfying the equations $(L-z)y = 0$ into functions $A[y]$ that satisfy the equations

$$(L-z)A[y] = -2(N'' + B')y' - \{2Nv' + 4N'(v-z) + B''\}y \ . \tag{4.1.6}$$

We choose the functions N and B so that the coefficients of y' in (4.1.6) will vanish, and that of y will not depend on z. These requirements are equivalent to the equalities

$$B = -N' \ , \quad -2Nv' - 4N'(v-z) + N''' = f(x,t) \ , \tag{4.1.7}$$

where $f(x,t)$ is an arbitrary function that does not depend on z. We look for N of the polynomial (in z) form

$$N = N_n = \sum_{j=0}^{n} a_{n-j}(x,t)z^j \ . \tag{4.1.8}$$

Substituting this expression into the left-hand side of equality (4.1.7), we see that this side does not depend on z if the coefficients $a_k = a_k(x,t)$ satisfy the system of equations

$$\left.\begin{aligned} -4a_0' &= 0 \ , \\ -4a_1' &= a_0''' - 4a_0'v - 2a_0v' \ , \\ &\vdots \\ -4a_n' &= a_{n-1}''' - 4a_{n-1}'v - 2a_{n-1}v' \ . \end{aligned}\right\} \tag{4.1.9}$$

If this is the case, then $f(x,t) = f_n(x,t)$ is given by the formula

$$f_n(x,t) = a_n''' - 4a_n'v - 2a_nv' = -4\frac{\partial}{\partial x}a_{n+1}(x,t) , \qquad ((4.1.9')$$

provided that a_{n+1} is determined from the equation

$$-4a_{n+1}' = a_n''' - 4a_n'v - 2a_nv' .$$

The successive integration of equations (4.1.9) yields

$$\left.\begin{aligned} a_0 &= p_0 , \\ a_1 &= \frac{1}{2}p_0v + p_1 , \\ a_2 &= -\frac{1}{8}p_0(v'' - 3v^2) + \frac{1}{2}p_1v + p_2 , \\ &\vdots \\ a_k &= p_0(ID)^k[1] + p_1(ID)^{k-1}[1] + \ldots + p_k , \end{aligned}\right\} \qquad (4.1.10)$$

where $p_k = p_k(t)$ $(0 \leq k \leq n)$ are arbitrary coefficients that may depend only on t, I is the integration operator, and

$$D = -\frac{1}{4}\frac{d^3}{dx^3} + v\frac{d}{dx} + \frac{1}{2}v' .$$

It follows from these formulas that the general solution of the system of equations (4.1.9) has the form

$$a_k(x,t) = \sum_{j=0}^{k} p_j(t)\hat{a}_{k-j}(x) , \qquad (4.1.10')$$

where $\hat{a}_0(x) = 1, \hat{a}_1(x), \ldots, \hat{a}_n(x)$ is an arbitrary particular solution of this system. Using the polynomial N_n with the coefficients $a_k(x,t)$ defined in this manner, we obtain the operators $A_n = 2N_n\frac{\partial}{\partial x} - N_n'$, for which equation (4.1.6) takes the form

$$(L-z)A_n[y] = f_n(x,t)y ,$$

where $f_n(x,t)$ is given by (4.1.9').

Upon differentiating the equation $-y'' + vy - zy = 0$ with respect to t, we find that the operator $\frac{\partial}{\partial t}$ takes the solutions of this equation into solutions $\dot{y} = \frac{\partial y}{\partial t}$ of $(L-z)\dot{y} = -\dot{v}y$. Hence, the operators

$$M_n = \frac{\partial}{\partial t} + A_n = \frac{\partial}{\partial t} + 2N_n\frac{\partial}{\partial x} - N_n' \qquad (4.1.11)$$

transform the solutions of the equation $(L-z)y = 0$ into functions $M_n[y]$ that satisfy the equations $(L-z)M_n[y] = (f_n-\dot{v})y$. We have thus proved the following assertion.

LEMMA 4.1.1. *If the coefficients of the polynomials* N_n *satisfy the system of equations* (4.1.9), *then the operators* M_n *of the form* (4.1.11) *transform the solutions* y *of the Sturm-Liouville equation* $(L-z)y = 0$ *into solutions* $M_n[y]$ *of the equations* $(L-z)M_n[y] = K_n[v]y$, *where*

$$K_n[v] = N_n''' - 4N_n'(v-z) - 2Nv' - \dot{v} \equiv -4\frac{\partial}{\partial x}a_{n+1}(x,t) - \frac{\partial v}{\partial t}.$$

□

The Sturm-Liouville equation $(L-z)y = 0$ is obviously equivalent to the system of first-order equations

$$y_1' = y_2 \quad , \quad y_2' = (v-z)y_1$$

for the functions $y_1 = y$ and $y_2 = y'$, which may be written in the matrix form

$$\mathcal{L}Y = 0 \quad , \quad \mathcal{L} = \frac{\partial}{\partial x} + \mathcal{V}(z,x,t) \ , \tag{4.1.12}$$

where

$$\mathcal{V}(z,x,t) = -\begin{pmatrix} 0 & 1 \\ v(x,t) - z & 0 \end{pmatrix}. \tag{4.1.12'}$$

[Warning: here special characters are used for matrices and operators with matrix coefficients, to distinguish them from their scalar counterparts. Later the special characters will be dropped; translator's note.]

Since

$$M_n[y] = \dot{y} + 2N_n y' - N_n' y = \dot{y}_1 + 2N_n y_2 - N_n' y_1$$

and

$$M_n[y]' = \dot{y}_1' + 2N_n' y_2 + 2N_n y_2' - N_n'' y_1 - N_n' y_1' = \dot{y}_2 + N_n' y_2 + (2N_n(v-z) - N_n'')y \ ,$$

we may, upon introducing the matrix operators

$$\mathcal{M}_n = \frac{\partial}{\partial t} + \mathcal{A}_n \quad , \quad \mathcal{C} = \begin{pmatrix} 0 & 0 \\ 1 & 0 \end{pmatrix} \tag{4.1.11'}$$

and

$$A_n = \begin{pmatrix} -N_n' & 2N_n \\ 2N_n(v-z) - N_n'' & N_n' \end{pmatrix} , \tag{4.1.11"}$$

reformulate Lemma 4.1.1 as follows.

LEMMA 4.1.1'. *If the coefficients of the polynomials* N_n *satisfy the system of equations* (4.1.9), *then the operators* $\mathcal{M}_n$ *of the form* (4.1.11') *transform the solutions* $\mathcal{Y}$ *of the equation* (4.1.12) *into vector-functions* $\mathcal{M}_n[\mathcal{Y}]$ *which satisfy the equations* $L\mathcal{M}_n[\mathcal{Y}] = K_n[v]C\mathcal{Y}$. □

COROLLARY. *The function* $v(x,t)$ *satisfies the equation* $K_n[v] = 0$ *if and only if the operators* M_n $(\mathcal{M}_n)$ *transform the solutions of the equation* $-y'' + v(x,t)y - zy = 0$ *(respectively,* $L\mathcal{Y} = 0$*) into solutions of the same equation.* □

In the following, an important role will be played by the operators M_n and $\mathcal{M}_n$ with $n = 0,1$ and special values of the coefficients $p_k(t)$. Setting $p_0 = -\frac{1}{2}$, we get

$$\left.\begin{aligned} &N_0 = -\frac{1}{2} , \quad M_0 = \frac{\partial}{\partial t} - \frac{\partial}{\partial x} , \quad K_0[v] = v' - v_t , \\ &\mathcal{M}_0 = \frac{\partial}{\partial t} - \begin{pmatrix} 0 & 1 \\ v-z & 0 \end{pmatrix} , \end{aligned}\right\} \tag{4.1.13}$$

whereas the choice $p_0 = -2$, $p_1 = 0$, yields

$$\left.\begin{aligned} &N_1 = -(v + 2z) , \quad K_1[v] = -v''' + 6vv' - v_t , \\ &M_1 = \frac{\partial}{\partial t} - 2(v+2z)\frac{\partial}{\partial x} + v' , \\ &\mathcal{M}_1 = \frac{\partial}{\partial t} + \begin{pmatrix} v' & -2(v+2z) \\ v'' - 2v^2 - 2vz + 4z^2 & -v' \end{pmatrix} . \end{aligned}\right\} \tag{4.1.14}$$

If v satisfies the KdV equation (4.1.2), then $K_1[v] = 0$, so that the operator M_1 transforms the solutions of the Sturm-Liouville equation $(L-z)y = 0$ into solutions of the same equation, i.e., it is a kind of transformation operator.

We next deduce explicit formulas for the operators M_n and $\mathcal{M}_n$ for arbitrary values of n . In view of (4.1.10'), it suffices to find a particular solution of system (4.1.9). Notice that, if

$$N_\infty = N_\infty(z,x) = \sum_{k=0}^{\infty} \hat{a}_k(x)z^{-k} \tag{4.1.15}$$

is a formal solution of the equation

$$N_\infty''' - 4N_\infty'(v-z) + 2N_\infty v' = 0 , \tag{4.1.16}$$

then the coefficients $\hat{a}_0(x), \hat{a}_1(x), \ldots$ satisfy the system (4.1.9). Therefore, in order to find a partial solution of the system (4.1.9), it suffices to find a formal solution of the form (4.1.15) to equation (4.1.16). According to (4.1.10), the general solution of (4.1.9) is provided by the coefficients of the nonnegative powers of z in the formal product $P_n(z,t)N_\infty(z,x)$, where

$$P_n(z,t) = \sum_{j=0}^{n} p_{n-j}(t)z^j$$

is an arbitrary polynomial. Consequently, the general form of the polynomials (4.1.18) for which the right-hand side of the equality (4.1.7) does not depend on z is

$$N_n = N_n(z,x,t) = \operatorname{Reg}\{P_n(z,t)N_\infty(z,x)\} ,$$

where the symbol Reg indicates that in the formal product $P_n(z,t)N_\infty(z,x)$, one retains only the nonnegative powers of z . Since $P_n(z,t)N_\infty(z,t) = \sum_{k=0}^{\infty} a_k(x,t)z^{n-k}$ is also a formal solution of equation (4.1.16), its coefficients satisfy the equalities

$$-4a_k' = a_{k-1}''' - 4a_{k-1}'v - 2a_{k-1}v' ,$$

from which we obtain, for $k = n+1$, by (4.1.9'),

$$f_n(x,t) = -4 \frac{\partial}{\partial x} a_{n+1}(x,t) = -4 \frac{\partial}{\partial x} \operatorname{Res}\{P_n(z,t)N_\infty(z,x)\} ,$$

where Res designates the coefficient of z^{-1} in the formal product $P_n(z,t)N_\infty(z,x)$.

Now let us find a formal solution (4.1.15) to equation (4.1.16). It is readily checked that the product y_1y_2 of two solutions of the Sturm-Liouville equation $-y'' + vy = zy$ satisfies equation (4.1.16). Hence, if $y(\sqrt{z},x)$ and $y(-\sqrt{z},x)$ are the solutions of the Sturm-Liouville equation, constructed in Lemma 1.4.2, then $y(\sqrt{z},x)y(-\sqrt{z},x)$ is a solution of equation

(4.1.16). But, by (1.4.21),

$$\frac{y(\sqrt{z},x)y(-\sqrt{z},x)}{2i\sqrt{z}+\sigma(\sqrt{z},0)-\sigma(-\sqrt{z},0)} = \{2i\sqrt{z}+\sigma(\sqrt{z},x)-\sigma(-\sqrt{z},x)\}^{-1} =$$

$$= 2i\sqrt{z}\left\{1 + 2\sum_{j=0}^{n}\sigma_{2j+1}(x)(-4z)^{-j-1} + o(z^{-n-2})\right\}^{-1} .$$

This shows that the formal expansion of the expression

$$N_\infty(z,x) = \left\{1 + 2\sum_{j=0}^{\infty}\sigma_{2j+1}(x)(-4z)^{-j-1}\right\}^{-1}$$

in negative powers of z, is a formal solution of equation (4.1.16). Therefore, all the polynomials $N_n(z,x,t)$, in which we are interested, and the corresponding functions $f_n(x,t)$, are given by the formulas

$$N_n(z,x,t) = \operatorname{Reg}\left\{P_n(z,t)\left[1 + 2\sum_{j=0}^{\infty}\sigma_{2j+1}(x)(-4z)^{-j-1}\right]^{-1}\right\}$$

and

$$f_n(x,t) = -4\frac{\partial}{\partial x}\operatorname{Res}\left\{P_n(z,t)\left[1 + 2\sum_{j=0}^{\infty}\sigma_{2j+1}(x)(-4z)^{-j-1}\right]^{-1}\right\} .$$

This shows, in particular, that $a_k(x,t)$ and $f_n(x,t)$ are polynomials in the potential v and its derivatives with respect to x, with coefficients which are independent of x. We have thus found the general form of the operators $A_n = 2N_n\frac{\partial}{\partial x} - N_n'$ that transform the solutions of the equation $(L-z)y = 0$ into solutions $A_n[y]$ of the equation $(L-z)A_n[y] = f_n(x,t)y$. On the space of solutions of the equation $(L-z)y = 0$, the operators of multiplication by the polynomials N_n and N_n' can be replaced by the differential operators that are obtained from N_n and N_n' upon substituting the Sturm-Liouville operator L for z. Hence, on the space of solutions y of $(L-z)y = 0$ (where z is an arbitrary complex parameter), the A_n's act as differential (with respect to x) operators $\hat{A}_n$ of order $2n+1$, and, in addition, the commutator $[\hat{A}_n,L]$ is equal to the operator of multiplication by the function $f_n(x,t)$. Since the linear span of all such solutions y is everywhere dense, $[\hat{A}_n,L] = f_n(x,t)$ on all sufficiently smooth functions. In other words, the differential operators $\hat{A}_n$, taken together with the Sturm-Liouville operator L, form Lax pairs. Reversing the argument, it is readily established that every differential operator $\hat{A}$, such that $L - \hat{A}$ is a Lax pair, is equal to one of the operators $\hat{A}_n$ built above.

Finally, we remark that

$$[L-zI,M_n] = 4N_n'L + K_n[v] \quad , \quad [\mathcal{L},M_n] = K_n[v]C$$

if z remains a complex parameter in the operators A_n and A_n .

PROBLEMS

The commutator $[L,M]$ of two differential operator L and M of the form

$$L = \frac{\partial}{\partial x} + V(z,x,t) \quad , \quad M = \frac{\partial}{\partial t} + A(z,x,t)$$

with operator-valued coefficients $V = v(z,x,t)$ and $A = A(z,x,t)$ belonging to the set $\mathcal{OH}$ (see Problem 4, Section 2, Chapter 1) has the form

$$[L,M] = LM - ML = A' - \dot{V} + [V,A] \ . \tag{4.1.17}$$

The problems proposed below are generalizations of Lemma 4.1.1': given the rules according to which the operators A and V depend on z , it is required to find their general form for which the commutator $[L,M]$ is independent of z . For example, if

$$V = z(V_{-1}z^{-1} + V_0) \quad , \quad A = z^N(A_N z^{-N} + A_{N-1}z^{-N+1} + \ldots + A_0) \ ,$$

we have, by (4.1.17),

$$[L,M] = \sum_{k=0}^{N} A_k' z^{N-k} - \dot{V}_1 - \dot{V}_0 z + \sum_{k=0}^{N} [V_1,A_k]z^{N-k} + \sum_{k=0}^{N} [V_0,A_k]z^{N-k+1} \ .$$

If the following equalities are satisfied:

$$\left.\begin{aligned} &[V_0,A_0] = 0 \\ &A_k' + [V_1,A_k] + [V_0,A_{k+1}] = 0 \quad (k = 0,1,\ldots,N-2) \\ &A_{N-1}' - \dot{V}_n + [V_1,A_{N-1}] + [V_0,A_N] = 0 \ , \end{aligned}\right\} \tag{4.1.18}$$

then

$$[L,M] = A_N' - V_1 + [V_1,A_N] \ , \tag{4.1.18'}$$

and hence the commutator $[L,M]$ does not depend on z . Thus, if A and V depend on z in the indicated way, the problem amounts to solving the system of equations (4.1.18) for $A_0,A_1,\ldots,A_N$.

1. The operator Dirac equation $By' + \Omega y - zy = 0$ is obviously equivalent to the equation $Ly = 0$, where

$$L = \frac{d}{dx} + V \ , \quad V = z(V_1 z^{-1} + V_0) \ , \quad V_1 = -B\Omega \ , \quad V_0 = B \ .$$

Find the general form of the operators

$$M_2 = \frac{\partial}{\partial t} + z^2(A_2 z^{-2} + A_1 z^{-1} + A_0)$$

for which the commutator $[L,M_2]$ is independent of z, and write the conditions under which it vanishes.

Hint. It follows from the equalities

$$B^2 = -I \ , \quad B\Omega + \Omega B = 0$$

that the operators

$$P = \frac{1}{2}(I + iB) \ , \quad Q = \frac{1}{2}(I - iB)$$

satisfy the relations

$$P^2 = P \ , \quad Q^2 = Q \ , \quad PQ = QP = 0 \ , \quad P+Q = I \ , \quad B = i(Q-P) \ ,$$

$$BP = PB = -iP \ , \quad BQ = QB = iQ \ , \quad P\Omega = \Omega Q \ , \quad Q\Omega = \Omega P \ .$$

Therefore, for arbitrary operators $X,Y \in \mathcal{O}H$:

$$[V_0,X] = [B,X] = 2i(QXP - PXQ) \ ,$$

$$[V_1,Y] = -[B\Omega,Y] = i\{P(\Omega QYP + PYQ\Omega)P - Q(\Omega PYQ + QYP\Omega)Q +$$
$$+ P(\Omega QYQ - PYP\Omega)Q - Q(\Omega PYP - QYQ\Omega)P\} \ ,$$

and hence (4.1.18) is equivalent to the system

$$PA_0Q = QA_0P = 0 \ ,$$

$$P\{A_k' + i(\Omega QA_kP + PA_kQ\Omega)\}P = 0 \ ,$$

$$Q\{A_k' - i(\Omega PA_kQ + QA_kP\Omega)\}Q = 0 \ ,$$

$$P\{A_k' + i(\Omega QA_kQ - PA_kP\Omega) - 2iA_{k+1}\}Q = 0 \ ,$$

$$Q\{A_k' - i(\Omega PA_kP - QA_kQ\Omega) + 2iA_{k+1}\}P = 0 \ .$$

The general solution to this system is found by means of the recursion formulas

$A_0 = C_0$ (with $[B,C_0] = 0$, $C_0 = \text{const}$) ,

$$PA_{k+1}Q = \frac{1}{2i} P\{A_k' + i[\Omega,A_k]\}Q ,$$

$$QA_{k+1}P = \frac{1}{2i} Q\{-A_k' + i[\Omega,A_k]\}P ,$$

$$PA_{k+1}'P = \frac{1}{2} P\{[\Omega,A_k'] - i[\Omega^2,A_k]\}P ,$$

and

$$QA_{k+1}'Q = \frac{1}{2} Q\{[\Omega,A_k'] + i[\Omega^2,A_k]\}Q ,$$

which for $N = 2$ yield

$$A_0 = C_0 \ , \quad A_1 = \frac{1}{2} [\Omega,C_0] + \frac{B}{2} [\Omega_2,C_0] + C_1 ,$$

and

$$A_2 = \frac{B}{4} [\Omega',C_0] + \frac{1}{4} [\Omega,B[\Omega_2,C_0] + 2C_1] + C_2(x,t) ,$$

where $C_0, C \in \mathcal{OH}$ are arbitrary constant operators, $C_2(x,t)$ is an arbitrary operator- ($\mathcal{OH}$-) valued function which commutes with B, and

$$\Omega_2(x,t) = \int_{x_0}^{x} \Omega^2(\xi,t)d\xi .$$

The commutator $[L,M_2]$ vanishes if

$$\frac{B}{4} [\Omega'',C_0] + \frac{1}{4} [\Omega,B[\Omega_2,C_0] + 2C_1]' + B\dot{\Omega} - B[\Omega,C_2] = 0$$

and

$$C_2' + \frac{1}{4} [\Omega,[C_0,\Omega'] + B[\Omega,B[\Omega_2,C_0] + 2C_1]] = 0 .$$

In particular, for $C_0 = C_2 = 0$ and $C_1 = B$, you find:

$$A_0 = 0 \ , \quad A_1 = B \ , \quad A_2 = -B\Omega ,$$

and if

$$M_1 = \frac{\partial}{\partial\tau} + Bz - B\Omega ,$$

then

$$[L,M_1] = -B\left(\frac{\partial}{\partial x} - \frac{\partial}{\partial\tau}\right)\Omega .$$

Also, for $C_0 = 2B$, $C_1 = 0$, and $C_2 = B\Omega^2$, you find:

$$A_0 = 2B \ , \quad A_1 = -2B\Omega \ , \quad A_2 = \Omega' + B\Omega^2 ,$$

and if

$$M_2 = \frac{\partial}{\partial t} + 2Bz^2 - 2B\Omega z + \Omega' + B\Omega^2 ,$$

then

$$[L,M_2] = \left(B\frac{\partial}{\partial t} + \frac{\partial^2}{\partial x^2}\right)\Omega - 2\Omega^3 .$$

Therefore, the operators M_1 and M_2 transform the solutions of the Dirac equation

$$By' + \Omega(x,\tau,t)y - zy = 0$$

into solutions of the same equation if the operator-valued potential $\Omega = \Omega(x,\tau,t)$ satisfy the equations

$$\left(\frac{\partial}{\partial x} - \frac{\partial}{\partial \tau}\right)\Omega = 0 \ , \quad \left(B\frac{\partial}{\partial t} + \frac{\partial^2}{\partial x^2}\right)\Omega - 2\Omega^3 = 0 .$$

We remark that in the case where the space H is of finite dimension $2n$ and

$$B = i\begin{pmatrix} I_n & 0 \\ 0 & -I_n \end{pmatrix} , \quad \Omega = \begin{pmatrix} 0 & P_n(x,\tau,t) \\ \pm P_n^*(x,\tau,t) & 0 \end{pmatrix} ,$$

the last equation is equivalent to

$$\left(i\frac{\partial}{\partial t} + \frac{\partial^2}{\partial x^2}\right)P_n \mp 2P_nP_n^*P_n = 0 .$$

2. Find the general form of the operators

$$M_n = \frac{\partial}{\partial t} + z^n \sum_{k=0}^{n} A_k z^{-k} ,$$

for which the commutator $[L,M_n]$ does not depend on z (the operator $L = \frac{d}{dx} - B\Omega + Bz$ is the same as in the preceding problem).

Hint. The operator coefficients T_k of the formal solution $T = \sum_{k=0}^{\infty} T_k z^{-k}$ of the equation

$$T' + [V,T] = 0 ,$$

where

$$V = -B\Omega + \Omega z ,$$

satisfy the same system of equations (4.1.18) as the sought-for operators A_k. Consequently,

$$M_n = \frac{\partial}{\partial t} + \operatorname{Reg}\{z^nT\} .$$

It is readily verified that the general solution of the equation $T' + [V,T] = 0$ is given by the formula $T = YCY^{-1}$, where Y is a nondegenerate solution of the equation $Y' + VY = 0$, and C is an arbitrary operator which does not depend on x. In the present case, the equation $Y' + VY = 0$ is equivalent to the Dirac equation

$$By' + \Omega y - zy = 0 ,$$

the nondegenerate solutions of which will be denoted by $y(z,x;B,\Omega)$. Since y^{-1} satisfies the equation

$$(y^{-1})' + (y^{-1})B\Omega - (y^{-1})Bz = 0 ,$$

you have

$$B^*\{(y^{-1})^*\}' + \Omega^*(y^{-1})^* - (-\bar{z})(y^{-1})^* = 0 ,$$

and hence

$$\{y(z,x;B,\Omega)\}^{-1} = \{y(-\bar{z},x;B^*,\Omega^*)\}^* .$$

Therefore,

$$T = y(z,x;B,\Omega)Cy(-\bar{z},x;B^*,\Omega^*) ,$$

where, for $y(z,x;B,\Omega)$ and $y(-z,x;B^*,\Omega^*)$, you may take the solutions constructed in Problems 1 and 2, Section 4, Chapter 1:

$$y(z,x;B,\Omega) = e^{izx}(I + v(z,x,\Omega))u_1(z,x,\Omega)P + e^{-izx}(I - v(-z,x,\Omega))u_1(-z,x,\Omega)Q$$

and

$$y(-\bar{z},x;B^*,\Omega^*) = e^{izx}Qu_1(-\bar{z},x,\Omega^*)^*(I + v(-z,x,\Omega^*))^* +$$
$$+ e^{-izx}Pu_1(\bar{z},x,\Omega^*)^*(I - v(z,x,\Omega^*)^*) .$$

In order for the solution T to admit a formal series expansion in negative powers of z, the operator C must commute with B. If this is the case, then

$$T = (I + v(z,x,\Omega))u_1(z,x,\Omega)PCPu_1(\bar{z},x,\Omega^*)^*(I - v(z,x,\Omega^*)^*) +$$
$$+ (I - v(-z,x,\Omega))u_1(-z,x,\Omega)QCQu_1(-\bar{z},x,\Omega^*)^*(I + v(-\bar{z},x,\Omega^*)^*) .$$

Taking $C = \sum_{k=0}^{n} C_k z^k$ with $[C_k,B] = 0$, you obtain the general form of the operators $M_n = \frac{\partial}{\partial t} + \text{Reg}\,\{z^n T\}$. Notice that $[L,M_n] = -V_{1t} - [V_0,\text{Res}\,T]$.

3. Find conditions under which the operators

$$L = \frac{d}{dx} + \sum_{k=-1}^{1} V_k z^k \quad , \quad M = \frac{d}{dx} + \sum_{k=-1}^{1} A_k z^k$$

commute, assuming that $V_k = \beta_k U_k$ and $A_k = \alpha_k U_k$, where α_k and β_k are numbers.

Hint. By (4.1.17),

$$[L,M] = [V_{-1},A_{-1}]z^{-2} + (A'_{-1} - \dot{V}_{-1} + [V_{-1},A_0] + [V_0,A_{-1}])z^{-1} +$$

$$+ (A'_0 - \dot{V}_0 + [V_{-1},A_1] + [V_0,A_0] + [V_1,A_{-1}]) +$$

$$+ (A'_1 - \dot{V}_1 + [V_0,A_1] + [V_1,A_1])z + [V_1,A_1]z^2 .$$

Taking note of the equalities $V_k = \beta_k U_k$ and $A_k = \alpha_k U_k$, you see that $[L,M] = 0$ if the operators U_k satisfy the following system of differential equations:

$$\begin{cases} \alpha_{-1}U'_{-1} - \beta_{-1}\dot{U}_{-1} + (\beta_{-1}\alpha_0 - \beta_0\alpha_{-1})[U_{-1},U_0] = 0 , \\ \alpha_0 U'_0 - \beta_0\dot{U}_0 + (\beta_{-1}\alpha_1 - \beta_1\alpha_{-1})[U_{-1},U_1] = 0 , \\ \alpha_1 U'_1 - \beta_1\dot{U}_1 + (\beta_0\alpha_1 - \beta_1\alpha_0)[U_0,U_1] = 0 . \end{cases}$$

In the most interesting case, where $U_1 = \text{const}$, $\alpha_0 = \alpha_1$, $\beta_0 = \beta_1$, this system reduces to the equations

$$\begin{cases} D_{-1}U_{-1} + [\delta U_0,U_{-1}] = 0 , \\ D_0 U_0 + [\delta U_1,U_{-1}] = 0 , \end{cases}$$

where

$$D_k = (\partial_k \frac{\partial}{\partial x} - \beta_k \frac{\partial}{\partial t}) \quad , \quad k = 0,-1 ,$$

and

$$\delta = \beta_0\alpha_{-1} - \beta_{-1}\alpha_0 .$$

For $\delta = 0$, the operators L and M commute if

$$D_{-1}U_{-1} = 0 \quad , \quad D_0 U_0 = 0 .$$

For $\delta = 0$, it is readily seen, just as in the preceding problem, that the operators U_{-1} and δU_0 are expressible through a nondegenerate solution of the equation

$$Y' + \delta U_0 Y = 0$$

via the formulas

$$\delta U_0 = -(D_{-1}Y)Y^{-1} \quad , \quad U_{-1} = YCY^{-1} \quad (\text{here } D_{-1}C = 0) .$$

Consequently, in this case $[L,M] = 0$ whenever the operator-valued function Y satisfies the equation

$$D_0\{(D_{-1}Y)Y^{-1}\} - \delta^2[U_1, YCY^{-1}] = 0 . \tag{4.1.19}$$

If, for example, the space H is two-dimensional, and

$$U_1 = \begin{pmatrix} 0 & -1 \\ 1 & 0 \end{pmatrix} \quad , \quad Y = \exp\left(\frac{1}{2} Bu\right) \quad , \quad B = \begin{pmatrix} i & 0 \\ 0 & -i \end{pmatrix} \quad , \quad C = \gamma U_1 \, ,$$

where $u = u(x,\tau,t)$ is a scalar function, then equation (4.1.19) reduces to

$$D_0(D_{-1}u) + 4\gamma\delta^2 \sin u = 0 .$$

Therefore, the operators

$$M_0 = \frac{\partial}{\partial\tau} + \sum_{k=-1}^{1} \beta_k U_k \quad , \quad M_1 = \frac{\partial}{\partial t} + \sum_{k=-1}^{1} \alpha_k U_k$$

(with $\alpha_0 = \alpha_1$ and $\beta_0 = \beta_1$), where

$$U_{-1} = \gamma e^{\frac{1}{2} Bu} U_1 e^{-\frac{1}{2} Bu} = \gamma(\cos uI + \sin uB)U_1 \quad , \quad U_0 = -\frac{D_{-1}u}{2\delta} B \, ,$$

transform the solutions of the equation

$$y' + \left(\sum_{k=-1}^{1} \beta_k U_k z^k \right) y = 0$$

into solutions of the same equation, provided that the function $u = u(x,\tau,t)$ satisfies the equations

$$\frac{\partial u}{\partial x} - \frac{\partial u}{\partial \tau} = 0 \quad , \quad D_0(D_{-1}u) + 4\gamma\delta^2 \sin u = 0 .$$

Notice that for $\beta_0 = \beta_{-1} = 1$ and $\alpha_0 = -\alpha_{-1}$, the last equation becomes

$$\left(-\alpha_0^2 \frac{\partial^2}{\partial x^2} + \frac{\partial^2}{\partial t^2} \right) u + 16\gamma\alpha_0^2 \sin u = 0 ,$$

with $\delta = -2\alpha_0$.

2. RAPIDLY DECREASING SOLUTIONS OF THE KORTEWEG-DE VRIES EQUATION

We say that the function $f(x,t)$ is rapidly decreasing if

$$\max_{|t|\leq T} \int_{-\infty}^{\infty} (1 + |x|)|f(x,t)|dx < \infty$$

for all nonnegative values of T . Also we say that the solution $v(x,t)$ of the KdV equation $\dot{v} - 6vv' + v''' = 0$ is rapidly decreasing if $v(x,t)$ and its derivatives of order ≤ 3 with respect to x are rapidly decreasing. We remark that the derivative of any rapidly decreasing solution with respect to t is automatically rapidly decreasing.

We consider the Cauchy problem

$$\dot{v}(x,t) - 6v(x,t)v'(x,t) + v'''(x,t) = 0 \quad , \quad v(x,0) = v_0(x) \tag{4.2.1}$$

with a real-valued rapidly decreasing initial function $v_0(x)$, and assume that it admits a real-valued, rapidly decreasing solution $v(x,t)$. To find this solution, we consider the family of Sturm-Liouville equations

$$-y'' + v(x,t)y - \lambda^2 y = 0 \quad (-\infty < x < \infty) \ , \tag{4.2.2}$$

generated by the sought-for solution $v(x,t)$. For each fixed value of t , this equation obviously satisfies condition (3.5.2), and hence all the results of Section 5, Chapter 3, apply.

For every value of the parameter λ in the closed upper half plane, equation (4.2.2) has solutions $e^+(\lambda,x;t)$ and $e^-(\lambda,x;t)$, which are representable in the form

$$e^{\pm}(\pm\lambda,x;t) = e^{\pm i\lambda x} \pm \int_x^{\pm\infty} K^{\pm}(x,y;t)e^{\pm i\lambda y}dy \ , \tag{4.2.3}$$

and are connected by the relations

$$\left.\begin{aligned} e^+(\lambda,x;t) &= b(\lambda,t)e^-(-\lambda,x;t) + a(\lambda,t)e^-(\lambda,x;t) \ , \\ e^-(-\lambda,x;t) &= -b(-\lambda,t)e^+(\lambda,x,t) + a(\lambda,t)e^+(-\lambda,x;t) \end{aligned}\right\} \tag{4.2.4}$$

for real λ , where

$$a(\lambda,t) = (2i\lambda)^{-1}\{e^+(\lambda,0;t)'e^-(-\lambda,0;t) - e^+(\lambda,0;t)e^-(-\lambda,0;t)'\}$$

and

$$b(\lambda,t) = (2i\lambda)^{-1}\{e^-(\lambda,0;t)'e^+(\lambda,0;t) - e^-(\lambda,0;t)e^+(\lambda,0;t)'\} .$$

The discrete eigenvalues of equation (4.2.2) are the squares of the zeros $i\kappa_k(t)$ of the function $a(\lambda,t)$, and the corresponding eigenfunctions are related by

$$e^-(-i\kappa_k(t),x;t) = c_k^-(t)e^+(i\kappa_k(t),x;t) . \tag{4.2.5}$$

Moreover,

$$(m_k^+(t))^{-2} = \int_{-\infty}^{\infty} |e^+(i\kappa_k(t),x;t)|^2 dx = i(c_k^-(t))^{-1} \left.\frac{\partial a(\lambda,t)}{\partial\lambda}\right|_{\lambda=i\kappa_k(t)} . \tag{4.2.6}$$

The potential $v(x,t)$ is uniquely recovered from the scattering data. Hence, to find the solution $v(x,t)$ of the Cauchy problem (4.2.1), it suffices to find the law governing the time evolution of the scattering data

$$r^+(\lambda,t) = -\frac{b(-\lambda,t)}{a(\lambda,t)} , \quad i\kappa_k(t) , \quad (m_k^+(t))^2 \tag{4.2.7}$$

of the family of equations (4.2.2). Since the potentials $v(x,t)$ of this family satisfy the KdV equation, then, as was shown in the preceding section, the operators

$$M_1 = \frac{\partial}{\partial t} - 2(v(x,t) + 2\lambda^2)\frac{\partial}{\partial x} + v'(x,t) \tag{4.2.8}$$

transform solutions (differentiable with respect to t) of equation (4.2.2) into solutions of the same equation.

Since the functions $v(x,t)$ and $v_t(x,t)$ are, by assumption, rapidly decreasing, it follows from the integral equations of the form (3.1.11) and (3.1.13), that the kernels $K^\pm(x,y;t)$ are differentiable with respect to t, and that

$$\left|\int_x^{\pm\infty} |\frac{\partial}{\partial t} K^\pm(x,y;t)\, dy\right| < \infty , \quad \left|\int_{x_1}^{\pm\infty} \left|\frac{\partial^2}{\partial t\partial x_i} K^\pm(x_1,x_2;t)\right| dx_2\right| < \infty ,$$

and

$$\lim_{x\to\pm\infty} \int_x^{\pm\infty} |\frac{\partial}{\partial t} K^\pm(x,y;t)|dy = 0 .$$

Consequently, the solutions $e^\pm(\pm\lambda,x;t)$, their derivatives $\frac{\partial}{\partial x} e^\pm(\pm\lambda,x;t)$, and hence the coefficients $a(\lambda,t)$, $b(\lambda,t)$, are differentiable with respect to t. Moreover, for every point λ in the closed upper half plane,

$$\left.\begin{array}{l}\lim\limits_{x\to\pm\infty} e^{\mp i\lambda x}\dot{e}^{\pm}(\pm\lambda,x;t) = 0 \ , \quad \lim\limits_{x\to\pm\infty} e^{\mp i\lambda x}e^{\pm}(\pm\lambda,x;t) = 1 \ , \\ \lim\limits_{x\to\pm\infty} e^{\mp i\lambda x}e^{\pm}(\pm\lambda,x;t)' = \pm i\lambda . \end{array}\right\} \qquad (4.2.9)$$

[Warning: in the preceding chapters, the dot was used to denote the derivative with respect to the spectral parameter λ ; translator's note.]

This permits us to apply the operator M_1 to every term of the equalities (4.2.4). On applying it to $e^+(\lambda,x;t)$, we obtain a solution $M_1[e^+(\lambda,x;t)]$ of equation (4.2.2) with the following asymptotics for $x \to \pm\infty$:

$$M_1[e^+(\lambda,x;t)] = -4i\lambda^3 e^{i\lambda x} + o(1) \quad (x \to +\infty) \ ,$$

and

$$M_1[e^+(\lambda,x;t)] = M_1[b(\lambda,t)e^-(-\lambda,x;t)] + M_1[a(\lambda,t)e^-(\lambda,x;t)] =$$

$$= \{\dot{b}(\lambda,t) + 4i\lambda^3 b(\lambda,t)\}e^{-i\lambda x} + \{\dot{a}(\lambda,t) - 4i\lambda^3 a(\lambda,t)\}e^{i\lambda x} + o(1) \quad (x \to -\infty) \ .$$

These follow from (4.2.9) upon taking note of the fact that $\lim\limits_{x\to\pm\infty} v(x,t) = \lim\limits_{x\to\pm\infty} v'(x,t) = 0$. But the only solutions of equation (4.2.2) with these asymptotics are

$$-4i\lambda^3 e^+(\lambda,x;t)$$

and

$$\{\dot{b}(\lambda,t) + 4i\lambda^3 b(\lambda,t)\}e^-(-\lambda,x;t) + \{\dot{a}(\lambda,t) - 4i\lambda^3 a(\lambda,t)\}e^-(\lambda,x;t) \ .$$

This implies that

$$-4i\lambda^3 e^+(\lambda,x;t) =$$

$$= \{\dot{b}(\lambda,t) + 4i\lambda^3 b(\lambda,t)\}e^-(-\lambda,x;t) + \{\dot{a}(\lambda,t) - 4i\lambda^3 a(\lambda,t)\}e^-(\lambda,x;t) \ .$$

Comparing this equality with (4.2.4), we obtain the identity

$$\{\dot{b}(\lambda,t) + 8i\lambda^3 b(\lambda,t)\}e^-(-\lambda,x;t) + \dot{a}(\lambda,t)e^-(\lambda,x;t) = 0 \ ,$$

from which we obtain, in view of the linear independence of the solutions $e^-(-\lambda;x,t)$ and $e^-(\lambda,x;t)$, the following differential equations for the coefficients $a(\lambda,t)$ and $b(\lambda,t)$:

$$\dot{a}(\lambda,t) = 0 \ , \quad \dot{b}(\lambda,t) + 8i\lambda^3 b(\lambda,t) = 0 \ .$$

Therefore,

$$a(\lambda,t) = a(\lambda,0) \quad , \quad b(\lambda,t) = b(\lambda,0)e^{-8i\lambda^3 t} ,$$

which implies, in particular, that the zeros $i\kappa_k(t)$ of the function $a(\lambda,t) = a(\lambda,0)$ are independent of t. Using this fact and applying the operator M_1 to both members of equality (4.2.5), we get, in just the same way as above,

$$4\kappa_k^3 e^-(-i\kappa_k,x;t) = \{\dot{c}_k^-(t) - 4\kappa_k^3 c_k^-(t)\}e^+(i\kappa_k,x;t) ,$$

whence

$$\dot{c}_k^-(t) - 8\kappa_k^3 c_k^-(t) = 0 ,$$

or

$$c_k^-(t) = c_k^-(0)e^{8\kappa_k^3 t} .$$

We have thus proved the following result.

Suppose that in the family of equations (4.2.2) *the potential* $v(x,t)$ *is a rapidly decreasing solution of the KdV equation. Then the time evolution of the coefficients* $a(\lambda,t)$, $b(\lambda,t)$, *and* $c_k^-(\lambda,t)$ *is given by the formulas*

$$a(\lambda,t) = a(\lambda,0) \quad , \quad b(\lambda,t) = b(\lambda,0)e^{-8i\lambda^3 t} \quad , \quad c_k^-(t) = c_k^-(0)e^{8\kappa_k^3 t} ,$$

respectively. In view of (4.2.6) *and* (4.2.7), *this means that the laws governing the time evolution of the scattering data of the given family of equations are*

$$r^+(\lambda,t) = r^+(\lambda,0)e^{8i\lambda^3 t} \quad , \quad \kappa_k(t) = \kappa_k(0) \quad , \quad m_k^+(t)^2 = m_k^+(0)e^{8\kappa_k^3 t} , \qquad (4.2.10)$$

where $\{r^+(\lambda,0), i\kappa_k(0), m_k^+(0)\}$ *are the scattering data corresponding to the initial potential* $v_0(x)$. *The solution* $v(x,t)$ *itself is found by the following recipe:*

— *construct the function*

$$F^+(x;t) = \sum_{k=1}^{n} m_k^+(0)^2 e^{-\kappa_k x + 8\kappa_k^3 t} + \frac{1}{2\pi}\int_{-\infty}^{\infty} r^+(\lambda,0)e^{i\lambda x + 8i\lambda^3 t}d\lambda ; \qquad (4.2.10')$$

— *for each fixed* x , *solve the equation*

$$F^+(x+y;t) + K^+(x,y;t) + \int_x^{\infty} F^+(y+\xi;t)K^+(x,\xi;t)d\xi = 0 \quad (x \leq y < \infty) ; \qquad (4.2.11)$$

— *apply the formula*

$$v(x,t) = -2 \frac{d}{dx} K(x,x;t) . \tag{4.2.12}$$

To this point, we have assumed that the Cauchy problem (4.2.1) admits a rapidly decreasing solution. However, one can drop this assumption by checking directly that the function $v(x,t)$ constructed above is indeed a solution of the KdV equation.

By the Corollary to Lemma 4.1.1, the function $v(x,t)$ satisfies the KdV equation if and only if the operator M_1 takes the solutions of equation (4.2.2) into solutions of the same equation. Hence, if the function $v(x,t)$, its t-derivative, and the first three x-derivatives are rapidly decreasing, thne the equation

$$M_1[e^+(\lambda,x;t)] = -4i\lambda^3 e^+(\lambda,x;t)$$

is equivalent to the KdV equation for the potentials $v(x,t)$. By (4.2.8) and (4.2.3), the last equation is equivalent to

$$\int_x^\infty \{K_t^+(x,y;t) + v_x(x,t)K^+(x,y;t) - 2v(x,t)K_x^+(x,y;t) - 4\lambda^2 K_x(x,y;t)\}e^{i\lambda y}dy +$$

$$+ \{-4i\lambda^3 - 2i\lambda v(x,t) + v_x(x,t) + 2v(x,t)K^+(x,x;t) + 4\lambda^2 K^+(x,x;t)\}e^{i\lambda x} =$$

$$= -4i\lambda^3 e^{i\lambda x} - 4i\lambda^3 \int_x^\infty K^+(x,y;t)e^{i\lambda y}dy .$$

Integrating by parts, we get

$$\int_x^\infty \{K_t^+(x,y;t) + v_x(x,t)K^+(x,y;t) - 2v(x,t)K_x^+(x,y;t) + 4K_{xyy}^+(x,y;t)\}e^{i\lambda y}dy +$$

$$+ \{-4i\lambda^3 - 2i\lambda v(x,t) + v_x(x,t) + 2v(x,t)K^+(x,x;t) + 4\lambda^2 K^+(x,x;t) +$$

$$+ [-4i\lambda K_x^+(x,y;t) + 4K_{xy}^+(x,y;t)]\big|_{y=x}\}e^{i\lambda x} =$$

$$= \{-4i\lambda^3 + [4\lambda^2 K^+(x,y;t) + 4i\lambda K_y^+(x,y;t) - 4K_{yy}^+(x,y;t)]\big|_{y=x}\}e^{i\lambda x} -$$

$$- 4 \int_x^\infty K_{yyy}^+(x,y;t)e^{i\lambda y}dy ,$$

or, equivalently,

$$\int_x^\infty \{K_t^+(x,y;t) + 4K_{xyy}^+(x,y;t) + 4K_{yyy}^+(x,y;t) -$$

$$- 2v(x,t)K_x^+(x,y;t) + v_x(x,t)K^+(x,y;t)\}e^{i\lambda y}dy =$$

$$= -\{v_x(x,t) + 4K^+_{xy}(x,y;t) + 4K^+_{yy}(x,y;t) -$$

$$- 2i\lambda[v(x,t) + 2K^+_x(x,y;t) + 2K^+_y(x,y;t)]\big|_{y=x}\}e^{i\lambda x} \equiv 0 .$$

since from (4.2.12) and the equation

$$K^+_{xx}(x,y;t) = K^+_{yy}(x,y;t) + v(x,t)K^+(x,y;t) , \tag{4.2.13}$$

which is always satisfied by the kernel of the transformation operator, it follows that

$$v(x,t) = -2 \frac{d}{dx} K^+(x,x;t) = -2 \left\{\frac{\partial}{\partial x} + \frac{\partial}{\partial y}\right\}K^+(x,y;t)\big|_{y=x} \tag{4.2.14}$$

and

$$v_x(x,t) = -2 \left\{\frac{\partial^2}{\partial x^2} + 2 \frac{\partial^2}{\partial x \partial y} + \frac{\partial^2}{\partial y^2}\right\}K^+(x,y;t)\big|_{y=x} =$$

$$= -2 \left\{2 \frac{\partial^2}{\partial x \partial y} + 2 \frac{\partial^2}{\partial y^2} + v(x,t)\right\} K^+(x,y;t)\big|_{y=x} . \tag{4.2.15}$$

Thus, a function $v(x,t)$ with the requisite number of derivatives satisfies the KdV equation if and only if the kernels $K^+(x,y;t)$ of the corresponding transformation operators satisfy the equation

$$\left\{\frac{\partial}{\partial t} + 4 \frac{\partial^3}{\partial y^3} + 4 \frac{\partial^3}{\partial x \partial y^2} - 2v(x,t) \frac{\partial}{\partial x} + v_x(x,t)\right\} K^+ = 0 . \tag{4.2.16}$$

Differentiating equation (4.2.13) with respect to x , we obtain

$$\frac{\partial^3}{\partial x \partial y^2} K^+(x,y;t) = \frac{\partial^3}{\partial x^3} K^+(x,y;t) - v_x(x,t)K^+(x,y;t) - v(x,t) \frac{\partial}{\partial x} K^+(x,y;t) .$$

This permits us to replace (4.2.16) by the equivalent equation

$$\left\{\frac{\partial}{\partial t} + 4 \frac{\partial^3}{\partial x^3} + 4 \frac{\partial^3}{\partial y^3} - 6v(x,t) \frac{\partial}{\partial x} - 3v_x(x,t)\right\} K^+ = 0 . \tag{4.2.16'}$$

Now let us show that the solutions $K^+(x,y;t)$ of the integral equations (4.2.11), in which the kernels $F^+(x+y;t)$ are defined by (4.2.10'), satisfy equation (4.2.16'), and hence that the function $v(x,t)$ is a solution of the Cauchy problem (4.2.1). From (4.2.10'), it obviously follows that $F^+(x;t)$ is a distribution solution of the equation

$$\frac{\partial}{\partial t} F^+(x;t) = -8 \frac{\partial^3}{\partial x^3} F^+(x;t) .$$

Therefore, if $F^+(x;t)$ is thrice continuously differentiable with respect to x and once with respect to t, then the identities

$$\frac{\partial}{\partial t} F^+(x+y;t) = -8 \frac{\partial^3}{\partial x^3} F^+(x+y;t) = -8 \frac{\partial^3}{\partial y^3} F^+(x+y;t) \tag{4.2.17}$$

hold in the ordinary sense. Moreover, if for every $a > -\infty$,

$$\int_a^\infty (1 + |x|) \left(\left|\frac{\partial}{\partial t} F^+(x;t)\right| + \left|\frac{\partial^k}{\partial x^k} F^+(x;t)\right| \right) dx < \infty \quad , \quad 0 \leqslant k \leqslant 3 , \tag{4.2.18}$$

then it follows from the equalities (4.2.11) that the solutions $K^+(x,y;t)$ are thrice continuously differentiable with respect to x and once with respect to t, that their derivatives, as functions of y, satisfy analogous inequalities, and that the equalities (4.2.11) can be differentiated three times with respect to x and y, and once with respect to t. Consequently, we can apply the operator

$$\frac{\partial}{\partial t} + 4 \frac{\partial^3}{\partial x^3} + 4 \frac{\partial^3}{\partial y^3} - 6v \frac{\partial}{\partial x} - 3v'$$

(where $v(x,t) = -2 \frac{d}{dx} K^+(x,x;t)$) to both sides of (4.2.11). Taking note of (4.2.13)-(4.2.15) and (4.2.17), and integrating once by parts, this yields

$$D(x,y;t) + \int_x^\infty F^+(y+\xi;t) D(x,\xi;t) d\xi = 0 ,$$

where

$$D(x,y;t) = \left\{ \frac{\partial}{\partial t} + 4 \frac{\partial^3}{\partial x^3} + 4 \frac{\partial^3}{\partial y^3} - 6v \frac{\partial}{\partial x} - 3v_x \right\} K^+(x,y;t) .$$

Since this integral equation for D has only the null solution (see Lemma 3.5.3), $D(x,y;t) \equiv 0$, as claimed.

Thus, the indicated method indeed provides the solution of the Cauchy problem (4.2.1) if the function $F^+(x,t)$, constructed from the initial data by the recipe (4.2.10'), meets the requirements (4.2.18). It is readily verified that these requirements are manifestly fulfilled whenever the initial function $v_0(x)$ is infinitely differentiable and has compact support. The applicability of the method to the case of an arbitrary thrice continuously differentiable initial function $v_0(x)$ with rapidly decreasing derivatives can be established by approximating v_0 with infinitely differentiable functions with compact support.

We conclude this section by examining an important particular case in which one can obtain explicit formulas for the solutions of the KdV equation. First of all, we remark that every collection of the form $\{r^+(\lambda) \equiv 0 \ , \ i\kappa_k \ (\kappa_k > 0) \ , \ m_k > 0 \ , \ k = 1,2,\ldots,n\}$ obviously satisfies all the conditions of Theorem 3.5.1. Hence, it serves as the scattering data of a Sturm-Liouville equation with real-valued rapidly decreasing potential $v(x)$. To find the latter, we must solve the integral equations

$$\sum_{k=1}^{n} m_k^2 e^{-\kappa_k(x+y)} + K^+(x,y) + \int_x^\infty \sum_{k=1}^{n} m_k^2 e^{-\kappa_k(y+\xi)} K^+(x,\xi)d\xi = 0 \ . \tag{4.2.19}$$

These are degenerate and reduce to algebraic equations. In fact, by (4.2.19),

$$K^+(x,y) = \sum_{k=1}^{n} e^{-\kappa_k y} P_k(x) \ ,$$

where

$$P_k(x) = -m_k^2 \left(e^{-\kappa_k x} + \int_x^\infty e^{-\kappa_k \xi} K^+(x,\xi)d\xi \right) .$$

Substituting this expression into equation (4.2.19), we obtain the identity

$$\sum_{k=1}^{n} e^{-\kappa_k y}\left(m_k^2 e^{-\kappa_k x} + P_k(x) + m_k^2 \sum_{l=1}^{n} \frac{e^{-(\kappa_k+\kappa_l)x}}{\kappa_k + \kappa_l} P_l(x) \right) \equiv 0 \ ,$$

which is equivalent to the following algebraic system of equations for the functions $P_k(x)$:

$$P_k(x) + \sum_{l=1}^{m} m_k^2 \frac{e^{-(\kappa_k+\kappa_l)x}}{\kappa_k + \kappa_l} P_l(x) = -m_k^2 e^{-\kappa_k x} \ .$$

By Cramer's rule,

$$P_l(x) = \Delta_l(x)[\Delta(x)]^{-1} \ ,$$

where

$$\Delta(x) = \mathrm{Det}\left[\delta_{kl} + m_k^2 \frac{e^{-(\kappa_k+\kappa_l)x}}{\kappa_k + \kappa_l} \right] ,$$

and the determinant $\Delta_l(x)$ is obtained from $\Delta(x)$ upon replacing the l-th column $m_k^2(\kappa_k + \kappa_l)^{-1} \exp\{-(\kappa_k + \kappa_l)x\}$ by the right-hand column $-m_k^2\exp\{-\kappa_k x\}$ Now observing that the right-hand column, multiplied by $\exp\{-\kappa_l x\}$, is equal to the derivative of the l-th column of the determinant $\Delta(x)$, we further get that

$$K^+(x,x) = \sum_{l=1}^{n} e^{-\kappa_l x} P_l(x) = \sum_{l=1}^{n} \tilde{\Delta}_l(x)[\Delta(x)]^{-1} ,$$

where $\tilde{\Delta}_l(x) = \Delta(x) \exp\{-\kappa_l x\}$ designates the determinant obtained from $\Delta(x)$ upon replacing the l-th column by its derivatives. Using the rule for differentiating determinants we finally obtain the formula for the sought-for potential:

$$v(x) = -2 \frac{d}{dx} K^+(x,x) = -2 \frac{d^2}{dx^2} \ln \Delta(x) .$$

In view of formulas (4.2.10), the solution of the Cauchy problem (4.2.1) with initial potential $v_0(x)$ of the form in question ($r^+(\lambda,0) \equiv 0$), can be expressed in the form

$$\left.\begin{aligned} &v(x,t) = -2 \frac{d^2}{dx^2} \ln \Delta(x,t) , \\ &\text{where} \\ &\Delta(x,t) = \mathrm{Det}\left[\delta_{kl} + m_k^2(0) \frac{e^{-(\kappa_k+\kappa_l)x+8\kappa_k^3 t}}{\kappa_k + \kappa_l}\right] . \end{aligned}\right\} \tag{4.2.20}$$

PROBLEMS

1. If the function $v(x,t)$ is infinitely differentiable with respect to x and all its derivatives are rapidly decreasing, then the solutions $e^+(\lambda,x;t)$ of the equations (4.2.2) can be represented in the form

$$e^+(\lambda,x;t) = \exp\left\{i\lambda x - \int_x^\infty \sigma(\lambda,\xi;t)d\xi\right\} .$$

Moreover, the function $\sigma(\lambda,\xi;t)$ admits the asymptotic series expansion

$$\sigma(\lambda,x;t) = \sum_{j=1}^{\infty} \frac{\sigma_j(x;t)}{(2i\lambda)^j} , \quad \lambda \to \pm\infty , \tag{4.2.21}$$

the coefficients $\sigma_j(x,t)$ of which are found from the recursion relations:

$$\sigma_1(x;t) = v(x,t) , \quad \sigma_{j+1}(x;t) = -\sigma_j'(x;t) - \sum_{l=1}^{j-1} \sigma_{j-l}(x;t)\sigma_l(x;t)$$

(see Problem 9, Section 1, Chapter 3).

Show that the integrals $\int_{-\infty}^{\infty} \sigma_j(x;t)dx$ are conserved (i.e., are independent of t) if $v(x,t)$ satisfies the equation $K_m[v] = 0$ (in particular, the KdV equation).

Hint. By the Corollary to Lemma 4.1.1, the operator

$$M_m = \frac{\partial}{\partial t} + 2N_m \frac{\partial}{\partial x} - N_m'$$

transforms the solution $e^+(\lambda,x;t)$ of equation (4.2.2) into a solution of the same equation. In the preceding section we showed that

$$N_m = N_m(\lambda^2,x,t) = \sum_{j=0}^{m} a_{m-j}(x;t)(2i\lambda)^{2j} =$$

$$= \operatorname{Reg}\left\{P_m(\lambda^2,t)\left[1 + 2\sum_{k=0}^{\infty} \sigma_{2k+1}(x;t)(-4\lambda^2)^{-k-1}\right]^{-1}\right\},$$

from which it follows that

$$\lim_{x\to\pm\infty} N_m(\lambda^2,x,t) = P_m(\lambda^2,t) \quad , \quad \lim_{x\to\pm\infty} N_m'(\lambda^2,x,t) = 0 \ ,$$

since $\lim_{x\to\pm\infty} \sigma_{2k+1}(x;t) = 0$. Consequently,

$$M_m[e^+(\lambda,x;t)] = 2i\lambda P_m(\lambda^2,t)e^{i\lambda x} + o(1)$$

as $x \to \infty$, and hence

$$M_m[e^+(\lambda,x;t)] = \left\{-\frac{\partial}{\partial t}\int_x^{\infty} \sigma(\lambda,\xi;t)d\xi + 2N_m(i\lambda + \sigma(\lambda,x;t)) - N_m'\right\}e^+(\lambda,x;t) =$$

$$= 2i\lambda P_m(\lambda^2,t)e^+(\lambda,x;t) \ ,$$

i.e.,

$$\frac{\partial}{\partial t}\int_x^{\infty} \sigma(\lambda,\xi;t)d\xi = 2i\lambda(N_m - P_m) + 2N_m\sigma(\lambda,x;t) - N_m' \ .$$

In view of the asymptotic formula (4.2.21),

$$\frac{\partial}{\partial t}\int_x^{\infty} \sigma_k(\xi;t)d\xi = 2\sum_{j=0}^{m} a_{m-j}(x;t)\sigma_{k+2j}(x;t) \quad , \quad k = 1,2,\ldots \ .$$

Since the coefficients $a_{m-j}(x;t)$ stay bounded and the function $\sigma_{k+2j}(x;t)$ tends to zero as $x \to \pm\infty$, you get

$$\frac{\partial}{\partial t} \int_{-\infty}^{\infty} \sigma_k(\xi;t)d\xi = 0$$

upon letting $x \to -\infty$ in the last set of equalities, as needed.

2. Show that the function (4.2.12) satisfies the KdV equation if K^+ is the unique solution of equation (4.2.11) and $F^+(x,t)$ satisfies equation (4.2.12). Investigate the behavior of the function (4.2.12) as $x \to \pm\infty$, when

$$F^+(x,t) = \int_0^1 m^2(\lambda)e^{-\lambda x+8\lambda^3 t}d\lambda .$$

3. Find the solution of the operator equation

$$\left(B \frac{\partial}{\partial t} + \frac{\partial^2}{\partial x^2}\right) \Omega - 2\Omega^3 = 0$$

using the inverse scattering method.

Hint. See [27] and Problem 1 to Section 1 of this chapter.

3. PERIODIC SOLUTIONS OF THE KORTEWEG-DE VRIES EQUATION

To explain better possible generalizations and to shorten the calculations we shall consider, instead of the family of Sturm-Liouville equations (4.2.2), the equivalent family of first-order matrix equations (4.1.12), and shall accordingly use Lemma 4.1.1' instead of 4.1.1.

We begin by reminding the reader of a few basic facts from the elementary theory of matrix differential equations of the form

$$Y' = A(x)Y .$$

The determinant of any solution $Y(x)$ of this equation satisfies the equation

$$\{\text{Det } Y(x)\}' = \text{Trace } A(x) \cdot \text{Det } Y(x) ,$$

from which one derives the Liouville formula

$$\text{Det } Y(x) = \text{Det } Y(x_1) \exp\left(\int_{x_1}^{x} \text{Trace } A(\xi)\right) d\xi .$$

A fundamental matrix of equation (4.3.1) is, by definition, a solution $\Phi(x)$ of (4.3.1) with nonvanishing determinant. The fundamental matrix which is equal to the identity matrix for $x = x_1$ will be denoted by $U(x,x_1)$. Obviously, $U(x,x_1) = \Phi(x)\Phi(x_1)^{-1}$, where $\Phi(x)$ is an arbitrary fundamental matrix. The method of variation of constants leads to the formula

$$Y(x) = U(x,x_1)Y_1 + U(x,x_1)\int_{x_1}^{x} U(\xi,x_1)^{-1}F(\xi)d\xi \tag{4.3.2}$$

for the solution of the Cauchy problem for the nonhomogeneous equation:

$$Y' = A(x)Y + F(x) \quad , \quad Y(x_1) = Y_1 \; .$$

Let $\Phi_1(x)$ and $\Phi_2(x)$ be fundamental matrices of the equations $\Phi_1' = A_1(x)\Phi_1$ and $\Phi_2' = -\Phi_2A_2(x)$, respectively. Then the general solution of the equation

$$Y' = A_1(x)Y - YA_2(x) \tag{4.3.3}$$

has the form

$$Y = \Phi_1(x)C\Phi_2(x) \; ,$$

where C is an arbitrary constant matrix. It follows from this formula and Liouville's formula that the determinant of any solution of equation (4.3.3) is independent of x provided that $\text{Trace } A_1(x) = \text{Trace } A_2(x)$. Since the trace of the commutator $[A,B] = AB - BA$ of two matrices is always equal to zero, the trace of any solution of equation (4.3.3) is independent of x if $A_1(x) = A_2(x)$. Hence, for the solutions of the equation

$$Y' = [Y,A(x)] \; , \tag{4.3.3'}$$

both the determinant and the trace are independent of x.

We consider now the family of operators

$$L = \frac{d}{dx} + V \quad , \quad V = V(z,x,t) = -\begin{pmatrix} 0 & 1 \\ v(x,t)-z & 0 \end{pmatrix} , \tag{4.3.4}$$

where $v(x,t)$ is an arbitrary sufficiently smooth function which is defined in a finite or infinite strip $x_1 \leqslant x \leqslant x_2$, $-\infty < t < \infty$.

We call a polynomial

$$N_n = N_n(z,x,t) = \sum_{j=0}^{n} a_{n-j}(x,t)z^j$$

admissible (for the given family (4.3.4)) if its coefficients $a_k(x,t)$ satisfy the system of equations (4.1.9). It follows from formulas (4.1.10) for the general solution of the system (4.1.9) that an admissible polynomial $N_n(z,x,t)$ is uniquely determined by its value $N_n(z,x_0,t)$ at some point x_0, which may be an arbitrary polynomial $B_n(z,t) = \sum_{j=0}^{n} b_{n-j}(t)z^j$. The explicit formula which expresses an admissible polynomial in terms of its value $B_n(z,t)$ at the point $x = x_0$ is readily established. In fact, as we saw earlier, every admissible polynomial can be written as

$$N_n(z,x,t) = \text{Reg}\,\{P_n(z,t)N_\infty(z,x)\},$$

where $N_\infty(z,x)$ denotes the formal series

$$N_\infty(z,x) = \left[1 + 2\sum_{j=0}^{\infty} \sigma_{2j+1}(x)(-4z)^{-j-1}\right]^{-1}.$$

Therefore, the formula

$$N_n(z,x,t) = \text{Reg}\,\{B_n(z,t)N_\infty(z,x_0)^{-1}N_\infty(z,x)\}$$

defines an admissible polynomial whose value at the point $x = x_0$ is $B_n(z,t)$. The nonlinear operator K_n,

$$K_n[v] = -\dot{v} + N_n''' - 4N_n'(v-z) - 2N_n v', \tag{4.3.5}$$

and the linear operators

$$M_n = \frac{\partial}{\partial t} + A_n, \quad A_n = A_n(z,x,t) = \begin{pmatrix} -N_n' & 2N_n \\ 2N_n(v-z) - N_n'' & N_n' \end{pmatrix}, \tag{4.3.6}$$

both of which figure in Lemma 4.1.1', are, obviously, also uniquely determined by the value $B_n(z,t)$ of the corresponding admissible polynomial N_n at $x = x_0$.

The foregoing discussion permits us to formulate Lemma 4.1.1' in the following sharper form.

LEMMA 4.3.1. *For any family of operators* L *of the form* (4.3.4), *and every point* x_0 *and arbitrary polynomial* $B_n(z,t) = \sum_{j=0}^{n} b_{n-j}(t)z^j$, *there corresponds a unique admissible polynomial* $N_n(z,x,t)$ *such that* $N_n(z,x_0,t) = B_n(z,t)$. *The operators* (4.3.5) *and* (4.3.6), *defined by* N_n, *satisfy the relations*

$$[L,M_n] = CK_n[v], \quad \text{where } C = \begin{pmatrix} 0 & 0 \\ 1 & 0 \end{pmatrix}. \tag{4.3.7}$$

PROOF. We only need to verify relation (4.3.7), which is elementary. □

Now consider the fundamental matrices $U = u(x,x_1)$ of the family of equations $L[Y] = 0$ corresponding to the operators (4.3.4). It follows immediatly from the form of these operators, that

$$U = \begin{pmatrix} c & s \\ c' & s' \end{pmatrix}, \quad U^{-1} = \begin{pmatrix} -s' & -s \\ -c' & c \end{pmatrix}, \tag{4.3.8}$$

where $c = c(\sqrt{z},x;x_1,t)$ and $s(\sqrt{z},x;x_1,t)$ is the fundamental system of solutions of the equation $-y'' + v(x,t)y = zy$, which is defined by the following initial conditions at the point $x = x_1$: $c' = s = 0$ and $c = s' = 1$. By Lemma 4.3.1, the operators M_n transform the matrix $U = U(x,x_1)$ into matrices

$M_n[U] = \dot{U} + A_nU$,

which satisfy the equations

$L\{M_n[U]\} = CK_n[v]U$

and the initial conditions $M_n[U]\big|_{x=x_1} = A_n(x_1)$, which follow from the equality $U\big|_{x=x_1} = I$. Hence, the $M_n[U]$ are solutions of the Cauchy problem

$L[Y] = CUK_n[v]$, $Y\big|_{x=x_1} = A_n(x_1)$,

and, by formula (4.3.2),

$$M_n[U] = \dot{U} + A_n(x)U = UA_n(x_1) + U\int_{x_1}^{x} U(\xi,x_1)^{-1}CU(\xi,x_1)K_n[v]d\xi ,$$

i.e.,

$$\dot{U} = UA_n(x_1) - A_n(x)U + U\int_{x_1}^{x} W(\xi)K_n[v]d\xi , \tag{4.3.9}$$

where

$W(\xi) = U(\xi,x_1)^{-1}CU(\xi,x_1)$.

Using formula (4.3.8), we obtain

$$W(\xi) = \begin{pmatrix} -cs & -s^2 \\ c^2 & cs \end{pmatrix}. \tag{4.3.9'}$$

Thus, Lemma 4.3.1 admits the following

COROLLARY. *The fundamental matrices* $U = U(x,x_1)$ *of the family of equations* $L[Y] = 0$ *associated with the operators* (4.3.4) *satisfy equalities* (4.3.9), *in which the operators* K_n *and the matrices* $A_n = A_n(z,x,t)$ *are given by formulas* (4.3.5) *and* (4.3.6), *respectively, and* $N_n = N_n(z,x,t)$ *are arbitrary polynomials which are admissible for the given family.* □

The value of a function $f(x)$ at the point $x = x_i$ will be denoted by $f(i)$, for short. For example, we shall write $U(x_2,x_1) = U(2,1)$, $N_n(z,x_i,t) = N_n(i)$, $A_n(z,x_i,t) = A_n(i)$, and so forth. We call the value of the fundamental matrix $U(x,x_1)$ of the equation $L[y] = 0$ at the point $x = x_2$, i.e., $U(x_2,x_1) = U(2,1)$, the transition matrix of this equation, or of the operator L itself (from the point x_1 to the point x_2). It is a function of the spectral parameter z and is directly connected with the spectral data of the boundary value problems generated by the equation $-y'' + v(x,t)y = zy$ on the segment $x_1 \leqslant x \leqslant x_2$. If $v(x,t)$ satisfies the equation $K_n[v] = 0$ in the strip $x_1 \leqslant x \leqslant x_2$, $-\infty < t < \infty$, then the transition matrices $U(2,1)$ of the family of operators (4.3.4) satisfy the equation

$$\dot{U}(2,1) = U(2,1)A_n(1) - A_n(2)U(2,1) , \tag{4.3.10}$$

which is an obvious consequence of formulas (4.3.9). We emphasize that in this equation the variable x does not appear, and the coefficients $A_n(i)$ depend polynomially on z. A very important fact is that the converse is also true.

THEOREM 4.3.1. *In order that the function* $v(x,t)$, *given in the strip* $x_1 \leqslant x \leqslant x_2$, $-\infty < t < \infty$, *satisfy the equation* $K_n[v] = 0$ *in this strip, which is defined by a polynomial* $N_n(z,x,t)$, *which is admissible for the family of operators* (4.3.4), *it is necessary and sufficient that there exist a pair of* 2×2 *matrices*

$$B^{(i)} = (B^{(i)}_{\alpha\beta}(z,t)) , \quad \text{Trace } B^{(i)} = 0 \quad (i = 1,2) ,$$

depending polynomially on z, *such that the transition matrix* $U(2,1)$ *of this family satisfies the equation*

$$\dot{U}(2,1) = U(2,1)B^{(1)} - B^{(2)}U(2,1) . \tag{4.3.11}$$

If a pair of matrices $B^{(i)}$ *with the indicated properties exist, then the polynomial* $N_n(z,x,t)$ *is uniquely determined by the condition* $2N_n(z,x_2,t) = B_{12}^{(2)}(z,t)$. *Moreover,* $B^{(1)} = A_n(1)$ *and* $B^{(2)} = A_n(2)$, *where the matrices* $A_n = A_n(z,x,t)$ *are constructed from the polynomial* $N_n(z,x,t)$ *by following the recipe* (4.3.6).

PROOF. The necessity of the conditions of the theorem was established earlier. To prove that these conditions are also sufficient, we remark that, by the Corollary to Lemma 4.3.1, the transition matrix $U(2,1)$ always satisfies the equalities

$$\dot{U}(2,1) = U(2,1)A_n(1) - A_n(2)U(2,1) + U(2,1)\int_{x_1}^{x_2} W(\xi)K_n[v]d\xi .$$

If the conditions of the theorem are satisfied, then substracting these equalities from (4.3.11), we obtain

$$0 = U(2,1)(B^{(1)} - A_n(1)) - (B^{(2)} - A_n(2))U(2,1) - U(2,1)\int_{x_1}^{x_2} W(\xi)K_n[v]d\xi ,$$

or, equivalently,

$$C^{(1)} = U(2,1)^{-1}C^{(2)}U(2,1) + \int_{x_1}^{x_2} W(\xi)K_n[v]d\xi , \tag{4.3.12}$$

where

$$C^{(i)} = B^{(i)} - A_n(i) = \left(C_{\alpha\beta}^{(i)}\right) , \quad \text{Trace } C^{(i)} = 0 \quad (i = 1,2) .$$

Moreover, (4.3.12) holds for every admissible polynomial $N_n(z,x,t)$ which intervenes in the definition of the operators K_n and the matrices A_n via formulas (4.3.5) and (4.3.6), respectively. By Lemma 4.3.1, the admissible polynomial $N_n = N_n(z,x,t)$ must be selected so that $2N_n(z,x_2,t) \equiv B_{12}^{(2)}(z,t)$, and hence $C_{12}^{(2)} \equiv 0$, too. The theorem will be proved if we show that, for such a choice of the admissible polynomial, $C^{(2)} \equiv 0$ and $K_n[v] \equiv 0$.

It follows from formulas (4.3.8) and (4.3.9'), upon taking note of the equalities Trace $C^{(2)} = 0$ and $C_{12}^{(2)} = 0$, that the 12 entries of the matrices intervening in (4.3.12) satisfy the relation

$$C_{12}^{(1)} = s(2)\left\{2s'(2)C_{11}^{(2)} - s(2)C_{21}^{(2)}\right\} - \int_{x_1}^{x_2} s^2(\sqrt{z},\xi)K_n[v]d\xi , \tag{4.3.13}$$

where $C_{12}^{(1)}$, $C_{11}^{(2)}$, and $C_{21}^{(2)}$ are polynomials in z , $s(2) = s(\sqrt{z},x_2)$, $s'(2) = s'(\sqrt{z},x_2)$, and

$$s(\sqrt{z},x) = s(\sqrt{z},x;x_1,t) = \frac{\sin\sqrt{z}\,(x-x_1)}{\sqrt{z}} + \int_{x_1}^{x} K(x,y;t)\,\frac{\sin\sqrt{z}\,(y-x_1)}{\sqrt{z}}\,dy\ .$$

The last formula leads readily to the following representation for the functions $s^2(\sqrt{z},x)$:

$$s^2(\sqrt{z},x) = (2z)^{-1}\left[1 - \cos 2\sqrt{z}\,(x-x_1) + \int_{x_1}^{x} H(x,y;t)\cos 2\sqrt{z}\,(y-x_1)dy\right] ,$$

which shows that they form a complete system in $L_2[x_1,x_2]$ (see Problem 1, Section 3, Chapter 1). Since the right-hand side of equality (4.3.13) tends to zero as z runs through the zeros of the function $s(2) = s(\sqrt{z},x_2)$, the polynomial $C_{12}^{(1)}$, appearing in its left-hand side, vanishes identically. Using this fact and considering the equality (4.3.13) at the zeros of $s'(2) = s'(\sqrt{z},x_2)$, we see that $C_{12}^{(2)} \equiv 0$. Finally, upon considering equality (4.3.13) for $2\sqrt{z}\,(x_1 - x_2) = \frac{\pi}{4} + 2n\pi$ $(n = 1,2,\ldots)$ (and taking into account the already established identities $C_{12}^{(1)} = C_{21}^{(2)} \equiv 0$), we conclude that

$$C_{11}^{(2)} \equiv 0 ,$$

and hence that

$$\int_{x_1}^{x_2} s^2(\sqrt{z},\xi)K_n[v]d\xi \equiv 0 .$$

In view of the completeness of the set of functions $s^2(\sqrt{z},x)$ in $L_2[x_1,x_2]$, this finally yields $K_n[v] \equiv 0$. □

Remark. The notations that were used for admissible polynomials may lead to missunderstandings when one deals with families of such polynomials that depend on parameters on which the potentials v may also depend. The general form of admissible polynomials,

$$N_n = \operatorname{Reg}\left\{P_n(z)\left[1 + 2\sum_{j=0}^{\infty}\sigma_{2j+1}(x)(-4z)^{-j-1}\right]^{-1}\right\} ,$$

shows that they are obtained by applying a certain nonlinear differential operator $N\{P_n\}$ to the potential v , regarded as a function of the variable x . The form of this operator determines the polynomial P_n . Hence, if the potentials $v = v(x,t)$ and the polynomials $P_n = P_n(z,t)$ depend on the parameter t , then the following notation for $N_n(z,x,t)$ is more appropriate:

$N_n(z,x,t) = N\{P_n(z,t)\}[v(x,t)]$.

For the same reasons, the following notations should be used for the formal series $N_\infty(z,x)$, the matrices A_n , and the operators K_n :

$N_\infty(z,x) = N_\infty(z,v(x,t))$, $A_n = A_n(P_n(z,t),v(x,t))$, $K_n[v] = K\{P_n(z,t)\}[v]$,

respectively. The equality

$K_n = K\{P_n(z,t)\}$

indicates that the operator K_n depends on t . For example, in Theorem 4.3.1 the operator K_n , defined by the matrices $B^{(i)}$, has the form

$K_n = K\{\text{Reg } [B_{12}^{(1)}(t,z)N_\infty(z,v(x_1,t))]\}$.

Therefore, it does not depend on t , provided that

$$\frac{d}{dt} \text{Reg } \{B_{12}^{(1)}(z,t)N_\infty(z,v(x_1,t))\} = 0 ,$$

and if this is the case, then $K_n[v] = 0$ is an equation with constant coefficients, as is, for example, the KdV equation.

Let us examine in more detail the particular case of Theorem 4.3.1 when the entries of the matrix $B^{(i)}$ are polynomials of degree at most two. By (4.1.10), the general form of admissible polynomials of first degree is

$$N_1 = p_0 z + \frac{1}{2} p_0 v + p_1 ,$$

while the corresponding operators K_1 and matrices A_1 look like

$$K_1[v] = -\dot{v} + \frac{1}{2} p_0\{v''' - 6vv'\} - 2p_1 v'$$

and

$$A_1 = \begin{pmatrix} -\frac{1}{2} p_0 v' & 2N_1 \\ 2N_1(v-z) - \frac{1}{2} p_0 v'' & \frac{1}{2} p_0 v' \end{pmatrix} ,$$

respectively, where $p_0 = p_0(t)$ and $p_1 = p_1(t)$ are arbitrary functions of t . Correspondingly, the matrices $B^{(i)}$ have the following structure:

$$B^{(i)} = \begin{pmatrix} -\frac{1}{2} p_0 b_i & B_{12}^{(i)} \\ B_{12}^{(i)}(a_i - z) - \frac{1}{2} p_0 d_i & \frac{1}{2} p_0 b_i \end{pmatrix} , \quad B_{12}^{(i)} = p_0(2z+a_i) + 2p_i, \quad (4.3.14)$$

where a_i, b_i and d_i are functions of t. According to Theorem 4.3.1, the function $v(x,t)$ satisfies the equation

$$-\dot{v} + \frac{1}{2} p_0\{v''' - 6vv'\} - 2p_1 v' = 0$$

in the strip $x_1 \leqslant x \leqslant x_2$, $-\infty < t < \infty$, if and only if there are matrices $B^{(i)}$ of the form (4.3.14) such that the transition matrix $U(2,1)$ of the family of operators (4.3.4) satisfies equation (4.3.11).

Now suppose that the function $v = v(x,\tau,t)$, which defines the family of operators (4.3.4), depends on the parameters τ and t. From the particular case of Theorem 4.3.1 considered above, we obtain the following important corollary.

COROLLARY. *In order that the function* $v(x,\tau,t)$ *be a joint solution of the equations*

$$\frac{\partial v}{\partial \tau} - \frac{\partial v}{\partial x} = 0 \ , \quad \frac{\partial v}{\partial t} = 6v\,\frac{\partial v}{\partial x} - \frac{\partial^3 v}{\partial x^3}$$

in the domain $x_1 \leqslant x \leqslant x_2$, $-\infty < \tau < \infty$, $-\infty < t < \infty$, *it is necessary and sufficient that there exist matrices* $B_0^{(i)}$ *and* $B_1^{(i)}$, $i = 1,2$, *of the form*

$$B_0^{(i)} = \begin{pmatrix} 0 & -1 \\ z - \tilde{a}_i & 0 \end{pmatrix} ,$$

and (4.3.15)

$$B_1^{(i)} = \begin{pmatrix} b_i & -2(a_i + 2z) \\ 4z^2 - 2a_i z + c_i & -b_i \end{pmatrix} ,$$

where $\tilde{a}_i$, a_i, b_i, *and* c_i *are functions of* τ *and* t, *such that the transition matrix* $U = U(2,1)$ *of the family of operators* (4.3.4) *satisfies the equations*

$$\frac{\partial}{\partial \tau} U = UB_0^{(1)} - B_0^{(2)}U \ , \quad \frac{\partial}{\partial t} U = UB_1^{(1)} - B_1^{(2)}U \ .$$

If matrices $B_0^{(i)}$ *and* $B_1^{(i)}$, *with the indicated properties exist, then*

$$\left.\begin{aligned} &\tilde{a}_i = a_i = v(x_i,\tau,t) \ , \quad b_i = v'_x(x_i,\tau,t) \ , \\ &c_i = v'_{xx}(x_i,\tau,t) - 2v^2(x_i,\tau,t) \ . \end{aligned}\right\} \qquad (4.3.15')$$

□

We now turn to the periodic Cauchy problem for the KdV equation

$$\dot{v} - 6vv' + v''' = 0 \quad , \quad v(x+\pi,t) = v(x,t) \quad (-\infty < x < \infty) \tag{4.3.16}$$

with a real-valued thrice continuously differentiable initial function

$$v(x,0) = v_0(x) = v(x+\pi) \ . \tag{4.3.16'}$$

Let $U_0(x)$ be the transition matrix (from $x_1 = 0$ to $x_2 = \pi$) of the operator (4.3.4) associated with the initial function $v_0(x)$. Let $U(t,\tau;z)$ be the transition matrices of the operators (4.3.4) associated with the functions $v(x+\tau,t)$, where $v(x,t)$ is a solution of problem (4.3.16), (4.3.16'). It is clear that the functions $v(x+\tau,t)$ satisfy simultaneously the two equations

$$\frac{\partial v}{\partial \tau} - \frac{\partial v}{\partial x} = 0 \quad , \quad \frac{\partial v}{\partial t} = 6v\,\frac{\partial v}{\partial x} - \frac{\partial^3 v}{\partial x^3}$$

in the strip $0 \leqslant x \leqslant \pi$, $-\infty < t < \infty$, and the periodic boundary conditions

$$\left.\frac{\partial^k v}{\partial x^k}\right|_{x=0} = \left.\frac{\partial^k v}{\partial x^k}\right|_{x=\pi} \quad , \quad k = 0,1,2 \ .$$

From this it follows, by the Corollary to Theorem 4.3.1, that there exist matrices $B_0^{(1)} = B_0^{(2)} = B_0$ and $B_1^{(1)} = B_1^{(2)} = B_1$ of the form

$$B_0 = \begin{pmatrix} 0 & -1 \\ z-a & 0 \end{pmatrix} , \quad B_1 = \begin{pmatrix} b & -2(a + 2z) \\ 4z^2 - 2az + c & -b \end{pmatrix} , \tag{4.3.17}$$

(with a , b , and c functions of τ and t), such that

$$\frac{\partial U}{\partial \tau} = [U,B_0] \quad , \quad \frac{\partial U}{\partial t} = [U,B_1] \ .$$

Conversely, suppose that from the transition matrix $U_0(z)$ one succeeds in constructing a matrix B_1 of the form (4.3.17) such that, for every value of t , the solution of the Cauchy problem

$$\frac{\partial U}{\partial t} = [U,B_1] \quad , \quad U\big|_{t=0} = U_0(z) \tag{4.3.18}$$

is the transition matrix $U = U(t,z)$ of an operator of the form (4.3.4). Then, by the same Corollary, the function $v(x,t)$ that defines this operator will be a solution of the boundary value problem for the KdV equation in the strip $0 \leqslant x \leqslant \pi$, $-\infty < t < \infty$, with initial boundary conditions

$$v(x,0) = v_0(x) \quad (0 \leqslant x \leqslant \pi) \ ,$$

$$v^{(k)}(0,t) = v^{(k)}(\pi,t) \quad , \quad k = 0,1,2 \; .$$

Since the equality $v'''(0,t) = v'''(\pi,t)$ is automatically satisfied here, the continuation of the function $v(x,t)$ by periodicity to the full real line $-\infty < x < \infty$ $(v(x+\pi,t) = v(x,t))$ will be a solution of the problem (4.3.16), (4.3.16'). Furthermore, if for the already available transition matrices $U(t;z)$, one succeeds in finding matrices B_0 of the form (4.3.17) such that the solutions of the Cauchy problem

$$\frac{\partial U}{\partial \tau} = [U,B_0] \quad , \quad U\big|_{\tau=0} = U(t;z) \tag{4.3.19}$$

are transition matrices $U(\tau,t;z)$ of operators of the form (4.3.4) for all values of τ, then the function $v(x,\tau,t)$ that defines the family of these operators is a solution of the initial boundary value problem

$$\frac{\partial v}{\partial \tau} - \frac{\partial v}{\partial x} = 0 \quad , \quad v(x,0,t) = v(x,t) \quad , \quad v(0,\tau,t) = v(\pi,\tau,t) \; .$$

But the unique solution of this problem is obviously the function $v(x+\tau,t)$. Hence, $v(x,\tau,t) = v(x+\tau,t)$, and in view of equalities (4.3.15'),

$$a(\tau,t) = v(0,\tau,t) = v(\tau,t) \; ,$$

where $a(\tau,t)$ is the function which defines the matrix B_0.

Therefore, if we succeed in effecting the two procedures indicated above, then we not only establish the existence of a solution to problem (4.3.16), (4.3.16'), *but also find it explicitly via the formula*

$$v(x,t) = a(x,t) \; . \tag{4.3.20}$$

In the beginning of this section, we mentioned that equations of the form $Y' = [Y,A]$ possess at least two first integrals: $\operatorname{Det} Y$ and $\operatorname{Trace} Y$. Accordingly, the equalities

$$\operatorname{Det} U(\tau,t;z) = \operatorname{Det} U(t;z) = \operatorname{Det} U_0(z)$$

and

$$\operatorname{Trace} U(\tau,t;z) = \operatorname{Trace} U(t;z) = \operatorname{Trace} U_0(z)$$

always hold for solutions of the Cauchy problem (4.3.18), (4.3.19). This permits us to express these solutions in the form

$$U(t,z) = \frac{1}{2} \operatorname{Trace} U_0(z) \cdot I + V(t;z) \quad ,$$

and

$$U(\tau,t;z) = \frac{1}{2}\,\text{Trace}\, U_0(z)\cdot I + V(\tau,t;z)\ ,$$

where $V(t;z)$ and $V(\tau,t;z)$ are solutions of the Cauchy problems

$$\frac{\partial V}{\partial t} = [V,B_1]\ ,\quad V(0;z) = V_0(z) = U_0(z) - \frac{1}{2}\,\text{Trace}\, U_0(z)\cdot I$$

and

$$\frac{\partial V}{\partial t} = [V,B_0]\ ,\quad V(0,t;z) = V(t,z)\ ,$$

respectively.

By formula (4.3.8), the transition matrix of any of the operators (4.3.4) has the form

$$U = \begin{pmatrix} c & s \\ c' & s \end{pmatrix} = u_+(\sqrt{z})I + \begin{pmatrix} u_-(\sqrt{z}) & s(\sqrt{z},\pi) \\ c'(\sqrt{z},\pi) & -u_-(\sqrt{z}) \end{pmatrix},$$

where $c = c(\sqrt{z},x)$, $s = s(\sqrt{z},x)$ is the fundamental system of solutions for the equation

$$-y'' + v(x)y = zy\ ,$$

$$u_+(\sqrt{z}) = \frac{1}{2}\,[c(\sqrt{z},\pi) + s'(\sqrt{z},\pi)] = \frac{1}{2}\,\text{Trace}\ U$$

is the Hill discriminant of this equation, and

$$u_-(\sqrt{z}) = \frac{1}{2}\,[c(\sqrt{z},\pi) - s'(\sqrt{z},\pi)]\ .$$

Therefore, if $v(x,t)$ is a solution of problem (4.3.16), (4.3.16'), then the Hill discriminant, and hence the spectra of the periodic and anti riodic boundary value problems on the segment $0 \leqslant x \leqslant \pi$ generated by the equations $-y'' + v(x+\tau,t)y = zy$, are independent of τ and t . In particular, if the initial potential $v(x,0) = v_0(x)$ is finite-zone, then the potentials $v(x+\tau,t)$ remain finite-zone for all τ and t , and the endpoints of their lacunae remain unchanged. In this case, the structure of the transition matrices is particularly simple. In fact, it follows from the asymptotic formula (4.3.26), that

$$1 - u_+(\sqrt{z})^2 = \pi^2(z-\mu_0)\prod_{k=1}^{\infty} k^{-4}(z - \mu_k^-)(z - \mu_k^+)\ ,$$

where $\mu_0 < \mu_1^- \leqslant \mu_1^+ < \mu_2^- \leqslant \mu_2^+ < \dots$ are the eigenvalues of the periodic and anti-periodic boundary value problems, while by (3.4.28') and (3.4.7),

$$s(\sqrt{z},\pi) = \pi \prod_{k=1}^{\infty} k^{-2}(\lambda_k - z) \quad , \quad \mu_k^- \leq \lambda_k \leq \mu_k^+ \; .$$

The fact that the potential is finite-zone means that the equation $1 - u_+(\sqrt{z})^2 = 0$ has only finitely many simple roots. We denote them by $\mu_0 < \mu_1^- < \mu_1^+ < \ldots < \mu_N^- < \mu_N^+$, modifying, if necessary, the labelling. Next, the inequalities $\mu_k^- \leq \lambda_k \leq \mu_k^+$ show that every double root of the equation $1 - u_+(\sqrt{z})^2 = 0$ is a simple root of the function $s(\sqrt{z},\pi)$. Therefore, for (N+1)-zone potentials

$$1 - u_+(\sqrt{z})^2 = T_{2N+1}(z)d^2(z) \quad , \quad s(\sqrt{z},\pi) = R_N(z)d(z) \; ,$$

where

$$T_{2N+1}(z) = (z - \mu_0) \prod_{j=1}^{N} (z - \mu_j^-)(z - \mu_j^+) \quad , \quad R_N(z) = \prod_{j=1}^{N} (z - \lambda_j) \; ,$$

$$d(z) = (-1)^N \pi(N!) \prod_{k=1}^{\infty} (\tilde{\mu}_k - z)(N + k)^{-2} \; ,$$

and the $\tilde{\mu}_k$ are the multiple roots of the equation $1 - u_+(\sqrt{z})^2 = 0$ and $\mu_j^- \leq \lambda_j \leq \mu_j^+$. Reexpressing the equality $\mathrm{Det}\, U = 1$ in the form

$$-u_-(\sqrt{z})^2 - c'(\sqrt{z},\pi)s(\sqrt{z},\pi) = 1 - u_+(\sqrt{z})^2 \; ,$$

we see that the functions $u_-(\sqrt{z})$ and $c'(\sqrt{z},\pi)$ are divisible by $d(z)$, and hence that

$$c'(\sqrt{z},\pi) = -W_{N+1}(z)d(z) \quad , \quad W_{N+1}(z) = \prod_{j=0}^{N} (z - \nu_j)$$

and

$$u_-(\sqrt{z}) = V_{N-1}(z)d(z) \quad , \quad V_{N-1}(z) = \sum_{k=0}^{N-1} v_k z^k \; ,$$

since

$$\lim_{z\to-\infty} \frac{c'(\sqrt{z},\pi)}{zs(\sqrt{z},\pi)} = -1 \quad , \quad \lim_{z\to-\infty} \frac{u_-(\sqrt{z})}{s(\sqrt{z},\pi)} = 0 \; .$$

Thus, the transition matrices of the operators of the form (4.3.4), generated by finite-zone potentials with the same Hill discriminant $u_+(\sqrt{z})$, have the form

$$U = u_+(\sqrt{z})I + d(z)P_{N+1}(z) \quad , \quad P_{N+1}(z) = \begin{pmatrix} U_{N-1}(z) & R_N(z) \\ -W_{N+1}(z) & -U_{N-1}(z) \end{pmatrix} , \qquad (4.3.21)$$

where $U_{N-1}(z)$, $R_N(z)$, and $W_{N+1}(z)$ are real polynomials which satisfy the relation

$$-U_{N-1}(z)^2 + W_{N+1}(z)R_N(z) = T_{2N+1}(z) ,$$

wherein

$$T_{2N+1}(z) = (1 - u_+(\sqrt{z})^2)d(z)^{-2} = (z - \mu_0) \prod_{j=1}^{N} (z - \mu_j^-)(z - \mu_j^+) .$$

Moreover, the coefficients of the highest-degree terms in the polynomials $R_N(z)$ and $W_{N+1}(z)$ are equal to 1, and every lacuna $[\mu_j^-,\mu_j^+]$ $(0 \leq j \leq N)$ contains exactly one root λ_j of the polynomial $R_N(z)$.

It follows from Theorem 4.3.2 that the converse is also true: if $u_+(\sqrt{z})$ is a Hill discriminant, then the matrices U of the form (4.3.21), in which the polynomials $U_{N-1}(z)$, $R_N(z)$, and $W_{N+1}(z)$ satisfy the conditions listed above, are transition matrices of operators of the form (4.3.4), generated by periodic finite-zone (and hence C^∞- smooth) potentials. Suppose that in problem (4.3.16), (4.3.16') the initial function $v_0(x)$ is a finite-zone potential, and let $v(x,t)$ be its solution. By the foregoing discussion, the transition matrices of the family of operators (4.3.4) associated with the functions $v(x+\tau,t)$, must be of the form

$$U(\tau,t;z) = u_+(\sqrt{z})I + d(z)P_{N+1}(\tau,t;z) ,$$

where the functions $u_+(\sqrt{z})$ and $d(z)$ are independent of t, and the matrix $P_{N+1}(\tau,t;z)$ is a polynomial in z and satisfies the equations

$$\frac{\partial P_{N+1}}{\partial t} = [P_{N+1},B_1] \quad , \quad \frac{\partial P_{N+1}}{\partial \tau} = [P_{N+1},B_0] , \tag{4.3.22}$$

with matrices B_0 and B_1 of the form (4.3.17). *The system formed by these equations is not closed in the sense that it contains, along with the sought-for solutions* P_{N+1}, *the functions* $a(t)$, $b(t)$, *and* $c(t)$, *which specify the matrices* B_0 *and* B_1. *However, the requirement that the solutions* P_{N+1} *be polynomials in* z *closes it and reduces it to systems of autonomous ordinary differential equations.*

In fact, by (4.3.17), equations (4.3.22) are equivalent to the systems

$$\left.\begin{aligned} \dot{U}_{N-1} &= \{c - 2az + 4z^2\}R_N - \{2a + 4z\}W_{N+1} , \\ \dot{R}_N &= -2\ (2a+4z)U_{N-1} - 2bR_N , \\ \dot{W}_{N+1} &= 2\{c - 2az + 4z^2\}U_{N-1} + 2bW_{N+1} , \end{aligned}\right\} \tag{4.3.23}$$

and

$$\left.\begin{aligned} U'_{N-1} &= (z-a)R_N - W_{N+1} , \\ R'_N &= -2U_{N-1} , \\ W'_{N+1} &= 2(z-a)U_{N-1} , \end{aligned}\right\} \tag{4.3.24}$$

for the entries U_{N-1}, R_N , and W_{N+1} of the matrix $P_{N+1}(\tau,t;z)$ (the dot and the prime denote differentiation with respect to t and τ , respectively). In order that these systems admit polynomials in z of the form

$$\left.\begin{aligned} U_{N-1} &= \sum_{k=0}^{N-1} u_k z^k , \\ R_N &= \sum_{j=0}^{N} r_j z^j , \\ W_{N+1} &= \sum_{l=0}^{N+1} w_l z^l , \end{aligned}\right\} \tag{4.3.25}$$

(with $r_N = w_{N+1} = 1$) as solutions, it is necessary and sufficient that the following equations be satisfied:

$$\left.\begin{aligned} \dot{u}_k &= cr_k - 2ar_{k-1} + 4r_{k-2} - 2aw_k - 4w_{k-1} , \\ \dot{r}_j &= -4au_j - 8u_{j-1} - 2br_j , \\ \dot{w}_l &= 2cu_l - 4au_{l-1} + 8u_{l-2} + 2bw_l , \end{aligned}\right\} \tag{4.3.23'}$$

and

$$\left.\begin{aligned} u'_k &= -ar_k + r_{k-1} - w_k , \\ r'_j &= -2u_j , \\ v'_l &= -2au_l + 2u_{l-1} , \end{aligned}\right\} \tag{4.3.24'}$$

(these are obtained by substituting the polynomials (4.3.25) into the systems (4.3.23) and (4.3.24), respectively). Since $u_k = w_l = r_j$ for negative values of k , j , and l , and also for $k > N-1$, $j > N$, and $l > N+1$, and since $w_{N+1} = r_N = 1$, the equations involving derivatives of u_N , u_{N+1} , u_{N+2} , r_N , and w_{N+1} , reduce to the algebraic equations:

$$\left.\begin{aligned} 0 &= c - 2a(r_{N-1} - w_N) + 4(r_{N-2} - w_{N-1}) , \\ 0 &= -4(a - r_{N-1} + w_N) , \\ 0 &= -a + r_{N-1} - w_N , \\ 0 &= -2(4u_{N-1} + b) . \end{aligned}\right\}$$

This shows that the systems (4.3.23), (4.3.24) have polynomial solutions of the form (4.3.25) if and only if

$$\left.\begin{aligned} a &= r_{N-1} - w_N , \\ b &= -4v_{N-1} , \\ c &= 2(r_{N-1}^2 - w_N^2) - 4(r_{N-2} - w_{N-1}) . \end{aligned}\right\} \tag{4.3.26}$$

Substituting these values for a , b , and c into the right-hand sides of systems (4.3.23') and (4.3.24'), we obtain certain autonomous systems of nonlinear ordinary differential equations for the functions u_k , r_j , w_l ($0 \leq k$, $j \leq N-1$, $0 \leq l \leq N$). Upon solving them and using formulas (4.3.26), we find the functions a , b , and c , and hence the matrices B_0 and B_1 for which the equations (4.3.22) have solutions which are polynomials in z .

Let $U_0(z) = u_+(\sqrt{z})I + d(z)P_{N+1}(z)$ be the transition matrix corresponding to the initial function $v_0(x)$. On solving the corresponding autonomous system for the initial data defined by the matrix $P_{N+1}(z)$, we find the real polynomials $U_{N-1}(t;z)$, $R_N(t;z)$, and $W_{N+1}(t;z)$, and a matrix B_1 of the form (4.3.17) such that the matrix $P_{N+1}(t;z)$, constructed from these polynomials, satisfies the equation $\dot{P}_{N+1} = [P_{N+1}, B_1]$. Hence, the matrix

$$U(t;z) = u_+(\sqrt{z})\cdot I + d(z)P_{N+1}(t;z)$$

solves the Cauchy problem (4.3.18). Since $\operatorname{Det} P_{N+1}(t;z) = -U_{N-1}(t;z)^2 + W_{N+1}(t;z)R_N(t;z)$ is conserved, the identity

$$-U_{N-1}(t;z)^2 + W_{N+1}(t;z)R_N(t;z) = T_{2N+1}(z) \tag{4.3.27}$$

holds on the entire domain of existence of the solutions to the autonomous system. Let $\lambda_k(t)$ ($k = 1,2,\dots,N$) denote the roots of the polynomial $R_N(t;z)$. At the initial moment ($t = 0$) they lie inside the lacunas: $\mu_k^- \leq \lambda_k(0) \leq \mu_k^+$. As t changes continuously, they move, but each one is trapped in its own lacuna, since by (4.3.27), the product

$$W_{N+1}(t;z)R_N(t;z) = T_{2N+1}(z) + U_{N-1}(t;z)^2$$

is strictly positive in the exterior of the lacunas. Therefore, the roots of the polynomial $R_N(t;z)$ remain simple, real, and they stay inside the corresponding lacunas, from which it follows, by the foregoing analysis, that $U(t;z)$ are transition matrices of operators of the form (4.3.4).

A similar argument, applied to the roots $\nu_k(t)$ of the polynomial $W_{N+1}(t;z)$, shows that they too, lie inside lacunas: $-\infty < \nu_0(t) \leq \mu_0$, $\mu_k^- \leq \nu_k(t) \leq \mu_k^+$ $(k = 1,2,\ldots,N)$. Upon taking note of identity (4.3.27), this shows that the coefficients of the three polynomials of interest are bounded by constants that do not depend on t, in the entire domain of existence of the solution to the autonomous system, and hence the solution of this system cannot blow-up (become infinite) for finite values of t, i.e., it can be continued throughout the real line $-\infty < t < \infty$. As we have seen earlier, this guarantees the existence and uniqueness of the solution $v(x,t)$ of the problem (4.3.16), (4.3.16'). Moreover, the matrices $U(t;z)$, found in the process, are the transition matrices of the operators (4.3.4) associated with this solution. Now we may solve the system obtained from (4.3.24) by replacing a by the function $r_{N-1} - w_N$, for the initial data specified by the matrix $P_{N+1}(t;z)$ found earlier, and thus obtain the function $a(\tau,t)$ that provides the solution $v(x,t)$ via formula (4.3.20).

Thus, we have proved the existence and uniqueness of the solution of the problem (4.3.16), (4.3.16') *for finite-zone initial data and, at the same time, we have reduced the task of solving it to that of solving the two systems of autonomous ordinary differential equations obtained from* (4.3.23'), (4.3.24').

The solution of the periodic Cauchy problem for the KdV equation with an arbitrary thrice continuously differentiable initial potential $v_0(x)$ may be obtained by approximating $v_0(x)$ with finite-zone potentials which, by Theorem 4.3.4, are dense in every space $\tilde{W}_2^n[0,\pi]$. For doing estimates, it is convenient to use the equalities

$$a(\tau,t) = \nu_0(\tau,t) + \sum_{k=1}^{N} (\nu_k(\tau,t) - \lambda_k(\tau,t)) , \qquad (4.3.28)$$

$$c(\tau,t) = -2\left\{\nu_0(t,\tau)^2 + \sum_{k=1}^{N} (\nu_k(\tau,t)^2 - \lambda_k(\tau,t)^2)\right\}, \tag{4.3.28}$$

which follow from Newton's formulas for the roots of polynomials and (4.3.26), and also the relation

$$\nu_0(t) + \sum_{k=1}^{N} (\nu_k(t) + \lambda_k(t)) = \mu_0 + \sum_{j=0}^{N} (\mu_j^- + \mu_j^+), \tag{4.3.29}$$

which is obtained upon comparing the coefficients of z^{2N} in (4.3.27). To estimate the function $b(\ ,t)$, one can use the interpolation formula

$$U_{N-1}(z) = R_N(z) \sum_{k=1}^{N} \frac{U_{N-1}(\lambda_k)}{(z - \lambda_k)R_N'(\lambda_k)}$$

and the equality

$$U_{N-1}(\lambda_k)^2 = -T_{2N+1}(\lambda_k), \tag{4.3.30}$$

which follows from identity (4.3.27) upon replacing z by the root λ_k of the polynomial R_N. Combining equalities (4.3.28) and (4.3.29), we obtain yet another useful formula:

$$a(\tau,t) = \mu_0 + \sum_{j=1}^{N} \{\mu_j^- + \mu_j^+ - 2\lambda_j(\tau,t)\} = 2[m + r_{N-1}(\tau,t)], \tag{4.3.31}$$

where $2m$ is a constant:

$$2m = \mu_0 + \sum_{j=1}^{N} \{\mu_j^- + \mu_j^+\}, \tag{4.3.31'}$$

and the function $r_{N-1}(\tau,t) = -\sum_{j=1}^{N} \lambda_j(\tau,t)$ is the coefficient of z^{N-1} in the polynomial R_N. Equalities (4.3.28), (4.3.29), and (4.3.31) are, of course, nothing but particular cases of the trace formulas, but they remain valid also in the case where U_{N-1}, R_N, and W_{N+1} are arbitrary polynomials which satisfy the systems of differential equations (4.3.23), (4.3.24).

From systems (4.3.23) and (4.3.24) we may also derive autonomous differential equations describing the motion of the roots λ_j and ν_l of the polynomials R_N and W_{N+1}, respectively. In fact, upon dividing both members of those equations that contain the derivatives of R_N by R_N, and then computing the residues of the resulting rational fractions at the points $z = \lambda_k$, we obtain

$$\dot{\lambda}_k = 4(a + 2\lambda_k)U_{N-1}(\tau,t;\lambda_k) \prod_{j\neq k} (\lambda_k - \lambda_j)^{-1} \tag{4.3.32}$$

and

$$\lambda_k' = 2U_{N-1}(\tau,t;\lambda_k) \prod_{j\neq k} (\lambda_k - \lambda_j)^{-1} , \tag{4.3.32'}$$

where

$$\prod_{j\neq k} (\lambda_k - \lambda_j) = \frac{d}{dz} R_N\Big|_{z=\lambda_k} .$$

These equalities are, in point of fact, systems of autonomous differential equations for the roots $\lambda_k(\tau,t)$, since the functions $a(\tau,\lambda)$ and $U_{N-1}(\tau,t;\lambda_k)$ can be expressed through these roots by the formulas (4.3.31) and (4.3.30), respectively. In exactly the same way, we can obtain differential equations for the roots $\nu_l(\tau,t)$ of the polynomial W_{N+1} :

$$\dot{\nu}_l = -2(c - 2a\nu_l + 4\nu_l^2)U_{N-1}(\tau,t;\nu_l) \prod_{j\neq l} (\nu_l - \nu_j)^{-1}$$

and

$$\nu_l' = -2(\nu_l - a)U_{N-1}(\tau,t;\nu_l) \prod_{j\neq l} (\nu_l - \nu_j)^{-1} ,$$

with

$$c = 4\{-m^2 + mw_N(\tau,t) - m_1 + w_{N-1}(\tau,t)\} ,$$

where $2m_1$ is the coefficient of z^{2N-1} in the polynomial T_{2N+1} , while $w_N(\tau,t)$ and $w_{N-1}(\tau,t)$ are the coefficients of z^N and z^{N-1} in the polynomial W_{N+1}, which are expressible through its roots by Newton's formulas.

In conclusion, we mention that from the method discussed above for solving the problem (4.3.16), (4.3.16') it obviously follows that the autonomous systems obtained from systems (4.3.23') and (4.3.24') are compatible, i.e., completely integrable. Hence, upon solving them for arbitrary initial polynomials $U_{N-1}(z)$, $R_N(z)$, and $W_{N+1}(z)$, which satisfy the sole requirement that coefficients of the highest power of z in $R_N(z)$ and $W_{N+1}(z)$ be equal to one, we obtain a function $a(x,t) = r_{N-1}(x,t) - w_N(x,t)$, which is necessarily a solution of the KdV equation. The reason behind this fact is that the KdV equation is obviously equivalent to the compatibility (integrability) condition for the systems (4.3.23) and (4.3.24). Needless to say, these assertions may also be proved directly (leading to the same results), i.e., making no use whatsoever of notions from spectral theory. Such an approach is particularly useful when the spectral properties of the operator L are difficult to investigate (for example, when L is non-self-adjoint, or the parameter z appears in it in a complicated, nonlinear way, and so forth). The basic guiding principle is that of "polynomial closure".

PROBLEMS

1. Let $v(x,t)$ be an infinitely differentiable periodic (i.e., $v(x,t) \equiv v(x+\pi,t)$) solution of the equation $K_m[v] = 0$ (for instance, of the KdV equation. Show that the integrals $\int_0^\pi \sigma_k(x;t)dt$ are conserved, i.e., are independent of t.

Hint. The transition matrices U of the family of operators (4.3.4), associated with the periodic solution $v(x,t)$ of the equation $K_m[v] = 0$, satisfy the equation $\dot{U} = [U,A_m]$, from which it follows that $\operatorname{Trace} U = u_+(\sqrt{z})$ does not depend on t. Expressing the Hill discriminant $u_+(\lambda) = \frac{1}{2}[c(\lambda,\pi) + s'(\lambda,\pi)]$ through the solutions $y(\lambda,x)$, $y(-\lambda,x)$ by means of equalities (1.4.24) and (1.4.24'), you get

$$2u_+(\lambda) = y(\lambda,\pi)\left[1 + \frac{\sigma(\lambda,\pi) - \sigma(\lambda,0)}{\omega(\lambda,0)}\right] + y(-\lambda,\pi)\left[1 + \frac{\sigma(-\lambda,0) - \sigma(-\lambda,\pi)}{\omega(\lambda,0)}\right] =$$

$$= y(\lambda,\pi)[1 + O(\lambda^{-n-1})] + y(-\lambda,\pi)[1 + O(\lambda^{-n-1})],$$

since $\sigma(\lambda,\pi) - \sigma(\lambda,0) = O(\lambda^{-n-1})$, in view of the periodicity of the potential. Consequently,

$$0 = \frac{d}{dt} 2u_+(\lambda) = y(\lambda,\pi)[1 + O(\lambda^{-n-1})]\frac{d}{dt}\int_0^\pi \sigma(\lambda,x;t)dx +$$

$$+ y(-\lambda,\pi)[1 + O(\lambda^{-n-1})]\frac{d}{dt}\int_0^\pi \sigma(-\lambda,x;t)dx,$$

where

$$\left.\begin{aligned} &\frac{d}{dt}\int_0^\pi \sigma(\lambda,x;t)dx = \frac{d}{dt}\sum_{j=1}^{n}\int_0^\pi \sigma_k(x;t)dx \cdot (2i\lambda)^{-k} + O(\lambda^{-n-1}) \\ &\text{and} \\ &y(\lambda,\pi) = \exp\left(i\lambda\pi + \int_0^\pi \sigma(\lambda,x;t)dx\right), \end{aligned}\right\} \quad (4.3.33)$$

which is possible only if

$$\frac{d}{dt}\int_0^\pi \sigma_k(x;t)dx = 0.$$

2. If the function $v(x,t) \equiv v(x)$ does not depend on t, then the equation $K_n[v] = 0$, which is generated by an admissible polynomial $N_n(z,x)$, becomes an ordinary differential equation of order $2n+1$ for $v(x)$:

$$\tilde{K}_n[v] \equiv N_n''' - 4N_n'(v-z) - 2N_n v' = 0 . \qquad (4.3.34)$$

In this particular situation, Theorem 4.3.1 says that $v(x)$ satisfies an equation of the form (4.3.34) on a segment $x_1 \leqslant x \leqslant x_2$ if and only if the transition matrix $U(2,1)$ of the corresponding operator (4.3.4) satisfies the equality $U(2,1)B^{(1)} - B^{(2)}U(2,1) = 0$, where $B^{(i)}$ are matrices which depend polynomially on z and $B^{(1)} = A_n(1)$, $B^{(2)} = A_n(2)$. Using this consequence of Theorem 4.3.1, show that the periodic function $v(x)$ is an m-zone potential if and only if it satisfies an equation of the form (4.3.34), in which case $m \leqslant n+1$.

<u>Hint</u>. If $v(x)$ is an m-zone potential, then, as was shown above, $U(2,1) = u_+(\sqrt{z})I + d(z)P_{m+1}(z)$, where $P_{m+1}(z)$ is a matrix polynomial in z. Consequently, $U(2,1)P_{m+1} - P_{m+1}U(2,1) = 0$, which, by the Corollary to Theorem 4.3.1, implies that $v(x)$ satisfies an equation of the form (4.3.34). Conversely, if the periodic function $v(x)$ satisfies (4.3.34), then the corresponding transition matrix $U(2,1)$ satisfies the equality $U(2,1)A_n(1) - A_n(2)U(2,1) = 0$, and by periodicity, $A_n(1) = A_n(2)$. Since

$$A_n(1) = \begin{pmatrix} -N_n' & 2N_n \\ 2N_n(v-z) - N_n'' & N_n' \end{pmatrix} = \begin{pmatrix} U_{m-1} & R_m \\ -W_{m+1} & U_{m-1} \end{pmatrix} q(z) = P_{m+1}q ,$$

where $q(z)$ is the common divisor of the entries of the matrix $A_n(1)$, you have $[V,P_{m+1}] = 0$, where $U(2,1) = u_+(\sqrt{z})I + V$, $m \leqslant n$, and

$$V = \begin{pmatrix} u_- & s \\ c' & -u_- \end{pmatrix} .$$

The equality $[V,P_{m+1}] = 0$ is equivalent to three scalar ones:

$$-sW_{m+1} = R_m c' , \quad u_- R_m = sU_{m-1} , \quad c'U_{m-1} = -u_- W_{m+1} ,$$

from which in turn it follows that

$$s = R_m d , \quad c' = -W_{m+1}d , \quad u_- = U_{m-1}d ,$$

and hence that

$$1 - u_+(\sqrt{z})^2 = -u_-^2 - c's = (-U_{m-1}^2 + W_{m+1}R_m)d^2 ,$$

where $d(z)$ is an entire function and $T_{2m+1}(z) = -U_{m-1}^2 + W_{m+1}R_m$ is a polynomial of degree $2m+1$. Therefore, the equation $1 - u_+(\sqrt{z})^2 = 0$ has only $2m+1$ simple roots and $v(x)$ is an $(m+1)$-zone potential.

3. Suppose that the periodic function $v_0(x)$ satisfies the equation $\tilde{K}_1[v] = 0$, and hence is a finite-zone potential. Show that the solution $v(x,t)$ of the periodic Cauchy problem for any of the equations $\tilde{K}_m[v] = 0$, with initial data $v(x,0) = v_0(x)$, satisfies the same equation $\tilde{K}_1[v] = 0$ for all t, i.e., the set of periodic solutions of the equation $\tilde{K}_1[v] = 0$ is an invariant manifold for any of the equations $K_m[v] = 0$.

Hint. The transition matrices of the operators (4.3.4) associated with the solution $v(x,t)$ have the form $U = u_+(\sqrt{z})\cdot I + d(z)P_{1+1}(z,t)$, where the functions $u_+(\sqrt{z})$ and $d(z)$ do not depend on t, and the 12-entries $R_1(z,t)$ of the matrices $P_{1+1}(z;t)$ are polynomials of degree 1; moreover, the solution $s(\sqrt{z},x;t)$ of the equation $-y'' + v(x,t)y = zy$ is such that $s(\sqrt{z},x;t) = R_1(z,t)d(z)$.

Obviously, $[U,P_{1+1}] = 0$, and by Theorem 4.3.1, the function $v(x,t)$ satisfies the equations

$$\tilde{K}_1[v] = \tilde{K}\{Q_1(z,t)\}[v] = 0 ,$$

where

$$Q_1(z,t) = \text{Reg}\ \{R_1(z,t)N_\infty(z,v(0,t))\} .$$

Hence, the claim you have to prove is equivalent to the equality

$$\frac{d}{dt}\text{Reg}\ \{R_1(z,t)N_\infty(z,v(0,t))\} = 0$$

(see the Remark to Theorem 4.3.1). By (1.4.24),

$$s(\sqrt{z}, ;t) = \frac{y(\sqrt{z},\pi;t) - y(-\sqrt{z},\pi;t)}{\omega(\sqrt{z},0;t)} = R_1(z,t)d(z) ,$$

and hence

$$R_1(z,t)\frac{\omega(\sqrt{z},0;t)}{2i\sqrt{z}} = \frac{y(\sqrt{z},\pi;t) - y(-\sqrt{z},\pi;t)}{y(\sqrt{z},\pi;0) - y(-\sqrt{z},\pi;0)} R_1(z,0)\frac{\omega(\sqrt{z},0;0)}{2i\sqrt{z}} . \qquad (4.3.35)$$

Next, it follows from formula (1.4.22) that for sufficiently large n

$$\operatorname{Reg}\left\{R_1(z,t)\,\frac{\omega(\sqrt{z},0;t)}{2i\sqrt{z}}\right\} = \operatorname{Reg}\,\{R_1(z,t)N_\infty(z,v(0,t))\}\ ,$$

whereas the equalities (4.3.33), (4.3.35) and the fact that $\int_0^\pi {}_k(x;t)dx$ does not depend on t imply that

$$\operatorname{Reg}\left\{R_1(z,t)\,\frac{\omega(\sqrt{z},0;t)}{2i\sqrt{z}}\right\} = \operatorname{Reg}\left\{R_1(0,z)\,\frac{\omega(\sqrt{z},0;0)}{2i\sqrt{z}}\right\}\ .$$

Therefore,

$$\operatorname{Reg}\,\{R_1(z,t)N_\infty(z,v(0,t))\} = \operatorname{Reg}\,\{R_1(z,0)N_\infty(z,v(0,0))\}\ ,$$

whence

$$\tilde{K}\{Q_1(z,t)\} \equiv \tilde{K}\{Q_1(z,0)\}\ ,$$

as claimed.

4. Reduce the integration of the operator partial differential equation

$$B\,\frac{\partial\Omega}{\partial t} + \frac{\partial^3\Omega}{\partial x^3} - 2\Omega^3 = 0\ , \tag{4.3.36}$$

(see Problem 1, Section 1, Chapter 4) to the integration of compatible systems of ordinary differential equations.

<u>Hint</u>. If $\Omega(x,t)$ is a periodic solution of equation (4.3.36), then the transition operators $U = U(2,1;\tau,t,z)$ of the family of equations $L[Y] = 0$, where

$$L = \frac{d}{dx} - B\Omega(x+\tau,t) + Bz\ ,$$

satisfy the compatible system of equations

$$\frac{\partial U}{\partial \tau} = [U,A_1]\quad ,\quad \frac{\partial U}{\partial t} = [U,A_2]\ ,$$

where

$$A_1 = Bz - B\Omega\quad ,\quad A_2 = 2Bz^2 - 2B\Omega z + \Omega' + B\Omega^2\ .$$

Using these equations as a guide, consider an arbitrary system of the form

$$\frac{\partial P}{\partial \tau} = [P,B_1]\quad ,\quad \frac{\partial P}{\partial t} = [P,B_2]\ , \tag{4.3.37}$$

where

$$B_1 = Bz - BC\quad ,\quad B_2 = 2Bz^2 - 2BCz + D + C^2\quad ,\quad C = C(\tau,t)\quad ,\quad D = D(\tau,t)\ ,$$

are operator-valued functions which satisfy the conditions

$$BC + CB = 0 \quad , \quad BD + DB = 0 \ .$$

The requirement that the system (4.3.37) admit a polynomial solution

$$P = \sum_{k=0}^{N} P_k(\tau,t)z^k$$

leads to the equalities

$$\frac{\partial P_k}{\partial \tau} = [P_{k-1},B] - [P_k,BC] \ , \quad \frac{\partial P_k}{\partial t} = [P_{k-2},2B] - [P_{k-1},2BC] + [P_k,D+BC^2] \quad (4.3.38)$$

which play the same role here as do the equalities (4.3.24'), (4.3.23') for the KdV equation. For $k = N+2$, $N+1$, N, (4.3.38) reduces to the algebraic equalities

$$0 = [P_N,B] \quad , \quad \frac{\partial P_N}{\partial \tau} = [P_{N-1},B] - [P_N,BC] \quad , \quad 0 = [P_N,2B] \ ,$$

$$0 = [P_{N-1},2B] - [P_N,2BC] \quad , \quad \frac{\partial P_N}{\partial t} = [P_{N-2},2B] \ ,$$

the fulfillment of which is necessary and sufficient for the existence of a solution which is a polynomial in z to each of the equations (4.3.37). These equalities may be fulfilled by choosing

$$P_N = B \quad , \quad C = \frac{1}{2}[B,P_{N-1}] \ ,$$

$$D = B\{[P_{N-2},B] - \frac{1}{2}[P_{N-1},B[B,P_{N-1}]]\} \ .$$

Substituting these expressions for P_N, C, and D into equation (4.3.38), you obtain two autonomous systems of ordinary differential equations for $P_{N-1}, P_{N-2}, \ldots, P_1, P_0$. Their compatibility may be checked directly: since the right-hand sides are simple polynomials in the unknowns, the proof of the compatibility reduces, by Frobenius' theorem, to checking a simple algebraic identity. The local solvability of the indicated systems is obvious. Finally, the simplest way to show that the operator-valued function $\Omega(\tau,t) = \frac{1}{2}[P,P_{N-1}(\tau,t)]$ satisfies equation (4.3.36) (with τ replaced by x), is to use the already established compatibility of equations (4.3.37).

5. Reduce the integration of the operator equation (4.1.20) to the integration of compatible systems of ordinary differential equations.

Hint. See Problem 3, Section 1 of this chapter.

4. EXPLICIT FORMULAS FOR PERIODIC SOLUTIONS OF THE KORTEWEG-DE VRIES EQUATION

In the preceding section it was established that a solution $v(x,t)$ of the Cauchy problem (4.3.16), (4.3.16') with a finite-zone initial potential exists, and that, according to (4.3.20) and (4.3.31),

$$v(x,t) = \mu_0 + \sum_{j=1}^{N} \{\mu_j^- + \mu_j^+ - 2\lambda_j(x,t)\} , \tag{4.4.1}$$

where the functions $\lambda_j(\tau,t)$ (i.e., the roots of the polynomial R_N) that must still be determined, satisfy the differential equations (4.3.32), (4.3.32'). The functions $U_{N-1}(\tau,t;\lambda_k(\tau,t))$ which appear in these equations must be expressed in terms of $\lambda_k(\tau,t)$ by means of equality (4.3.30), which however only permits us to find them up to a sign. To get rid of this indeterminacy, we bring the Riemann surface Γ of the function $\sqrt{-T_{2N+1}(z)}$ into picture, and replace equations (4.3.32) and (4.3.32') by the equations which are derived from them for the points of the surface Γ lying over the points $\lambda_k(\tau,t)$. The Riemann surface Γ can be realized by taking two copies of the z-plane, slit along the segments $[\mu_0,\mu_1^-],[\mu_1^+,\mu_2^-],\dots,[\mu_{N-1}^+,\mu_N^-],[\mu_N^+,\infty)$, the boundaries of which are pasted crosswise. On this two-sheeted surface the function $\sqrt{-T_{2N+1}}$ is single-valued. Now let us associate to each pair $\lambda_k(\tau,t)$, sign $U_{N-1}(\lambda_k(\tau,t);\tau,t)$ the point $\gamma_k(\tau,t) \in \Gamma$ lying over $\lambda_k(\tau,\lambda)$ on that sheet on which sign $U_{N-1}(\tau,t;\lambda_k(\tau,t)) = \text{sign} \sqrt{-T_{2N+1}(\gamma_k(\tau,t))}$. It follows from formula (4.3.30) that under this correspondence

$$U_{N-1}(\tau,t;\lambda_k(\tau,t)) = \sqrt{-T_{2N+1}(\gamma_k(\tau,t))} ,$$

and equations (4.3.32), (4.3.32') turn into the differential equations

$$\dot{\gamma}_k = 4(a + 2\gamma_k) \sqrt{-T_{2N+1}(\gamma_k)} \prod_{j\neq k} (\gamma_k - \gamma_j) \tag{4.4.2}$$

and

$$\gamma_k' = 2 \sqrt{-T_{2N+1}(\gamma_k)} \prod_{j\neq k} (\gamma_k - \gamma_j)^{-1} \tag{4.4.2'}$$

for the points γ_k of Γ lying over the points λ_k .

Equations (4.4.2), (4.4.2') can be integrated by means of Abel's substitution. To this end, we introduce canonical sections (cycles) a_k , b_k $(1 \leq k \leq N)$ on Γ : a_k is a closed contour on the upper sheet which encircles the cut $[\mu_{k-1}^+,\mu_k^-]$, whereas b_k is a closed contour which starts on the upper lip of the slit $[\mu_{k-1}^{+},\mu_k^{-}]$, continues on the upper sheet until it reaches the upper lip of the slit $[\mu_N^{+},\infty)$, and then passes to the lower sheet on which it return to the starting point (see Fig. 6).

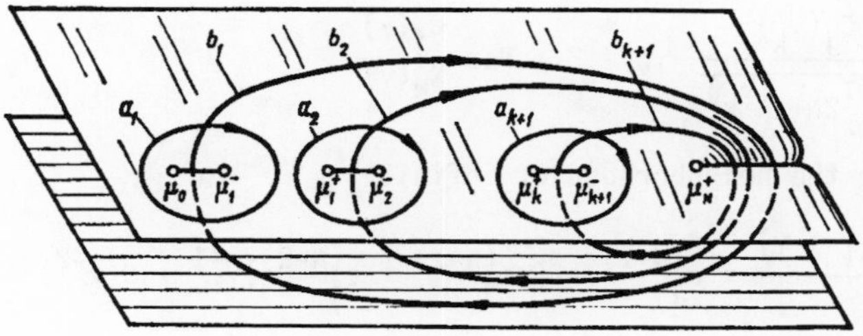

Figure 6

On the Riemann surface Γ of the function $\sqrt{-T_{2N+1}(\gamma)}$, there exist N linearly independent Abelian differentials

$$dU_j = \frac{C_j(\gamma)}{\sqrt{-T_{2N+1}(\gamma)}}\,d\gamma \quad , \text{ with } C_j(\gamma) = \sum_{p=0}^{N-1} c_j^{(p)}\gamma^p \ , \tag{4.4.3}$$

which can be chosen so that

$$\oint_{a_k} dU_j = \delta_{jk} \quad , \quad b_{jk} = \oint_{b_k} dU_j \ . \tag{4.4.4}$$

Moreover, by Riemann's theorem, the matrix $B = (b_{jk})$ of b-periods of these differentials is symmetric ($b_{jk} = b_{kj}$) and its imaginary part is positive definite (see, e.g.,[25]). The Abel mapping $(\gamma_1,\ldots,\gamma_N) \to (e_1,\ldots,e_N)$ is defined by the rule

$$e_j(\gamma_1,\ldots,\gamma_N) \equiv \sum_{k=1}^{N} \int_{\infty}^{\gamma_k} dU_j = \sum_{k=1}^{N} U_j(\gamma_k) \ , \tag{4.4.5}$$

in which the symbol $\equiv$ designates congruence modulo the periods of the Abelian differentials dU_j. The functions $e_j(\gamma_1(\tau,t),\ldots,\gamma_N(\tau,t))$, obtained via this mapping, depend linearly on τ and t if $\gamma_1(\tau,t),\ldots,\gamma_N(\tau,t)$ satisfy equations (4.4.2), (4.4.2'). In fact, by (4.4.3), (4.3.31), and (4.4.5),

$$\frac{\partial e_j}{\partial t} = \sum_{k=1}^{N} \frac{C_j(\lambda_k)}{\sqrt{-T(\gamma_k)}} \dot{\gamma}_k = 4 \sum_{k=1}^{N} C_j(\gamma_k)[a + 2\gamma_k] \prod_{j \neq k} (\gamma_k - \gamma_j)^{-1} =$$

$$= 4 \sum_{k=1}^{N} \frac{C_j(\gamma_k)}{R_N'(\gamma_k)} = 8 \sum \operatorname{Res} \frac{C_j(k)[m + r_{N-1} + \gamma]}{R_N(\gamma)}$$

and

$$\frac{\partial e_j}{\partial \tau} = \sum_{j=1}^{N} \frac{C_j(\gamma_k)}{\sqrt{-T_{2N+1}(\gamma_k)}} \gamma_k' = 2 \sum \operatorname{Res} \frac{C_j(\gamma)}{R_N(\gamma)} ,$$

and, since in the neighborhood of infinity,

$$\frac{C_j(\gamma)[m + r_{N-1} + \gamma]}{R_N(\gamma)} = c_j^{(N-1)} + (mc_j^{(N-1)} + c_j^{(N-2)})\gamma^{-1} + O(\gamma^{-2})$$

and

$$\frac{C_j(\gamma)}{R_N(\gamma)} = \frac{c_j^{(N-1)}\gamma^{N-1} + c_j^{(N-2)}\gamma^{N-2} + \ldots}{\gamma^N + r_{N-1}\gamma^{N-1} + \ldots} = c_j^{(N-1)}\gamma^{-1} + O(\gamma^{-2}) ,$$

the partial derivatives

$$\frac{\partial e_j}{\partial t} = 8(mc_j^{(N-1)} + c_j^{(N-2)}) \quad , \quad \frac{\partial e_j}{\partial \tau} = 2c_j^{(N-1)}$$

do not depend on the variables τ and t, and the functions $e_j(\gamma_1(\tau,t),\ldots,\gamma_N(\tau,t))$ depend on them linearly.

Hence, the points $\gamma_k(\tau,t)$ $(1 \leqslant k \leqslant N)$ satisfy the system of equations

$$e_j(\gamma_1,\ldots,\gamma_N) \equiv g_j\tau + v_jt + p_j \quad (1 \leqslant j \leqslant N) , \tag{4.4.6}$$

in which, for the sake of brevity, we denoted

$$g_j = 2c_j^{(N-1)} \quad , \quad v_j = 8(mc_j^{(N-1)} + c_j^{(N-2)}) \quad , \quad p_j = \sum_{j=1}^{N} \int_{\infty}^{\gamma_k(0,0)} dU_j .$$

Thus, solving the system of differential equations (4.4.2), (4.4.2') amounts to determining the functions $\gamma_k(\tau,t)$ from the system of equations (4.4.6) or, in other words, to the inversion of the Abel mapping. This is known as the Jacobi inversion problem. It was solved by Riemann by means of the (Riemann) θ-function, which he introduced, and which is constructed as follows. First, one considers the θ-function of N complex variables

$$\theta(\vec{z}) = \theta(z_1,\ldots,z_N) = \sum_{\vec{k}} \exp\{i\pi[(B\vec{k},\vec{k}) + 2(\vec{z},\vec{k})]\} ,$$

where $\vec{z} = (z_1,\ldots,z_N)$, $\vec{k} = (k_1,\ldots,k_N)$ is an N-dimensional vector with integer coordinates $k_j = 0,\pm1,\pm2,\ldots$, B is the matrix (4.4.4') of b-periods of the system dU_j of Abelian differentials considered above, and $(\vec{x},\vec{y})$ designates the usual inner product in $\mathbb{C}^N$. The positive-definitness of the matrix $\operatorname{Im} B$ guarantees that this series converges, and that its sum is an entire function of the N complex variables $z_1,\ldots,z_N$. The function $\theta(\vec{z})$ is even: $\theta(\vec{z}) = \theta(-\vec{z})$, and enjoys the following readily verifiable properties:

$$\theta(z_1,\ldots,z_p+1,\ldots,z_N) = \theta(z_1,\ldots,z_p,\ldots,z_N) , \tag{4.4.7}$$

and

$$\theta(z_1+b_{1p},\ldots,z_N+b_{Np}) = \theta(z_1,\ldots,z_N) \exp\{-i\pi(b_{pp}+2z_p)\} . \tag{4.4.7'}$$

If in this θ-function we now replace the arguments z_j by the Abelian integrals

$$U_j(\gamma) - e_j = \int_\infty^\gamma dU_j - e_j \quad (1 \leqslant j \leqslant N) ,$$

where $e_1,\ldots,e_N$ are arbitrary constants, then the resulting function

$$\theta_1(\gamma) = \theta(\vec{u}(\gamma) - \vec{e}) = \theta(U_1(\gamma) - e_1,\ldots,U_N(\gamma) - e_N) , \tag{4.4.8}$$

will obviously be holomorphic and single-valued on the Riemann surface Γ , cut along the canonical sections a_k and b_k , $k = 1,\ldots,N$. The function (4.4.8) is called Riemann's θ-function. Its fundamental value in solving the Jacobi inversion problem is explained by the following result of Riemann.

THEOREM. *If the θ-function does not vanish identically, then it has exactly* N *zeros* $\gamma_1,\ldots,\gamma_N$, *and for every* p , $1 \leqslant p \leqslant N$,

$$\sum_{j=1}^N U_p(\gamma_j) \equiv e_p - k_p ,$$

where

$$k_p = \frac{1}{2}\left(\sum_{j=1}^N b_{pj} - p\right) .$$

For the proof, see, for example, [28, 25]. □

It follows from this theorem and the relations (4.4.6) that the points $\gamma_k(\tau,t)$ will be the zeros of Riemann's function $\theta_1(\gamma) = \theta(\vec{u}(\gamma) - \vec{e})$ if we put

$$e_j = g_j\tau + v_jt + p_j - k_j \quad , \quad 1 \leqslant j \leqslant N . \tag{4.4.9}$$

To find the solutions $v(x,t)$ of the Cauchy problem (4.3.16), (4.3.16'), we actually need not the points $\gamma_k(\tau,t)$ themselves, but the sum $\sum_{k=1}^{N} \lambda(\gamma_k(\tau,t))$ of their projections $\lambda(\gamma_k(\tau,t)) = \lambda_k(\tau,t)$ onto the complex plane. By the residue theorem,

$$\sum_{k=1}^{N} \lambda_k(\tau,t) = \frac{1}{2\pi i}\int_{\partial\tilde{\Gamma}} \lambda(\gamma)d\ \ln\ \theta_1(\gamma) - \mathrm{Res}\ \{\lambda(\gamma)d\ \ln\ \theta_1(\gamma)\}\Big|_{\gamma=\infty} ,$$

where $\tilde{\Gamma}$ is the domain obtained from the Riemann surface Γ by cutting it along the canonical sections a_k and b_k , $1 \leqslant k \leqslant N$, and

$$\partial\tilde{\Gamma} = \sum_k (a_k^+ - a_k^- + b_k^+ - b_k^-)$$

is its boundary, which consists of the clockwise oriented left lips (a_k^+,b_k^+) and the counter-clockwise oriented right lips $(-a_k^-,-b_k^-)$ of the canonical sections. Integrating the differentials dU_j along the cycles a_k^+ and b_k^+ , and invoking (4.4.4), we obtain the equalities

$$U_j(\gamma)\Big|_{a_k^+} - U_j(\gamma)\Big|_{a_k^-} = -b_{kj} \quad , \quad U_j(\gamma)\Big|_{b_k^+} - U_j(\gamma)\Big|_{b_k^-} = \delta_{kj} ,$$

respectively, from which in turn it follows, upon taking note of (4.4.7) and (4.4.7'), that

$$\oint_{a_k^+} \lambda(\gamma)d\ \ln\ \theta_1(\gamma) + \oint_{-a_k^-} \lambda(\gamma)d\ \ln\ \theta_1(\gamma) = 2\pi i \oint_{a_k^+} \lambda(\gamma)dU_k(\gamma)$$

and

$$\oint_{b_k^+} \lambda(\gamma)d\ \ln\ \theta_1(\gamma) + \oint_{-b_k^-} \lambda(\gamma)d\ \ln\ \theta_1(\gamma) = 0 .$$

Therefore,

$$\sum_{k=1}^{N} \lambda_k(\tau,t) = \sum_{k=1}^{N} \oint_{a_k^+} \lambda(\gamma)dU_k(\gamma) - \mathrm{Res}\ \{\lambda(\gamma)d\ \ln\ \theta_1(\gamma)\}\Big|_{\gamma=\infty} . \tag{4.4.10}$$

In the neighborhood of infinity in the domain $\tilde{\Gamma}$, the Abelian integrals

$$U_j(\gamma) = i\int_{\infty}^{\gamma}\left(\sum_{p=0}^{N-1} c_j^{(p)} z^{p-N+1}\right)(1 - 2mz^{-1} + \dots)^{-1/2} z^{-3/2} dz$$

can be developed in series in the odd powers of $s = \lambda(\gamma)^{-1/2}$:

$$U_j(\gamma) = -2i\{c_j^{(N-1)} s + \dots\} . \tag{4.4.11}$$

Hence, the function $\ln \theta_1(\gamma)$ admits the series expansion

$$\ln \theta_1(\gamma) = \sum_{j=0}^{\infty} \frac{q_j s^j}{j!} ,$$

where $q_j = \left.\dfrac{d^j \ln \theta_1(\gamma)}{ds^j}\right|_{s=0}$, and in the neighborhood of infinity,

$$\lambda(\gamma) d \ln \theta_1(\gamma) = s^{-2} \sum_{j=1}^{\infty} \frac{q_j s^{j-1}}{(j-1)!} ds ,$$

so that

$$\operatorname{Res} \{\lambda(\gamma) d \ln \theta_1(\gamma)\}\big|_{\gamma=\infty} = q_2 = \left.\frac{d^2 \ln \theta_1(\gamma)}{ds^2}\right|_{s=0} . \tag{4.4.12}$$

It follows from the definition of Riemann's θ-function in terms of the multi-dimensional θ-function and formulas (4.4.11) that

$$\left.\frac{d^2 \ln \theta_1(\gamma)}{ds^2}\right|_{s=0} =$$

$$= \left\{ \sum_{j=1}^{N} \sum_{l=1}^{N} \frac{\partial^2 \ln \theta(\vec{u} - \vec{e})}{\partial U_j \partial U_l} \frac{dU_j}{ds} \frac{dU_l}{ds} + \sum_{j=1}^{N} \frac{\partial \ln \theta(\vec{u} - \vec{e})}{\partial U_j} \frac{d^2 U_j}{ds^2} \right\}\Bigg|_{s=0} =$$

$$= - \sum_{j=1}^{N} \sum_{l=1}^{N} \frac{\partial^2 \ln \theta(\vec{e})}{\partial e_j \partial e_l} \frac{\partial e_j}{\partial \tau} \frac{\partial e_l}{\partial \tau} = - \frac{d^2}{d\tau^2} \ln \theta(\vec{e}) ,$$

since $\theta(\vec{e}) = \theta(-\vec{e})$, and

$$\frac{\partial e_k}{\partial \tau} = g_k = 2c_k^{(N-1)} ,$$

by (4.4.9) and (4.4.6). Now, combining the equalities (4.4.1), (4.4.6), (4.4.9), (4.4.10), and (4.4.12), we obtain the final formula

$$v(x,t) = -2 \frac{d^2}{dx^2} \ln \theta(\vec{g}x + \vec{v}t + \vec{w}) + C , \tag{4.4.13}$$

in which

$$g_k = 2c_k^{(N-1)} ,$$

$$v_k = 8(mc_k^{(N-1)} + c_k^{(N-2)}) ,$$

$$w_k = \sum_{j=1}^{N} \int_{\infty}^{\gamma_k(0,0)} dU_k - \frac{1}{2}\left(\sum_{j=1}^{N} b_{kj} - k\right) ,$$

and

$$C = \mu_0 + \sum_{j=1}^{N} \{\mu_j^- + \mu_j^+\} - 2 \sum_{j=1}^{N} \oint_{a_j^+} \lambda(\gamma)dU_j .$$

This formula was obtained for solutions with finite-zone initial functions. However, from the remark made at the end of the preceding section it follows that the function (4.4.13) satisfies the KdV equation throughout its domain of existence for every choice of the polynomial $T_{2N+1}(z)$ and the points $\gamma_k(0,0)$. If the roots of the polynomial $T_{2N+1}(z)$ are real and simple, and $\mu_k^- \leqslant \lambda(\gamma_k(0,0)) \leqslant \mu_k^+$, then the solution exists everywhere and is an almost-periodic function. In other cases, there may exist a wide variety of solutions. For example, the solutions (4.2.20) are limit cases of (4.4.13) for $\mu_k^+ = \mu_{k-1}^-$ (see [11]).

CHAPTER 5

STABILITY OF INVERSE PROBLEMS

1. PROBLEM FORMULATION AND DERIVATION OF MAIN FORMULAS

It was shown in the previous chapters that the Sturm-Liouville boundary-value problem can be completely reconstructed either from its spectral function or from the scattering data and that the reconstruction procedures are quite efficient. In particular, they allowed us to find necessary and sufficient conditions for spectral functions and scattering data of the boundary-value problems under consideration. These conditions show that the symmetric boundary-value problem is uniquely reconstructed from its spectral function $\rho(\mu)$ given for all μ.

The same is valid for the reconstruction of the boundary-value problem from its scattering data.

At the same time, the physical sense of the inverse problems indicates that neither the spectral function nor the spectral data can be completely known. This is clearly seen in the case of the inverse quantum scattering problem. Indeed, in this problem the parameter λ^2 is proportional to the system energy (see (3.3.2)), so in order to know the scattering data for all values of λ one has to conduct experiments with particles of arbitrarily large energy. But for sufficiently large, though finite, values of energy, the scattering process is not described anymore by the Schrödinger equation (3.3.1) with the potential $V(x) = \frac{\hbar^2}{2M}q(x)$. Therefore, even allowing, ideally, the possibility to experiment with particles of arbitrarily large energies, we would obtain, starting from a certain energy, data relevant to a process, which has certainly nothing to do with the equation that we want to reconstruct. Hence, a principal question is as follows: what information about the function $q(x)$ or the boundary-value problem in general can be obtained, if the spectral function or the scattering data are known (generally speaking, approximately) only on a finite interval of values of the spectral parameter? To answer this question, one has to know to what extend can two boundary-value problems differ from each other, if it is known that their spectral functions or scattering data differ slightly for λ^2 varying on a finite interval. It is evident that if nothing is known a priori about these problems, then they can differ as much as you want. For example, for any boundary-value problems with $q(x) \geq N$ and $h > 0$, the corresponding spectral functions vanish for all $\mu < N$ and, therefore, they coincide on the interval $(-\infty, N)$. Therefore, a meaningful (and natural, from the physical point of view) question about the stability of inverse spectral problems is the following: how much can two boundary-value problems differ from each other, if their spectral functions differ a little on a given interval of values of the spectral parameter λ^2, under the condition that certain estimates for $|h| + \int_0^x |q(t)|dt$ are known a priori?

In a similar way one can formulate the stability problem for the inverse problem of the quantum scattering theory. Notice the similarity of these questions with the question that is typical for the approximation theory: what can one say about a function, for which a finite part of its Fourier series is known? Or (which is actually the same): how can one estimate a function whose few first Fourier coefficients are equal to 0? It is known that one can answer this question (with the help of the G. Bohr inequality) only assuming that the function under consideration belongs to a certain functional class (for example, its derivative is bounded by a prescribed number, etc.)

Now let us introduce definitions and notations, which will be used in what follows. Denote the symmetric boundary-value problems (2.2.1), (2.2.2) by $\{h, q(x)\}$ and the boundary-value problem (3.1.1), (3.1.2), that satisfies condition (3.1.3), by $\{q(x)\}$. The function $q(x)$ is called the potentials.

Let $a(x)$ be an arbitrary nondecreasing continuous function with $a(0) = 0$ and let A be an arbitrary nonnegative number. Denote by $V\{A, a(x)\}$ the set of all boundary-value problems $\{h, q(x)\}$ such that

$$|h| \leq A, \quad \int_0^x |q(t)|dt \leq a(x) \quad (0 \leq x < \infty), \tag{5.1.1}$$

and denote by $V\{\alpha(x)\}$ the set of boundary-value problems $\{q(x)\}$ such that

$$\int_x^\infty |q(t)|dt \leq \alpha(x) \qquad (0 < x < \infty), \tag{5.1.2}$$

where $\alpha(x)$ is a continuous nonincreasing function integrable on $(0, \infty)$. We will study the accuracy of the reconstruction of the boundary-value problem $\{h, q(x)\}$ from a part of its spectral function in the set $V\{A, a(x)\}$ while the accuracy of the reconstruction of the boundary-value problem $\{q(x)\}$ from a part of its scattering data will be studied in the set $V\{\alpha(x)\}$.

Our primary interest is the accuracy of reconstruction of the solutions $\omega(\lambda, x; h)$ or $e(\lambda, x)$ of the corresponding equations; this is because their reconstruction is more stable.

In this section we derive convenient representations for the differences of such solutions in terms of the differences of the corresponding spectral functions (or the scattering data).

Consider two boundary-value problems $\{q_1(x)\}$ and $\{q_2(x)\}$ from the set $V\{\alpha(x)\}$. Subtracting the main integral equations for the corresponding inverse problems gives the equation

$$F_1(x+y) - F_2(x+y) + K_1(x,y) - K_2(x,y) + \int_x^\infty F_1(t+y)\times$$

$$\times\{K_1(x,t) - K_2(x,t)\}dt + \int_x^\infty \{F_1(t+y) - F_2(t+y)\}K_2(x,t)dt = 0,$$

where $K_1(x,y)$, and $K_2(x,y)$ are the kernels of the corresponding transformation operators and the functions $F_1(x)$ and $F_2(x)$ are constructed from the scattering data by (3.2.7). To shorten notations, define

$$K_{1,2}(x,y) = K_1(x,y) - K_2(x,y), \qquad F_{1,2}(x) = F_1(x) - F_2(x).$$

Then the equation above can be written as

$$K_{1,2}(x,y) + \int_x^\infty F_1(t+y)K_{1,2}(x,t)dt = -\{F_{1,2}(x+y) + \int_x^\infty F_{1,2}(t+y)K_2(x,t)dt\}.$$

For each fixed $x \geq 0$, this equality is an equation with respect to the function $K_{1,2}(x,y)$. Solving it we find

$$K_{1,2}(x,y) = (\mathbf{I} + \mathbf{F}_{1x})^{-1}\{F_{1,2}(x,y) + \int_x^\infty F_{1,2}(t+y)K(x,t)dt\}, \tag{5.1.3}$$

where, in view of (3.2.16),

$$(\mathbf{I} + \mathbf{F}_{1x})^{-1} = (\mathbf{I} + \mathbf{K}_{1x}^*)(\mathbf{I} + \mathbf{K}_{1x}) \tag{5.1.4}$$

and the operators $\mathbf{K}_{1x}^*$, $\mathbf{K}_{1x}$ are defined by (3.2.16').

Let $\{S_j(\lambda),\ \lambda_k,\ m_k(j)\}$ $(j = 1, 2)$ be the scattering data and let $e_j(\lambda, x)$ be the solutions of the considering problems, defined in Lemma 3.1.1. To shorten notations we omit the index j at λ_k and assume that $m_k(j) = 0$ if $\mathrm{i}\lambda_k$ is not a zero of $e_j(\lambda, 0)$. Then

$$F_{1,2}(x) = \sum_k \mathrm{e}^{-\lambda_k x}\{m_k^2(1) - m_k^2(2)\} + \frac{1}{2\pi}\int_{-\infty}^\infty \{S_2(\lambda) - S_1(\lambda)\}\mathrm{e}^{\mathrm{i}\lambda x}d\lambda.$$

Thus

$$F_{1,2}(x+y) + \int_x^\infty K_2(x,t)F_{1,2}(t+y)dt = \sum_k \mathrm{e}^{-\lambda_k y}e_2(\mathrm{i}\lambda_k, x)\times$$

$$\times\{m_k^2(1)-m_k^2(2)\}+\frac{1}{2\pi}\int_{-\infty}^{\infty}\{S_2(\lambda)-S_1(\lambda)\}e^{i\lambda y}e_2(\lambda,x)d\lambda,$$

$$\phi(x,y)=(\mathbf{I}+\mathbf{K}_{1x})\{F_{1,2}(x,y)+\int_x^\infty K_2(x,t)F_{1,2}(t+y)dt\}=\sum_k e_1(i\lambda_k,y)\times$$

$$\times e_2(i\lambda_k,x)\{m_k^2(1)-m_k^2(2)\}+\frac{1}{2\pi}\int_{-\infty}^{\infty}\{S_2(\lambda)-S_1(\lambda)\}e_1(\lambda y)e_2(\lambda,x)d\lambda. \quad (5.1.5)$$

Equations (5.1.3) and (5.1.4) imply that

$$K_{1,2}(x,y)=-(\mathbf{I}+\mathbf{K}_{1x}^*)\phi(x,y) \quad (5.1.6)$$

and, therefore, for $\operatorname{Im}\mu>0$ we have

$$e_1(\mu,x)-e_2(\mu,x)=\int_x^\infty K_{1,2}(x,y)e^{i\mu y}dy=-\left(\{\mathbf{I}+\mathbf{K}_{1x}^*\}\phi(x,y),\exp(-i\overline{\mu}y)\right)_x=$$

$$=-\left(\phi(x,y),\{\mathbf{I}+\mathbf{K}_{1x}\}\exp(-i\overline{\mu}y)\right)_x=-\left(\phi(x,y),\ \overline{e_1(\mu,y)}\right)_x,$$

where $(f,\,g)_x$ denotes the scalar product in the space $L^2(x,\infty)$:

$$(f,\,g)_x=\int_x^\infty f(t)\overline{g(t)}dt.$$

Thus we obtain the equation

$$e_1(\mu,x)-e_2(\mu,x)=-\int_x^\infty \phi(x,y)e_1(\mu,y)dy, \quad (5.1.7)$$

where $\operatorname{Im}\mu>0$ and the function $\phi(x,y)$ is defined by the r.h.s. of (5.1.5).

From the equations for $e_1(\lambda,y)$ it follows that

$$\int_x^\infty e_1(\lambda,y)e_1(\mu,y)dy=\frac{e_1'(\lambda,x)e_1(\mu,x)-e_1(\lambda,x)e_1'(\mu,x)}{\lambda^2-\mu^2}.$$

Using this equality and (5.1.5), we have

$$e_1(\mu,x)-e_2(\mu,x)=\sum_k E_{1,2}(i\lambda_k,\mu,x)\{m_k^2(1)-m_k^2(2)\}+$$

$$+\frac{1}{2\pi}\int_{-\infty}^{\infty}\{S_2(\lambda)-S_1(\lambda)\}E_{1,2}(\lambda,\mu,x)d\lambda, \quad (5.1.8)$$

where

$$E_{j,i}(\lambda, \mu, x) = \frac{e_i(\lambda, x)\{e_j(\lambda, x)e_j'(\mu, x) - e_j'(\lambda, x)e_j(\mu, x)\}}{\lambda^2 - \mu^2}.$$

Now we notice that one can interchange the indices 1 and 2 in (5.1.8), which gives

$$e_2(\mu, x) - e_1(\mu, x) = \sum_k E_{2,1}(\mathrm{i}\lambda_k, \mu, x)\{m_k^2(2) - m_k^2(1)\}+$$

$$+\frac{1}{2\pi}\int_{-\infty}^{\infty}\{S_1(\lambda) - S_2(\lambda)\}E_{2,1}(\lambda, \mu, x)d\lambda. \tag{5.1.9}$$

Multiply the both sides of (5.1.8) by $e_2(\mu, x)$ and the both sides of (5.1.9) by $e_1(\mu, x)$ and sum them up. As a result, in the l.h.s. we obtain $-\{e_1(\mu, x) - e_2(\mu, x)\}^2$. In order to represent the r.h.s. in a form convenient for further considerations, we perform the following transformations:

$$(\lambda^2 - \mu^2)\{E_{1,2}(\lambda, \mu, x)e_2(\mu, x) - E_{2,1}(\lambda, \mu, x)e_1(\mu, x)\} =$$

$$= e_2(\mu, x)e_2(\lambda, x)e_1(\lambda, x)e_1'(\mu, x) - e_2(\mu, x)e_2(\lambda, x)e_1'(\lambda, x)e_1(\mu, x)-$$

$$-e_1(\mu, x)e_1(\lambda, x)e_2(\lambda, x)e_2'(\mu, x) + e_1(\mu, x)e_1(\lambda, x)e_2'(\lambda, x)e_2(\mu, x) =$$

$$= e_1(\lambda, x)e_2(\lambda, x)\{e_1'(\mu, x)e_2(\mu, x) - e_1(\mu, x)e_2'(\mu, x)\}-$$

$$-e_1(\mu, x)e_2(\mu, x)\{e_1'(\lambda, x)e_2(\lambda, x) - e_1(\lambda, x)e_2'(\lambda, x)\} =$$

$$= \int_x^{\infty}\{q_1(t) - q_2(t)\}\{e_1(\mu, x)e_2(\mu, x)e_1(\lambda, t)e_2(\lambda, t)-$$

$$-e_1(\lambda, x)e_2(\lambda, x)e_1(\mu, t)e_2(\mu, t)\}dt.$$

Here we have used the identities

$$\int_x^{\infty}\{q_1(t) - q_2(t)\}e_1(\nu, t)e_2(\nu, t)dt = -\{e_1'(\nu, x)e_2(\nu, x) - e_1(\nu, x)e_2'(\nu, x)\},$$

that follow from the Sturm-Liouville equations for $e_1(\nu, x)$ and $e_2(\nu, x)$. Hence, the r.h.s. of the resulting equation can be written as

$$\int_x^{\infty}\{q_1(t) - q_2(t)\}\{A_{1,2}(\mu, x, t) - A_{1,2}(\mu, t, x)\}dt,$$

where

$$A_{1,2}(\mu,x,t)=e_1(\mu,x)e_2(\mu,x)\sum_k \frac{m_k^2(2)-m_k^2(1)}{\lambda_k^2-\mu^2}e_1(\mathrm{i}\lambda_k,t)e_2(\mathrm{i}\lambda_k,t)+$$

$$+\frac{e_1(\mu,x)e_2(\mu,x)}{2\pi}\int_{-\infty}^{\infty}\frac{S_2(\lambda)-S_1(\lambda)}{\lambda^2-\mu^2}e_1(\lambda,t)e_2(\lambda,t)dt. \quad (5.1.10)$$

Thus we have proved the following

LEMMA 5.1.1. *For all μ from the open upper half plane, for which $Im\mu^2\neq 0$, the following identity holds:*

$$\{e_1(\mu,x)-e_2(\mu,x)\}^2=\int_x^{\infty}\{q_1(t)-q_2(t)\}\{A_{1,2}(\mu,t,x)-A_{1,2}(\mu,x,t)\}dt, \quad (5.1.11)$$

where the function $A_{1,2}(\mu,x,t)$ is defined by the r.h.s. of (5.1.10).

Now consider two boundary-value problems, $\{h_j,q_j(x)\}$ $(j=1,2)$, from the set $V\{A,a(x)\}$ and denote by $\omega_j(\sqrt{\lambda},x)$ the solutions of the following equations:

$$-y''+q_i(x)y=\lambda y,\quad y(0)=1,\quad y'(0)=h_i\quad (i=1,2). \quad (5.1.12)$$

LEMMA 5.1.2. *For all μ with $\operatorname{Im}\mu\neq 0$, the following identity holds:*

$$\{\omega_1(\sqrt{\mu},x)-\omega_2(\sqrt{\mu},x)\}^2=(h_1-h_2)\{B_{1,2}(\mu,0,x)-B_{1,2}(\mu,x,0)\}+$$

$$+\int_x^{\infty}\{q_1(t)-q_2(t)\}\{B_{1,2}(\mu,t,x)-B_{1,2}(\mu,x,t)\}dt, \quad (5.1.13)$$

where

$$B_{1,2}(\mu,x,t)=\omega_1(\sqrt{\mu},x)\omega_2(\sqrt{\mu},x)\int_{-\infty}^{\infty}\frac{\omega_1(\sqrt{\lambda},t)\omega_2(\sqrt{\lambda},t)}{\lambda-\mu}d\{\rho_1(\lambda)-\rho_2(\lambda)\}, \quad (5.1.14)$$

and $\rho_j(\lambda)$ $(j=1,2)$ are the spectral functions of the boundary-value problems $\{h_j,q_j(x)\}$.

PROOF. Let $\mathbf{I}+\mathbf{K}_j^l$ be the transformation operators which transform the solution $\omega_j(\sqrt{\lambda},x)$ to the solution $\omega_l(\sqrt{\lambda},x)$. Consider the integrals

$$I_j(\mu,x)=\int_{-\infty}^{\infty}\omega_2(\sqrt{\lambda},x)d\rho_j(\lambda)\int_0^x\omega_1(\sqrt{\lambda},t)\omega_1(\sqrt{\mu},t)dt,$$

where $\rho_j(\lambda)$ $(j=1,2)$ are the spectral functions of the associated boundary-value problems. Then we have

$$I_1(\mu,x)=\int_{-\infty}^{\infty}\omega_1(\sqrt{\lambda},x)d\rho_1(\lambda)\int_0^x\omega_1(\sqrt{\lambda},t)\omega_1(\sqrt{\mu},t)dt+$$

$$+\int_{-\infty}^{\infty}\left\{\int_0^x K_1^2(x,t)\omega_1(\sqrt{\lambda},t)dt\right\}\left\{\int_0^x\omega_1(\sqrt{\lambda},t)\omega_1(\sqrt{\mu},t)dt\right\}d\rho_1(\lambda),$$

where $K_1^2(x,t)$ is the kernel of the operator $\mathbf{K}_1^2$. By the equiconvergence theorem, the first integral in this equality exists and is equal to $\frac{1}{2}\omega_1(\sqrt{\mu},x)$. From the Parceval equality we find the second summand:

$$\int_0^x K_1^2(x,t)\omega_1(\sqrt{\mu},t)dt=\omega_2(\sqrt{\mu},x)-\omega_1(\sqrt{\mu},x).$$

This implies the existence of $I_1(\mu,x)$ and the equality

$$I_1(\mu,x)=\omega_2(\sqrt{\mu},x)-\frac{1}{2}\omega_1(\sqrt{\mu},x).$$

Applying the equiconvergence theorem to $I_2(\mu,x)$, we get

$$I_2(\mu,x)=\int_{-\infty}^{\infty}\omega_2(\sqrt{\lambda},x)d\rho_2(\lambda)\int_0^x\omega_2(\sqrt{\lambda},t)\omega_1(\sqrt{\mu},t)dt+$$

$$+\int_{-\infty}^{\infty}\omega_2(\sqrt{\lambda},x)d\rho_2(\lambda)\int_0^x\int_0^t K_2^1(t,\xi)\omega_2(\sqrt{\lambda},\xi)\omega_1(\sqrt{\mu},t)d\xi dt=\frac{1}{2}\omega_1(\sqrt{\mu},x)+$$

$$+\int_{-\infty}^{\infty}\omega_2(\sqrt{\lambda},x)\left\{\int_0^x\int_\xi^x K_2^1(t,\xi)\omega_1(\sqrt{\mu},t)\omega_2(\sqrt{\lambda},\xi)dtd\xi d\rho_2(\lambda)\right\}=$$

$$=\frac{1}{2}\omega_1(\sqrt{\mu},x)+\frac{1}{2}\left\{\int_\xi^x K_2^1(t,\xi)\omega_1(\sqrt{\mu},t)dt\right\}|_{\xi=x}=\frac{1}{2}\omega_1(\sqrt{\mu},x).$$

Therefore

$$\int_{-\infty}^{\infty}\omega_2(\sqrt{\lambda},x)\int_0^x\omega_1(\sqrt{\lambda},t)\omega_1(\sqrt{\mu},t)dtd\{\rho_1(\lambda)-\rho_2(\lambda)\}=$$

$$=I_1(\mu,x)-I_2(\mu,x)=\omega_2(\sqrt{\mu},x)-\omega_1(\sqrt{\mu},x),$$

and, using the equality

$$\int_0^x \omega_1(\sqrt{\lambda},t)\omega_1(\sqrt{\mu},t)dt = -\frac{\omega_1'(\sqrt{\lambda},x)\omega_1(\sqrt{\mu},x)-\omega_1(\sqrt{\lambda},x)\omega_1'(\sqrt{\mu},x)}{\lambda-\mu},$$

we arrive at the formula

$$\omega_1(\sqrt{\mu},x)-\omega_2(\sqrt{\mu},x) = \int_{-\infty}^{\infty} \frac{\omega_1'(\sqrt{\lambda},x)\omega_1(\sqrt{\mu},x)-\omega_1(\sqrt{\lambda},x)\omega_1'(\sqrt{\mu},x)}{\lambda-\mu}\times$$

$$\times\omega_2(\sqrt{\lambda},x)d\{\rho_1(\lambda)-\rho_2(\lambda)\},$$

which is similar to (5.1.8). Further transformations, which are in complete analogy with those given above, are based on the parity of indices 1 and 2. □

2. STABILITY OF THE INVERSE SCATTERING PROBLEM

Let the scattering data of a boundary value problem $\{q(x)\} \in V\{\alpha(x)\}$ be defined for all $\lambda^2 < N^2$ only, with an error δ. How accurately can one reconstruct this boundary-value problem? To fix ideas, we will assume that $\delta = 0$; the general case can be treated in a completely similar way, but the resulting formulas turn out to be much more cumbersome.

To answer the question above, one has to understand how much can differ two boundary-value problems $\{q_j(x)\} \in V\{\alpha(x)\}$ $(j=1,2)$ having the scattering data $\{S_j(\lambda;\ \lambda_k(j);\ m_k(j)\}$ coinciding for $\lambda^2 \in (-\infty, N^2)$, that is,

$$S_1(\lambda) = S_2(\lambda),\quad -N<\lambda<N,\quad \lambda_k(1)=\lambda_k(2),\quad m_k(1)=m_k(2)\quad (k=1,..,n).$$

THEOREM 5.2.1. *If the scattering data of two boundary-value problems* $\{q_j(x)\} \in V\{\alpha(x)\}$ $(j=1,2)$ *coincide for all* $\lambda^2 \in (-\infty, N^2)$, *then for all* $\mu^2 \in (-\infty, N^2)$ *the following inequalities hold:*

$$|e_1(\mu,x)-e_2(\mu,x)|^2 \le \frac{8\alpha(x)\exp\{4\alpha_1(x)\}}{\pi N\left(1-\frac{|\mu|^2+\mu^2}{2N^2}\right)}, \tag{5.2.1}$$

$$\int_0^\infty |e_1(\mu,x)-e_2(\mu,x)|^2 dx \le \frac{4e^{2\alpha_1(0)}\sinh\{2\alpha_1(0)\}}{\pi N\left(1-\frac{|\mu|^2+\mu^2}{2N^2}\right)}, \tag{5.2.2}$$

where

$$\alpha_1(x) = \int_x^\infty \alpha(t)dt. \tag{5.2.3}$$

PROOF. Let us first suppose that μ belongs to the upper half plane and that $\operatorname{Im}\mu^2 \neq 0$. Then one can use formula (5.1.11), where, since the scattering data of the considered problems coincide for all $\lambda^2 \in (-\infty, N^2)$, one has

$$A_{1,2}(\mu, x, t) = \frac{e_1(\mu, x)e_2(\mu, x)}{2\pi} \int\limits_{|\lambda|>N} \frac{S_2(\lambda) - S_1(\lambda)}{\lambda^2 - \mu^2} e_1(\lambda, t)e_2(\lambda, t)d\lambda. \qquad (5.2.4)$$

Hence, formulas (5.1.11) and (5.2.4) are valid also for $\mu^2 \in (-\infty, N^2)$, which can be verified by passing to the limit. From the estimate (3.1.19) and the inequality (5.1.2) it follows that for all real ν and $x \geq 0$ one has $|e_j(\nu, x)| \leq \exp \alpha_1(x)$, $j = 1, 2$, where $\alpha_1(x)$ is defined by formula (5.2.3). Thus for all $N^2 > 0$ and $\mu^2 < N^2$ we have

$$| A_{1,2}(\mu, x, t) | \leq \frac{\exp\{2(\alpha_1(x) + \alpha_1(t))\}}{2\pi} \int\limits_{|\lambda|\geq N} \frac{|S_1(\lambda) - S_2(\lambda)|}{\lambda^2 \left(1 - \frac{|\mu|^2+\mu^2}{2N^2}\right)} d\lambda \leq$$

$$\leq \frac{2\exp\{2(\alpha_1(x) + \alpha_1(t))\}}{\pi N \left(1 - \frac{|\mu|^2+\mu^2}{2N^2}\right)},$$

and hence, in view of (5.1.11), we obtain the estimate

$$|e_1(\mu, x) - e_2(\mu, x)|^2 \leq \frac{4\exp\{2\alpha_1(x)\}}{\pi N \left(1 - \frac{|\mu|^2+\mu^2}{2N^2}\right)} \int\limits_x^\infty |q_1(t) - q_2(t)| \exp\{2\alpha_1(t)\} dt \leq$$

$$\leq \frac{8\alpha(x)\exp\{4\alpha_1(x)\}}{\pi N \left(1 - \frac{|\mu|^2+\mu^2}{2N^2}\right)}.$$

Integrating now the inequality (5.2.1) along the whole positive semiaxis and taking into account that $\alpha(x) = -\alpha_1'(x)$, we obtain (5.2.2). □

The estimate (5.2.1) holds for all positive N, x, and $\mu^2 < N^2$. Nevertheless, an evident estimate $|e_1(\mu, x) - e_2(\mu, x)|^2 \leq 4\exp\{2\alpha_1(x)\}$, that follows from the inequalities (3.1.9), shows that the estimate (5.2.1) is nontrivial for

$$N > 2\pi^{-1}\alpha(x)\exp\{2\alpha_1(x)\}\left(1 - \frac{|\mu|^2 + \mu^2}{2N^2}\right)^{-1}$$

only. Therefore, it is desirable to get more precise estimates for large N, which are valid for all x.

For the further convenience, let us introduce the function

$$\Delta(x,N)=\alpha_1(x)-\alpha_1(x+N^{-1})=\int\limits_x^{x+N^{-1}}\alpha(t)dt.$$

It is defined for all $x\geq 0$ and $N>0$, is positive and nonincreasing with respect to the both variables.

Evidently, the potentials of the considered boundary-value problems satisfy the inequalities

$$\int\limits_x^{\infty}|q(t)|dt=\sigma(x)\leq\alpha(x),\quad \sigma_1(x)=\int\limits_x^{\infty}\sigma(t)dt\leq\int\limits_x^{\infty}\alpha(t)dt=\alpha_1(x),$$

$$\sigma_1(x)-\sigma_1(x+N^{-1})\leq\Delta(x,N)\leq\Delta(0,N)=\int\limits_0^{N^{-1}}\alpha(t)dt,$$

from which the following inequalities follow:

$$\int\limits_x^{\infty}\left|\frac{\sin\lambda(t-x)}{\lambda}\right||q(t)|dt\leq\sigma_1(x)-\sigma_1(x+|\lambda|^{-1})\leq\Delta(x,|\lambda|), \tag{5.2.5}$$

$$\int\limits_{|\lambda|\geq N}\left(\int_x^{\infty}\frac{\sin^2\lambda(t-x)}{\lambda^2}|q(t)|dt\right)d\lambda\leq 2\pi(\sigma_1(x)-\sigma_1(x+N^{-1}))\leq 2\pi\Delta(x,N). \tag{5.2.6}$$

Namely, since for all real values of λ and y,

$$|\lambda^{-1}\sin\lambda y|\leq|y|,\quad |\lambda^{-1}\sin\lambda y|\leq|\lambda|^{-1}, \tag{5.2.7}$$

it follows that

$$\int\limits_x^{\infty}\left|\frac{\sin\lambda(t-x)}{\lambda}\right||q(t)|dt\leq\int\limits_x^{x+|\lambda|^{-1}}(t-x)|q(t)|dt+|\lambda|^{-1}\int\limits_{x+|\lambda|^{-1}}^{\infty}|q(t)|dt=$$

$$=-\int\limits_x^{x+|\lambda|^{-1}}(t-x)d\sigma(t)+|\lambda|^{-1}\sigma(x+|\lambda|^{-1})=-|\lambda|^{-1}\sigma(x+|\lambda|^{-1})+$$

$$+\int\limits_x^{x+|\lambda|^{-1}}\sigma(t)dt+|\lambda|^{-1}\sigma(x+|\lambda|^{-1})=\sigma_1(x)-\sigma_1(x+|\lambda|^{-1})\leq\Delta(x,|\lambda|),$$

and

$$\int\limits_{|\lambda|\geq N}\left(\int\limits_x^\infty \frac{\sin^2\lambda(t-x)}{\lambda^2}|q(t)|dt\right)d\lambda=\int\limits_x^\infty |q(t)|\left(\int\limits_{|\lambda|\geq N}\frac{\sin^2\lambda(t-x)}{\lambda^2}d\lambda\right)dt\leq$$

$$\int\limits_x^{x+N^{-1}}(t-x)|q(t)|\left(\int\limits_{|\lambda|\geq 0}\frac{\sin^2\lambda(t-x)}{\lambda^2(t-x)}d\lambda\right)dt+\int\limits_{x+N^{-1}}^\infty |q(t)|dt\int\limits_{|\lambda|\geq N}|\lambda|^{-2}d\lambda=$$

$$=-2\pi\int\limits_x^{x+N^{-1}}(t-x)d\sigma(t)+2N^{-1}\sigma(x+N^{-1})=-2\pi N^{-1}\sigma(x+N^{-1})+$$

$$+2\pi\int\limits_x^{x+N^{-1}}\sigma(t)dt+2N^{-1}\sigma(x+N^{-1})\leq 2\pi(\sigma_1(x)-\sigma_1(x+N^{-1}))\leq 2\pi\Delta(x,N).$$

Now introduce the functions

$$\phi(\lambda,x)=e(\lambda,x)-e^{i\lambda x}, \tag{5.2.8}$$

$$\psi(\lambda,x)=\phi(\lambda,x)-\phi(-\lambda,x)=e(\lambda,x)-e(-\lambda,x)-2i\sin\lambda x.$$

In view of (3.1.20), these functions satisfy the inequalities

$$|\phi(\lambda,x)|\leq(\sigma_1(0)-\sigma_1(|\lambda|^{-1}))e^{\sigma_1(0)}\leq\Delta(0,|\lambda|)e^{\alpha_1(0)},$$

$$|\psi(\lambda,x)|\leq 2\Delta(0,|\lambda|)e^{\alpha_1(0)},$$

which implies that for all $x\geq 0$ and $\lambda\in(-\infty,\infty)$,

$$m(\lambda,x)=\sup_{t\geq x}|\phi(\lambda,t)|<\infty,$$

and for all $N_1>N>0$,

$$J(N,N_1)=\sup_{t\geq 0}\int\limits_{N<|\lambda|<N_1}\left|\frac{\psi(\lambda,t)}{\lambda}\right|d\lambda<\infty.$$

In the next lemma we will make these rough estimates more precise.

LEMMA 5.2.1. *If $\Delta(0,N) < 1$, then for all $x \geq 0$ and $|\lambda| \geq N$,*

$$|\phi(\lambda,x)| \leq \frac{\Delta(x,|\lambda|)}{1-\Delta(x,|\lambda|)} \leq \frac{\Delta(0,N)}{1-\Delta(0,N)}, \tag{5.2.9}$$

and for all $x \geq 0$,

$$\int\limits_{|\lambda|\geq N}^{\infty} \left|\frac{\psi(\lambda,x)}{\lambda}\right| d\lambda \leq \frac{2\pi\Delta(0,N)}{1-\Delta(0,N)}. \tag{5.2.10}$$

PROOF. In view of (3.1.7), the functions $\phi(\lambda,x)$ and $\psi(\lambda,x)$ satisfy the equations

$$\phi(\lambda,x) = \int_x^\infty \frac{\sin\lambda(t-x)}{\lambda} e^{i\lambda t} q(t)dt + \int_x^\infty \frac{\sin\lambda(t-x)}{\lambda} q(t)\phi(\lambda,t)dt,$$

$$\psi(\lambda,x) = 2i\int_x^\infty \frac{\sin\lambda(t-x)}{\lambda} \sin\lambda t q(t)dt + \int_x^\infty \frac{\sin\lambda(t-x)}{\lambda} q(t)\psi(\lambda,t)dt,$$

which imply the following inequalities

$$|\phi(\lambda,x)| = \int_x^\infty \left|\frac{\sin\lambda(t-x)}{\lambda} q(t)\right| dt(1+m(\lambda,x)),$$

$$\left|\frac{\psi(\lambda,x)}{\lambda}\right| \leq \int_x^\infty \frac{\sin^2\lambda(t-x)+\sin^2\lambda t}{\lambda^2}|q(t)|dt+$$

$$+\int\limits_x^{x+N^{-1}} (t-x)|q(t)| \left|\frac{\psi(\lambda,t)}{\lambda}\right| dt + \frac{1}{N}\int\limits_{x+N^{-1}}^{\infty} |q(t)| \left|\frac{\psi(\lambda,t)}{\lambda}\right| dt \tag{5.2.11}$$

(to derive the second inequality, we use (5.2.6) and the inequality $|2\sin\lambda(t-x)\sin\lambda t| \leq \sin^2\lambda(t-x)+\sin^2\lambda t$). From the first inequality, in view of (5.2.5) it follows that

$$|\phi(\lambda,x)| \leq \Delta(x,|\lambda|) + m(\lambda,x)\Delta(x,|\lambda|). \tag{5.2.12}$$

Integrating the second inequality over $|\lambda| \in [N,N_1]$, we obtain, in view of (5.2.6), the estimate

$$\int\limits_{N<|\lambda|<N_1} \left|\frac{\psi(\lambda,x)}{\lambda}\right| d\lambda \leq 2\pi\{\Delta(x,N)+\Delta(0,N)\}+J(N,N_1)\left\{\int\limits_x^{x+N^{-1}} (t-x)|q(t)dt+\right.$$

$$\left.+N^{-1}\int\limits_{x+N^{-1}}^{\infty} |q(t)|dt\right\} = 2\pi\{\Delta(x,N)+\Delta(0,N)\}+J(N,N_1)\{(x-t)\sigma(t)\Big|_x^{x+N^{-1}}+$$

$$+\int_x^{x+N^{-1}} \sigma(t)dt + N^{-1}\sigma(x+N^{-1})\} \leq 4\pi\Delta(0,N) + J(N,N_1)\{\sigma_1(x) - \sigma_1(x+N^{-1})\}.$$

Since $\sigma_1(x) - \sigma_1(x+N^{-1}) \leq \Delta(x,N) \leq \Delta(0,N,)$, then

$$\int_{N<|\lambda|<N_1} \left|\frac{\psi(\lambda,x)}{\lambda}\right| d\lambda \leq 4\pi\Delta(0,N) + J(N,N_1)\Delta(0,N). \tag{5.2.13}$$

The r.h.s. of (5.2.12) is a nonincreasing function of x, and the r.h.s. of (5.2.13) is independent of x. Thus we have

$$\sup_{t\geq x} |\phi(\lambda,t)| = m(\lambda,x) \leq \Delta(x,|\lambda|) + m(\lambda,x)\Delta(x,|\lambda|),$$

$$\sup_{x\geq 0} \int_{N<|\lambda|<N_1} \left|\frac{\psi(\lambda,x)}{\lambda}\right| d\lambda = J(N,N_1) \leq 4\pi\Delta(0,N) + J(N,N_1)\Delta(0,N).$$

Since $\Delta(x,|\lambda|) \leq \Delta(0,N) < 1$ for $|\lambda| \geq N$, it follows that

$$|\phi(\lambda,x)| \leq m(\lambda,x) \leq \frac{\Delta(x,|\lambda|)}{1-\Delta(x,|\lambda|)} \text{ and } J(N,N_1) \leq \frac{4\pi\Delta(0,N)}{1-\Delta(0,N)}.$$

Thus

$$\int_{|\lambda|>N} \left|\frac{\psi(\lambda,x)}{\lambda}\right| d\lambda \leq \lim_{N_1\to\infty} J(N,N_1) \leq \frac{4\pi\Delta(0,N)}{1-\Delta(0,N)}.$$

□

COROLLARY. *If $2\Delta(0,N) < 1$, then*

$$\int_{|\lambda|\geq N} \left|\frac{S(\lambda)-1}{\lambda}\right| d\lambda \leq \frac{4\pi\Delta(0,N)}{1-2\Delta(0,N)}. \tag{5.2.14}$$

Namely, according to (5.2.8) and (5.2.9),

$$|e(\lambda,0)| \geq 1 - \frac{\Delta(0,N)}{1-\Delta(0,N)} = \frac{1-2\Delta(0,N)}{1-\Delta(0,N)}.$$

Therefore,

$$|S(\lambda)-1| = \left|\frac{e(-\lambda,0)-e(\lambda,0)}{e(\lambda,0)}\right| = \left|\frac{\psi(\lambda,0)}{e(\lambda,0)}\right| \leq \frac{1-\Delta(0,N)}{1-2\Delta(0,N)}\,|\psi(\lambda,0)|\,,$$

which together with $(5.2.12)$ yields the estimate

$$\int\limits_{|\lambda|\geq N} \left|\frac{S(\lambda)-1}{\lambda}\right| d\lambda \leq \frac{1-\Delta(0,N)}{1-2\Delta(0,N)} \int\limits_{|\lambda|\geq N} \left|\frac{\psi(\lambda,0)}{\lambda}\right| \leq$$

$$\leq \left(\frac{1-\Delta(0,N)}{1-2\Delta(0,N)}\right)\left(\frac{4\pi\Delta(0,N)}{1-\Delta(0,N)}\right) = \frac{4\pi\Delta(0,N)}{1-2\Delta(0,N)}.$$

THEOREM 5.2.2. *If the scattering data of two boundary-value problems* $\{q_j(x)\} \in V\{\alpha(x)\}$, $(j=1,2)$ *coincide for all* $\lambda^2 \in (-\infty, N^2)$ *and*

$$2\Delta(0,N) = 2\int_0^{N^{-1}} \alpha(t)dt < 1, \tag{5.2.15}$$

then for all $\mu^2 \in (-\infty, N^2)$, *the following estimates hold:*

$$|e_1(\mu,x)-e_2(\mu,x)|^2 \leq \frac{16\Delta(0,N)\alpha(x)\sup_{t\geq x}|e_1(\mu,t)e_2(\mu,t)|}{N(1-2\Delta(0,N))^2\left(1-\frac{|\mu|^2+\mu^2}{2N^2}\right)} \leq$$

$$\leq \frac{16\Delta(0,N)\alpha(x)e^{2\alpha_1(x)}}{N(1-2\Delta(0,N))^2\left(1-\frac{|\mu|^2+\mu^2}{2N^2}\right)}, \tag{5.2.16}$$

and

$$\int_0^\infty |e_1(\mu,x)-e_2(\mu,x)|^2dx \leq \frac{16\Delta(0,N)e^{\alpha_1(0)}\sinh\alpha_1(0)}{N(1-2\Delta(0,N))^2\left(1-\frac{|\mu|^2+\mu^2}{2N^2}\right)}. \tag{5.2.17}$$

PROOF. Due to (3.1.19), for all $t \geq x$ and $\lambda \in (-\infty,\infty)$ we have $|e(\lambda,t)| \leq \exp\alpha_1(x)$; then from the inequality $(5.2.9)$ and condition $(5.2.15)$ it follows that for all $x \geq 0$ and $|\lambda| \geq N$,

$$|\phi(\lambda,x)| \leq \frac{\Delta(x,N)}{1-\Delta(x,N)} \quad \text{and} \quad e(\lambda,x)| \leq 1+\frac{\Delta(x,N)}{1-\Delta(x,N)} = \frac{1}{1-\Delta(x,N)}.$$

These inequalities and formula $(5.2.4)$ imply that

$$|A_{1,2}(\mu,t,x)-A_{1,2}(\mu,x,t)| \leq \frac{\sup_{t\geq x}|e_1(\mu,t)e_2(\mu,t)|}{N(1-\Delta(0,N))^2\left(1-\frac{|\mu|^2+\mu^2}{2N^2}\right)}\times$$

$$\times \int\limits_{|\lambda|\geq N} \left|\frac{S_2(\lambda)-S_1(\lambda)}{\lambda}\right| d\lambda.$$

Using the inequality $|S_1(\lambda) - S_2(\lambda)| \le |S_1(\lambda) - 1| + |S_2(\lambda) - 1|$ and Corollary of Lemma 5.2.1., we get the estimate

$$\int\limits_{|\lambda|\ge N} \left|\frac{S_2(\lambda) - S_1(\lambda)}{\lambda}\right| d\lambda \le \frac{8\pi\Delta(0,N)}{1 - 2\Delta(0,N)}$$

and, therefore,

$$|A_{1,2}(\mu,t,x) - A_{1,2}(\mu,x,t)| \le \frac{8\Delta(0,N)\sup_{t\ge x}|e_1(\mu,t)e_2(\mu,t)|}{N(1-2\Delta(0,N))(1-\Delta(0,N))^2\left(1 - \frac{|\mu|^2+\mu^2}{2N^2}\right)} \le$$

$$\le \frac{8\Delta(0,N)e^{2\alpha_1(x)}}{N(1-2\Delta(0,N))(1-\Delta(0,N))^2\left(1 - \frac{|\mu|^2+\mu^2}{2N^2}\right)}.$$

The inequality (5.2.16) follows now directly from this formula, formula (5.1.11), and the simple inequalities

$$(1-\Delta(0,N))^2 = 1-2\Delta(0,N)+\Delta(0,N)^2 \ge 1-2\Delta(0,N), \quad \int_x^\infty |q_1(t)-q_2(t)|dt \le 2\alpha(x).$$

Finally, the inequality (5.2.17) can be obtained from (5.2.16) by integration over the positive semiaxis $x \in [0,\infty)$. □

Let us estimate now the difference of potentials $q_1(x) - q_2(x)$. To do this, recall (5.1.3) and set there $y = x$. Then, according to the definition of the operator K_{1x}^* (formula (3.2.16')), we get $K_{1,2}(x,x) = -\phi(x,x)$, from which, in view of (5.1.5) and (3.1.6), it follows that

$$\frac{1}{2}\int_x^\infty \{q_1(t) - q_2(t)\}dt = \sum e_1(\mathrm{i}\lambda_k,x)e_2(\mathrm{i}\lambda_k,x)\{m_k^2(2) - m_k^2(1)\}+$$

$$+\frac{1}{2\pi}\int_{-\infty}^\infty \{S_1(\lambda) - S_2(\lambda)\}e_1(\lambda,x)e_2(\lambda,x)d\lambda.$$

In particular, if the conditions of Theorem 5.2.1. are fulfilled, then

$$\frac{1}{2}\int_x^\infty \{q_1(t) - q_2(t)\}dt = \frac{1}{2\pi}\int\limits_{|\lambda|>N} \{S_1(\lambda) - S_2(\lambda)\}e_1(\lambda,x)e_2(\lambda,x)d\lambda. \quad (5.2.18)$$

THEOREM 5.2.3. *If the conditions of theorem 5.2.2. are fulfilled and the potentials $q_j(x)$ $(j = 1,2)$ are differentiable, then*

$$|q_1(x) - q_2(x)| \le 2\pi hD(x,h) + \frac{16\Delta(0,N)(1+9h\alpha(x))}{9h^2(1-2\Delta(0,N))},$$

where

$$D(x,h) = \max_{j=1,2} \sup_{x \le y \le x+\pi h} |q_j'(y)|, \tag{5.2.19}$$

for any $h > N^{-1}$.

PROOF. Let us rewrite formula (5.2.18) in a more convenient form, taking into account that $2\Delta(0,N) < 1$. From the definition of $\phi(\lambda,x)$ it follows that

$$e_1(\lambda,x)e_2(\lambda,x) = \mathrm{e}^{2\mathrm{i}\lambda x} + \mathrm{e}^{\mathrm{i}\lambda x}(\phi_1(\lambda,x) + \phi_2(\lambda,x)) +$$

$$+\phi_1(\lambda,x)\phi_2(\lambda,x) = \mathrm{e}^{2\mathrm{i}\lambda x} + r(x,\lambda), \tag{5.2.20}$$

where, by Lemma 5.2.1., the function

$$r(x,\lambda) = \mathrm{e}^{\mathrm{i}\lambda x}(\phi_1(\lambda,x) + \phi_2(\lambda,x)) + \phi_1(\lambda,x)\phi_2(\lambda,x)$$

satisfies, for all $x \ge 0$ and $|\lambda| > N$, the inequality

$$|r(x,\lambda)| \le \frac{2\Delta(x,|\lambda|)}{1-\Delta(x,|\lambda|)} + \left(\frac{\Delta(x,|\lambda|)}{1-\Delta(x,|\lambda|)}\right)^2 =$$

$$= \frac{\Delta(x,|\lambda|)}{1-\Delta(x,|\lambda|)}\left(2 + \frac{\Delta(x,|\lambda|)}{1-\Delta(x,|\lambda|)}\right).$$

Since $2\Delta(x,|\lambda|) \le 2\Delta(0,N) < 1$ and the function $\alpha(t)$ does not increase, we have $\Delta(x,|\lambda|)(1-\Delta(x,|\lambda|))^{-1} < 1$ and

$$\Delta(x,|\lambda|) = \int_x^{x+|\lambda|^{-1}} \alpha(t)dt \le \frac{\alpha(x)}{|\lambda|},$$

which implies that

$$|r(x,\lambda)| \le \frac{6\alpha(x)}{|\lambda|}. \tag{5.2.21}$$

Set in (5.2.18) $x = y + h\xi$; then, in view of (5.2.20), we have

$$\int\limits_{y+h\xi}^{\infty} \{q_1(t) - q_2(t)\}dt = \frac{1}{\pi} \int\limits_{|\lambda|>N} (S_1(\lambda) - S_2(\lambda)) \left(\mathrm{e}^{2\mathrm{i}\lambda(y+h\xi)} + r(y+h\xi,\lambda)\right) d\lambda.$$

Multiply the both parts of this equality by $\cos\xi$ and integrate over $\xi \in (0,\pi)$. In the l.h.s., integrating by parts leads to the expression $h\int_0^\pi \{q_1(y+h\xi) - q_2(y+h\xi)\}\sin\xi d\xi$. Taking into account that $\int_0^\pi \sin\xi d\xi = 2$, rewrite this expression as

$$2h(q_1(y) - q_2(y)) + h\int_0^\pi Q_{1,2}(y,h,\xi)\sin\xi d\xi,$$

where $Q_{1,2}(y,h,\xi) = (q_1(y+h\xi) - q_1(y)) - (q_2(y+h\xi) - q_2(y))$. In the r.h.s., changing the integration order gives the expression

$$\frac{1}{\pi}\int\limits_{|\lambda|>N} (S_1(\lambda) - S_2(\lambda)(A(y,\lambda) + B(y,\lambda))d\lambda,$$

where

$$A(y,\lambda) = \int_0^\pi \mathrm{e}^{2\mathrm{i}\lambda(y+h\xi)}\cos\xi d\xi = \frac{\mathrm{i}\mathrm{e}^{2\mathrm{i}\lambda y}(\mathrm{e}^{2\mathrm{i}\lambda h\pi}+1)}{2\lambda h\,(1-\frac{1}{4\lambda^2h^2})}, \tag{5.2.22}$$

$$B(y,\lambda) = \int_0^\pi r(y+h\xi,\lambda)\cos\xi d\xi. \tag{5.2.23}$$

Thus,

$$q_1(y) - q_2(y) = -\frac{1}{2}\int_0^\pi Q_{1,2}(y,h,\xi)\sin\xi d\xi +$$

$$+\frac{1}{2\pi h}\int\limits_{|\lambda|>N} (S_1(\lambda) - S_2(\lambda))(A(y,\lambda) + B(y,\lambda))d\lambda$$

and

$$|q_1(y) - q_2(y)| \le \frac{1}{2}\int_0^\pi |Q_{1,2}(y,h,\xi)|\sin\xi d\xi +$$

$$+\frac{1}{2\pi h}\int\limits_{|\lambda|>N} |S_1(\lambda) - S_2(\lambda)|(|A(y,\lambda)| + |B(y,\lambda)|)d\lambda. \tag{5.2.24}$$

For $|\lambda| > N$ and $h > N^{-1}$, in view of (5.2.22) we have

$$|A(y,\lambda)| \le \frac{1}{|\lambda|h\,(1-\frac{1}{4\lambda^2h^2})} \le \frac{4}{3|\lambda|h}$$

and, according to (5.2.21) and (5.2.23),

$$|B(y,\lambda)| \le \int_0^\pi |r(y+h\xi,\lambda)||\cos\xi|d\xi \le \frac{6\alpha(y)}{|\lambda|}\int_0^\pi |\cos\xi|d\xi = \frac{12\alpha(y)}{|\lambda|}.$$

Therefore,

$$|A(y,\lambda)|+|B(y,\lambda)|\le \frac{4}{3|\lambda|h}\left(1+9h\alpha(y)\right),$$

and formula (5.2.14) implies that

$$\frac{1}{2\pi h}\int\limits_{|\lambda|>N}|S_1(\lambda)-S_2(\lambda)|(|A(y,\lambda)|+|B(y,\lambda)|)d\lambda \le \frac{2}{3\pi h^2}\left(1+9h\alpha(y)\right)\times$$

$$\times\int\limits_{|\lambda|>N}\left(\left|\frac{S_1(\lambda)-1}{\lambda}\right|+\left|\frac{S_2(\lambda)-1}{\lambda}\right|\right)d\lambda\le \frac{16\Delta(0,N)\left(1+9h\alpha(y)\right)}{3h^2(1-2\Delta(0,N))}. \qquad (5.2.25)$$

Next, by the mean-value theorem, we have

$$Q_{1,2}(y,\lambda,\xi)=h\xi\left(q_1'(y+h\tilde{\xi})-q_2'(y+h\tilde{\xi})\right),$$

where $0\le\tilde{\xi}\le\xi$, and formula (5.2.19) implies that $Q_{1,2}(y,h,\xi)|\le 2h\pi D(y,h)$ and

$$\frac{1}{2}\int_0^{\pi}|Q_{1,2}(y,h,\xi)|\sin\xi d\xi\le h\pi D(y,h)\int_0^{\pi}\sin\xi d\xi=2h\pi D(y,h). \qquad (5.2.26)$$

Now the statement of the theorem follows directly from the inequalities (5.2.24), (5.2.25), and (5.2.26). □

3. ERROR ESTIMATE FOR THE RECONSTRUCTION OF A BOUNDARY VALUE PROBLEM FROM ITS SPECTRAL FUNCTION GIVEN ON THE SET $(-\infty,N^2)$ ONLY.

We will need estimates for the spectral functions $\rho(\lambda)$ and the solutions $\omega(\sqrt{\lambda},x)=\omega(\sqrt{\lambda},x,h)$ of boundary-value problems $\{h,q(x)\}\in V\{A,a(x)\}$. Let us introduce the following notations:

$$a_1(x)=\int_0^x a(t)dt,\quad b(x)=A+a(x),\quad b_1(x)=\int_0^x b(t)dt=Ax+a_1(x),$$

and note, that the boundary value problems $\{h,q(x)\}\in V\{A,a(x)\}$ satisfy inequalities

$$\sigma_0(x)=\int_0^x|q(t)|dt\le a(x),\quad \sigma_1(x)=\int_0^x\sigma_0(t)dt\le a_1(x),\quad |h|\le A.$$

It is convenient to normalize the spectral functions of these problems by $\rho(-\infty)=0$.

LEMMA 5.3.1. *The spectral functions of the boundary value problems* $\{h, q(x)\} \in V\{A, a(x)\}$ *satisfy the inequality*

$$\rho(\lambda) \leq \frac{3}{2}\sqrt{\lambda} \exp 2b_1(\lambda^{-1/2}), \qquad \lambda > 0. \tag{5.3.1}$$

PROOF. It follows from formula (2.4.1) and equality $1 - \cos 2\sqrt{\lambda}x = 2\sin^2\sqrt{\lambda}x$, that for all $x > 0$

$$2x^2 \int_{-\infty}^{\infty} \left(\frac{\sin\sqrt{\lambda}x}{\sqrt{\lambda}x}\right)^2 d\rho(\lambda) = 2x + \int_0^{2x} (2x - t)L(t, 0, h)dt.$$

Since for all $t^2 \in (-\infty, 1)$

$$\left(\frac{\sin t}{t}\right)^2 \geq \left(1 - \frac{t^2}{6}\right)^2 \geq \frac{25}{36} \geq \frac{2}{3},$$

then

$$2x + \int_0^{2x} (2x-t)L(t,0,h)dt \geq 2x^2 \int_{-\infty}^{x^{-2}} \left(\frac{\sin\sqrt{\lambda}x}{\sqrt{\lambda}x}\right)^2 d\rho(\lambda) \geq \frac{4}{3}x^2 \int_{-\infty}^{x^{-2}} d\rho(\lambda) = \frac{4}{3}x^2\rho(x^{-2}).$$

So,

$$\rho(x^{-2}) \leq \frac{3}{2}x^{-1}\left(1 + \int_0^{2x}\left(1 - \frac{t}{2x}\right)L(t,0,h)dt\right) \leq \frac{3}{2}x^{-1}\left(1 + \int_0^{2x} L(t,0,h)dt\right),$$

and since by (1.2.34')

$$|L(t,0,h)| \leq \left(|h| + \sigma_0\left(\frac{t}{2}\right)\right)\exp 2\left(|h|\frac{t}{2} + \sigma_1\left(\frac{t}{2}\right)\right) \leq$$

$$\leq b\left(\frac{t}{2}\right)\exp 2b_1\left(\frac{t}{2}\right) = \frac{d}{dt}\exp 2b_1\left(\frac{t}{2}\right),$$

then

$$\rho\left(x^{-2}\right) \leq \frac{3}{2}x^{-1}\exp 2b_1(x).$$

Inequality (5.3.1) follows by putting here $\lambda = x^{-2}$. □

COROLLARY.

$$\int_{N^2}^{\infty} \frac{d\rho(\lambda)}{\lambda} = -\frac{\rho(N^2)}{N} + \int_{N^2}^{\infty} \frac{\rho(\lambda)}{\lambda^2}d\lambda \leq \frac{3}{2}\exp 2b_1(N^{-1}) \int_{N^2}^{\infty} \frac{d\lambda}{\lambda^{3/2}} = 3N^{-1}\exp 2b_1(N^{-1}).$$

LEMMA 5.3.2. *The solutions* $\omega(\sqrt{\lambda}, x) = \omega(\sqrt{\lambda}, x, h)$ *of the boundary value problems* $\{h, q(x)\} \in V\{A, a(x)\}$ *satisfy inequalities*

$$|\omega(\sqrt{\lambda}, x)| \leq (1 + |h|x) \exp\left(\sigma_1(x) + \left|\operatorname{Im}\sqrt{\lambda}\right| x\right) \leq \exp\left(b_1(x) + \left|\operatorname{Im}\sqrt{\lambda}\right| x\right),$$

moreover, in the domain $a(x) < \sqrt{\lambda}$ *they also satisfy inequalities*

$$|\omega(\sqrt{\lambda}, x)| \leq \frac{\sqrt{\lambda} + |h|}{\sqrt{\lambda} - \sigma_0(x)} \exp\left(\left|\operatorname{Im}\sqrt{\lambda}\right| x\right) \leq \frac{\sqrt{\lambda} + A}{\sqrt{\lambda} - a(x)} \exp\left(\left|\operatorname{Im}\sqrt{\lambda}\right| x\right).$$

PROOF. The functions $\phi(\lambda, x) = \omega(\sqrt{\lambda}, x) \exp\left(-\left|\operatorname{Im}\sqrt{\lambda}\right| x\right)$ satisfy equations

$$\phi(\lambda, x) = \left(\cos\sqrt{\lambda}x + h\frac{\sin\sqrt{\lambda}x}{\sqrt{\lambda}}\right) \mathrm{e}^{\left(-|\operatorname{Im}\sqrt{\lambda}|x\right)} +$$

$$+ \int_0^x \frac{\sin\sqrt{\lambda}(x-t)}{\sqrt{\lambda}} \mathrm{e}^{\left(-|\operatorname{Im}\sqrt{\lambda}|(x-t)\right)} q(t)\phi(\lambda, t)dt. \tag{5.3.2}$$

The latter can be solved by the method of successive approximations:

$$\phi(\lambda, x) = \sum_{k=0}^{\infty} \phi_k(\lambda, x),$$

where

$$\phi_0(\lambda, x) = \left(\cos\sqrt{\lambda}x + h\frac{\sin\sqrt{\lambda}x}{\sqrt{\lambda}}\right) \mathrm{e}^{\left(-|\operatorname{Im}\sqrt{\lambda}|x\right)},$$

$$\phi_k(\lambda, x) = \int_0^x \frac{\sin\sqrt{\lambda}(x-t)}{\sqrt{\lambda}} \mathrm{e}^{\left(-|\operatorname{Im}\sqrt{\lambda}|(x-t)\right)} q(t)\phi_{k-1}(\lambda, t)dt.$$

Putting $m_k(x) = \max\limits_{0<t\leq x} |\phi_k(t)|$ we find, that $m_0(x) \leq 1 + |h|x$ and

$$m_k(x) \leq \int_0^x (x-t)|q(t)|m_{k-1}(t)dt = \int\limits_0^x (x-t)d\left(\int_0^t |q(\xi)|m_{k-1}(\xi)d\xi\right) =$$

$$= \int_0^x \left(\int_0^t |q(\xi)|m_{k-1}(\xi)d\xi\right) dt \leq \int_0^x m_{k-1}(t)\sigma_0(t)dt,$$

which implies

$$m_1(x) \leq (1 + |h|x) \int_0^x \sigma_0(t)dt \leq (1 + |h|x)\sigma_1(x),$$

$$m_2(x) \leq (1+|h|x)\int_0^x \sigma_1(t)\sigma_0(t)dt = (1+|h|x)\frac{(\sigma_1(x))^2}{2},$$

and by induction

$$m_k(x) \leq (1+|h|x)\frac{(\sigma_1(x))^k}{k!}.$$

Therefore,

$$|\omega(\sqrt{\lambda},x)| = |\phi(\lambda,x)|\mathrm{e}^{|\operatorname{Im}\sqrt{\lambda}|x} \leq \left(\sum_{k=0}^{\infty} m_k(x)\right)\mathrm{e}^{|\operatorname{Im}\sqrt{\lambda}|x} \leq (1+|h|x)\exp\left(\sigma_1(x)+\right.$$

$$\left.+\left|\operatorname{Im}\sqrt{\lambda}\right|x\right) \leq \exp\left(|h|x+\sigma_1(x)+\left|\operatorname{Im}\sqrt{\lambda}\right|x\right) \leq \exp\left(b_1(x)+\left|\operatorname{Im}\sqrt{\lambda}\right|x\right).$$

Next, it follows immediately from equation (5.3.2), that for $\lambda > 0$

$$\max_{0<t\leq x}|\phi(\lambda,t)| \leq \left(1+\frac{|h|}{\sqrt{\lambda}}\right)+\frac{\max_{0<t\leq x}|\phi(\lambda,t)|}{\sqrt{\lambda}}\int_0^x |q(t)|dt.$$

When $\sqrt{\lambda} > a(x) \geq \int_0^x |q(t)|dt < \sqrt{\lambda}$, it implies

$$\max_{0<t\leq x}|\phi(\lambda,t)| \leq \frac{\sqrt{\lambda}+|h|}{\sqrt{\lambda}-\sigma_0(x)} \leq \frac{\sqrt{\lambda}+A}{\sqrt{\lambda}-a(x)}$$

and

$$|\omega(\sqrt{\lambda},x)| \leq \frac{\sqrt{\lambda}+A}{\sqrt{\lambda}-a(x)}\exp\left(\left|\operatorname{Im}\sqrt{\lambda}\right|x\right).$$

□

THEOREM 5.3.1. *If the spectral functions $\rho_j(\lambda)$ of two boundary value problems $\{h_j, q_j(x)\} \in V\{A, a(x)\}$, $j=1,2$, coincide on the semiaxis $(-\infty, N^2)$, then the solutions $\omega_j(\sqrt{\mu},x)$ satisfy inequality*

$$|\omega_1(\sqrt{\mu},x)-\omega_2(\sqrt{\mu},x)|^2 \leq \frac{24b(x)\exp\left(4b_1(x)+2\left|\operatorname{Im}\sqrt{\mu}\right|x+b_1(N^{-1})\right)}{N(1-N^{-2}\mu)}$$

for all $\mu \in (-\infty, N^2)$ and $x \in (0,\infty)$. If $a(x) < N$ they also satisfy the inequality

$$|\omega_1(\sqrt{\mu},x)-\omega_2(\sqrt{\mu},x)|^2 \leq \frac{24b(x)\exp\left(2\left(b_1(x)+\left|\operatorname{Im}\sqrt{\mu}\right|x+b_1\left(\frac{1}{N}\right)\right)\right)}{N(1-N^{-2}\mu)}\left(\frac{N+A}{N-a(x)}\right)^2.$$

PROOF. From formula (5.1.14), Lemma 5.3.2. and corollary to Lemma 5.3.1. it follows, that for all $\mu \in (-\infty, N^2)$ and $x \geq t \geq 0$

$$|B_{1,2}(\mu,x,t) - B_{1,2}(\mu,t,x)| \leq \frac{2\exp\left(4b_1(x) + 2\left|\operatorname{Im}\sqrt{\lambda}\right| x\right)}{(1-N^{-2}\mu)} \int_{N^2}^{\infty} \frac{d\rho_1(\lambda) + d\rho_2(\lambda)}{\lambda} \leq$$

$$\leq \frac{12\exp\left(4b_1(x) + 2\left|\operatorname{Im}\sqrt{\mu}\right| x + b_1(N^{-1})\right)}{N(1-N^{-2}\mu)},$$

and when $a(x) < N$

$$|B_{1,2}(\mu,x,t) - B_{1,2}(\mu,t,x)| \leq \frac{12\exp\left(2\left(b_1(x) + \left|\operatorname{Im}\sqrt{\mu}\right| x + b_1(N^{-1})\right)\right)}{N(1-N^{-2}\mu)} \left(\frac{N+A}{N-a(x)}\right)^2.$$

The result now follows immediately from these inequalities and formula (5.1.13). □

In conclusion we find the bound for the difference of potentials of two boundary value problems $\{h_j, q_j(x)\}$, $j = 1,2$, whose spectral functions coincide on the interval $(-\infty, N^2)$.

For compactly supported functions $f(x)$ the transformation operators $\mathbf{I} + \mathbf{K}_j^i$ connect the Fourier transforms

$$\omega_j(\sqrt{\lambda}, f) = \int_0^{\infty} f(x)\omega_j(\sqrt{\lambda}, x)dx$$

by equalities

$$\omega_i(\sqrt{\lambda}, f) = \omega_j(\sqrt{\lambda}, f) + \int_0^{\infty}\left(\int_0^{\infty} f(x)K_j^i(x,t)dx\right)\omega_j(\sqrt{\lambda}, t)dt.$$

Here the functions $K_j^i(x,t)$ (the kernels of operators $\mathbf{K}_j^i$) are absolutely continuous and

$$K_j^i(x,t) = h_i - h_j + \frac{1}{2}\int_0^x (q_i(t) - q_j(t))\,dt. \tag{5.3.3}$$

The Parceval equality implies

$$\int_{-\infty}^{\infty} \omega_i(\sqrt{\lambda}, f)\omega_j(\sqrt{\lambda}, f)d\rho_j(\lambda) = \|f\|^2 + \int_0^{\infty}\int_0^{\infty} f(x)K_j^i(x,t)f(t)dtdx,$$

$$\int_{-\infty}^{\infty} \omega_i(\sqrt{\lambda}, f)\omega_j(\sqrt{\lambda}, f)d\left(\rho_j(\lambda) - \rho_i(\lambda)\right) = \int_0^{\infty}\int_0^{\infty} Q_j^i(x,t)f(t)f(x)dtdx,$$

where

$$Q_j^i(x,t) = K_j^i(x,t) - K_i^j(x,t) \tag{5.3.4}$$

In particular, if the spectral functions $\rho_1(\lambda)$ and $\rho_2(\lambda)$ coincide on the interval $(-\infty, N^2)$ and if we choose f to be the Steklov function

$$\delta(x) = \begin{cases} \delta^{-1} & x \in (y-\delta, y) \\ 0 & x \notin (y-\delta, y), \end{cases}$$

then

$$\int\limits_{N^2}^{\infty} \omega_1(\sqrt{\lambda}, \delta)\omega_2(\sqrt{\lambda}, \delta) d\left(\rho_2(\lambda) - \rho_1(\lambda)\right) = \delta^{-2} \int\limits_{y-\delta}^{y} \int\limits_{y-\delta}^{x} Q_2^1(x,t)\,dt\,dx. \qquad (5.3.5)$$

By putting

$$R(x,t;y) := Q_2^1(y,y) - Q_2^1(x,t) = \left(Q_2^1(y,y) - Q_2^1(x,x)\right) + \left(Q_2^1(x,x) - Q_2^1(x,t)\right) \qquad (5.3.6)$$

we find, that

$$\delta^{-2} \int\limits_{y-\delta}^{y} \int\limits_{y-\delta}^{x} Q_2^1(x,t)\,dt\,dx = \delta^{-2} \int\limits_{y-\delta}^{y} \int\limits_{y-\delta}^{x} \left\{Q_2^1(y,y) - R(x,t;y)\right\} dt\,dx =$$

$$\frac{1}{2} Q_2^1(y,y) - \delta^{-2} \int\limits_{y-\delta}^{y} \int\limits_{y-\delta}^{x} R(x,t;y)\,dt\,dx.$$

Due to formulas (5.3.3), (5.3.4) and (5.3.6)

$$Q_2^1(y,y) = 2\left(h_1 - h_2 + \frac{1}{2}\int_0^y \left(q_1(t) - q_2(t)\right) dt\right),$$

$$R(x,t;y) = \int_x^y \left(q_1(t) - q_2(t)\right) dt + \int_t^x \frac{\partial Q_2^1(x,\xi)}{\partial \xi} d\xi.$$

For all $y - \delta \le t \le x \le y$ we have $|R(x,t;y)| \le \delta D(y,\delta)$, where

$$D(y,\delta) = \sup_{y-\delta \le t \le x \le y} \left\{ |q_1(t)| + |q_2(t)| + \left| \frac{\partial}{\partial t}\left(K_2^1(x,t) - K_1^2(x,t)\right)\right| \right\}.$$

Now we can rewrite equality (5.3.5) as

$$h_1 - h_2 + \frac{1}{2}\int_0^y \left(q_1(t) - q_2(t)\right) dt = \int\limits_{N^2}^{\infty} \omega_1(\sqrt{\lambda}, \delta)\omega_2(\sqrt{\lambda}, \delta) d\left(\rho_2(\lambda) - \rho_1(\lambda)\right) +$$

$$+\delta^{-2}\int\limits_{y-\delta}^{y}\int\limits_{y-\delta}^{x} R(x,t;y)dtdx, \qquad (5.3.7)$$

where

$$\delta^{-2}\left|\int\limits_{y-\delta}^{y}\int\limits_{y-\delta}^{x} R(x,t;y)dtdx\right| \leq \delta^{-1}D(y,\delta)\int\limits_{y-\delta}^{y}\int\limits_{y-\delta}^{x} dtdx = \frac{\delta D(y,\delta)}{2}. \qquad (5.3.8)$$

Next, it follows from Lemma 5.3.2. that for $\lambda > 0$

$$\omega(\sqrt{\lambda},x) = \cos\sqrt{\lambda}x + h\,\frac{\sin\sqrt{\lambda}x}{\sqrt{\lambda}} + r(\sqrt{\lambda},x),$$

where

$$\left|r(\sqrt{\lambda},x)\right| \leq \frac{1}{\sqrt{\lambda}}\left(1+\frac{|h|}{\sqrt{\lambda}}\right)\sigma_0(x)\mathrm{e}^{\sigma_1(x)}.$$

Therefore, for $\lambda > N^2$

$$\left|\omega(\sqrt{\lambda},\delta)\right| = \delta^{-1}\left|\int\limits_{y-\delta}^{y}\omega(\sqrt{\lambda},x)dx\right| \leq \frac{\delta^{-1}}{\sqrt{\lambda}}\left\{2\left(1+\frac{|h|}{\sqrt{\lambda}}\right)+\right.$$

$$\left.+\left(1+\frac{|h|}{\sqrt{\lambda}}\right)\left(\mathrm{e}^{\sigma_1(y)}-\mathrm{e}^{\sigma_1(y-\delta)}\right)\right\} \leq \frac{2\delta^{-1}}{\sqrt{\lambda}}\left(1+\frac{A}{N}\right)\mathrm{e}^{a_1(y)},$$

and according to the corollary to Lemma 5.3.1

$$\int_{N^2}^{\infty}\omega_1(\sqrt{\lambda},\delta)\omega_2(\sqrt{\lambda},\delta)d\left(\rho_2(\lambda)-\rho_1(\lambda)\right) \leq 4\delta^{-2}\left(1+\frac{A}{N}\right)^2\mathrm{e}^{2a_1(y)}\times$$

$$\times\int_{N^2}^{\infty}\frac{d\left(\rho_2(\lambda)+\rho_1(\lambda)\right)}{\lambda} \leq \frac{24}{\delta^2 N}\left(1+\frac{A}{N}\right)^2\exp 2\left(a_1(y)+b_1(N^{-1})\right). \qquad (5.3.9)$$

From formula (5.3.7) and estimates (5.3.8), (5.3.9) with $\delta = N^{-\frac{1}{3}}$ we obtain the main inequality

$$\left|h_1-h_2+\frac{1}{2}\int_0^y\left(q_1(t)-q_2(t)\right)dt\right| \leq 24N^{-\frac{1}{3}}\left\{(1+AN^{-1})^2\exp 2\left(a_1(y)+b_1(N^{-1})\right)+\right.$$

$$\left.+\frac{1}{48}D(y,N^{-\frac{1}{3}})\right\}. \qquad (5.3.10)$$

One can estimate the value $D(y, N^{-\frac{1}{3}})$ via $\sup_{0\le t\le y}|q(t)|$ using the integral equations for the kernels $K_j^i(x,t)$. □

PROBLEMS

1. Prove the formula

$$h_1 - h_2 + \frac{1}{2}\int_0^x \{q_1(t) - q_2(t)\}dt = \int_{-\infty}^{\infty} \omega_1(\sqrt{\mu}, x)\omega_2(\sqrt{\mu}, x)d\{\rho_1(\mu) - \rho_2(\mu)\}.$$

2. Prove that if the boundary-value problems $\{q_j(x)\}$ $(j = 1, 2)$ belong to the set $V\{\alpha(x)\}$, then in the domain $\{x : 2\nu > \alpha(x)\}$, the following inequality holds:

$$\left| e_1(\mathrm{i}\nu, x) - e_2(\mathrm{i}\nu, x) - \frac{\mathrm{e}^{-\nu x}}{2\nu}\int_x^{\infty}\left(1 - \mathrm{e}^{-2\nu(t-x)}\right)(q_1(t) - q_2(t))\,dt\right| \le \frac{\mathrm{e}^{-\nu x}\alpha^2(x)}{\nu^2}.$$

Hint. Integrate equation (3.1.7) once.

3. Prove the following generalization of Theorem 5.2.3.: If the scattering data of the two boundary problems $\{q_j(x)\} \in V\{\alpha(x)\}$ $(j = 1, 2)$ for all $\lambda^2 \in (-\infty, N^2)$ differ by no more than δ, i.e.

$$\sum_k |m_k^2(2) - m_k^2(1)| + \frac{1}{2\pi}\int_{-\infty}^{\infty} |S_1(\lambda) - S_2(\lambda)|\,d\lambda < \delta,$$

then

$$|q_1(x) - q_2(x)| \le 2\delta\mathrm{e}^{\alpha_1(x)} + 2\pi h D(x, h) + \frac{16\Delta(0, N)(1 + 9h\alpha(x))}{9h^2(1 - 2\Delta(0, N))}.$$

4. Obtain a similar generalization for formula (5.3.10).

5. Generalize theorems 5.2.2. and 5.3.1. for the case when the scattering data or spectral functions are not equal but are only slightly different on the interval $(-\infty, N^2)$.

REFERENCES

1. Akhiezer, N. I.: *A continuous analogue of orthogonal polynomials on a system of intervals*, Dokl. Akad. Nauk SSSR 141, No. 2 (1961), 263-266; English transl.: Soviet Math. Dokl., 2 (1961), 1409-1412.

2. Borg, G.: *Eine Umkehrung der Sturm-Liouvilleschen Eigenwertaufgabe*, Acta Math., 78, fasc. 1 (1946), 1-96.

3. Dubrovin, B. A., Matveev, V. B., and Novikov, S. P.: *Non-linear equations of the Korteweg-de Vries type, finite-zone linear operators, and Abelian varieties*, Usp. Mat. Nauk 31, No. 1 (1976), 55-136; English transl.: Russian Math. Surveys 31, No. 1 (1976), 59-146.

4. Dubrovin, B. A. and Novikov, S. P.: *A periodicity problem for the Korteweg-de Vries and Sturm-Liouville equations*, Dokl. Akad. Nauk SSSR 219, No. 3 (1974), 531-534; English transl.: Soviet Math. Dokl., 15, No. 6 (1974), 1597-1601.

5. Faddeev, L. D.: *Properties of the S-matrix of the one-dimensional Schrödinger equation*, Trudy Mat. Inst. Steklov 73 (1974), 314-336; English transl.: Amer. Math. Soc. Transl. (2) 65 (1967), 139-166.

6. Gardner, C., Green, J., Kruskal, M., and Miura, R.: *A method for solving the Korteweg-de Vries equation*, Phys. Rev. Lett., 19 (1967), 1095-1098.

7. Gasymov, I. M. and Levitan, B. M.: *Determination of a differential equation by two of its spectra*, Usp. Mat. Nauk 19, No. 2 (1964), 3-63; English transl.: Russian Math. Surveys 19, No. 2 (1964), 1-63.

8. Gelfand, I. M. and Levitan, B. M.: *On the determination of a differential equation from its spectral function*, Izv. Akad. Nauk SSSR Ser. Mat., 15, No. 4 (1951), 309-360; English transl.: Amer. Mat. Soc. Transl. (2) 1 (1955), 253-304.

9. Gelfand, I. M. and Levitan, B. M.: *On a simple identity for the eigenvalues of a second-order differential operator*, Dokl. Akad. Nauk SSSR 88, No. 4 (1953), 593-596. (Russian)

10. Its, A. R. and Matveev, V. B.: *On Hill operators with a finite number of lacunae*, Funkts. Anal. Prilozhen., 9, No. 1 (1975), 69-70; English transl.: Funct. Anal. Appl., 9, No. 1 (1975), 65-66.

11. Its, A. R. and Matveev, V. B.: *On a class of solutions of the KdV equation*, Problemy Mat. Fiz., No. 8 (1976), 70-92. (Russian)

12. Lax, P. D.: *Integrals of nonlinear equations of evolution and solitary waves*, Comm. Pure Appl. Math., 21 (1968), 467-490.

13. Lax, P. D.: *Periodic solutions of the KdV equation*, Lect. Appl. Math., 15 (1974), 85-96.

14. Levin, B. Ya.: *Fourier- and Laplace-type transformations by means of solutions of a second-order differential equation*, Dokl. Akad. Nauk SSSR 106, No. 2 (1956), 187-190. (Russian)

15. Levitan, B. M.: *On the asymptotic behavior of the spectral function of a self-adjoint differential equation of second order and on expansion in eigenfunctions. I,II*, Izv. Akad. Nauk SSSR Ser. Mat., 17, No. 4 (1953), 331-364; 19, No. 1 (1955), 33-58; English transl.: Amer. Math. Soc. Transl. (2) 102 (1973), 191-229; 110 (1977), 165-188.

16. Levitan, B. M.: *Theory of Generalized Translation Operators, 2nd ed.*, "Nauka", Moscow, 1973; English transl. of 1st ed.: Israel Program for Scientific Translations, Jerusalem, 1964.

17. Marchenko, V. A.: *Some problems in the theory of second-order differential operators*, Dokl. Akad. Nauk SSSR 72, No. 3 (1950), 457-460. (Russian)

18. Marchenko, V. A.: *On inversion formulas generated by a linear second-order differential operator*, Dokl. Akad. Nauk SSSR 74, No. 4 (1950), 657-660. (Russian)

19. Marchenko, V. A.: *Reconstruction of the potential energy from the phases of the scattered waves*, Dokl. Akad. Nauk SSSR 104, No. 5 (1955), 695-698. (Russian)

20. Marchenko, V. A.: *Spectral Theory of Sturm-Liouville Operators*, "Naukova Dumka", Kiev, 1972. (Russian)

21. Marchenko, V. A.: *The periodic Korteweg-de Vries problem*, Mat. Sb., 95, No. 3 (1974), 331-356; English transl.: Math. USSR Sbornik 24, No. 3 (1974), 319-344.

22. Marchenko, V. A. and Ostrovskii, I. V.: *A characterization of the spectrum of Hill's operator*, Mat. Sb., 97, No. 4 (1975), 540-606; English transl.: Math. USSR Sbornik 26, No. 4 (1975), 493-554.

23. Novikov, S. P.: *The periodic problem for the Korteweg-de Vries equation. I*, Funkts. Anal. Prilozhen., 8, No. 3 (1974), 54-66; English transl.: Funct. Anal. Appl., 8, No. 3 (1974), 236-246.

24. Povzner, A. Ya.: *On differential equations of Sturm-Liouville type on a half-axis*, Mat. Sb., 23(65) (1948), 3-52; English transl.: Amer. Math. Soc. Transl. (1) 4 (1962), 24-101.

25. Springer, G.: *Introduction to the Theory of Riemann Surfaces*, Addison-Wesley, Reading, Mass., 1957.

26. Zakharov, V. E. and Faddeev, L. D.: *Korteweg-de Vries equation: a completely integrable Hamiltonian system*, Funkt. Anal. Prilozhen., 5, No. 4 (1971), 18-27; English transl.: Funct. Anal. Appl., 5, No. 4 (1971), 280-287.

27. Zakharov, V. E. and Shabat, A. B.: *Exact theory of two-dimensional self-focusing and one-dimensional self-modulation of waves in nonlinear media*, Zh. Eksper. Teor. Fiz., 61, No. 1 (1971), 118-134; English transl.: Soviet Physics JETP 34 (1972), 62-69.

28. Zverovich, E. I.: *Boundary value problems in the theory of analytic functions in Hölder classes on Riemann surfaces*, Usp. Mat. Nauk 26, No. 1 (1971), 113-179; English transl.: Russian Math. Surveys 26, No. 1 (1971), 117-192.

SUPPLEMENTARY REFERENCES*

1. Ablowitz, M. J. and Segur, H.: *Solitons and the Inverse Scattering Transform*, SIAM Studies in Appl. Math., no. 4, SIAM, Philadelphia, 1981.

2. Beals, R.: *The inverse problem for ordinary differential operators on the line*, Amer. J. Math., to appear.

3. Beals, R. and Coifman, R.R.: *Scattering, transformations spectrales, et equations d'evolutions nonlineaires I,II*, Seminaire Goulalouic-Meyer-Schwartz 1980-1981, exp. 22, 1981-1982, exp. 21, Ecole Polytechnique, Palaiseau.

4. Beals, R. and Coifman, R.R.: *Scattering and inverse scattering for first order systems*, Comm. Pure Appl. Math., 37 (1984), 39-90.

5. Beals, R. and Coifman, R.R.: *Inverse scattering and evolution equations*, Comm. Pure Appl. Math., to appear.

6. Birnir, B.: *Complex Hill's equation and the complex Korteweg-de Vries equation*, Comm. Pure Appl. Math., 39 (1986), 1-49.

7. Bullough, R. K. and Caudrey, P. J. (editors): *Solitons*, Topics in Current Physics, Springer-Verlag, Berlin, Heidelberg, New York, 1980.

---

* Translator's note: In the last 15 years the theory of nonlinear evolution equations of the "Korteweg-de Vries type" and the allied spectral and scattering theories have undergone an explosive development, which still continues to this day. This very short supplementary list of references (a complete list would include hundreds if not a few thousand titles) has been compiled by the translator in order to introduce the interested reader to the literature dealing with some aspects of these theories.

8. Calogero, F. and Degasperis, A.: *Spectral Transform and Solitons, Vol. 1*, Studies in Math. Appl., no. 13, North-Holland, Amsterdam, 1982.

9. Chadan, K. and Sabatier, P. C.: *Inverse Problems in Quantum Scattering Theory*, Springer-Verlag, Berlin, Heidelberg, New York, 1977.

10. Dahlberg, B. E. and Trubowitz, E.: *The inverse Sturm-Liouville problem III*, Comm. Pure Appl. Math., to appear.

11. Deift, P. and Trubowitz, E.: *Inverse scattering on the line*, Comm. Pure Appl. Math., 32 (1979), 121-251.

12. Deift, P., Tomei, C., and Trubowitz, E.: *Inverse scattering and the Bousinesq equation*, Comm. Pure Appl. Math., 35 (1982), 567-628.

13. Dodd, R. K., Eilbeck, J. C., Gibbon, J. D., and Morris, H. C.: *Solitons and Nonlinear Wave Equations*, Academic Press, London, 1982.

14. Eastham, M. S. P.: *The Spectral Theory of Periodic Differential Equations*, Scottish Academic Press, Edinburg and London, 1973.

15. Eckhaus, W. and van Harten, A.: *The Inverse Scattering Transformation and the Theory of Solitons*, Math. Studies no. 50, North-Holland, Amsterdam, 1981.

16. Faddeev, L. D.: *The inverse problem in the quantum theory of scattering*, Usp. Mat. Nauk 14, No. 4 (1959), 57-119; English transl.: J. Math. Phys., 4, No. 1 (1963), 72-104.

17. Faddeev, L. D.: *The inverse problem of quantum scattering theory II*, Itogi Nauki Tekhn. Sovrem. Probl. Mat., 3 (1974), 93-180; English transl.: J. Soviet Math., 5, No. 1 (1976), 334-396.

18. Flaschka, H.: *On the inverse problem for Hill's operator*, Arch. Rat. Mech. Anal., 59 (1975), 293-308

19. Folland, G. B.: *Spectral analysis of a singular nonselfadjoint boundary value problem*, J. Diff. Equations 37, No. 2 (1980), 206-224.

20. Folland, G. B.: *Spectral analysis of a nonselfadjoint operator*, J. Diff. Equations 39, No. 2 (1981), 151-185.

21. Garnett, J. and Trubowitz, E.: *Gaps and bands of one-dimensional Schrödinger operators*, Comment. Math. Helv., 59 (1984), 258-312.

22. Hochstadt, H.: *On the determination of a Hill's equation from its spectrum*, Arch. Rat. Mech. Anal., 19 (1965), 353-362.

23. Isaacson, E. L., McKean, H. P., and Trubowitz, E.: *The inverse Sturm-Liouville problem. II*, Comm. Pure Appl. Math., 37 (1984), 1-11.

24. Isaacson, E. L. and Trubowitz, E.: *The inverse Sturm-Liouville problem. I*, Comm. Pure Appl. Math., 36 (1983), 767-783.

25. Lamb, G. L., Jr.: *Elements of Soliton Theory*, J. Wiley & Sons, New York, 1980.

26. Levitan, B. M.: *Inverse Sturm-Liouville Problems*, "Nauka", Moscow, 1984. (Russian)

27. Levitan, B. M. and Sargsjan, I. S.: *Introduction to Spectral Theory: Selfadjoint Ordinary Differential Operators*, "Nauka", Moscow, 1970; English transl.: Transl. of Math. Monographs Vol. 39, Amer. Math. Soc., Providence, R. I., 1975.

28. McKean, H. P.: *Integrable systems and algebraic curves*, in: Global Analysis, Proc. Calgary 1978, Lect. Notes Math. 755, Springer-Verlag, Berlin, Heidelberg, New York (1979), pp. 83-200.

29. McKean, H. P.: *The sine-Gordon and sinh-Gordon equations on the circle*, Comm. Pure Appl. Math., 34 (1981), 197-257.

30. McKean, H. P.: *Boussinesq's equation on the circle*, Comm. Pure Appl. Math., 34 (1981), 599-691.

31. McKean, H. P. and van Moerbeke, P.: *The spectrum of Hill's equation*, Invent. Math., 30 (1975), 217-274.

32. McKean, H. P. and Trubowitz, E.: *Hill's operator and hyperelliptic function theory in the presence of infinitely many branch points*, Comm. Pure Appl. Math., 26 (1976), 143-226.

33. McKean, H. P. and Trubowitz, E.: *Hill's surfaces and their theta functions*, Bull. Amer. Math. Soc., 84, No. 6 (1978), 1042-1085.

34. Melin, A.: *Operator methods for inverse scattering on the real line*, Comm. Partial Diff. Equations 10, No. 7 (1985), 677-766.

35. Miura, R. M.: *The Korteweg-de Vries equation: a survey of results*, SIAM Review 18, No. 3 (1976), 412-459.

36. Newell, A. C.: *Solitons in Mathematics and Physics*, SIAM Reg. Conf. Series Appl. Math., no. 48, SIAM, Philadelphia, 1985.

37. Newton, R. G.: *Inverse scattering. I. One dimension*, J. Math. Phys., 21, No. 3 (1980), 493-505.

38. Novikov, S. P. et al.: *Integrable Systems. Selected Papers*, London Math. Soc. Lect. Notes Series no. 60, Cambridge Univ. Press, Cambridge, 1981.

39. Novikov, S. P. (editor): *Theory of Solitons*, "Nauka", Moscow, 1980; English transl.: Plenum Press, New York, 1984.

40. Poschel, J. and Trubowitz, E.: *Lectures on Inverse Spectral Theory*, Bonn Univ. and ETH Zurich, 1985, to appear.

41. Trubowitz, E.: *The inverse problem for periodic potentials*, Comm. Pure Appl. Math., 30 (1977), 321-337.

ISBN 978-0-8218-5316-0
9 780821 853160

About this book

The spectral theory of Sturm-Liouville operators is a classical domain of analysis, comprising a wide variety of problems. Besides the basic results on the structure of the spectrum and the eigenfunction expansion of regular and singular Sturm-Liouville problems, it is in this domain that one-dimensional quantum scattering theory, inverse spectral problems, and the surprising connections of the theory with nonlinear evolution equations first become related. The main goal of this book is to show what can be achieved with the aid of transformation operators in spectral theory as well as in their applications. The main methods and results in this area (many of which are credited to the author) are for the first time examined from a unified point of view.

The direct and inverse problems of spectral analysis and the inverse scattering problem are solved with the help of the transformation operators in both self-adjoint and nonself-adjoint cases. The asymptotic formulae for spectral functions, trace formulae, and the exact relation (in both directions) between the smoothness of potential and the asymptotics of eigenvalues (or the lengths of gaps in the spectrum) are obtained. Also, the applications of transformation operators and their generalizations to soliton theory (i.e., solving nonlinear equations of Korteweg-de Vries type) are considered.

The new Chapter 5 is devoted to the stability of the inverse problem solutions. The estimation of the accuracy with which the potential of the Sturm-Liouville operator can be restored from the scattering data or the spectral function, if they are only known on a finite interval of a spectral parameter (i.e., on a finite interval of energy), is obtained.